Multimodal Narratives in Research and Teaching Practices

J. Bernardino Lopes
Universidade de Trás-os-Montes e Alto Douro, Portugal

Maria Clara Viegas
Politécnico do Porto, Portugal

José Alexandre Pinto
Politécnico do Porto, Portugal

A volume in the Advances in Educational Technologies and Instructional Design (AETID) Book Series

IGI Global
DISSEMINATOR OF KNOWLEDGE

Published in the United States of America by
IGI Global
Information Science Reference (an imprint of IGI Global)
701 E. Chocolate Avenue
Hershey PA, USA 17033
Tel: 717-533-8845
Fax: 717-533-8661
E-mail: cust@igi-global.com
Web site: http://www.igi-global.com

Library of Congress Cataloging-in-Publication Data

Names: Lopes, J. Bernardino, editor. | Viegas, Maria Clara, 1968- editor. | Pinto, Jose Alexandre, 1968- editor.
Title: Multimodal narratives in research and teaching practices / J. Bernardino Lopes, Maria Clara Viegas, and Jose Alexandre Pinto, editors.
Description: Hershey, PA : Information Science Reference, [2019] | Includes bibliographical references.
Identifiers: LCCN 2018055422| ISBN 9781522585701 (hardcover) | ISBN 9781522585718 (ebook) | ISBN 9781522585732 (softcover)
Subjects: LCSH: Science--Study and teaching--Research.
Classification: LCC Q181 .M95165 2019 | DDC 507.1--dc23 LC record available at https://lccn.loc.gov/2018055422

This book is published in the IGI Global book series Advances in Educational Technologies and Instructional Design (AE-TID) (ISSN: 2326-8905; eISSN: 2326-8913)

British Cataloguing in Publication Data
A Cataloguing in Publication record for this book is available from the British Library.

For electronic access to this publication, please contact: eresources@igi-global.com.

Advances in Educational Technologies and Instructional Design (AETID) Book Series

Lawrence A. Tomei
Robert Morris University, USA

ISSN:2326-8905
EISSN:2326-8913

Mission

Education has undergone, and continues to undergo, immense changes in the way it is enacted and distributed to both child and adult learners. In modern education, the traditional classroom learning experience has evolved to include technological resources and to provide online classroom opportunities to students of all ages regardless of their geographical locations. From distance education, Massive-Open-Online-Courses (MOOCs), and electronic tablets in the classroom, technology is now an integral part of learning and is also affecting the way educators communicate information to students.

The **Advances in Educational Technologies & Instructional Design (AETID) Book Series** explores new research and theories for facilitating learning and improving educational performance utilizing technological processes and resources. The series examines technologies that can be integrated into K-12 classrooms to improve skills and learning abilities in all subjects including STEM education and language learning. Additionally, it studies the emergence of fully online classrooms for young and adult learners alike, and the communication and accountability challenges that can arise. Trending topics that are covered include adaptive learning, game-based learning, virtual school environments, and social media effects. School administrators, educators, academicians, researchers, and students will find this series to be an excellent resource for the effective design and implementation of learning technologies in their classes.

Coverage

- Online Media in Classrooms
- Collaboration Tools
- Social Media Effects on Education
- Curriculum Development
- Higher Education Technologies
- Adaptive Learning
- E-Learning
- Bring-Your-Own-Device
- Instructional Design
- Virtual School Environments

IGI Global is currently accepting manuscripts for publication within this series. To submit a proposal for a volume in this series, please contact our Acquisition Editors at Acquisitions@igi-global.com or visit: http://www.igi-global.com/publish/.

The Advances in Educational Technologies and Instructional Design (AETID) Book Series (ISSN 2326-8905) is published by IGI Global, 701 E. Chocolate Avenue, Hershey, PA 17033-1240, USA, www.igi-global.com. This series is composed of titles available for purchase individually; each title is edited to be contextually exclusive from any other title within the series. For pricing and ordering information please visit http://www.igi-global.com/book-series/advances-educational-technologies-instructional-design/73678. Postmaster: Send all address changes to above address.

Titles in this Series

For a list of additional titles in this series, please visit: www.igi-global.com/book-series

Handbook of Research on Assessment Literacy and Teacher-Made Testing in the Language Classroom
Eddy White (University of Arizona, USA) and Thomas Delaney (University of Oregon, USA)
Information Science Reference • copyright 2019 • 410pp • H/C (ISBN: 9781522569862) • US $255.00 (our price)

Interdisciplinary and International Perspectives on 3D Printing in Education
Ieda M. Santos (Emirates College for Advanced Education, UAE) Nagla Ali (Emirates College for Advanced Education, UAE) and Shaljan Areepattamannil (Emirates College for Advanced Education, UAE)
Information Science Reference • copyright 2019 • 346pp • H/C (ISBN: 9781522570189) • US $195.00 (our price)

Bayesian Networks for Managing Learner Models in Adaptive Hypermedia Systems Emerging Research and Opportunities
Mouenis Anouar Tadlaoui (Abdelmalek Essaadi University, Morocco) Mohamed Khaldi (Abdelmalek Essaadi University, Morocco) and Rommel Novaes Carvalho (University of Brasília, Brazil)
Information Science Reference • copyright 2019 • 199pp • H/C (ISBN: 9781522574132) • US $165.00 (our price)

Handbook of Research on TPACK in the Digital Age
Margaret L. Niess (Oregon State University, USA) Henry Gillow-Wiles (Southern Oregon University, USA) and Charoula Angeli (University of Cyprus, Cyprus)
Information Science Reference • copyright 2019 • 490pp • H/C (ISBN: 9781522570011) • US $255.00 (our price)

Educational and Social Dimensions of Digital Transformation in Organizations
Paula Peres (Polytechnic of Porto, Portugal) Fernando Moreira (Universidade Portucalense, Portugal & Universidade de Aveiro, Portugal) and Anabela Mesquita (Polytechnic of Porto, Portugal)
Information Science Reference • copyright 2019 • 372pp • H/C (ISBN: 9781522562610) • US $185.00 (our price)

Handbook of Research on Curriculum Reform Initiatives in English Education
Christopher Denman (Sultan Qaboos University, Oman) and Rahma Al-Mahrooqi (Sultan Qaboos University, Oman)
Information Science Reference • copyright 2019 • 372pp • H/C (ISBN: 9781522558460) • US $225.00 (our price)

Global Perspectives on Work-Based Learning Initiatives
Jon Talbot (University of Chester, UK)
Information Science Reference • copyright 2019 • 344pp • H/C (ISBN: 9781522569770) • US $185.00 (our price)

IGI Global
DISSEMINATOR OF KNOWLEDGE

701 East Chocolate Avenue, Hershey, PA 17033, USA
Tel: 717-533-8845 x100 • Fax: 717-533-8661
E-Mail: cust@igi-global.com • www.igi-global.com

Editorial Advisory Board

List of Reviewers

Table of Contents

Detailed Table of Contents

Section 1
Multimodal Narratives: A Tool and an Open Resource for Practice, Research, and Professional Development

In Section 1, the reader can get a complete idea of what a multimodal narrative (MN) is, how to produce and validate a MN, and its different purposes. This section also serves as a reference part where readers can go to address any doubt regarding specifics aspects of the MN. In the end of Section 1, there are indications on how to proceed to access the MN collection (Chapter 1: Appendix 1). Furthermore, it provides two complete examples of MN in order to better understand its characteristics and potentialities (Chapter 1: Appendix 2).

It is acknowledged that to improve the value of the learning process and outcomes in areas such as science, technology, engineering and math, the teaching quality needs to be enhanced. Therefore, it is crucial to have access to real teaching practices. The multimodal narrative (MN) tool allows teaching practices to become public, sharable, and usable (open science perspective), preserving their holistic, complex, and ecological nature. This tool has characteristics and a structure that enable an in-depth study of teaching practices, in different contexts, with several purposes (e.g., teacher education, professional development, and research). This chapter presents MNs and the necessary steps involved in collecting multimodal data, structuring the narrative, and validating the document. MNs can be used by teachers and researchers, or other professionals, with multiple specific objectives, globally contributing to improving professional practices.

Section 3
Professional Development of Science and Technology Teachers Using Multimodal Narratives

In this section of the book, the use of multimodal narrative (MN) in the practice of Science and Technology teachers is contemplated including a view to their professional development. Throughout the chapters, situations in which action and object centered reflection potentiates professional knowledge and sharing of experiences are presented. Despite these cases being referent to specific themes, since MNs focus on classroom actions and practices, they may prove to be rich in strategies and approaches. They are thus useful in any context, particularly in Science and Technology teaching. The reading of this section can be done from two perspectives: first, from the perspective of teachers interested in performing MN of their classes as a way to better understand their practice to reflect on their work or even to use them for didactic purposes; second, by the reading of MN performed by other teachers (even in different realities) to identify strengths and weaknesses of classroom practices and to better understand the outcome of certain strategies. Regardless of the perspective, it will be possible to use the MN to enrich the practice of teaching.

In this chapter, the authors discuss professional development (PD), self- and hetero-directed based on cases of physics education teachers in secondary and higher education and the role that multimodal narratives have in both processes. They adopt the PD perspective, which assumes that PD should be evaluated by its impact on teaching practices improvement. In this context, the PD aims to improve, incrementally, teaching practices. This process is difficult and complex and involves both teachers themselves and other professionals. Multimodal narratives (MN) are a tool that makes it possible to make teaching practices public and thus open to analysis by the teacher, other professionals, and researchers. The reported PD uses MNs as a tool that supports the teacher's collaborative work with peers and researchers. This collaborative work results in important incremental improvements in teaching practices.

Teacher training processes must incorporate a reflective dimension as a strategy for professional development. The pursuit of a professional identity and the need to give personal meaning to theoretical principles grounds the emergence of young teachers' reflection. In this chapter, multimodal narrative (MN) is presented as a tool to support reflexive approaches in the development of teachers' professional knowledge. The data collected about the perspectives of student teachers and supervisors who experienced the use of MN show the interest of these actors about the tool and about the processes of its use. This chapter presents and discusses constraints identified throughout the pilot study of using a MN in teacher training that the authors developed. It also presents a proposal for the use of MN in the context of initiation of the professional practice that includes an adapted version of the MN tool and a phased process of its use.

This chapter presents how multimodal narratives were employed as a self-reflection tool within an online professional development program for in-service teacher training at Universidade Aberta, Portugal during two editions of a pedagogic practice course. The chapter includes the aspects that raised issues and those that trainees performed correctly. This is done in three stages: beforehand, upon initial contact with multimodal narratives, and after providing feedback to trainees. The most relevant issues were in aspects directly required to enrich the narrative. Aspects related to multimodal narrative structure and features were completed successfully. It is recommended that future attempts to employ multimodal narratives in this context adapt learning resources and pedagogic support practices by employing formative feedback and continual support during the trainees' process of exploring and exploiting multimodal narratives.

This chapter focuses on the impact that multimodal narratives (MN), elaborated to a first intervention (MN1) and to a succession of interventions (MN2), had in the professional development of a teacher to increase the impact of visual representations in the teaching and learning of mathematical content. The study follows an interpretative paradigm, with design science research and case study investigation methods. The participants are two groups of students (mean age 10.7 years and 11.3 years) and their Mathematics teacher. The data were collected from the students' written records, lesson plans, and teacher's notes. The data collected were organized in a MN. The results show that the MN is a tool that facilitates the process of professional development at the self-reflection level and in combination with researchers. The authors conclude that MN is an efficient and useful tool for professional development that aims to improve teaching practices and student learning.

This chapter focuses on teachers' professional development, on the task design and on experimental work, as well as on the role of the teachers' mediation in the quality of student learning. The research problem was how the teacher can promote self-directed professional development, namely, improving the quality of teaching practices to influence the quality of students' learning, in their engagement in experimental tasks and epistemic practices. A longitudinal research methodology was followed during 10 years, based on a qualitative case study, from a curricular approach in secondary education. The analysis of data collection on teaching practices and students' learning over time and the teachers' professional pathways allow to formulate the following conclusions: (1) new traits of teaching practices were identified that promote students' productive engagement; (2) changes to the task design were enough to trigger differences in teachers' mediation, with consequences for students' epistemic practices and their productive engagement.

Section 4
Researching Science and Technology Practices Using Multimodal Narratives

Section 4 of this book reports research on Science and Technology teaching practices based on academic research criteria using as empirical base the multimodal narratives (MNs) in their final public and shareable form, available in a collection of MNs (see Section 1). The set is representative, but not exhaustive, of the several research studies that have already been carried out with MNs.

This chapter begins by presenting how multimodal narratives were developed and how they have evolved in time as a research tool. Their characteristics are explained, as well as their potential for use in research studies. One important aspect discussed in this chapter is the versatility of this tool, which allows for many possible investigative approaches. In particular, it addresses how multimodal narratives can be used to carry out studies in science and technology education research. Several examples are provided of such studies that were developed within a research community which has been working with multimodal narratives for over a decade.

This chapter describes an empirical study using multimodal narratives for research into students' development of epistemic practices in the classroom. Multimodal narratives can give access to classroom events, preserving their complex and holistic nature. Through content analysis, they allow a good comprehension of the multimodal nature of teaching and learning practices. The results of this work highlight the importance of multimodal narratives as a research instrument. Their importance is based on the richness of elements they contain that allow the identification, categorization, and characterization of teacher mediation actions that promote, scaffold, and enlarge students' epistemic practice development. This chapter seeks to describe both their multiple potentialities as an instrument and their limitations when researching the development of students' epistemic practices in the physical sciences classroom.

Section 5
Using Multimodal Narratives in Other Contexts

In this section of the book we try to blur boundaries and to point out contexts that are less exploited in the initial moments of multimodal narrative (MN) tool development. Seeking to go beyond familiar circumstances, we are moving forward with the conviction that the characteristics of MN are sufficient grounds to support our confidence that they will be usable and useful in several professional contexts. The two approaches we bring in this section of the book are, first, illustrations of the diversity we foresee for the use of MN. In the following two chapters, we present work done using MN in two very distinct contexts. The first context presents the work done on the Science Museum Educators' Discourse and the second on Nursing Education. It is intended that these examples, in addition to the intrinsic and relevant issues they present, constitute a challenge to the reader to reflect on the MN's potentialities in multiple contexts.

Museum educators play an important role in science communication, as they connect elements of an exhibit with visitors through emotion-driven experiences that are meaningful to them. Language is their main modus operandi in face-to-face interactions, but little is known of how they use it to communicate science, in part because little attention has been given to their practices and professional development. Nowadays, museums are changing, and science communication has become more demanding as these institutions exhibit hot themes of science. In this context, it is important that museum educators become aware of how they communicate science with an intended audience and reflect on how their practice can be improved. In this chapter, the authors focus on the way multimodal narratives can be used as a tool to access museum educators' discourse as well as how to promote museum educators' reflection about their practice and their professional development.

Nursing education is driven by emerging challenges of scientific, technological, and professional advances that require the use of strategies that promote students' development of critical thinking for decision making in different contexts. It also requires that teachers constantly reflect on their pedagogical practices and (re)think them using strategies that allow their enhancement. The use of multimodal narratives (MNs) can be an important tool for teachers' professional development, namely to improve their classroom practices. Given the novelty of the use of MNs in nursing education, this chapter presents an analysis concerning the experience of making a MN and how it has been reflected in the authors' pedagogical practices. With this experience, potentialities of continuing to use MNs in nurse education are explored.

Foreword

Here is the new book of the UTAD's team! A natural question is: how do they do, to be able to publish so many books? Simple: they work. They have worked hard, for almost twenty years, many people together, so now they have a huge amount of accumulated experience, results and knowledge.

This book is an introduction to the method, which, accompanying a strong collective spirit, allowed them to produce these results and knowledge: the multimodal narratives (MNs). This method has been described briefly in scientific articles, and as such, is validated by the scientific community in Science Education. But articles are always too short, and the format of a book is quite necessary to understand the whole richness of this method of collecting and analysing classroom data, to perceive all the opportunities it offers. It is by the way a quality of this book to give to the reader detailed insights of many various MNs.

We must pay attention to both the words in the expression: "multimodal narratives". In two different dimensions, they manifest the complexity of classroom activity. "Multimodal" because the reality in the classroom mobilises a range of semiotic registers, verbal, gestural, graphical etc., at a given time; if you want to grasp the state of the interaction, you must collect data at these different planes. "Narrative", because the classroom activity is a story, coherent along time, with a history and an aim, and you cannot understand one moment of the story without knowing what happened before and what happened after.

I have just used the word "complexity"; not in its common meaning: complicated! Following Edgar Morin's theory, by "a complex reality" I mean an entity constructed from intertwined elements, which are linked by relationships of complementarity, competition, and antagonism – with a different mixt of each, depending on the case. These relationships between elements allow the complex reality to progress, through the mechanism of recursion, installing constructive loops.

Just as education, research on education is a complex process, with many elements and many loops. Among those elements, students of course, teachers of course – and the book pays a great tribute to the study of interactions between those two kinds of elements. But a long-term characteristics of UTAD's team is to aggregate, without identifying them, the researchers to the social activity of teachers that is, not to put them on a separate (and implicitly higher) level. These characteristics is theoretically grounded on a reflection, developed in Chapter 7, about the nature of professional knowledge, which recognises it from a different nature than scientific knowledge, and which makes it necessary to take into account both perspectives, researchers and teachers. That is why this book is of interest both for researchers and for teachers.

MNs are the tool for making this kind of complexity alive. First, the research practice of UTAD's team is that a teacher, alone in his/her class, can create MNs valid for research. It is the proof of a deep confidence in the professional rigour of the teacher; but is also the affirmation that subjectivity has a scientific value, far from the hard-science perspective trying to rule human sciences. But the subjectivity is carefully controlled – by researchers – through the unified protocol of MNs. The MNs install a well-balanced relationship between research and professional practice.

Second, MNs install the loop. When a teacher, alone in his/her class, elaborates a MN, s/he is not a pure observer; of course, s/he describes what s/he has wanted and done, what the answer of the class was; but this elaboration changes him or her, by the reflection it demands, and the process of collective validation of the MN will deepen the changes as well. Chapter 5 and chapter 7 are eloquent to this regard. After a MN, the teacher will not be the same person, and will not teach in the same way. The product has changed the process. This is a starting point towards long-term professional development. The interested reader can find examples in a former and recent book of the team, *Teaching Science. Contributions of Research for Planning, Practice and Professional Development.*

I do not want to end these notes without emphasizing the great importance of Chapter 12. You fully understand a human or social reality only when you know its history. That is why Chapter 12 recalls the development of MNs' technology. The examples in this chapter in particular give to see the development of a loop at the research level too, due to the use of MNs: when you engage using the MNs with their preformatted framework, you change the nature of the data you collect.

Finally, we must be grateful to the team to have created a website, giving access to a lot of MNs, and giving the possibility to other researchers to enrich this database. This offer is consistent with the adoption of the perspective of the Open Science movement, and also with the choice to address both teachers and researchers.

And that also allows understanding a reason, among others, why they publish a book about the method *after* some books about the results: the method has permitted to obtain some results, through a work that has refined and developed and extended the method, up to a step where other researchers or teachers can appropriate the method, solid enough, and participate in the recursion, enriching the results and the method, in other countries, other specific knowledge domains, other fields of interactive activities, beyond science education.

Enter the loop!

Christian Buty
Université Lyon 2, France
2018, November 2

Foreword

Research on educational practices is receiving a lot of attention currently. Whether for the purpose of a deep reflection on such practices, or simply due to the requirement for teachers to do research as part of their professional performance, studies have truly multiplied. However, the remaining problem is the socialization of them in their entirety. While the Open Science movement has expanded and institutional repositories have progressively become daily resources, authentic records of empirical data still remain mostly hidden. Neither journal articles nor books can include them due to space limitations.

On the other hand, researchers are facing an increasing difficulty entering the field due to several reasons including transportation costs in some cases, but primarily the educational institutions' reluctance to see their routine invaded or affected by academics (trained or in training process), who do not always return to share the results of their studies. Another obstacle is the intrinsic difficulty in collecting field data without altering the natural course of events. Currently methodological discussions in the literature often cover ethical concerns (e.g., how to guarantee total anonymity), and technological issues (how to take advantage of technologies to capture reality), but thus far solutions have not been found.

These reasons, among others, point to the need for a radical shift in empirical research; specifically, to fully utilize the empirical materials collected by teachers and researchers. How could this be done? By developing data collection techniques that capture the complex nature of educational settings, and portray the findings with multimodal elements that transform the narrative into a vivid experience. Diversified types of data, such as audio or audiovisual recordings, teachers' and students' notes, photographs of blackboards, classroom plans, diagrams illustrating the spatial positions and movements of teachers and students, educational resources used in class (posters, slide presentations), students' tests, among others, could possibly more thoroughly capture this complexity. In this way, it might be possible to "transport" the reader to the class.

Through such techniques, we could examine specific phenomena like the teacher's proposals and the students' reactions, or the intended work and the work actually performed by the students, and expand them in turn to the context (for a complete transversal view) and/or to the line story (for a longitudinal perspective), overcoming the reductionism of critical incidents or specific moments in class.

Although this approach seems to be a solution, it is not a simple one. It requires mechanisms capable of guaranteeing the fidelity of the empirical records, and making them complete and standardized so that they can later be used and compared with others. Those mechanisms, in turn, require a guide for achieving a rigorous description which can be openly shared; basically, a protocol to elaborate a Multimodal Narrative (MN). Finally, this tool requires the hard work of creating the necessary database to compile the MNs; validated documents by the academic community in an open source webpage. Thus, MN can be a tool and a collection, a process and a product, and the participants can be MN users and contributors.

The current book addresses these requirements, framed as the useful formulaic concept of Multimodal Narrative (MN), with ample background information throughout the chapters. Complemented with up-to date references, the book allows readers to build and complete their knowledge on the topic of MN. Overall, the strength of the book is based on two factors: the topic itself -i.e., the importance of MN-, and multiple explanations that show its benefits. Due to the clarity of language, this work is appropriate for a wide audience, including both researchers and educators. The possibilities of MN for research, teaching, and professional development in different contexts are clearly presented and explained. Furthermore, due to the particularities of each kind of MN practice, the book is organized in a first section that addresses them together and then devotes a section to each one.

In this dialectic of the specific profiles and the shared roles between teachers and researchers, existing barriers are removed. On the one hand, teachers act like researchers by collecting their own field material, taking advantage of their complete exposure to natural settings. Likewise, the methodological rigor, usually attributed to researchers, becomes the *modus operandi* of both the teachers and researchers. On the other hand, the researchers have the possibility to engage directly in the analysis of the material constructed by the teachers, suppressing the traditional hierarchies in which the former "own" the data. In this way, the authors fulfill a need for real collaboration between teaching and research, and link the different investigative processes across several contexts, countries and educational levels.

On the part of academics, the scientific research is made accessible to everyone, sharing the results of their work to promote professional development, and thus, contributing to the improvement of educational practices. This objective is imperative in Science, Technology, Engineering and Math, in which there is a need to increase the number of professionals and their skills to deal with a more complex society. In this context, this book is an invitation, a challenge, a proposal, a guide, and a goal.

María Isabel Pozzo
Universidad Nacional de Rosario, Argentina
2018, October 18

Preface

This book is dedicated to all teachers and Science and Technology (S&T) education researchers who want to improve teaching practices from a professional or a theoretical point of view.

It introduces a tool already validated by the scientific community – Multimodal Narratives (MN) – to a vaster audience.

This tool was developed by a research team as they felt the necessity of being able to capture the holistic nature of a classroom practice. A MN intends to be the class "story", combining subjective and objective data, producing a written genuine document able to transport the reader to within the class itself. The prior validation of each NM by a research team gives the reader the assurance of its truthfulness.

The reader may immerse in the MN "world" with different levels of commitment and this book will help the understanding of advantages in each one. For instance, the reader can find the necessary indications in order to begin to produce MN on its own. Or the reader may decide to simply read some MNs and reflect upon other teachers' practice or compare with personal experiences.

On the other hand, the reader may want to use MNs as a research tool in education research. Again, the reader will find different levels in which MN proves useful. From the data collection phase to the study and development of the theoretical framework in any area, by analyzing several MN produced in different contexts. As it will be explained, due to MN main characteristic of focusing on the teaching practice and not on the subjects, this diversity of contexts becomes a major advantage. As a *cherry on the top*, MNs can be found in an open source webpage, in an Open Science perspective, allowing any researcher to access MNs as a reader or as a contributor to enlarge this collection.

In this sense, the book is organized into these different perspectives of usage, following a gradual increase of commitment of the reader with MNs' usage: as a teacher to acknowledge different practices inside the classroom (Section 2), as a teacher or teacher educator using MNs as a tool to reflect and improve teaching practices (Section 3), as a teacher-researcher or researcher using MNs as a research tool (Section 4).

Following this reader-alignment, in Section 1, the reader can get a complete idea of what a MN represents, how to produce a MN and its different purposes. This section also serves as a reference part, where readers can go to address any doubt regarding structural or methodological aspects of the MN. It also provides indications on how to proceed in the access to the MN collection (http://multimodal.narratives.utad.pt) and an index of all MN present as their context and level of education. In the end of Section 1 there are two full examples of complete MN in order to better understand its potentialities. In Section 2, the reader can see particular aspects of teaching practices through MNs, how teachers exploit them and the corresponding developments in class. In Section 3, MN are used and explored in professional development, that is, how teachers may use MNs of their own classes or from other teachers to improve their own practice. Finally, in Section 4, some research studies using MNs are presented, allowing better understandings of the complexity of teacher mediation in the classroom. The Section 5 intends to show that MN can also be applied beyond S&T teaching practices or contexts. As long as there is a role of an educator, MN can be useful to better understand and improve that practice.

All through this book the reader can have different viewpoints: educator or researcher. In each one, can opt to read about it or to produce work with MNs. Also, with an open science perspective, using the MN collection, the reader may choose to work upon other teachers' MNs or produce some of his own.

All authors had directly or indirectly contacted with MN and their singular perspectives allowed this book to consolidate MN advantages from different angles. Every chapter in this book went through a process of double blinded review by at least two members of our Scientific Board. These experts from 29 universities and from 12 countries never worked with MN and the great majority never even heard about it before. This was intentional and provided an impartial scrutiny of this work. The thorough analysis produced in each chapter allowed the authors to improve their works and better capture the essence of their contributions. Complementary to this particular analysis, a methodologic and structural scrutiny was produced by the Advisors. Their insights allowed us to successfully improve the readability and flow of the book.

The massive positive opinion about MN capacities, both from the Scientific and Advisory Boards, is already a highly encouraging step towards the success of this book.

J. Bernardino Lopes
Universidade de Trás-os-Montes e Alto Douro, Portugal

Maria Clara Viegas
Politécnico do Porto, Portugal

José Alexandre Pinto
Politécnico do Porto, Portugal

Acknowledgment

This book represents a combined effort between a large team of teachers and researchers of Science and Technology Education to whom we are profoundly grateful and proud of having work with.

Even though we cannot name them all, we must express our gratitude to all researchers, teachers, students and school staff of all the involved institutions. Each one of them was invaluable in their cooperation and enthusiasm.

In particular, we address a word of thankfulness to each author of the chapters. Their belief in this project from day one, the quality of the presented work and their timely response to all challenges, allowed us to successfully complete this errand.

A special appreciation to our Scientific Board, who certified the quality of each work and gave precious inputs to improve their value and legibility.

An unlimited gratitude to the Advisory Board. We believe that with their help and expertise through the editing process and namely their advices upon the book structure, coherence and flow between the different parts, led to crucial improvements which, in our opinion, helped this book become an important contribution to the scientific community.

At last, we must acknowledge our gratefulness to the research center: *Centro de Investigação em Didática e Tecnologia na Formação de Formadores* – CIDTFF.

This work was financially supported by Portuguese National Funds through FCT – *Fundação para a Ciência e a Tecnologia*, I.P., project UID/CED/00194/2013.

Section 1

Multimodal Narratives: A Tool and an Open Resource for Practice, Research, and Professional Development

In Section 1, the reader can get a complete idea of what a multimodal narrative (MN) is, how to produce and validate a MN, and its different purposes. This section also serves as a reference part where readers can go to address any doubt regarding specifics aspects of the MN. In the end of Section 1, there are indications on how to proceed to access the MN collection (Chapter 1: Appendix 1). Furthermore, it provides two complete examples of MN in order to better understand its characteristics and potentialities (Chapter 1: Appendix 2).

Chapter 1
The Importance of Making Teaching Practices Public, Shareable, and Usable:
The Role of Multimodal Narratives

J. Bernardino Lopes
Universidade de Trás-os-Montes e Alto Douro, Portugal

Maria Clara Viegas
Politécnico do Porto, Portugal

José Alexandre Pinto
Politécnico do Porto, Portugal

ABSTRACT

It is acknowledged that to improve the value of the learning process and outcomes in areas such as science, technology, engineering and math, the teaching quality needs to be enhanced. Therefore, it is crucial to have access to real teaching practices. The multimodal narrative (MN) tool allows teaching practices to become public, sharable, and usable (open science perspective), preserving their holistic, complex, and ecological nature. This tool has characteristics and a structure that enable an in-depth study of teaching practices, in different contexts, with several purposes (e.g., teacher education, professional development, and research). This chapter presents MNs and the necessary steps involved in collecting multimodal data, structuring the narrative, and validating the document. MNs can be used by teachers and researchers, or other professionals, with multiple specific objectives, globally contributing to improving professional practices.

DOI: 10.4018/978-1-5225-8570-1.ch001

AN OPEN SCIENCE CONTRIBUTION: MAKING TEACHING PRACTICES PUBLIC, SHAREABLE, AND USABLE

In the context of our societies' strong immersion in Science and Technology (S&T) in everyday life, it is fundamental to ensure that the advances in S&T support and promote well-being both for individuals and for society from a holistic perspective. This principle not only involves the challenge of providing people with the capacity to understand and interact in scientific and technological contexts but also encompasses the recognition of the need to shorten the time between the creation of new knowledge and its dissemination and incorporation. This book can contribute to the Open Science movement.

The Open Science movement seeks to make scientific research accessible to everyone, particularly the research community, by taking advantage of the technological communication networks currently available. It is a concept that is widely recognized among researchers (e.g. Altunay et al., 2011; Fecher & Friesike, 2014) and institutions. In this regard, the European Union recognized the value of Open Science in the program Horizon 2020: EU Research and Innovation. It is also acknowledged by the Group of Seven (G7). In a statement (2017, p. 7), the G7 pronounced:

First, the incentives for the openness of the research ecosystem: the evaluation of research careers should better recognise and reward Open Science activities. Secondly, the infrastructures for an optimal use of research data: all researchers should be able to deposit, access and analyse scientific data across disciplines and at the global scale, and research data should adhere to the FAIR principles of being findable, accessible, interoperable, and reusable.

This book shows how Science and Technology (S&T) teaching practices can be public and shareable (the open science perspective), preserving their holistic, complex, and multimodal nature using a tool—Multimodal Narrative (MN)—validated by the academic community (Lopes et al., 2014). This tool has two components: a protocol to elaborate an MN and a validated document. As a document, the MN can be used in education research, in professional development, and in the articulation between S&T teaching practices and S&T education research to improve the quality of both. This tool allows researchers to overcome the difficulty of entering a classroom and obtaining reliable data on teaching practices that can be used and compared by teachers and researchers.

The characteristics and structure of MNs allow for comparability between them (even in different contexts) to achieve purposes related to the improvement of S&T teaching, research, and professional development.

MNs, both as a tool and as a collection, occupy an open science perspective, since (a) professionals can elaborate an MN of their own professional practices, based on data, thus participating in the research process and gradually increasing the MN collection available, and (b) researchers can access and analyze scientific data about S&T teaching practices in several contexts and countries and at different educational levels.

MNs open new possibilities for:

- Research, because they allow the comparison of teaching practices at different levels of education and in different countries and contexts, and even professional profiles, allowing a deeper understanding;

- Teacher education and professional development, because researching professional practices may influence the quality of teaching practices.

The team of researchers supporting this book, in tune with this movement, was sensitive early on to the dimensions of dissemination and the sharing of data and results. On one hand, it was considered to be very important to disseminate an instrument (MNs) that, from our perspective, could be useful or even fundamental for other teams. Its characteristics allow the preservation of the holistic, complex, and ecological nature of teaching practices. This is highly significant both for research and for professional development processes. On the other hand, through this work, the team publishes and makes available an extensive corpus of data that have already been produced and validated, which can be exploited by other teams of researchers in other places and with other perspectives in mind. With this sharing, it is hoped that these "ready-to-use" resources will provide an incentive for structuring new or expanded research networks. It is also our goal to increase the number of available MNs, which will be achieved if new partnerships and networks emerge. In this sense, this work is a contribution to concretizing the concept of open science.

In this book, we also present a set of scholarly works that show how MNs can be used for different purposes: seeing teaching practices in S&T through MNs, aiding the professional development of S&T teachers using MNs, researching S&T practices using MNs, and using MNs in other professional contexts. The use of MNs was initially confined to science teaching. It becomes evident in this book that the open science perspective is assumed by presenting, explaining, using, and exploring the possibilities of use (researching, teaching, and professional development) of MNs, which may:

1. Enhance the possibilities of researching S&T teaching practices by other teams, using the currently available MN collection in a database articulated with the book; the portal http://multimodal.narratives.utad.pt includes the MN collection and information about the project (Appendix 1);
2. Expand the MN collection with other MNs from different professionals and contexts;
3. Disseminate the MNs' potential to larger audiences, which can include not only academics but also teachers and other professionals who have daily contact with other people in their practice.

BACKGROUND

One of the major difficulties in educational research is accessing classrooms to obtain data on real teaching practices. There are several instruments for collecting data from teaching practices, such as video or audio recordings (Kung, Kung, & Linder, 2005), diaries (Hundhausen, 2002), and narratives (Craig, 2011). However, our research concluded that these instruments do not thoroughly capture crucial features, namely: descriptions of teachers' decisions and their intentions and context and descriptions of students' efforts and the work actually undertaken. Moreover, it was found that these instruments did not have a characteristic that was revealed to be essential to our work—a given focus and structure—to allow more rigorous, precise, and powerful comparability between different situations and cases (Lopes et al., 2014, p. 416).

Several approaches have to studying the teaching practice of teachers in the classroom been identified: inferring teaching characteristics by linking them to the school culture, curriculum material, teacher knowledge and beliefs, and students' learning (Borko, Stecher, & Kuffner, 2007; Tytler, 2003) and using artefacts in the classroom combined with self-reporting (Martínez, Borko, & Stecher, 2012).

During the pursuit of a tool that made those combined characteristics explicit, we came across an important book, *Researching your own Practice: The Discipline of Noticing* (Mason, 2002). This work was very inspiring and gave us the idea of mixing the class recordings with the first-hand teachers' intentions, remarks, or complementary information. We continued to search for inspiration on reports (Alsop, Bencze, & Pedretti, 2005), narratives (Craig, 2011), and structured daily records with information about events or interactions in the classroom (Rowan, Camburn, & Correnti, 2004). However, we were not merely interested in critical incidents or specific moments in class; we also wanted to acknowledge their context and line story. Still, our team could not find a tool that combined all these characteristics.

Thus, we identified the need for a tool that is not limited to collecting data about the class but is also capable of gathering, as completely as possible, the context of the lesson, the teacher's intentions and his/her perception, the students' reactions, the planned work, the work actually performed by the students, and so on. Consequently, we began to collect a more thorough and diversified type of data, such as audio recordings, teachers' notes, pictures of blackboards, diagrams, figures, software and movies, classroom plans and the predominant spatial position of teachers and students, students' notebooks, other materials used in class (slide presentations, tasks, etc.), students' perception tests (about teaching and learning), teachers' and students' attitudes or teachers' objectives, and work performed in class to develop skills and concepts.

Finally, the Multimodal Narrative tool began to take shape within the science education research team, headed by J. B. Lopes, and targeted the need for access to real classroom practices without jeopardizing its integrity.

The first narratives started to take place without the multimodal elements. The first reference to this approach was made by Lopes et al. (2008). After being analyzed systematically by the teachers and researchers, our research team felt the need to refine the narratives. At first, their characteristics strongly evidenced the different personal choices of the authors. Since the objective was to research teacher mediation in the classroom, independent from its context, the next step was to standardize the narratives' features to ensure the same characteristics and allow comparability. One of the major problems was understanding that it was not possible to focus on the issues that teachers wanted to highlight, because this would bring too much subjectivity into the narratives. This would not only jeopardize their comparability but would also lose the timeline of the class and everything else that happened. Thus, the focus should be on the class narrative as a whole, as a story, with a timeline and without the omission of any steps.

To overcome these difficulties, the Multimodal Narrative structure was consolidated into two different parts: first, a succinct narrative of the whole class, with a storyline filled with contextual information; second, a detailed description of the actions and the teachers' and students' discourse during the work developed in the classroom. Each activity should be described thoroughly from the beginning to the end. Afterwards, some characteristics were established as being independent of the class context or the narrator. These included being a true story with a chronological timeline that could be verifiable (linking to audio recordings, student-produced documents, and other data). The final touch was the inclusion of the multimodal elements that would make the Multimodal Narrative (MN) form unique, transporting the reader to the class. This tool was finally validated by the scientific community (Lopes et al., 2014).

In sum, this MN tool focuses on the substantive description of what the teacher and the students do and say during the proposed tasks and should describe in practical terms the teacher's intentions and decisions during the class, all substantiated with rich multimodal elements that transform the narrative into a vivid experience.

Several Multimodal Narratives (MNs) were then produced by various authors from different levels of teaching and from different science and technology contexts. Their structure, characteristics, and focus were analyzed to ensure their comparability. Separate meetings with each author were conducted to assess particular issues and identify ways to solve them. At that moment, it became clear that (a) there should be a set of procedures for collecting the data; (b) it would be possible to solve the problem of objectivity and the systematization of the MNs without prejudicing the inclusion of subjective aspects of teaching (e.g. teacher intentions and justifications of actions) if a set of clearly identified steps was followed; and (c) the narrative would constitute a document linked to the remaining data and therefore acquire its multimodal status.

External expertise was consulted during this phase of consolidation as a research tool. It helped in establishing the MN protocol to elaborate and validate the tool. This tool was then validated by the research community. Our team used the expression multimodal narration in Lopes et al. (2010) and the expression Multimodal Narrative in the international literature of Santos, Lopes, and Cravino (2011).

The term Multimodal Narrative (MN) can refer to two different entities:

1. The MN document, which constitutes a validated resource of classroom practice through a self-contained descriptive narrative (albeit linked to independent data) incorporating all the multimodal elements. This document is usable, public, and sharable.
2. The MN protocol for constructing an MN, which implies:
    a. A set of indications of how to collect data from and about the classroom;
    b. A set of follow-up steps to construct an MN that helps the narrator to transform the collected data into a written document;
    c. A set of steps to validate an MN that includes the certification of its major characteristics and internal consistency.

These two entities will be explained fully in the following sections. The section "The MN Document" will explain the characteristics of the final document. The following section, "The Multimodal Narrative Protocol," will explain in detail the whole process of constructing an MN, from the data collection to the validation of the final document. Finally, in the "Discussion" section, the roles of Multimodal Narratives discussed in this book will be debated, providing some insights into their application.

THE MN DOCUMENT

As a document, an MN is a rich chronological narration of what actually happened in a specific professional practice context (e.g. classroom teaching practice), aggregating several types of data. It is a multimodal and self-contained description of teachers' and students' actions (Lopes et al., 2014). MNs are characterized by their reliability in representing public descriptions of professional practices in the real-work context. It may report teachers' practices in the classroom or other professionals' practices. Its elaboration is performed according to a protocol with indications to collect data, narrate, and validate the final document. It is important to stress that an MN does not have the purpose of representing paradigmatic classes. It merely reports how the events took place based on described decisions and reactions and avoids making judgments.

In this way, different MNs, in spite of their different contexts, can be used for multiple objectives due to their similar features and structure:

- As an object of research. Since their foci are not the "teaching subjects" but the "teaching practices," MNs from different contexts can be used to compare different practices and serve wider purposes related to the improvement of S&T teaching or other professional practices.
- To provide an integrative vision of an S&T teacher in action, showing how events may evolve in the classroom and be addressed in different ways.
- To be used by teachers in the context of their professional development or even in their initial training.

The MN document aggregates all the collected data into a written story of a class, with a timeline and without the omission of any important facts. It prevents the difficulty of addressing multiple sources or the exposure of teachers' identification. It is a single document that is ready to be used by several communities, such as educational research, teacher professional development, teacher education, or the education of other professionals.

The structure of an MN contains two distinct parts. The first part consists of (a) contextual elements (the specific time, space, and circumstances of the classroom and the teacher's and students' characteristics), a classroom plan, and the predominant spatial position of the teacher and students; and (b) a brief description of the storyline between the distinct classroom tasks. These units around the tasks are called episodes, and they chronologically mark the time of the class. The second part consists of a descriptive, detailed, and multimodal report of each episode, in the order in which they occurred in the class. Each episode is vividly narrated from its beginning until its end (that being the closure of the task or the teacher's decision to end it).

MNs' characteristics are summarized below:

- **Modularity** – based on a contextualized description of the class and a detailed description of distinct episodes; each MN can report one or more episodes in the order in which they were accomplished in the class;
- **Accountability** – an honest and self-contained account of the events from the beginning to the end of each episode (usually associated with a task);
- **Wholeness:** A chronological narrative focused on the teacher's and the students' actions and discourse, whether their nature is pedagogical, didactical, sociological, or epistemological, during the development of each task;
- **Singularity:** A genuine history (with the mark of the author) and singular nature (occurring in times, spaces, and particular circumstances);
- **Multimodality:** Incorporating a variety of multimodal elements, including actions, representations, photographs, schematics, space organization, the students' reactions, excerpts from dialogues, silences, the teacher's decisions and intentions, the students' work, the types of language and discourse, and so on (this multimodal component is fundamental because it is not registrable otherwise);

- **Verifiability:** Through its connection to other data: audio recordings, video, documents produced by the students, tasks set by the teacher, and so on; previous validation accomplished by an expert team of researchers who have cross-analyzed all the data and attested to their truthfulness;
- **Objectivity:** Avoiding judgments and justification (predominance of nouns and verbs, avoidance of subjectivity, with the minimum possible interpretation);
- **Usefulness:** Represents a single written document that can be made public, shareable, and useful for research, teaching, and professional development;
- **Comparability:** Allowed by the set of characteristics mentioned above and by its structure;

These characteristics are assured by the protocol for preparing MNs and by their validation before being considered finished and becoming public and usable.

At this Section 1 of the book, shortly after this chapter, the reader can find two examples of Multimodal Narratives with comments showing its structure and main characteristics. Epistemologically, MNs represent a validated description of an unrepeatable reality (what happened in the classroom), which is reliable (congruent with the data) and public. It is the first degree of abstraction in the treatment of the reality of what happened in class (which is, by nature, singular and unrepeatable)—an essential step towards the construction of scientific knowledge (Mugur-Schächter, 2006).

THE MULTIMODAL NARRATIVE PROTOCOL

The process of constructing an MN, although complex, helps to structure the collected data. With some practice, the process becomes more methodical and easier to accomplish. It consists of three phases, and each one will now be explained in detail.

Collecting Relevant Data for a Multimodal Narrative

To be able to produce a rich MN, special care must be taken to prepare for the gathering of data. The crucial multimodal elements need to be identified as well as the means for acquiring them. The key factor is to collect as many elements as possible. Each class is unique and will occur only once; we must ensure that we have all the necessary elements to attest without a doubt to what happened in the class.

There are two kinds of data: independent data and teacher-dependent data. The first group refers to the kind of data that stand on their own and can be worked by anyone; this includes audio recordings, written documents that served as support for the class, photographs of the blackboard, photographs of the students' layout in the room, documents produced by the students, and the teacher's description of the context (identification of the resources used, characteristics of the school, students, and teacher, etc.). On the contrary, the second group, dependent data, requires the teacher's participation to be described. This group may include the teacher's intentions, actions (which may or may not be on record), reactions, attitudes, silences, gestures, decisions, and so on.

The purpose is to bring these two kinds of data together throughout the narrative, using them to create the class story. What was said, asked, or discussed, how and when, how difficulties were overcome, what was changed relative to the initial plan and why (immediately, after the class, the registration of the teacher's opinions, perceptions, and reactions), students' perceptions and reactions (complemented by the report on the teacher's observations), and so on. Regardless of how insignificant they may seem at the time, all the identified and collected elements can make a difference.

Thus, the MN tool has also become a tool for organizing, systematizing, and collecting data, which is useful not only in research but also in didactic work.

Writing the Multimodal Narrative

The second phase of the process of constructing an MN is the longest. It may be accomplished by the teacher himself or by a researcher, as long he/she stays in contact with the teacher who led the class.

Most narrators find it helpful methodologically to follow this sequence:

- Gather all the data and compare and organize the teacher-independent and teacher-dependent data;
- Listen to the audio recording (or watch the video) and identify the moments into which the multimodal data can be inserted;
- Identify all the episodes and, with the help of the audio recording timer, register the time of their beginning and end (for longer episodes also somewhere in between);
- Elaborate a succinct narrative of the whole class, with its storyline, and identify all the episodes;
- Fill it with all the contextual information (about the school, the students, the resources, the background, etc.);
- Write each episode, narrating what happened as if telling a story (listening to the audio recording in more detail) and focusing on the teacher and student discourse, interactions, or actions;
- Progressively enrich the MN and each episode with all the multimodal elements.

A complete written transcript of the recording is not mandatory (and, in most cases, is not recommended). However, the narrator must carefully select the relevant segments of each episode and transcribe in full the parts that he/she considers to be relevant (including dialogues). However, to avoid too much subjectivity, the narrator should follow the rule of considering to be important any aspect (remotely linked to the class thematic) that caught the students' attention. It may also be the case that some segments are synchronous, that is, referring to the same moment in the class (for instance in the case of having multiple audio recorders scattered around the room in team work). To enable a better understanding (and to assist the validation process), the time in the audio recording should be registered on a regular basis.

The MN should be able to transport the reader to that class experience; that is, it should show how actions were linked and recognize them as a genuine practice. Its descriptive and honest character, together with the multimodal aspects, helps to make it rich in expressiveness and identify it as an object of reliable study. However, for this to be possible, the narrator must assure that all the aspects are verifiable; that is, all the narrated elements must be substantiated by the collected data. This constitutes a hermeneutic unit in which the written document represents the public face of the MN (Lopes et al., 2014).

Naturally, if the narrator is the teacher himself, the whole process of writing the MN will be easier and less time consuming. The narrator will automatically remember the details of what happened inside the classroom while listening to the recording, especially if he/she listens to it promptly. However, in some cases, that is not possible, and a researcher (not present in the class) can become the narrator. In these cases, the narrator must remain in close contact with the teacher throughout the process to gain a better understanding of the events and the introduction of the teacher-dependent data (Branco, Magalhães, Ferreira, Lopes, & Cravino, 2009). For example, special multimodal elements, such as class dynamics, students' faces, or silent reactions, can only be transmitted by those in the class. Furthermore, the teacher's intentions or the contextual information can only be explained by him/her. In these cases, it will be necessary to add a step to the procedure: the teacher must read the written document of the MN and feel comfortable with it or suggest the necessary modifications.

Validation of the Multimodal Narrative

The beauty of an MN is that, for the public reader (a researcher, a teacher, or a professional in any other context), it appears as though it is only a written document. However, this written document has the advantage of having been certified as a truthful document, holistically expressing the occurrences inside a classroom. This certification occurs in the last phase of the process of the MN construction.

The analysis is conducted by expert researchers, who attest to the precision, reliability, legibility, and internal coherence. In parallel, they assure that the MN's characteristics were achieved, and their focus is on the teacher's and students' actions and discourse during a specific task. Individually, these researchers must consult the data collection while reading the written document. In the case of any doubt, reformulation is requested, specifying the issues to which the narrator must attend. Afterwards, re-evaluation is conducted. When all the MN characteristics are guaranteed by the external researchers, the MN is considered to be valid and can become a public document only consisting of the written narrative enriched by the multimodal elements.

The whole process of the MN construction is schematically shown in Figure 1.

Once validated, the MN remains unchanged, can become public, and can be used for numerous purposes. The difference from a classroom video, an audio recording, or simply a transcription is that, in those cases, they are traditionally analyzed in isolation and do not represent an integrated, intersubjective, and articulate description of what happened in the classroom.

PERSPECTIVES ON THE ROLE OF MULTIMODAL NARRATIVES

As mentioned before, the MN tool was developed due to the need apparent in an educational research context. This context, from which this tool emerges, is therefore particularly familiar and constitutes a privileged field of use. However, it was soon realized that this tool could offer much more. Teachers who used the MN tool reported the experiences themselves as being rich. It was common for teachers to acknowledge that the reality of their classes was sometimes very different from their immediate (post-class) perception. As they attested, it improved their awareness about their intentionality and decision-making processes, increased their experience and practical knowledge, and helped them to grow professionally.

Consequently, it was realized that the MN tool could serve different purposes and be considered useful to several communities of practice: pre-service teachers, teachers in practice, and researchers. More recently, its usefulness for further professional practices/contexts (other than formal classes), in which communication with others is relevant to pursue a certain goal of achievement, also became clear.

Figure 1. Constructing an MN: The process and its phases.
(adapted from Lopes et al., 2014)

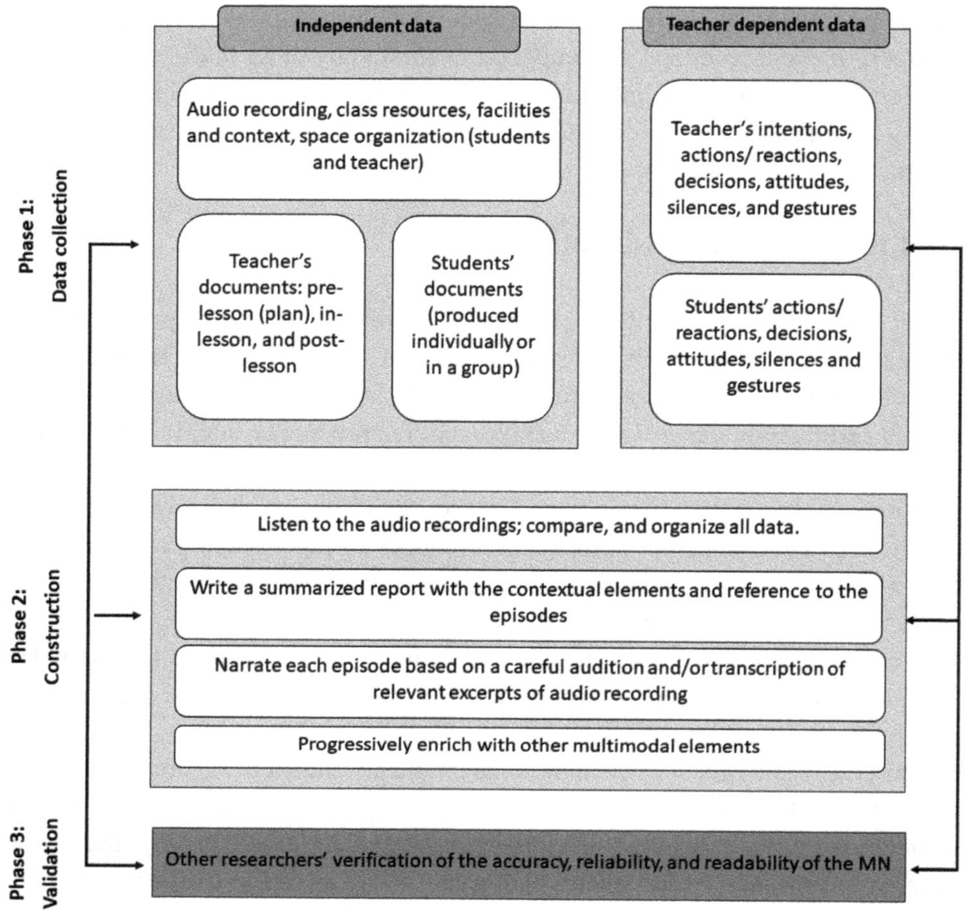

Seeing Teaching Through Multimodal Narratives

While preserving classes in their holistic, complex, and multimodal nature, teachers can use MNs to help them to reflect on their teaching practices. In essence, an MN is a multimodal, self-contained, reliable, and public description of professional practices in the real-work context. Thus, the study of different MNs allows teachers to see and compare different approaches (even in different contexts) and reflect on their own.

On the other hand, teachers who write their own MN can benefit from the natural self-reflection process that seeing their own process will demand. That will help them to evolve and/or consolidate their own practice. This can be useful in different circumstances. For example:

- Providing pre-service teachers with wider contact with real teaching practices, enabling them to benefit from live examples of dealing with the natural situations in a classroom.

- Aiding young teachers in analyzing their own first practices by constructing their own MN of their classes. The simple act of constructing the MN can help them to change some personal beliefs and grow professionally by increasing their experience and practical knowledge.
- Enabling in-service teachers to reflect on other teachers' practices and experiences. The reflection on different approaches may aid changes or consolidation of their own practices.
- Helping teachers to reflect on their own practice and gain more consciousness of their teaching classes.

Section 2 of this book presents some studies showing how MNs can be used to address specific issues related to teaching practices. These chapters will allow the reader to understand better how MNs can help them to see a classroom from the inside and to observe, for instance, particular dimensions of the teacher's mediation, how a particular didactic resource works in interaction with students, or even characteristics of students' interactions with each other.

Professional Development with Multimodal Narratives

Regardless of their nature, educational research approaches' goals are mostly related to improving teaching and, consequently, learning. In this sense, by nature, MNs' characteristics support their relevance to professional teacher development. The use of MNs allows different perspectives, presents facts for observation and discussion, and shares dynamics with a beginning and an end. Within a limit, every situation described in the previous subsection can be a potential case of professional development. In particular, this can be useful in the framework of ongoing teacher training, in which reflection on teachers' action is as important as core competences for professional development. However, these competences are still equally relevant in an initial teacher training framework. Firstly, the use of MNs is easily adjustable to the initiation processes of professional practice and internship. Returning to the centrality of reflection on the contexts and practices, MNs can not only foster the adjusted and necessary distance from the young teachers while providing them with the opportunity to consider what happened and perform a retrospective analysis of the class. Lopes and Cunha (2017) showed that MNs can work as a tool for professional development in the double sense that they support the interaction and reflection among teachers based on selected excerpts of MNs and they can form the base for collaborative work with researchers to analyze, from a research perspective, a specific aspect of teaching practices, creating the conditions to improve it.

We intend to emphasize the MN tool's capabilities of supporting professional teacher development processes, independently or linked to research dynamics (Lopes, 2017). As we mentioned earlier, within the research team, we have experiences of using the MN in a context of collaboration between teachers. Some of these studies, addressing general professional development and initial teacher training, will be presented in Section 3 of this book.

Research with Multimodal Narratives

Educational contexts are known for their diversity and complexity; it is impossible to categorize them all together. All teachers know that every school is different, every scholastic year is different, every student is different, and every class—even when repeating the subject—is different. In this sense, researching education is a challenge, and the comprehension of what happens inside the classroom is crucial. The need for detailed studies that show the intersubjective and interactive dimensions of these phenomena led to the development of qualitative studies (Bogdan & Biklen, 1997).

Researching education using the MN tool can easily fall into this category. By systematically focusing on teacher mediation inside the classroom and addressing and organizing relevant and diversified data, the MN is a powerful tool that can help researchers to understand this complexity better. An MN not only provides information about how the events took place in the classroom but also preserves the relationships and their context surroundings, and that can be determinant in a study of this nature.

However, MNs also have great potential for use in quantitative or combinations of research methodologies. Their ability to preserve large data samples, their comparability, and their cross-analysis of cause/effect situations may be some examples, for instance the identification of patterns between teachers' mediation actions and students' outcomes.

Some research studies using MNs are presented in detail in Section 4 of this book. It is our belief that MNs' potential is exponentially and qualitatively growing with the consistent increase in the collection of MNs. The developing richness of different educational contexts, levels of education, countries, teachers' backgrounds, students' backgrounds, resources, approaches, and so on multiplies the opportunities for research that can be addressed. In Section 1 of this book, two examples of MNs and the MN collection structure are presented (Appendix 2). Our goal is to enrich this collection's potentiality by inviting researchers/teachers/other professionals not only to use the existent MNs but also to contribute to their enlargement.

Multimodal Narratives' Usage in Other Contexts

In most professions in which the access is preceded by a formal period of training, this initial period before fully embracing the profession provides young professionals with a high number of challenges and uncertainties. The outcomes of this period may have a tremendous effect on these young professionals. The analysis of this initial period has been attracting growing attention. Usually it is accomplished through an experienced person helping less experienced ones to take their first steps in the profession. This mentoring process seems to be an area in which the MN tool can also be extremely useful for opening areas that are not directly related to formal education to other professionals.

This can be the case in several professions in which interaction with others is fundamental, such as the examples reported in Section 5, in which the MN tool has already proved itself to be useful: nurse-related practices, apprenticeships, or scientific museums' visit guides (see chapters 17 and 18).

According to its characteristics, the MN is a robust tool, able to preserve and organize data on events and flexible realities, potentiating reflection centered on practices and interaction with people. We believe that many other fields or scenarios will be revealed in the future.

REFERENCES

G7. (2017). *Group of Seven Science Ministers Meeting | Declaration*. Retrieved from http://www.g7italy.it/sites/default/files/documents/G7%20Science%20Communiqu%C3%A9.pdf

Alsop, S., Bencze, L., & Pedretti, E. (Eds.). (2005). *Analysing exemplary science teaching*. Maidenhead, UK: Open University Press.

Altunay, M., Avery, P., Blackburn, K., Bockelman, B., Ernst, M., Fraser, D., ... Livny, M. (2011). A science driven production cyberinfrastructure—The open science grid. *Journal of Grid Computing*, *9*(2), 201–218. doi:10.100710723-010-9176-6

Bogdan, R., & Biklen, S. K. (1997). *Qualitative research for education*. Boston, MA: Allyn & Bacon.

Borko, H., Stecher, B., & Kuffner, K. (2007). *Using artifacts to characterize reform-oriented instruction: The scoop notebook and rating guide* (CSE Technical Report 707). Los Angeles, CA: The Regents of the University of California.

Branco, M. J., Magalhães, R., Ferreira, A., Lopes, J., & Cravino, J. (2009). Narrações de aulas: um estímulo ao desenvolvimento profissional? [Narratives from lessons: A stimulus to professional development?]. In F. Paixão & F. Jorge (Eds.), *Proceedings XIII Encontro Nacional de Educação em Ciências* (pp. 740–749). Castelo Branco: Instituto Politécnico de Castelo Branco.

Craig, C. J. (2011). Narrative inquiry in teaching and teacher education. In J. Kitchen, D. C. Parker, & D. Pushor (Eds.), *Advances in research on teaching* (Vol. 13, pp. 19–42). Bingley, UK: Emerald Group Publishing.

Fecher, B., & Friesike, S. (2014). Open science: One term, five schools of thought. *Opening Science*, 17–47.

Hundhausen, C. D. (2002). Integrating algorithm visualization technology into an undergraduate algorithms course: Ethnographic studies of a social constructivist approach. *Computers & Education*, *39*(3), 237–260. doi:10.1016/S0360-1315(02)00044-1

Kung, R. L., Kung, P., & Linder, C. (2005). Equipment issues regarding the collection of video data for research. Physical review special topics. *Physics Education Research, 1*(1), 010105-1/010105-9. doi:10.1103/PhysRevSTPER.1.010105

Lopes, J., Silva, A., Cravino, J., Santos, C., Cunha, A., Pinto, A., ... Branco, M. (2014). Constructing and using Multimodal Narratives to Research in Science Education: Contributions Based on Practical Classroom. *Research in Science Education*, *44*(3), 415–438. doi:10.100711165-013-9381-y

Lopes, J. B. (2017). Researching science teaching practices in the classroom over two decades: What matters? In J. B. Lopes, J. P. Cravino, E. Cruz, & A. Barbot (Eds.), *Teaching Science: Contributions of Research for Planning, Practice and Professional Development* (pp. 1–25). New York: Nova Science Publishers.

Lopes, J. B., Cravino, J. P., Branco, M., Saraiva, E., & Silva A. A. (2008). Mediation of student learning: Dimensions and evidences in science teaching. *Problems of Education in the 21st Century, 9*(9), 42–52.

Lopes, J. B., & Cunha, A. E. (2017). Self-directed professional development to improve effective teaching: Keys points for a model. *Teaching and Teacher Education, 68*, 262–274. doi:10.1016/j.tate.2017.09.009

Lopes, J. B., Silva A. A., Cravino, J. P., Viegas, C., Cunha, A. E., Saraiva, E., … Santos, A. (2010). *Investigação sobre a Mediação de professores de Ciências Físicas em sala de aula* [Research on Teacher Mediation of Physical Science on classroom]. Vila Real: UTAD.

Martínez, J. F., Borko, H., & Stecher, B. M. (2012). Measuring instructional practice in science using classroom artifacts: Lessons learned from two validation studies. *Journal of Research in Science Teaching, 49*(1), 38–67. doi:10.1002/tea.20447

Mason, J. (2002). *Researching your own practice: The discipline of noticing*. New York: Routledge.

Mugur-Schächter, M. (2006). *Sur le tissage des connaissances* [On the weaving of knowledge]. Paris: Hermès Science Publishing – Lavoisier.

Rowan, B., Camburn, E., & Correnti, R. (2004). Using teacher logs to measure the enacted curriculum: A study of literacy teaching in third-grade classrooms. *The Elementary School Journal, 105*(1), 75–101. doi:10.1086/428803

Santos, C. A., Lopes J. B., & Cravino, J. P. (2011). Patterns of didactic decisions made by teachers in physical sciences classrooms. *Problems of Education in the 21st Century, 36*, 76–89.

Tytler, R. (2003). A window for a purpose: Developing a framework for describing effective science teaching and learning. *Research in Science Education, 33*(3), 273–298. doi:10.1023/A:1025423704068

APPENDIX 1

Multimodal Narratives Collection: Information and Access

The collection of validated Multimodal Narratives (MNs) currently exceeds 100 MNs which are accessible on the website http://multimodal.narratives.utad.pt/

In this section we intend to provide information on how MNs are organized and the diversity of educational contexts and countries (up to now: Angola, Brazil, Portugal) where these teaching practices occurred. There are MNs narrated by the teacher himself and others narrated by a researcher. The MNs relate to teaching practices that have taken place in Basic Education (Natural Sciences, Mathematics, Physical and Chemical Sciences, ICT), Secondary Education (Physics and Chemistry, ICT), Professional Courses (ICT, Informatics), Higher Education (Physics, Mathematics) and Adult Education and Training.

The MNs' collection is open, that is, after validated, it can receive MNs of professional practices of teachers or other professionals in science and technology education, or other professionals who are based on interaction with people such as doctors, nurses, caregivers, etc. The submission process is described on the same website. In this collection all the names of people and institutions are fictitious as other ethical aspects which are described in detail on the website.

The MNs' collection may be used for educational or investigative purposes, by registering on the website and providing that it is correctly referenced:

Lopes, J. B., Cravino, J. P. (2017). Práticas de Ensino de Ciências e Tecnologia - Acervo de Narrações Multimodais [Science and Technology Teaching Practices – Multimodal Narratives Collection]. Vila Real: Universidade de Trás-os-Montes e Alto Douro. http://multimodal.narratives.utad.pt ISBN (electronic version): 978-989-704-253-9 (PDF/A).

The portal and the pdf book are under Creative Commons Attribution-NonCommercial-NoDerivatives 4.0 International Public License (CC BY-NC-ND 4.0). http://creativecommons.org/licenses/by-nc-nd/4.0/

The Current state of the MNs' Collection is divided in three major levels of Education:

Basic Education

Section 1 – Natural Sciences or Physical and Chemical Sciences (37 MN)
Section 2 – ICT (4 MN)
Section 3 – Mathematics Teaching (1 MN)

Figure 2.

Secondary Education

Section 4 – Physics and Chemistry (25 MN)
Section 5 – ICT (1 MN)
Section 6 – Professional course – ICT (15 MN)

Higher Education

Section 7 – Physical Sciences – ICT (15 MN)
Section 8 – Mathematics Teaching (5 MN)
Section 9 – ICT Professional Training (1 MN)
Section 10 – Adult Education and Training (1 MN)
Section 11 – Other Contexts (2 MN)

As an example, two of these MN are transcript in full in the end of Section 1 (MNs from two different contexts in terms of level of Education and thematic).

These two MNs are identified in in *italics* in the Tables 1-11.

APPENDIX 2

Multimodal Narrative Examples

Multimodal Narrative 1

Context: Formal education
Country: Portugal
Professional Code: Sofia
Professional Activity: Teacher
Narrator: Teacher who taught the class
Narrator Code: Sofia
Context of Education: Physical and Chemical Sciences
Course: Chemistry
Topics: Acid-base chemical reactions; atmospheric pollution
Level of Education: Basic education—Eighth grade
Age Group: 13 years
School Year: 2007/2008
First Class on the subject (05 May 2008)
Total Class Time: 90 min

Table 1. Current state of the Structure of the MNs' Collection and brief information about each MN (105 MNs): Section 1

Section 1 **Ensino Básico [Basic education]** **Natural Sciences or Physical and Chemical Sciences**			
Narrated by the Teacher Himself			
Teacher Maria	**4MN**	2MN 2008	1st MN of 2008: Electric receivers; Series and parallel circuits; Equipment protection organs; Electrical installations at home.
			2nd MN of 2008: Electrical resistance of a conductive wire; Electric circuit; Relation between the potential difference at the terminals of a copper wire, and the current intensity; Ohm's Law.
		2MN 2009	1st MN of 2009: Properties and Applications of Light. Simulation.
			2nd MN of 2009: Reflection of Light. Simulation.
Teacher Sofia	**2MN**	2MN 2008	*1st MN of 2008: Acid - base chemical reactions; Atmospheric pollution.*
			2nd MN of 2008: Base acid chemical reactions (continuation); Chemical precipitation reactions; Atmospheric pollution: Formation of stalactites and stalagmites.
Teacher Flor	**1MN**	1MN 2014	1st MN of 2014: Water treatment; Decantation; Filtration; Disinfection.
Narrated by a Researcher			
Lessons from Teacher Isabel narrated by the researcher Diana	**7MN**	1MN 2008	1st MN of 2008: Atoms and molecules; Elemental substance and compound substance; Atomic structure.
		4MN 2009	1st MN of 2009: Electric circuits; Electrical circuits in series, and in parallel.
			2nd MN of 2009: Electric circuits; Conductors; intensity of electric current.
			3rd MN of 2009: Electric circuits; Resistance; Potential difference.
			4td MN of 2009: Electric circuits; Ohm's Law.
		2MN 2010	1st MN of 2010: Properties and Applications of Light; Parallel, convergent and divergent light beams; Translucent body; transparent body; opaque body.
			2nd MN of 2010: Properties and Applications of Light; Bodies opaque, transparent and translucent; Reflection laws.
Lessons from Teacher Joana narrated by the researcher Diana	**8MN**	2MN 2008	1st MN of 2008: Properties and Applications of Light; What is light; Propagation of light; Shadows; Electromagnetic Spectrum.
			2nd MN of 2008: Properties and Applications of Light; Reflection of light; Reflection laws.
		6MN 2009	1st MN of 2009: Atomic structure; Evolution of the atomic model.
			2nd MN of 2009: Atomic structure; Constitution of the atom; Eletronic distribution.
			3rd MN of 2009: Atomic structure; Isotopes; Ions; Periodic table.
			4th MN of 2009: Atomic structure; Periodic table.
			5th MN of 2009: Atomic structure; Periodic table; Ions.
			6th MN of 2009: Atomic structure; Covalent bond.

Contextual Information

This is a group of 22 students aged between 13 and 16 years old, who are attending the eighth year of schooling for the first time and of whom only 5 have retentions throughout their school year.

Section 1 Ensino Básico [Basic education] Natural Sciences or Physical and Chemical Sciences			
Lessons from Teacher Ricardo narrated by the researcher Diana	11 MN	2MN 2008	1st MN of 2008: Energy; Source; Receptor; energy transfers.
			2nd MN of 2008: Energy; Renewable and non-renewable energy sources; Kinetic energy and potential energy.
		6MN 2009	1st MN of 2009: Atomic structure; The evolution of the atomic model.
			2nd MN of 2009: Atomic structure; Atomic model.
			3rd MN of 2009: Atomic structure; Ion formation; Location of elements in Periodic Table.
			4th MN of 2009: Atomic structure; Size of atoms and ions.
			5th MN of 2009: Atomic structure; Covalent bond.
			6th MN of 2009: Atomic structure; Electrical conductivity of aqueous solutions, ionic bonding.
		3MN 2010	1st MN of 2010: Properties and Applications of light; Reflection of light; Characteristics of the images formed by the mirrors.
			2nd MN of 2010: Properties and Applications of light; Refraction; Total reflection; Characteristics of images formed through lenses.
			3rd MN of 2010: Properties and Applications of light; The color of light and objects.
Lessons from Angolan's Teachers narrated by the researcher Gil	4MN	4MN 2007	1st MN of 2007: Interaction Forces.
			2nd MN of 2007: Force and mass; Interaction between material systems.
			3rd MN of 2007: Force and mass; Interaction between material systems.
			4th MN of 2007: Force and mass; Interaction between material systems.
TOTAL MN from this section: 37 MN			

Table 2. Current state of the Structure of the MNs' Collection and brief information about each MN (105 MNs): Section 2

Section 2 Ensino Básico [Basic education] ICT			
Narrated by the Teacher Himself			
Teacher Nelson	1MN	1MN 2011	1st MN of 2011: Microsoft Excel - Graphics Types and Formatting; Creating MS Excel charts from tables with data.
Teacher Wilson	1MN	1MN 2012	1st MN of 2012: Microsoft Excel; Management of sales and stocks of a warehouse; Drawing and printing reports.
Teacher Jack	1MN	1MN 2013	1st MN of 2013: Insert and format charts. Graph the table data.
Teacher Paulo	1MN	1MN 2014	1st MN of 2014: Use and manipulation of the Microsoft Word program; Word processor and its use in day-to-day.
TOTAL MN from this section: 4 MN			

Table 3. Current state of the Structure of the MNs' Collection and brief information about each MN (105 MNs): Section 3

Section 3 Ensino Matemática [Mathematics Teaching]			
Narrated by the Teacher Himself			
Teacher Priscila	1MN	1MN 2017	1st MN of 2017: polygons and circles.
TOTAL MN from this section: 1 MN			

Table 4. Current state of the Structure of the MNs' Collection and brief information about each MN (105 MNs): Section 4

<table>
<tr><th colspan="4">Section 4
Ensino Secundário [Secondary education]
Physics and Chemistry</th></tr>
<tr><th colspan="4">Narrated by the Teacher Himself</th></tr>
<tr><td rowspan="11">Teacher Isabel</td><td rowspan="11">11MN</td><td rowspan="4">4MN
2006</td><td>1st MN of 2006: Wave properties; Waves, frequency and amplitude modulation.</td></tr>
<tr><td>2nd MN of 2006: Wave properties; Waves, frequency and amplitude modulation; Modulation and demodulation.</td></tr>
<tr><td>3rd MN of 2006: Wave properties; Waves, frequency and amplitude modulation; Wave transmission; Phenomena of reflection refraction, Diffraction; Snell's Laws.</td></tr>
<tr><td>4th MN of 2006: Wave properties; Analog signal and digital signal; Signal conversion; Transmission of information; Frequencies used by each device; Waves, frequency and amplitude modulation; Wave transmission.</td></tr>
<tr><td rowspan="3">3MN
2008</td><td>1st MN of 2008: Long distance communications. Wave properties.</td></tr>
<tr><td>2nd MN of 2008: Long distance communications. Wave properties .</td></tr>
<tr><td>3rd MN of 2008: Long distance communications. Wave properties.</td></tr>
<tr><td rowspan="4">4MN
2011/2012</td><td>1st MN of 2011: Fotelétrico effect; Ionization energy, electron withdrawal energy, Current generation. Simulation.</td></tr>
<tr><td>2nd MN of 2011: Blackbody radiation; Displacement of Wien; Radiation; Absorption; Emission of energy.</td></tr>
<tr><td>3rd MN of 2012: Blackbody radiation; Displacement of Wien; Radiation; Absorption; Emission of energy.</td></tr>
<tr><td>4th MN of 2012: Blackbody radiation; Displacement of Wien; Radiation; Absorption; Emission of energy.</td></tr>
<tr><td>Teacher Sofia</td><td>1MN</td><td>1MN
2012</td><td>1st MN of 2012: Electrochemical cells. Simulation.</td></tr>
<tr><td rowspan="2">Teacher Francisca</td><td rowspan="2">2MN</td><td>1MN
2011</td><td>1st MN of 2011: Computational Simulation - Photoelectric Effect. Simulation.</td></tr>
<tr><td>1MN
2012</td><td>1st MN of 2012: From the sun to the "warming"; Black Body Radiation; Law of Displacement of Wien. Simulation.</td></tr>
<tr><th colspan="4">Narrated by a Researcher</th></tr>
<tr><td rowspan="3">Lessons from Teacher Joana narrated by the researcher Diana</td><td rowspan="3">3MN</td><td rowspan="3">3MN
2010</td><td>1st MN of 2010: Wave phenomena; Reflection and refraction.</td></tr>
<tr><td>2nd MN of 2010: Wave phenomena; Refraction of light.</td></tr>
<tr><td>3rd MN of 2010: From Atmosphere to Ocean: Solutions on Earth for the Earth; Acid rain; Oxidation-reduction reactions.</td></tr>
<tr><td rowspan="8">Lessons from Teacher Ricardo narrated by the researcher Diana</td><td rowspan="8">8MN</td><td rowspan="8">8MN
2011</td><td>1st MN of 2011: Atomic models; Electrical conductivity of materials.</td></tr>
<tr><td>2nd MN of 2011: Atomic models; Electrical conductivity of materials.</td></tr>
<tr><td>3rd MN of 2011: Chemical Bonding - Ionic and Covalent Bonding.</td></tr>
<tr><td>4th MN of 2011: Chemical Bonding - Ionic Bonding.</td></tr>
<tr><td>5th MN of 2011: Chemical Bonding - Covalent Bonding.</td></tr>
<tr><td>6th MN of 2011: Periodic table.</td></tr>
<tr><td>7th MN of 2011: Periodic table.</td></tr>
<tr><td>8th MN of 2011: Periodic table.</td></tr>
<tr><td colspan="4">TOTAL MN from this section: 25 MN</td></tr>
</table>

Table 5. Current state of the Structure of the MNs' Collection and brief information about each MN (105 MNs): Section 5

Section 5 Ensino Secundário [Secondary education] ICT			
Narrated by the Teacher Himself			
Teacher Ernesto	**1MN**	1MN 2012	1st MN of 2012: Xcode Object Library. Use other types of Xcode objects to compose / design the interface of an iPhone application.
TOTAL MN from this section: 1 MN			

Table 6. Current state of the Structure of the MNs' Collection and brief information about each MN (105 MNs): Section 6

Section 6 Ensino Secundário [Secondary education] – Professional course ICT			
Narrated by the Teacher Himself			
Teacher Michael	**1MN**	1MN 2014	1st MN of 2014: Constituent elements of PLC's. CPU, power supply, and input and output modules.
Teacher Cindy	**4 MN**	4MN 2017 – 3 are synthetic	1st MN of 2017: Low-tech prototyping test session; introduction of the theme and construction of a game.
			2nd MN of 2017: Computer applications (Kodu).
			3rd MN of 2017: Computer applications (Kodu).
			4th MN of 2017: Computer applications (Kodu).
Teacher Nero	**5MN**	5MN 2017 – 3 are synthetic	1st MN of 2017: Display of HTML commands; Webpage.
			2nd MN of 2017: General concepts of the HTML language; Viewing an HTML page in a browser; Prototyping session; OSI Model.
			3rd MN of 2017: HTML Commands, HTML Command Features.
			4th MN of 2017: HTML Commands, HTML Command Features.
			5th MN of 2017: Construction and formatting of tables; Building a website.
Teacher Ryan	**5MN**	5MN 2017 – 3 are synthetic	1st MN of 2017: Language C, Arduino Board.
			2nd MN of 2017: Language C, Arduino Board, low –tech prototyping.
			3rd MN of 2017: S4A - (Scratch) + Arduino.
			4th MN of 2017: S4A - (Scratch) + Arduino.
			5th MN of 2017: Programming in the Arduino IDE.
TOTAL MN from this section: 15MN			

Table 7. Current state of the Structure of the MNs' Collection and brief information about each MN (105 MNs): Section 7

<table>
<tr><th colspan="4">Section 7
Ensino Superior [Higher education]
Physical Sciences</th></tr>
<tr><th colspan="4">Narrated by the Teacher Himself</th></tr>
<tr><td rowspan="6">Teacher Luís</td><td rowspan="6">6MN</td><td rowspan="4">4MN 2009</td><td>1st MN of 2009: The nature of science.</td></tr>
<tr><td>2nd MN of 2009: Candle of the "Horror vacui".</td></tr>
<tr><td>3rd MN of 2009: "The longest fountain"</td></tr>
<tr><td>4thMN of 2009: "Shower Efficiency".</td></tr>
<tr><td rowspan="2">2MN 2012</td><td>1st MN of 2012: Simulators 1st Phase. Reach of a fountain - Hydrostatic pressure and launch of a projectile.</td></tr>
<tr><td>2nd MN of 2012 Simulators 2nd Phase. Reach of a fountain - Hydrostatic pressure and launch of a projectile.</td></tr>
<tr><td rowspan="6">Teacher Mara</td><td rowspan="6">6MN</td><td rowspan="5">5MN 2008</td><td>1st MN of 2008: Temperature concept.</td></tr>
<tr><td>2nd MN of 2008: Heat transfer by conduction; Thermal conductivity of a material and comparison with electrical conductivity; Heat conduction: establishment of the Fourier law; Thermal flow: amount of energy flowing per unit of time; Permanent and non-permanent thermal flow.</td></tr>
<tr><td>3rd MN of 2008: Thermal flow; Temperature gradient. Relationship with the slope: linear variation due to the imposition of the permanent regime within the discipline.</td></tr>
<tr><td>4th MN of 2008: Analogy of Ohm's law for "thermal circuits"; Determination of the thermal resistance of one and several layers of different materials; Thermal resistors in series; Coefficient of global thermal transmission, used in civil engineering; Heat transfer: convection; Convection currents; Natural and Forced Convection.</td></tr>
<tr><td>5th MN of 2008: Convection currents; Process of transfer of heat involved: radiation; Relation of emission and absorption with the color of the object and its temperature; Infrared radiation as part of the electromagnetic spectrum; Electromagnetic waves: wavelength, frequency, and velocity; Emission of a blackbody, the power curves and their dependence on temperature; Stefan-Boltzmann's law and temperature to the fourth; Balance between the emission and absorption of a body; Destination of the solar radiation that reaches the Earth and call; Different types of radiation between absorbed and reflected and that which is re-emitted by the Earth; Thermal flow; Surface thermal resistance; Total thermal resistance; Notion of thermal inertia.</td></tr>
<tr><td>1MN 2012</td><td>1st MN of 2012: Study of elastic collisions (one dimension) and their characteristics; derivation of the principle of conservation of linear momentum; Coefficient of elasticity and its calculation through simulation data; Principle of conservation of linear momentum in another type of collisions (inelastic) in one dimension; Characteristics of the inelastic collisions and of the intermediate elasticity coefficient. Simulation.</td></tr>
<tr><td rowspan="2">Teacher Mário</td><td rowspan="2">2MN</td><td rowspan="2">2MN 2011/2012</td><td>1st MN of 2011: Law of Archimedes and fluctuation; Gravitational Potential Energy. Simulation.</td></tr>
<tr><td>2nd MN of 2012: Law of Archimedes and fluctuation; Gravitational Potential Energy. Simulation.</td></tr>
<tr><td colspan="4">TOTAL MN from this section: 14 MN</td></tr>
</table>

Table 8. Current state of the Structure of the MNs' Collection and brief information about each MN (105 MNs): Section 8

Section 8 **Ensino Superior [Higher education]** **Mathematic**			
Narrated by the Teacher Himself			
Teacher Rafael	**5MN**	5MN 2011/2012	*1st MN of 2011: Matrix language; operations with matrices.*
			2nd MN of 2011: Systems of linear equations; Gauss elimination method.
			3rd MN of 2012: Systems of linear equations.
			4th MN of 2012: Inverse matrix.
			5th MN of 2012: Linear combination; subspace spanned.
TOTAL MN from this section: 5 MN			

Table 9. Current state of the Structure of the MNs' Collection and brief information about each MN (105 MNs): Section 9

Section 9 **Professional training** **ICT**			
Narrated by the Teacher Himself			
Teacher Mauro	**1MN**	1MN 2014	1st MN of 2014: MS Excel - Editing and formatting cells in tables with data.
TOTAL MN from this section: 1 MN			

Table 10.

Section 10 **Adult Education and Training**			
Narrated by the Teacher Himself			
Teacher Cindy	**1MN**	1MN 2017	1st MN of 2017: Word text processor: contents: text boxes, drawing, importing images, inserting clipart and automatic shapes and wordart.
TOTAL MN from this section: 1 MN			

Table 11.

Section 11 **Other Contexts**			
Narrated by the Teacher Himself			
Teacher Mário	**1MN**	1MN 2012	1st MN 2008: Electrolysis and Fuel Cells. Simulation.
Teacher Leonor	**1MN**	1 MN 2012	1st MN 2012: Professional development in a Lusophone country
TOTAL MN from this section: 2 MN			

The school has little availability of space and therefore there is no duplication of classrooms for practical classes.

Since I needed students to come to the laboratory, I switched rooms with another colleague. This was the first class in this environment and with this work organization, which motivated some agitation and lack of concentration among the students. In the room where I had worked with the class in the previous periods, the students had never sat in groups, because there was not enough room to move (it was a room that is part of a prefabricated building improvised by the school to face the problem of overcrowding, which is designated as "os palheiros" given the manifest lack of conditions of these rooms). The students had previously worked in pairs.

Another fact that I think is relevant is the lack of contact that these students had had with the material, equipment, and reagents normally used in chemistry. Students in the seventh year participated in few practical activities (they did not take classes in the laboratory and the class was larger) and in the first and second periods had studied only physics.

These students have a workload of 90 minutes per week for the discipline of physical–chemical sciences, and there are no shifts, so there are always 22 students in the room.

The classroom does not offer many possibilities for space organization, since it is a laboratory with 3 fixed workbenches (with a large dimension), and I asked for extra tables to be placed in the room to be able to organize more work groups.

In this way, the 22 students were distributed into work groups with 4 or 5 members each, as can be observed in the classroom plan shown in Figure 3.

Figure 3. Classroom plan

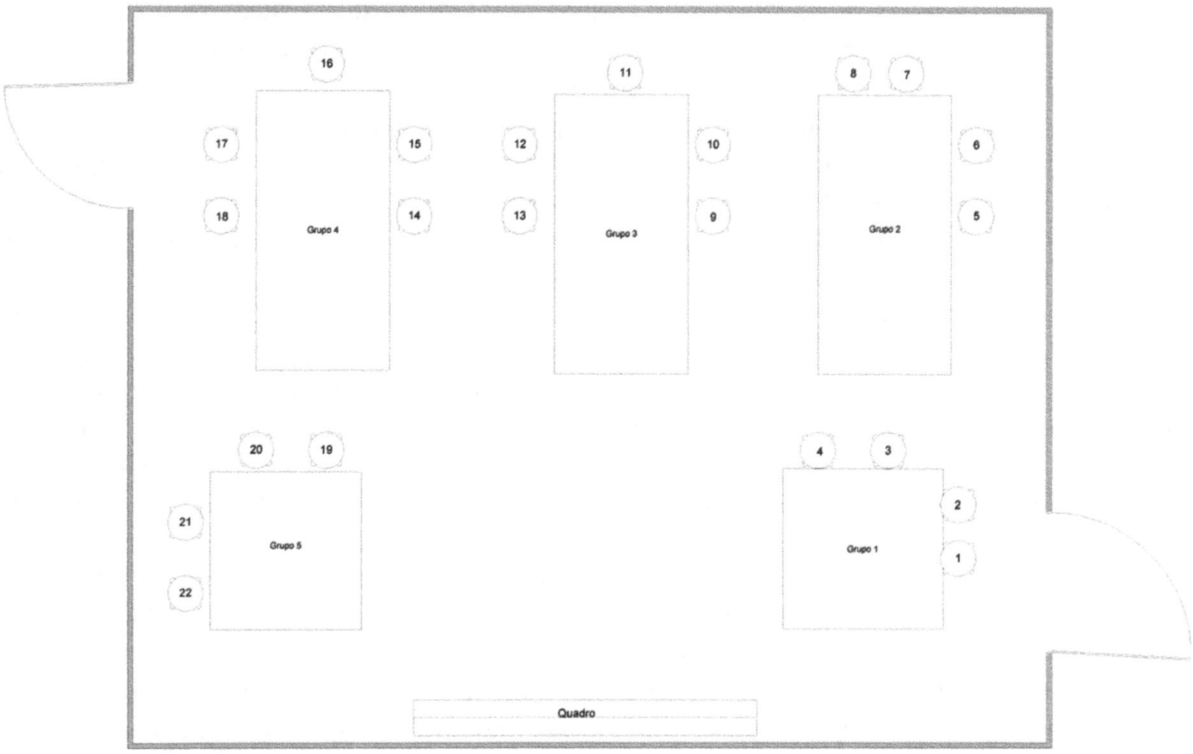

I used the free top on the center bench to place and organize the material to distribute to the students.

Synthetic Narration of the Whole Lesson

I started by organizing the students into five working groups (see Figure 3) and called attention to the need to keep the worktables organized, which took about 8 minutes.

Then, I distributed the grid for the self-assessment of students' performance in the tasks, one per group, giving a brief summary of the fulfillment of the same and noting the importance of its correct completion.

Then, I showed a photograph of dead fish floating on a lake as a result of the acidification of the water (physical situation) and questioned the students, based on the knowledge already acquired in the discipline of natural sciences, about this phenomenon. Almost all the students tried to respond at the same time: "pollution," said a student; "acid rain," said another; "acidification," another added. As all the students wanted to answer, I asked them to try to speak one at a time. Taking advantage of these answers, I asked whether they knew how acid rain is created. Only one student reported that acid rains forms through carbon dioxide.

Nine minutes after the beginning of the lesson, I distributed a questionnaire about the physical situation containing five questions on this topic, printed on strips of paper. As I was distributing it, I asked the students to transfer to their notebook all the questions that they would have to answer. I emphasized the importance of reading the issue well, discussing it as a group, and reaching the answer together, which would later be presented by the group's spokesperson. As the students attempted to respond, I moved around the room, trying to ensure that they had understood the task and checked for doubts, encouraging collective participation and the exchange of ideas within the groups. During this task, two students showed little commitment: Sandro, who has difficulty in remaining seated in his group, and Pedro, who has a tendency to discuss other matters and has been repeatedly called to pay attention.

During the accomplishment of the task on the questionnaire, I was not asked much by the students, since all of them demonstrated sufficient knowledge to answer the questions. In general, all of the students were committed and working, although the groups showed different rhythms.

When all of the students had indicated that they had finished (after 15 minutes), I started by calling a member of the first group to the board to write their answer. I did this successively with everyone for all the questions, which in the end were complemented by the answers of the other groups, in an iterative way. I helped the students with the final synthesis of each question.

After the students had finished writing the answers on the board and in their notebooks, I handed out the necessary material for the execution of the second task, which consisted of the neutralization of an aqueous solution of sodium hydroxide through the action of the CO_2 released by the students, as they spoke into the Erlenmeyer containing the solution. This task was presented on a strip of paper containing an explanation of the task (exemplary type) and questions regarding the interpretation of the phenomenon in question. Once again, I asked the students to transfer the questions to their notebook. I explained the task that they would carry out and exemplified it by speaking into the Erlenmeyer.

During the execution of the task by the students, some confusion was generated due to the failure to obtain the desired results. The previously prepared solution was highly concentrated, and the students could not see the change of color despite successive attempts. As such, I had to dilute the solution and ask the students to repeat the procedure. Even so, the students continued to struggle to complete the task. Consequently, they often called for me and I had to move between groups, questioning them and reassuring them in the meantime. In the end, I asked the students to record all their observations in their daily notebooks and to respond in groups to the question: "Why did the color of the solution change?" The recording of observations and the understanding of them were always accompanied by me, as I was asking questions to help the students to identify the reagents for the elaboration of the word scheme that translated the chemical reaction. In almost all the groups, the students showed some difficulty in understanding that the initial solution was basic and, as such, in identifying the phenomenon observed as a neutralization reaction. I explained several times what it meant to "neutralize," and I had to refer to the name of the reaction products, since no group could write the word equation. In some groups, I found that students referred to phenolphthalein as a reagent. In general, all the groups concluded the existence of a chemical reaction, although they could not identify it. Thus, based on the students' responses, I had to present the word equation for this reaction on the board, because the students were not able to identify it on their own.

As it was only about 20 minutes before the end of the class, the third task, initially intended to be carried out by the students, had to be demonstrated by me. Using universal indicator paper and three acidic solutions (carbonic, sulfuric, and nitric acid) and with the students' help, the pH of these solutions was verified using the respective color code. I asked the students to record the pH values of these three solutions in a table and explained that the increased rain acidity was due to the presence of carbonic acid, sulfuric acid, and nitric acid in the atmosphere.

Finally, I showed the figure in the fourth task and recalled the pH values obtained in the previous solutions. I also mentioned the influence of these acids on the atmosphere, as regards the formation of acid rain. However, since we did not have any more time, I asked the students to complete task 4 at home or in the Accompanied Study session.

At the end of the lesson, I asked the students not to forget to complete the sheet distributed at the beginning, regarding the evaluation of the performance in the tasks.

Episode(s) Relative to This Lesson

All the episodes of this class will be presented, since they include aspects related to the investigation of mediation, namely with regard to the choice, presentation, and management of the tasks. These three aspects were decisive for the events reported in the episodes. Regarding the use of aspects concerning the graphic language, images and diagrams were presented to the students and they were asked to elaborate representations (tables and word schemes). Thus, there was no choice or selection of episodes for this multimodal narrative.

First Episode

Beginning: 8 min 42 s

End: 42 min 28 s

After the students had observed the image (Figure 4) on A4 paper (one copy for each group and another one in my hand) showing dead fish floating in a lake as a result of the acidification of the water (physical situation), I told the students that it was a phenomenon that they had already studied in natural sciences and asked whether they knew what it was.

After observing the image, the students began to give their answers without respecting their turn, all trying to speak at the same time:

"Pollution," said Joel, Manuel, Rúben, Pedro, and Miguel, talking over each other at the same time.

"Aquatic," added Manuel.

Figure 4. Image provided to the students on A4 paper

And I continued:

"But this pollution has a specific name ... and that causes acidity ... of the waters?!"

"Acid rain," said Rúben.

"Acidification!" said Miguel.

"Pollution of rivers," added Joel.

"Acidification of aquifers," insisted Miguel.

Everyone tried to respond at the same time, arguing with each other, while Miguel insisted on acidification. At this point, the colleagues wanted to "beat" Miguel, but he insisted ...

"Acidification of rivers!"

"One at a time, please!" I added, trying to keep order in the interventions.

"Acidification of aquifers", continued Miguel.

"And how does that happen, Paulo?" I addressed one of the students who stubbornly said acid rain.

"It's from acid rain," replied the student.

"And how does acid rain form? Anybody know?" I questioned the crowd by looking around.

Once again, the students responded at the same time.

"Through carbon dioxide," answered Rita and Rúben.

"My friends, one at a time!"

At this point I was "abrupt" with Rúben because he had always spoken at the same time as his colleagues in all his speeches. In fact, this student has difficulty respecting his colleagues' turn to talk and has an old "bad habit" of intervening instead of his colleagues, so he said that it was Rita's turn to speak.

"Oh, Teacher!" Reuben reacted by sulking at not being allowed to speak.

Rita followed with the answer:

"Gases released by cars and industries go into the atmosphere and in contact with water acidify and then precipitate."

"Okay. Anybody else know anything about it?" I tried to get some more students to add some more ideas, but Rita's response left her colleagues without much more to add. Besides, Reuben was still sulky ...

"That's what I meant!"

"That's exactly ... or in other words? Say, what did you have to say ...." I tried to smile, but Rúben was silent.

From this moment, given the incident with Reuben, the debate of ideas was interrupted, and I proceeded to the task of the questionnaire.

The excess of disorderly interventions and the noise that was generated caused the end of the discussion. Rita, given a chance to respond, calmly exposed the causes of acid rain, putting an end to the dispute with her colleague for the best answer but preventing the other students from continuing to explore their ideas. There was an abrupt cut in the dialogue, and I closed the question, moving to the distribution of questions (Figure 5), on paper strips, by the different groups.

I handed out a strip of paper to every two students, since it was a way of saving paper and photocopies, given that there is a limit to how much we can do in school. I did not choose to project the issues on an acetate or through a multimedia projector because this room faces south in one of the corners of the building and has two glass walls, without blinds (only curtains), which prevent any quality viewing projection on the whiteboard, given the excess luminosity. In addition, the class begins at 1:30 p.m., which aggravates the problem.

I began to distribute the strips of paper to the groups; meanwhile, I was saying:

"So, you've heard so much about the subject—you're in a position to answer these questions."

They moved to the notebook that was there.

Meanwhile, Miguel asked: "All?"

"Everyone goes to the notebook, to read well," I replied.

"Since they were so enthusiastic to talk about the subject ... they know perfectly how to answer that one there!" I insisted.

Once again, I showed the grid that allows the self-assessment of students' performance during the tasks. I had already explained how they should fill it in, but, as they were using it for the first time, I decided to show them where to place the cross in this case:

"That is, in relation to this record ... have heard about the subject," I said while showing the grid.

Meanwhile, Manuel called me to ask what the date was, as I was starting to organize the notebook to write the questions.

I noticed that there were students who had not yet begun to transfer the questions to their notebook ....

"Boys, everyone must use the notebook."

Figure 5. Questions presented to the students

Translation - Task 1: *Based on your previous knowledge about acid rain e its effects, discuss with your peers: a) which are the effects of acid rain on living beings? b) How does acid rain affects aquatic ecosystems? c) How does acid rain affects terrestrial plants? d) How does acid rain affects animal life?*

Tarefa 1: Com base no vosso conhecimento sobre as chuvas ácidas e os seus efeitos, debate com os teus colegas de modo a serem capazes responder às seguintes questões:

a) Qual é o efeito das chuvas ácidas nos seres vivos?
b) Como é que as chuvas ácidas afectam os ecossistemas aquáticos?
c) Como podem as chuvas ácidas afectar a vida das plantas terrestres?
d) Como podem as chuvas ácidas afectar a vida animal?

"The idea is to read the question, talk to each other, and try to give a group response. Please note that I do not want individual answers now. They will talk to each other and the responses they write are the response of the group."

I tried to ensure that this question about the need to exchange ideas among themselves had been understood by all. In the meantime, I went around the room and found that in group V only the two girls were trying to answer the first question and the boys were distracted. So, I addressed this group:

"The answer is in group ... you must talk about it, see what you already know and what you do not know, and try to give a group answer. In the end, a student spokesperson will present the group's response on the board."

I circled the room again to see how the students were working and reinforced the need to work in groups.

"That's why I do not want individual answers. Any questions you have ...."

I tried to assure, in each group, that there was no doubt about the task.

Because it was a small set of questions, the students easily achieved what they were intended to achieve. Quickly, some students started transferring only the first question to the notebook and trying to respond. Other students (in particular the girls) started by transferring all the questions to the notebook so that they could start trying to respond to them. In any case, I noticed that the task itself was not a problem.

I let them work, but quickly realized that the students hardly spoke to each other. At 1 min 30 s, some students called me individually to ask questions like:

"Do I have to give a big answer?" asked Miguel.

"No, it has to be complete. Try to answer based on what you already know."

I stressed that they should talk to each other to find the answers.

Group I called for help to clarify the sense of the questions. This group showed some commitment, so I went there ...

"I do not understand question a)," Miguel and Rita said.

I approached and just read the question out loud.

"What is the effect of acid rain on living things?"

"Death," Miguel said with an air of astonishment.

"Of living things," I insisted.

But Miguel continued to repeat:

"Death!"

"They are dead now, but they were alive!" I replied, smiling as I reached the true sense of the students.

All the students in the group laughed and began to write in the notebook without saying anything else. Apparently, they were able to move forward.

I decided to circle the room and remain silent. I found that the students seemed to be working on the answers to the questions, despite Sandro's sporadic attempts to distract students from other groups, since in the meantime he had stood up to sharpen his pencil. At the end of 4 minutes, I had to call attention to him, and, for that reason, I stopped being silent. Meanwhile, I told Sandro and Daniel that it would be better to sit on the other side of the table to be closer to the other group members. However, my real intention was for Sandro to sit with his back to his colleagues in the other groups so there would be less cross-talk!

After that, I just watched ....

I noticed that group III worked very well, even with one of the students having complicated behavior (Manuel). It was a group with a student (Joana) who shows great maturity for her age and manages to involve her colleagues in working with her, although she is a student who intervenes little in the extended discussions in class. They did not call me once, and, in the next phase (presentation and discussion), I realized that this was the group with the best answers (Figure 6). I also noticed that she did not just dictate the answers to her colleagues, because she tried to talk to them and asked what they thought of their answers.

The students were working, and only 2 minutes after calling Sandro to pay attention, the two students of his group IV, who were working in a committed way but without great collaboration on the part of the colleagues (Sandro and Bruno), called for me. At that moment, the two boys tried to pay attention to the dialogue I had with their colleagues, pretending to be engaged.

"The effect of acid rain on plants is that it eats them away?" asked Irene.

"Not only! Those died, but in relation to all the others ...?" I said, trying not to answer the question.

"No more plants will be born in these places!" Irene concluded.

"Because?" I asked.

"Acid rain affects the soil," Irene and Cátia answered.

As additional information, I told the group that there are plants that do not like acid soils and that if the acidity level is high, few plants will develop there.

The students seemed to have understood what was intended, and I let them remain alone. However, I noticed that Bruno and Sandro simply copied what their two colleagues (Irene and Cátia) tried to undertake together. Sandro is a critical case and distracts Bruno.

Meanwhile, I was called over by group II regarding the same question.

Joel, pointing to the paper, said:

"The c)?"

I just read the question aloud:

"How can acid rain affect the life of terrestrial plants?"

"They affect your development," replied Telmo.

"How?"

Figure 6. Responses presented by group V

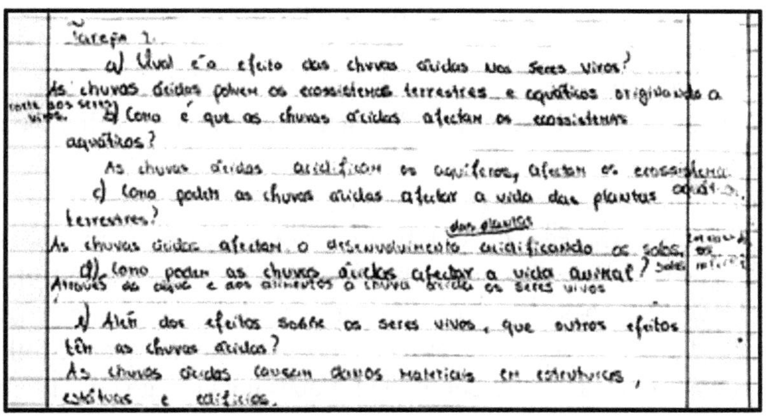

Tarefa 1.

a) Qual é o efeito das chuvas ácidas nos seres vivos?

As chuvas ácidas poluem os ecossistemas terrestres e aquáticos originando a morte aos seres vivos.

b) Como é que as chuvas ácidas afectam os ecossistemas aquáticos?

As chuvas ácidas acidificam os aquíferos, afectam os ecossistemas aquáticos.

c) Como podem as chuvas ácidas afectar a vida das plantas terrestres?

As chuvas ácidas afectam o desenvolvimento das plantas acidificando os solos.

d) Como podem as chuvas ácidas afectar a vida animal?

Através da água e dos alimentos a chuva ácida [illegible] os seres vivos

e) Além dos efeitos sobre os seres vivos, que outros efeitos têm as chuvas ácidas?

As chuvas ácidas causam danos materiais em estruturas, estátuas e edifícios.

The students looked at each other but did not say anything. Meanwhile, Telmo began writing and I stood, watching him. Then he dictated to his colleagues what he had written in his notebook.

The students continued to work, but, in this final phase (about 3 minutes), there was already a greater buzz in the room, which indicated that some groups were already finishing, including group III, which was the first one (about 4 minutes before all the others).

Group IV took the longest time. This group, although it seemed to be committed while I was in the room, did not work as much as it seemed at first sight. With the exception of the student Pedro (whom I had to call to twice for talking to Sandro), all the others seemed to be working and talking about the issues. It was only when they came to expose their answers on the board that I had the notion that they had worked less.

At this stage, while waiting for the students who were finishing the task, I suggested that they completed the self-assessment grid.

The execution phase of this task took about 15 minutes.

(Intermediate time: 27 min 10 s)

After answering all the questions that were included in the task, the students were in a position to perceive the effects of acid rain on living beings and on the abiotic world. As such, after everyone had answered the questions, one student from each group came to the board to answer a question. The students from the other groups made their contributions iteratively to improve and/or clarify the answer. In this way, the contributions of each group helped to enrich the final answer to each question.

Miguel was the first to come to the board and presented the answer of group I to the question about the effects of acid rain on living beings:

"So, Miguel, according to the opinion of his group, is going to tell you what the effect of acid rain is on living beings. You will listen to everyone and if there are different things to add ... we will add them at the end! So, come on!"

"I think it can cause death and diseases," said Miguel, before beginning to write it on the board.

As the student wrote on the board, I repeated it so that everyone would hear what Miguel had said. Then, addressing group II:

"What did you say?"

Students Rúben and Telmo tried to speak at the same time. Once again, Rúben could not resist speaking!

"One at a time or no one will understand," I said, trying to keep order in the room. Then I suggested:

"Start there ... one of you!" I said, looking at Telmo and Paulo.

"It can cause the death of ecosystems," answered Telmo.

Then I approached group III:

"And here? What did you answer?"

"It can cause the destruction of terrestrial and aquatic ecosystems, causing the death of living beings," replied Manuel.

"Okay … and there?" I asked the next group, looking at them.

In this group, no students answered at once, as they had not yet defined their spokesperson... there was excitement in the room because the other groups were inciting Pedro to respond, but he was not sure of the answer, since his colleagues' answers were different from his. This was indicative that the members of this group had worked on their own. I also checked this in the answers to the following questions.

"One of you ... we will not make a fuss out of this! It does not hurt, go on!" I said, prompting Pedro not to hesitate to answer.

"Acid rain can cause death of living beings and acidification of aquifers," Pedro finally answered.

Then I went to the last group:

"Acid rain can cause the death of living beings," replied Bruno.

Addressing the student standing on the board, I said that we could complete the answer given by group III.

"No need to erase the answer ... let's add what the colleagues said. Complete your answers to look like the one on the board," I said.

The answer was corrected on the board by Miguel and in the notebooks of each student.

The intervention of the spokesperson of group II followed, with Paulo writing on the board:

"Acidify the aquifers."

By questioning the students, I found out that practically everyone had given the same answer. However, when questioning Sandra (group IV), she answered:

"Acid rain pollutes aquatic ecosystems and causes the death of living beings."

I then asked the student to go to the board and complete the answer, which she did, adding:

"Originating the death of ecosystems."

The answer to the third question, by Manuel (group III), was immediately interrupted by Sandro, who was manifesting difficulty in remaining seated.

While Manuel gave his answer, it was interrupted by me asking my colleagues to listen carefully.

"Acid rain, polluting the soil," Manuel said, until he was interrupted.

"Listen to Manuel ... if you do not listen, nothing can be accomplished."

In the meantime, I had to call out to Sandro again so that he would not interrupt his colleagues' work. Then, Manuel resumed his response:

"Acid rains when polluting the soil make it infertile, making it difficult for the plants to develop."

"Can anyone present any reason for it to be infertile?" I asked, addressing the class.

"Soil acidification," added Miguel.

I was questioning the other groups. In the end, I suggested:

"Your answer is fine. I would only add: 'make soils infertile because of acidification.' Can you come and write it down, Manuel? You will only add in your answer 'to make soils infertile because of acidification.'"

"Next question ... here ... someone," I continued, addressing group IV.

"Who's coming to the board?"

It was Carolina's turn:

"Animals, by eating and drinking it, can catch diseases and cause their death."

In the meantime, I had to call for the students to pay attention to Carolina. Once again, I had to ask Sandro not to bother his colleagues.

Groups I and III, represented by Miguel and Joana, respectively, had the same answer. Therefore, I only suggested that they replace the expression "by eating and drinking" with "through food and water."

Irene had to answer the last question, but, as the class was a little agitated by then and the student spoke very quietly, they could not quite hear what she said. Moreover, as my attention was focused on what Irene was saying and I was standing next to her, the students sitting in the farthest places took the opportunity to talk about extra subjects and did not pay attention to their colleague's response.

I had to ask the students to keep quiet to hear Irene say:

"Acid rain can cause deforestation, respiratory diseases, and skin cancer."

"Right, but that is about ... skin cancer is about living beings, deforestation on plants ... The question was: 'In addition to the effects we have already mentioned, from previous questions, what other effects can have acid rains have?'" I tried to help her understand what was at stake in this point, since this group did not seem to have answered it right.

"Material damages," says Miguel.

I tried to involve the whole class in the matter, and I looked at the several groups, waiting for more answers:

"Material damages, namely ...?" I asked.

"Buildings, cars," Joana tried.

"More? Basically, everything that is metal structures, structures of rocks," I tried to finish, with some difficulty due to the agitation of the class.

As the students had mostly mentioned the effects on living things in their answers, I ended up dictating to Irene so that she would write it on the board:

"They cause material damages to structures, statues, and buildings, etc."

Meanwhile, Rita added:

"In cars!"

Then Pedro was heard saying:

"The stone melts!"

"Not melts!" I answered Pedro.

Then I addressed the class and asked:

"What happens to the stone?"

"Gets destroyed," some students answered in chorus.

"Wears out. It is attacked by acids," I finished.

This presentation and discussion of the responses was very time consuming (about 15 minutes), as it was often necessary to interrupt with calls for attention to students who were disturbing the work of their colleagues. In addition, students took a long time to write on the board and copy the answers.

During this presentation of the answers, to improve them, I was concerned with guiding the interventions of each one, as I had verified from the beginning of the lesson that oral interventions became more difficult because of the novelty of being seated in groups of four or five elements and facing each other, potentiating cross-over comments on extra-class subjects. In the case of students like Sandro and Bruno, this caused them constantly to challenge their classmates from other groups through gestures and facial expressions, which disturbed them and the course of the lesson.

(End of First Episode: 42 min 28 s)

Multimodal Narrative 2

Context: Formal education
Country: Portugal
Professional code: Rafael
Professional activity: Teacher
Narrator: Teacher teaching class
Narrator code: Rafael

Context of Education: Degree in Electrical Engineering (first year)—Discrete Mathematics and Linear Algebra
Course: Linear Algebra
Topics: Matrix language; operations with matrices
Level of education: Higher education
Age group: Adults (> 18 years)
School year: 2011/2012
Second Lesson on the subject (03 November 2011)
Total class time: 95 min

Contextual Information

This is a class of the curricular unit of Discrete Mathematics and Linear Algebra, at a public institution of polytechnic higher education, with students in the first year of the Electrical Engineering course. To date, the discrete mathematics component has already been taught, taking place during the first eight classes, as well as a class on the linear algebra component. It should be noted that this class precedes an interim moment of the written evaluation.

Thirty-one students are enrolled, of whom sixteen are attending this curricular unit for the first time. Since few of the repeating students attend classes, the group of students who regularly attend classes is almost entirely the first one. Nineteen students were present in this class.

The classes take place alternately in two rooms, one of which is an electronic laboratory with capacity for 25 students, with individual tables arranged in a line. On each side, there are two workbenches, with different equipment and devices. The board is near the door, and the room has only small windows in the background, with opaque glass. Figure 7 shows the layout of the room and the seat layout of the students present.

Figure 7. Classroom plan

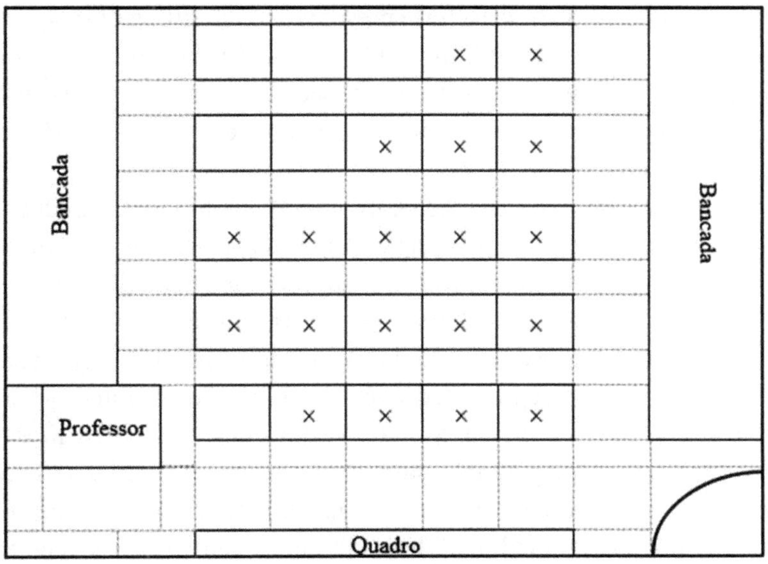

In the room, there is a video projector, which is placed on the teacher's desk, the image being projected onto the wall next to the board. It is used in all classes to project the supporting text slides and at certain times to show how to use the software involved and its exploration.

In the context of the sequence of subjects' exploration, this class continued the presentation and exemplification of basic definitions of matrices started in the previous class and evolved to the introduction of algebraic operations with matrices. A methodological innovation centered on the proposal for solving a task as an introduction to the multiplication of matrices, being the first moment at which students performed work of this nature.

Synthetic Narration of the Whole Class

I started the class by announcing a scheme construction as a way to review the matrix concepts addressed in the previous class. I also informed the students that the scheme would continue on the board throughout the class, rather than being erased, with the purpose of completing two new concepts that would be introduced in this class. I then added various concepts concerning matrices to the scheme, either by talking about their meaning and the relations between concepts or by requiring information from the students about what they remembered about a certain concept. In the end, transversal to all the concepts, I added to the scheme the reference to Scilab software.

The class continued with the introduction of the concepts of echelon form and row reduced echelon form of a matrix. By designing an example of supporting text, containing matrices written in echelon form and canonical form reduced by row reduced echelon form, I established a dialogue with the students about the requirements needed to arrive at those concepts. In the meantime, I turned to the computer to show Scilab's ref command, which reduces a matrix to a row reduced echelon form, and to inform them that this software does not reduce a matrix to the echelon form. Then I wrote some more matrices into the table for the students to conclude their writing in the echelon form or the row reduced echelon form and clarified some doubts.

Since about 28 minutes had elapsed since the beginning of the lesson, I proposed the resolution of 2 exercises related to the different definitions of matrices already seen. I ended up not giving the students time to try its resolution alone. As the first exercise presupposed the filling of spaces, I continually asked the answer in each case and, once accepted, the students would wrote the answers in their notebooks. By questioning the students, we found the answer to the second proposed exercise together. Meanwhile, focusing again on the previously constructed scheme, I completed it with the reference to the two new concepts of matrices addressed.

After nearly 39 minutes had passed from the beginning of the lesson, I introduced the study of subchapter 1.2—Operations with matrices—beginning with the algebraic addition of matrices. The approach that I outlined was to transpose students' knowledge about vector operations, which they had already studied in previous education cycles, to algebraic operations with matrices. The purpose was to show that the subject at hand was not entirely new and disconnected from what they already knew.

Using the software GeoGebra, I represented two vectors in the plan and questioned the students about their coordinates. Considering the sum of the vectors that I executed in the software and the mention that I made of the rule of the parallelogram, the students advanced that the vector sum was the result of adding the respective coordinates of the vectors, already speaking of the elements' sum of the corresponding lines. I then wrote on the board the addition of the two vectors and the result, using the matrices' language. I conveyed to the students that they already knew how to add matrices, since they were processed as the addition of vectors. Next, through a proposal to add the two matrices that I wrote on the board, the students were able to establish that, to add two matrices, it was necessary to have the same dimension and proceed individually to their sum. To conclude this topic, I also drew up an example of the supporting text and, using gestures to mark the homologous terms, repeated the example of the addition of matrices.

In an analogous way, returning to GeoGebra, I explored the multiplication of a scalar by a vector on the plane. The students were able to indicate the coordinates of the resulting vector in each case that I proposed, having registered one of the cases in the table, using the matrix notation. Then I wrote a matrix and the proposal for its multiplication by a scalar, via which the students established that the multiplication was performed "one by one," that is, the scalar by each element of the matrix.

In the second half of the class, announcing the study of the multiplication of matrices, I began by saying that I would propose "something different." I designed task 1 of the supporting text, indicating that, before knowing how to multiply matrices and without presenting any algorithm, the students would perform two subtasks, the final discussion of which would lead to the formalization of an algorithm. I also emphasized that the first subtask was not relevant to those who already knew the matrix multiplication algorithm. Suggesting that students organize themselves into groups of two or three members, I then distributed the task statement to those who did not have the supporting text.

Observing students' difficulty in beginning to solve the task, I tried to help them with the interpretation of the statement. I added that they had information about the multiplication of two matrices performed in Scilab and that the challenge was to interpret and apply the information provided in each subtask. I even stated that the multiplication of an element by the homologous element would not translate into the result presented. I also briefly reviewed the concept of internal products.

Then I followed the work of the different groups, encouraging them and continuing to clarify students' doubts when requested. As time passed, questions about the statement began to give rise to questions about possible answers, which either I would accept or I would point out the error, suggesting its reformulation. About 12 minutes after the task had started, I proposed a discussion of the results, even though I realized that the students had not yet made much progress in responding to subtask 2.

When I asked for a response to subtask 1, the students remained silent. The impasse was unblocked when, at my request, Filipe proposed that each element of the first row of the first matrix should be multiplied with each element of the first column of the second matrix. I went on to describe the process, using the signage of the elements that I was considering, and the students began interacting more and more naturally. At this point, I took the opportunity to address the basic requirement for the possibility of multiplying two matrices in light of the size of the matrices. I completed this subtask, performing the calculations until I was confronted with the result presented at the beginning.

As for the second subtask, and given that no group was able to find an answer, I then proposed to exemplify another possibility for the multiplication of matrices, highlighting the choice for the rows and columns that students should consider and the operations with matrices learned previously, until Catia and Filipe began to say that it involved "so much work." I emphasized the importance of the comments, as this was a clue to the choice of an algorithm for multiplying matrices that we would perform in the end. After we had attained the desired result for subtask 2, I raised the question: "What is the best algorithm in terms of time saving?" The answers were not unanimous, which was surprising to me, since I thought that everyone would think that the algorithm underlying subtask 1 would lead to the product of the matrices being obtained faster. In any case, I left it open, stating that they could then multiply arrays by whichever algorithm they preferred.

At this point, the students were already feeling restless due to the lack of time until the end of the class (about 15 minutes), since they asked whether they were still practicing and whether that subject would be on the test. I continued the class, formalizing an algorithm for the multiplication of matrices, the one underlying subtask 1, from an image that I projected and that was in the supporting text.

Then, by presenting an example with two matrices, I began by reviewing the criteria for the possibility of multiplying matrices and proceeded with a suggestion for students to perform the intermediate calculations mentally. Then, listening to the students, I wrote the result of the multiplication of each row by each column. Adding another matrix, I asked them to solve an exercise involving the operations learned during the lesson.

I finished the class with information about the content that I would contemplate for the written evaluation and some exercises that the students should solve as training. In the final minute, I structured the resolution of an exercise that I thought was important, related to the resolution of a matrix equation and the application of algebraic operations with matrices.

Episode(s) Relative to This Lesson

First Episode

Beginning: 00 h 00 min 20 s
End: 00 h 38 min 18 s

While constructing the scheme, first I wrote the term vector and referred to it as underlying the language of the matrices. Then, I added the terms square matrix, row matrix, and column matrix. As for the latter, I emphasized again that it corresponds to an alternative writing of the coordinates of a vector. I continued with the reference to a zero matrix and notation $0_{m \times n}$. This matrix has "all zeros," just as Joana said. Continuing the scheme, Filipe came forward with a designation:

"Diagonal matrix!"

"Well done! And what is a diagonal matrix?" I asked, addressing the question.

"When above and below the line it gives zero," Hugo anticipated.

"They are all null elements," I added, referring to the elements above and below the diagonal.

"And if by chance all the elements of this main diagonal are equal, how do you name the matrix?" I continued.

"Identity," said Joana.

"Not quite ... Identity is more particular," I corrected.

"Scalar," Joel suggested.

"Scalar!" I confirmed immediately.

I continued:

"When the elements of the diagonal are equal. And now, yes ... if they are all one it is when you can call matrix ... are you seeing the arrows going from the most general to the most particular?" I said, pointing to the sequence that I had just put into the scheme (diagonal matrix → scalar matrix → identity matrix).

After talking about the notation of the identity matrix, even having mentioned the case of the matrix I_2, I continued the scheme with reference to the transposed matrix and the symmetric matrix, as related concepts. In the end, Pedro recalled that the concept of a triangular matrix was still lacking and advanced with its definition, given my request for it. I completed the scheme with the reference to the Scilab software, giving the commands that would allow the writing or the calculations associated with most of those concepts approached in the previous class. After 6 minutes, I finished the construction of the scheme (Figure 8), saying that it would continue on the board and it would be completed with the next concepts that would be discussed (the ones that are indicated in a different color [at the bottom]).

For the study of the concepts of echelon form and row reduced echelon form of a matrix, I began by writing these designations as a title in the table. Next, I projected Example 1.9 of the supporting text (Figure 9) and questioned the students about what they saw and what they thought of the colored spot in each of the matrices.

Faced with the suggestions of being diagonal matrices or triangular matrices, I had to remind them that these would also have to be square matrices, which was not the case. Hugo even added the occurrence of zeros, before I, without giving more time, introduced the intuitive designation "on a ladder." Then,

Figure 8. Scheme built on the language of the matrices

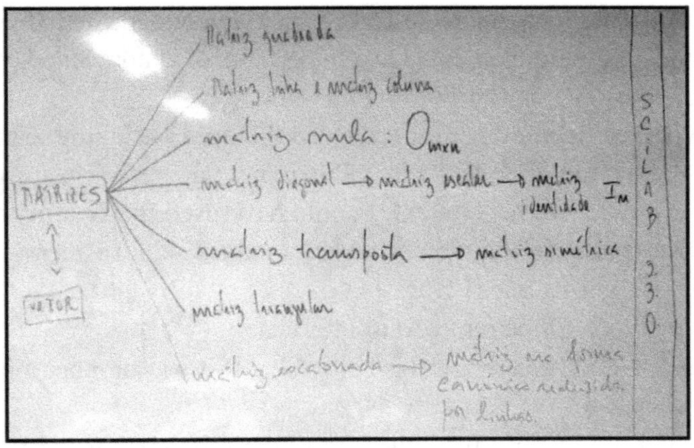

Figure 9. Excerpt from the supporting text of Example 1.9

Exemplo 1.9	• Matrizes na forma escalonada:
	$\begin{bmatrix} 2 & 1 & 0 & -2 \\ 0 & 1 & 3 & 1 \\ 0 & 0 & -5 & 0 \\ 0 & 0 & 0 & 2 \end{bmatrix}$, $\begin{bmatrix} 2 & 4 & -1 & 1 \\ 0 & 0 & 2 & -3 \end{bmatrix}$, $\begin{bmatrix} 8 & 2 & 7 & 3 \\ 0 & 0 & 0 & 2 \\ 0 & 0 & 0 & 0 \end{bmatrix}$, $\begin{bmatrix} 0 & 2 & 4 \\ 0 & 0 & -3 \\ 0 & 0 & 0 \end{bmatrix}$.
	• Matrizes na forma canónica reduzida por linhas:
	$\begin{bmatrix} 1 & 0 & 0 \\ 0 & 1 & 0 \\ 0 & 0 & 1 \end{bmatrix}$, $\begin{bmatrix} 1 & 2 & 0 & 0 & 0 \\ 0 & 0 & 1 & 0 & 0 \\ 0 & 0 & 0 & 1 & 0 \\ 0 & 0 & 0 & 0 & 1 \end{bmatrix}$, $\begin{vmatrix} 1 & 0 & 4 \\ 0 & 1 & -3 \\ 0 & 0 & 0 \\ 0 & 0 & 0 \end{vmatrix}$.

marking the case of matrices in the echelon form, I proceeded with the search for requirements for the identification of matrices of this nature. I started by asking whether any of the four matrices contained zero rows, and, after the students answered positively, I indicated the last two. However, the students remained silent when I questioned them about the zero rows' position. I then played with them by saying, between smiles, that I would have to enlarge the image and repeat the question. The students still hesitated to answer, until Hugo said they were "always on stairs" and Pedro said "last row." Therefore, I continued:

"May I say that they are at the bottom of the matrix? This is a requirement! To be a matrix in the echelon form, if there are rows of null elements ... all null ... they have to be ... they have to be the last rows of the row."

Looking for another requirement, I pointed to the third matrix of the example and tried to interact with the students about the position of the first nonzero element on the second row, but I kept noticing that the students were not very spontaneous with their contributions, although they were correct. I went on to say that it had to be element 2, all null elements preceding it. He said that the next row was null. I then transcribed the matrix on the board (Figure 10), while reinforcing the intuitive idea of a ladder and the requirements:

"When there are zero rows, they are at the base of the matrix ... and then the first nonzero input of the next row ... has to be to the right," I added.

"It does not say anything about having to be zero, but I know or deduce that if the first nonzero entry has to be on the right, then all these elements have to be ... zero," I concluded, pointing to the element below the pivot and the elements that followed on this row.

Then Filipe asked whether "it could be just a zero on the right," alluding to the circumstance that in a row only the element below the pivot was zero. This issue may have arisen from the fact that such a situation had not been explored. In this context, I could have used the first matrix of the example, but I decided to write another matrix in the table. I added a matrix with two rows, with the second pivot immediately to the right of the first one (Figure 11):

When identifying the pivots, Filipe returned to intervene:

"Because there is zero ... and because zero is below 2 ... if it was another number then it would not be…?"

I immediately validated the remark, recalling one of the requirements previously reported. Then I asked the students to help me to add a third row (Figure 11) so that the matrix would continue in the echelon form. At this point, the students showed spontaneity and understanding of the subject. Then I repeated the requirements according to what was written in the projected support text and further clarified Hugo's doubt about the possible positions of the pivot on the last row of the matrix that the students

Figure 10. Excerpt from the notebook of a student of the matrix written on the board

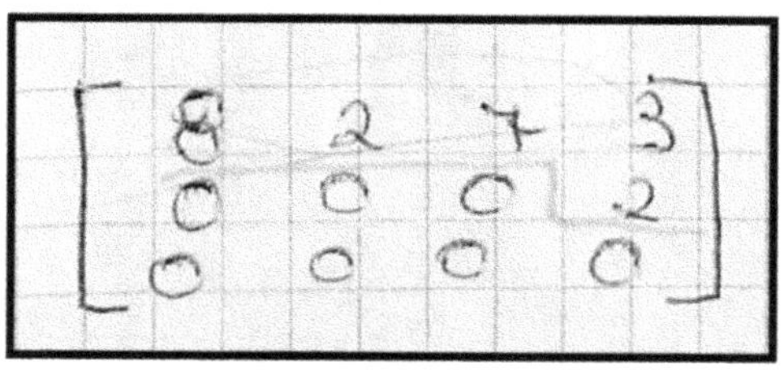

Figure 11. Excerpt from the notebook of a student of the matrix written on the board, initially only with the first two rows

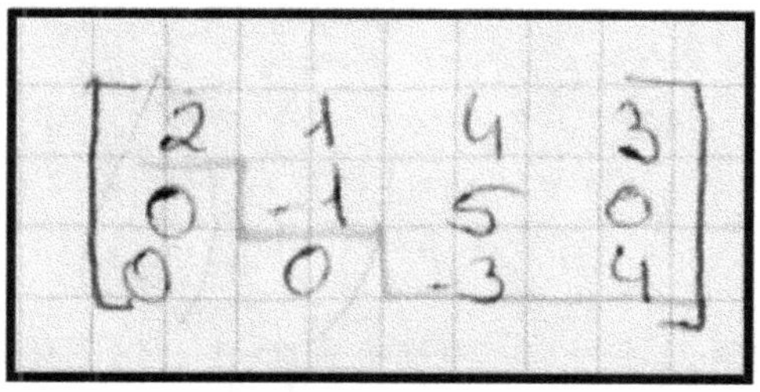

Figure 12. Excerpt from the notebook of a student of the matrix written on the board

had helped to complete. It was at this point in the class that I referred to the designation "pivot" for the first time. Finally, I decided to give an example in the negative, writing on the board a matrix that was not written in the echelon form (Figure 12).

Easily, students identified that there could not be two pivots in the same column. Representing the ladder that defined the position of the pivots, it occurred to me to say in a more informal way that "it costs a lot to climb this step," as an allusion to the indicated failure, and added that "if the steps were by levels in all the rows, we could climb them."

Advancing to the concept of a matrix in row reduced echelon form, first I established that the two requirements for obtaining a matrix in the echelon form seen above were still valid and that we would have to look for other complementary requirements. To do so, I took advantage of the matrix in the echelon form that I had previously written on the board (Figure 10) to introduce such requirements:

"Continuing from a in the echelon form, to be row reduced, the pivots have to be 1," I reported and exemplified this substitution in the matrix.

"If there are some ... as pivots ... in this column, everything else is zero," I continued, referring to the second requirement and returning to exemplify the necessary exchange in some elements of the matrix.

Then, using Scilab, I showed the students how to find the row reduced echelon form of this matrix and warned that this program does not have a command to find the echelon form of a matrix (Figure 13). I told the students that presently they could only recognize whether a given matrix was written in echelon form or row reduced echelon form. Afterwards, they would learn to perform the calculations, as Scilab did. I pointed out that such a procedure was crucial in learning the subjects of linear algebra.

Then I corroborated Abel's remark about paying attention to the columns rather than the rows to see whether a matrix is written in one of the two forms studied. I continued, writing on the board and discussing two more examples of matrices in row reduced echelon form with the students. Among them, I pointed again to the matrices of Example 1.9 alluding to this topic (Figure 9).

Figure 13. Scilab's environment in the search for the row reduced echelon form of a matrix

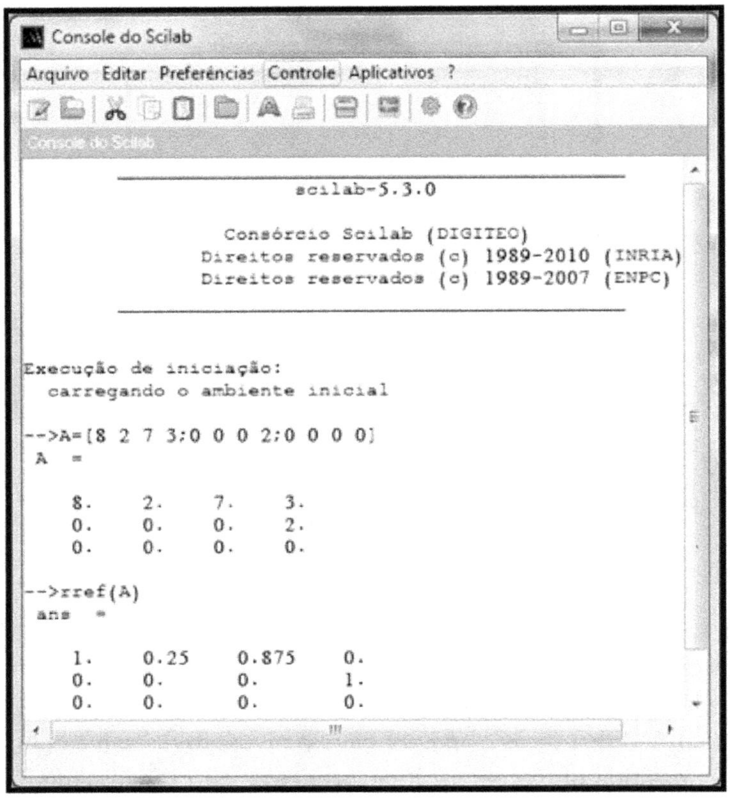

In the second matrix that I wrote on the board, I tried to recreate a difficulty with the placement of element 1 in the last column (Figure 14), questioning the students about whether this matrix was also written in row reduced echelon form.

The students answered affirmatively but were somehow apprehensive and lacked justification. After a few seconds, Antonio replied:

"The element of the first row and first column is 1."

"Well done! We call this a pivot," I added.

"Then we have ... so ... the right, which is also 1, right?"

"Another pivot. Very well ... and the fact that here is a 1 and there is a 3," I asked, interrupting Antonio's reasoning.

Figure 14. Excerpt from a student's notebook with other examples of matrices written in row reduced echelon form

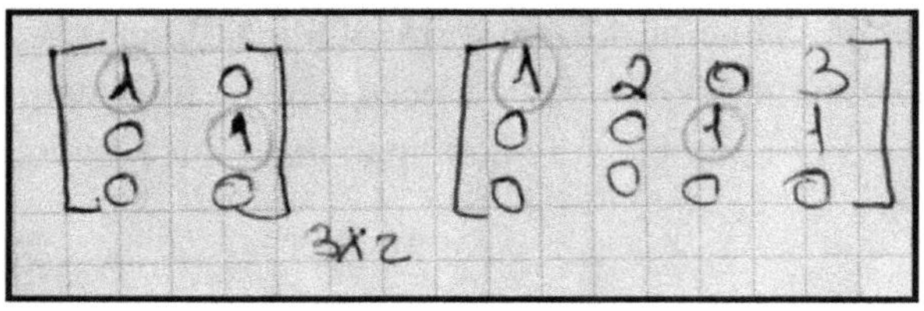

"That's not a pivot," Filipe said quickly.

"That's wrong," said Cátia.

"It is a requirement not to be in row reduced echelon form, or it is not ... for not being a pivot," I continued, ignoring the answers of Filipe and Cátia and addressing Antonio.

"Not a pivot, no," Antonio continued.

"That's it!" I said.

As I summarized this last example, Daniel interrupted me to ask whether the matrix was written in row reduced echelon form. I told him he could not have been paying attention. However, I repeated it all over again, with Daniel interacting with me and wondering about the effect of altering the values of some elements.

Closing this, I pointed to the scheme constructed at the beginning of the class and said that it was now possible to add other designations of matrices, which we would do later.

Intermediate time: 27 min 22 s

To apply the knowledge about the definitions of the matrices seen so far, I proposed the joint resolution of Exercise 1.2 (filling in a table indicating whether each matrix presented was square, zero, diagonal, triangular, scalar, and/or symmetric) and of Exercise 1.6 (Figure 15). As the students looked for the exercise page and started to solve them, I reiterated that the solutions to the exercises presented in the supporting text were listed at the end of each chapter, a fact that the students did not seem to know. I also said that it would be convenient not to consult the solutions immediately, not without thinking about the exercises' resolution first.

Almost without giving time to the students, I pointed to the board alluding to Exercise 1.2, which was projected, and questioned the students about their choices in each case. In this way, we solved this exercise together. Adopting the same strategy, I also guided a joint resolution of Exercise 1.

When, in the first paragraph, Cátia said that the matrix was simultaneously written in echelon and row reduced echelon form, I joked with her that I had not taught it. In this case, the matrix was only

Figure 15. Excerpt from the text supporting the statement of Exercise 1.6

Exercício 1.6

Indique se as seguintes matrizes estão na forma escalonada, forma canónica reduzida por linhas ou nenhuma das situações:

a. $\begin{bmatrix} 2 & 3 & 0 & 0 \\ 0 & 0 & 0 & 1 \end{bmatrix}$;

b. $\begin{bmatrix} 3 & 2 & 0 & 2 & -1 \\ 0 & -4 & 1 & -1 & 8 \\ 0 & 0 & 1 & 5 & -3 \\ 0 & 0 & 2 & 3 & 1 \end{bmatrix}$;

c. $\begin{bmatrix} 1 & 0 & 0 & 0 \\ 0 & 0 & 1 & 0 \end{bmatrix}$;

d. $\begin{bmatrix} 0 & 1 \\ 0 & 0 \\ 0 & 0 \end{bmatrix}$;

e. $\begin{bmatrix} a & 0 & -5 \\ 0 & b & 2 \\ 0 & 0 & 0 \end{bmatrix}$.

written in echelon form. The suggestions made by her colleagues were inconsistent with each other, so I reiterated the reference to the requirements previously discussed. Moving forward, I asked whether I was moving too fast, having received feedback from Filipe, saying "yes." This student, who is the most playful, said that he was from *Alentejo* [a Portuguese region where people are known to do things slowly due to the heat—editor's remark], but I devalued the comment. Regarding the answer to the second item, I heard Cátia say that it was not written in any of the known forms because "it had a very large step"; I immediately focused my attention on this circumstance and repeated to the rest of the class what the student had said. It was a way of highlighting the informal idea that I had launched a few minutes ago and I appreciated that it was being used by the student to achieve the correct answer.

In the next three minutes, we answered two more paragraphs together and I explained how much the last row was different, since it involved discussing unknown parameters, but I left the students to solve them as homework. Some students started to ask questions and make suggestions, and I ended up giving them some clues:

"Think about which values you may attribute to the a and b for the matrix to be written in an echelon form and in a row reduced echelon form."

I added, giving some examples:

"If a and b value 1, it becomes a matrix in row reduced echelon form. But what if it is all zero, for example?"

"It's nothing! —They must be greater than zero!" Filipe said promptly.

"What if one of them is one and the other zero?" I continued to question, not paying attention to the comments that were arising.

"There are several possibilities!" I finished.

At the end, I completed the scheme constructed at the beginning of the class, purposely left on the board. Using a different color, I added the two new designations of matrices studied (Figure 8).

Section 2

Seeing Teaching Practices in Science and Technology Through Multimodal Narratives

Section 2 is intended to show multimodal narratives (MNs) in action, that is, to use MN excerpts to transport readers into a classroom and understand in first-hand the teaching practices in action. As already explained in Section 1, these MN excerpts are not chosen as exemplary ones, simply as samples upon which the author and the reader may want to reflect upon.

Different aspects of teaching practices are addressed in this section, again with no intention of stating them as the most important, but merely as examples.

Chapter 2 addresses an important aspect of any teaching practice: students' assessment and teachers' feedback. By tackling this thematic through MN excerpts, the authors intend to show the reader teachers intended curricular choices may transform themselves in the classroom due to different teacher mediation or class developments. Different outcomes are illustrated in those excerpts, aiding to better understand some important theoretical aspects of teaching education research.

Chapter 3 tackles an important aspect of a class: how teachers deal with unexpected situations during class. through MN the authors show how students are able to overcome difficulties or end a specific task much depending on how the teacher treated the situation. Th authenticity characteristic reported in MN allows teachers to identify themselves in similar situations. The development of the teachers' actions and its repercussions on students learning are certainly interesting for all readers.

Chapter 4 shows in detail, through excerpts from MN, how a task was being worked on and developed by teachers during an intensive teacher training course in Angola. In this case, the MN documents the professional development of the teachers in particular showing how they appropriated a tool (a specific didactic model) that was object of the training.

Chapter 5 reports how MN contributed to the developing of conceptual teaching and learning. Along this chapter the reader may understand how this teacher used and linked two scientific concepts. The richness of the MN allows to better understand the teaching practice developed and the students' outcomes.

Chapter 6 is dedicated to show particularities and constrains of teaching math in college for non-math majors. The level of abstraction usually involved in these classes brings this thematic as a major concern in education. Through MN excerpts, readers may better understand, for instance, how different ways of asking students to participate in class produces different outcomes in terms of involvement and understanding.

Chapter 2
Understanding the Importance of Students' Assessment and Feedback With Multimodal Narratives

Maria Clara Viegas
Politécnico do Porto, Portugal

Eliane de Souza Cruz
Universidade Federal de São Paulo - Diadema, Brazil & Centro de Formação de Educadores da Escola Básica, Brazil

ABSTRACT

Multimodal narratives give insight of a classroom, teacher mediation, and students' work. In this chapter, the authors use this facility to better understand the importance of assessment and feedback in classroom practice. The analysis of excerpts from two complete multimodal narratives made it possible to highlight the importance of teacher intentions regarding the assessment process, the chosen tasks, and the subsequent feedback. It also provided an inside perspective of what was actually implemented in the classroom and how different teacher mediation leads to different results. These identifications in teaching mediation and students' involvement emerge transversally regarding teaching levels or contexts as objects of reflection, illustration, or simply inspiration. This shows the importance this tool can have in education by addressing parallel and complementary aspects of a classroom, which makes it a powerful tool in teaching.

INTRODUCTION

The learning assessment has always been part of the teaching process. It is usually used to enhance student learning – called formative assessment or assessment for learning, as well as to judge and certify learning achievements – called summative assessment or assessment of learning (Nicol, 2008; Black & William, 1998; Sadler, 1989). Since the "teaching objective is that students would learn" (Lopes, 2004),

DOI: 10.4018/978-1-5225-8570-1.ch002

teachers have increasingly been devoting their attention to the formative role of assessment, which aims to underpin students' learning in their ongoing effort to build knowledge, helped by their teacher's feedback (Dufresne & Gerace, 2004) and their peers' feedback (Cruz, 2012; Nicol, 2008). Simultaneously, teachers are also using learning technologies to aid them in the process (Laurillard, 2002).

According to Sambell (2016), the challenge involves viewing assessment and feedback as dialogic, dynamic and open to negotiation, with an emphasis on process as well as product.

The mediation between teacher and students is a crucial analysis factor that will influence the learning outcomes. The ways teachers organize the curriculum, idealize the tasks and present them to students, guide class discussions or monitor their progress can lead to substantially different results (Cruz, 2012; Lopes, 2004; Viegas, Lopes & Cravino, 2010). Teachers may also have an important role in helping students to understand their achievements toward the learning goals by introducing strategies of learning target identification that they can complete with their peers (Sindelar, 2015).

Inherently, studies using classroom practices are very important to gain understanding of the dynamics of interaction between the teacher and the students and strategies to develop the didactical aspects into practice. The literature presents several studies on teaching, learning, teacher reflections or classroom research (Earley, 2014), but usually as separate studies.

Unfortunately, classroom practices and the link between teachers' actions and students' reactions are not easy to study from the inside. Multimodal narratives (MN) is a tool that can fulfill this request and give the reader insight into a classroom, teacher mediation and students' work (Lopes et al., 2014). This tool, as fully presented in chapter 1, *provides a succinct narrative of the whole class with its storyline filled with contextual information* and *the detailed description of the actions and the teachers' and students' discourse during the work developed in the classroom.* The great advantage of its usage is that the complex and thorough work of analyzing all the classroom data is validated by the scrutiny of experts; i.e., the MN is a simple text document, easy to read, that will transport the reader inside the classroom. MN makes it possible to visualize how teacher mediation in the classroom can in fact affect student development (Lopes et al., 2014). In addition to the class narratives or even the class video recording, MN allows readers to understand teacher intentions, takes into consideration the class's background and conditions, and shows the way the teacher addresses situations by observing the class's development. Hence, MN may aid in highlighting different aspects of the assessment, such as the design, the chosen tasks and the subsequent feedback. On the other hand, it makes it possible to comprehend not only the teachers' intentions but also the way they are able to implement them in classroom – as well as to infer the results with regard to the classroom environment and students' involvement and comprehension.

The objective of the chapter is the use of MN excerpts to better understand the importance of assessment of and feedback on teaching practices and student development.

BACKGROUND

This section is divided according to the three identified aspects of assessment: the assessment process, assessment tasks and assessment feedback.

Assessment Process

Currently, consensus exists as to the importance of the choice of the assessment process to students' learning (Laurillard, 2002). It is known that students put their effort into what they know to be required (Christoforou & Yigit, 2008). The formative assessment should, therefore, be articulated with the summative assessment. Nicol (2008) emphasizes the ordinary example of students who do not get much involved in formative assessment activities when they know that they will be assessed only through a final exam. This can be particularly relevant in the first year of higher education, where summative assessments are usually single events and have an excessive effect on individual results. This may interfere with collaborative peer learning in an unwanted way. In this sense, an alternative might be the use of several summative assessments with formative purpose (Cruz, 2012; Viegas, Alves & Lima, 2015), e.g., portfolios, projects focusing on real problems and graded summative tasks, where weighting may increase as the students become more and more involved with frequent activities. It would also be important to design assessments that enable partial reformulation of the final work/project based on feedback from the teacher and peers (Cruz, 2012).

In addition, the assessment design should contain some characteristics that encourage students' participation, involvement and responsibility and that occur at several instances during the course, giving credit to students for their participation (Crouch & Mazur, 2001) but considering their performance firstly, so that they will take the tasks seriously. Studies of students' performance have been conducted that address students who complete formative assessment tasks merely to gain credit, instead of focusing on learning. They report that credit earned for mere participation will not help them to accomplish the following steps successfully. Thus, this credit may simply work as an incentive for the student to engage in learning (Arnold, 2016; Viegas et al., 2015).

In order to foster students' involvement, different productive activities should be proposed to students during the course, both individually and collectively (Crisp, 2012). The set of tasks should be diversified, not only to stimulate the participation of students with different learning styles or privileged forms of learning (Felder, Woods, Stice, & Rugarcia, 2000; Mazur, 1997) but also to create different moments in class (avoiding monotony) and the development of different competences, knowledge and attitudes (Lopes, 2004).

Many researchers have argued that the relationship between assessment processes and motivation is a neglected consideration in research and teaching practice (Laurillard, 2002; Nicol, 2008; Sindelar, 2015). Research in this area suggests that motivation is not a fixed attribute of the student, nor is it completely determined by any context. Students "build motivation" based on their own appraisal of both teaching and learning processes in a given context (Nicol, 2008). Thus, it is important to create strategies that could focus on how to:

- Structure tasks in progressive levels of difficulty and, if possible, individualize them;
- Foster mutual respect and responsibility;
- Use dynamic feedback and real formative scenarios, providing greater validation of developed competences; and
- Express performance levels in terms of "aspects in need of improvement" (rather than evidencing negative aspects), which may contribute to motivation (Black & William, 2005).

Assessment Tasks

As mentioned above, tasks may be summative or formative. They should be aligned with learning objectives (Biggs, 1999; Lopes, 2004; Redish, 2003) in order to develop the desired outcomes. It is important to complement the work done in class with out-of-class work (e.g., homework, e-learning platforms, project work) (Sondergaard & Mulder, 2012).

In addition, tasks should foster the development of different kinds or levels of competences (Redish, 2003), providing the mobilization of knowledge in an integrated way, applications to new contexts and the consolidation of knowledge (Sabella & Redish, 2007). However, these tasks should be achievable challenges, that is to say, neither too easy nor too difficult, both of which could undermine students' commitment. Tasks should promote student engagement in productive authentic work, avoiding "recipe tasks" or complete guidance from teachers; they should compel students to address their doubts; and they should support students' reflection in order to help students overcome their doubts by themselves (Lopes & Cunha, 2017).

The research has shown that fostering self-assessment tasks contributes to the improvement of students' achievements in final exams, their autonomy, their intrinsic motivation (McDonald & Boud, 2003) and their critical thinking (Nicol, 2008). Particularly in classes with a large number of students, the use of techniques such as the "one-minute paper" or automatic voting systems (currently available for free) can be motivating and useful to provide reflection moments during classes.

The ability to monitor, manage and self-assess learning is a key requirement in professions and life-long learning (Black & William, 1998; Nicol, 2008).

Research on learning assessment has been focusing on the teacher's role, the assessment process and feedback as being the teacher's appreciation of students' performance and/or work. However, there is an increasing need to share this responsibility with students: Students are constantly monitoring and self-assessing their own work, as has been reported by Black and William (2005) and Sindelar (2015). Students use the teacher's feedback to evaluate their own work and act on this reflection, which also implies an important self-assessment process (Nicol & Macfarlane-Dick, 2004).

On the other hand, since learning is essentially a social act, it is important to provide group work that stimulates collaborative learning among peers (Lladó, 2014) and peer feedback. Peer assessment contributes not only to the development of critical thinking (Boud, Cohen, & Sampson, 1999) but also to the improvement of peer performance (presenting students with a different perspective from that provided by the teacher) and increases their learning acknowledgment. Peer assessment and peer feedback are considered crucial for the regulation of learning (Nicol, 2008). Hence, it becomes important to present some strategies that enhance peer interaction and feedback: (i) providing open spaces/moments for students to discuss and study with peers and (ii) encouraging learners to give feedback on work before they submit it to the teacher.

Assessment Feedback

Feedback can center on: (i) *learning* (called *assessment feedback*) – providing information that can students can use to enhance their own learning and achievement and (ii) *teaching* (feedback *for teachers*) – giving information to teachers that they can use to help shape their practice (Nicol, 2008).

Cruz (2012), based in Hattie and Timperley (2007), report that assessment feedback can be:

- **Regulatory:** Focusing on the task process (the strategies used for its resolution);
- **Corrective:** Focused on task performance (relative to task accomplishment);
- **Self-Regulatory:** Centered on each student's self-regulation processes (autonomy, discipline and attitudes); and
- **Personal:** Focused on individual performance (with a view to enhancing students' ability to complete tasks).

The formative character of assessment is reinforced through feedback about students' learning (Black & William, 1998; Dufresne & Gerace, 2004; Etkina, 2000; Gibbs, 2003).

Thus, the effectiveness of formative assessment is related to the way feedback on students' achievements becomes important for their progress in knowledge and completion of the next steps (Gibbs, 2003; Mestre, 2001). Hence, these moments should occur regularly (several times during the course) and should be timely – that is, the time between when the students deliver their work and when they receive feedback should not be too long; otherwise, the answers to their doubts may no longer make much sense to them (Cruz, 2012; Rodrigues & Oliveira, 2014; Viegas et al., 2010). Feedback can focus on specific features of a task – the what, how and why of a problem – rather than simply indicating whether the work is correct or not (Evans, 2013).

The quality of assessment feedback can be assessed by how much it contributes to the progression of students' learning (Black & William, 2005). Two important ways of implementing feedback mediation should be highlighted: (i) giving students autonomy and (ii) providing aids based on relevant and pertinent clues and only at key moments – never before the students' awareness of their doubts (Lopes, 2004; Viegas et al., 2010, 2015).

The process of formative assessment should also find ways to generate continuous information for teachers about student learning. This feedback *for teachers* can be used to manage the curriculum in a timely way, taking into account students' difficulties or progressive steps that have not been foreseen (Dufresne & Gerace, 2004). There are several ways in which teachers can acquire continuous feedback; examples include regular training tasks and opportunities for regular classroom and non-class discussions through learning management systems, e.g., e-learning platforms (Nicol, 2008).

All of these aspects can contribute to the involvement and motivation of the majority of students. Inherently they affect students' academic success and the development of scientific and social skills.

METHODS

Assessment and feedback will be analyzed through MN excerpts, identifying relevant aspects mentioned in the literature following an interpretive and inferential analysis. The usage of multimodal narratives allows a holistic treatment of this issue, contributing to the understanding of the complexity of this aspect of teaching.

The analyzed data consisted of the two MN presented in full in the beginning of the book (see Part I). In no way are the chosen MN or the excerpts meant to be addressed as outstanding examples: They were chosen because they illustrate some real class events related to the topic addressed in this chapter, upon which it is possible to reflect and draw conclusions.

The analyzed dimensions are the aspects identified in the previous section: the assessment process, assessment tasks and assessment feedback, which will be analyzed and illustrated with MN excerpts. To do so, for each dimension, each presented excerpt will be analyzed methodologically in order to differentiate and better understand what the teacher planned (design), what occurred in the classroom (class implementation) and the outcome of the situation, e.g., student and teacher reactions, student involvement, peer collaboration or learning progression (results). Design, class implementation and results will constitute the units of analysis.

STUDENT ASSESSMENT AND FEEDBACK ANALYZED THROUGH MULTIMODAL NARRATIVES

In this section, by using excerpts from the two multimodal narratives presented in Part I, different outcomes from initial similar class designs will be illustrated, reinforcing the importance of how different aspects of teacher mediation can influence student motivation. The first MN represents a science class in a high school, from which six excerpts are analyzed (E1, E2, E7, E8, E9 and E10), and the second MN is from a university math class, with five excerpts (E3, E4, E5, E6 and E11). All of these excerpts will be analyzed, showing those aspects in terms of their design, class implementation and results.

Following the systematization of the "Background" section, where some particularly important aspects were addressed, the same three dimensions will be analyzed: the assessment process, assessment tasks and assessment feedback. In order for a curriculum design to be coherent, these three dimensions should be designed together. However, as often happens, due to various circumstances (e.g., class restrictions, lack of time, lack of resources, last-minute changes, different teachers), the idealized design is not always actually implemented. Several didactical studies show how to delineate the curriculum and how it can be possible to take the best options, in general all focusing on the prime objective of any teacher: improving students' learning (Biggs, 1999; Lopes, 2004; Redish, 2003; Sadler, 1989). However, is rarer to encounter studies based on real classroom events, teachers' intentions versus their decisions in class, and students' reactions. These might focus, for instance, on understanding how a task may or may not change its purpose, how a teacher might change his/her mind in the middle of a task or how students' collaborative work might lead to better or worse developments. Or, to give another example, teachers are used to asking questions, but every experienced teacher knows that the way of delivering those questions to the class can lead to substantially different results (Pedrosa-de-Jesus & Watts, 2014). These are just a few examples of what a MN might have to offer, showing how these events have happened. Often, teachers learn from their good or bad experiences. Studying MN also makes it possible to learn from other teachers' practices. Regardless of their context, successful or unsuccessful examples make them generalizable or of situated generalization (Cruz, 2012).

This work does not pretend to be exhaustive in terms of covering every teaching issue related to assessment and feedback. It merely serves the purpose of showing how one can use MN to better understand the importance of some aspects related to student assessment and teacher feedback.

Assessment Process Observed With Multimodal Narratives

The usage of formative assessment during class can foster student attention and involvement. The design of these moments in a course curriculum helps teachers to proceed and execute several actions in order to deliver classes with rich and profitable moments for students' learning. This involves not only the type of tasks and type of feedback but also the approaches that teachers should use to launch these challenges and mediate the discussions.

The teacher's intentions are established in the first part of a MN, where a thorough contextualization of the background classes and resources is given, along with a description of the approach to be taken and a summary of outcomes.

In excerpt 1 (E1), the teacher makes explicit her intention of doing team work and organizes the students in groups around the benches. She tries to foster student discussion by showing them a picture.

Multimodal Narrative Excerpt 1: Teacher A (Classroom Discussion Centered on a Starting Question)

Then, I showed a photograph of dead fish floating on a lake as a result of the acidification of the waters (physical situation) and questioned the students, based on the knowledge already acquired in the course of natural sciences, about this phenomenon. Almost all the students tried to answer at the same time: "pollution", said a student; "acid rain", said another; "acidification", another added. As all the students wanted to answer, I asked them to try to speak one at a time. Taking advantage of these answers, I asked whether they knew how acid rain is created. Only one student reported that acid rains forms through carbon dioxide.

As becomes clear by the teacher's words, when the teacher poses the question to the class, the faster students tend to answer it straight away. There are a few students who struggle to be heard, and it can be imagined that the rest of the students, who were not fast enough, were unable to think about the question since answers were already being given. Every teacher has experienced this kind of situation. This illustration encourages teachers to reflect upon what can be done to avoid this situation.

Systematizing this analysis of E1: design – formative assessment, fostering students' participation and responsibility, creating a different moment in class; implementation – showed a picture and asked a question to the entire class; results – some students tried to answer all at once; not all students had the opportunity to think about the question.

The class continues, and the teacher has the chance to pursue her prior objective of engaging the students in a group discussion, illustrated in E2:

Multimodal Narrative Excerpt 2: Teacher A (Group Discussion on Some Issues Before Classroom Discussion)

Nine minutes after the beginning of lesson, I [the teacher] distributed a questionnaire about the physical situation containing five questions on this topic printed on strips of paper. [...] I emphasized the importance of reading the issue well, discussing it as a group and reaching the answer together, which would later be presented by the group's representative. As the students attempted to respond, I moved around the room, trying to ensure that they had understood the task and checked for doubts, encouraging collective participation and the exchange of ideas within the groups. [...] During the accomplishment of the task on the questionnaire, I was not asked much by the students, since all of them demonstrated sufficient knowledge to answer the questions. In general, all of the students were committed and working, although the groups showed different rhythms. When all of the students had indicated that they had finished (after 15 minutes), I started by calling a member of the first group to the board to write their answer. I did this successively with everyone for all the questions, which in the end were complemented by the answers of the other groups, in an iterative way. I helped the students with the final synthesis of each question.

The teacher's effort in meeting the need of pushing students into discussing with their peers is visible. She also makes an effort to guarantee that the students have understood the questions. In this excerpt: design – formative assessment, fostering students' participation and responsibility, creating a different moment in class, fostering collaborative work and allowing for the development of different competences; implementation – distributed the questions to the groups of students and asked them to discuss them before writing the answers; someone from each group would transmit their answers to the class; results – students were committed to the work, and the final answers were iteratively constructed on the blackboard with the participation of each group.

When comparing the excerpts E1 and E2:

- In terms of the design, both excerpts show the teacher's intention of encouraging students to participate and think about a situation/question.
- In terms of the implementation, in E1, the question is posed to the class and their participation is not controlled, whereas in E2, the teacher asks them to discuss the question with their peers first.
- As for results, E1 reveals that not all students had the opportunity and time to think about the question, while in E2 all students participate in small work groups before participating in the classroom group.
- In the second example, it becomes evident how a class plan can change as the teacher realizes he is beginning to run out of time. This can be observed in the following excerpt, E3:

Multimodal Narrative Excerpt 3:Teacher B (Resolution of Exercises in Classroom Group)

Since about 28 minutes had elapsed since the beginning of the lesson, I [the teacher] proposed the resolution of 2 exercises related to the different definitions of matrices already seen. I ended up not giving the students time to try its resolution alone. As the first exercise presupposed the filling of spaces, I continually asked the answer in each case and, once accepted, the students wrote the answers in their notebooks.

Teachers understand the pressure of fulfilling course subjects, but even so, the teacher in E3 knew the class needed a break from routine and chose to continue with the class plan of proposing a problem to the students. However, due to lack of time, the prior objective was cut short. Reflecting on the actual value this task might have brought to the majority of students may help teachers make their own decisions in the future. Here, the decision was to address those students who were responding, which is understandable under the circumstances. However, the MN can help teachers reflect on what they might have done differently. Again, this excerpt presents a situation where a question is posed to the class and only a few students interact with the teacher to solve the exercise. It is implicit that the majority of students only copied the answers into their notebooks. Often, these situations appeal to students (because they collect the answers in hopes that they will understand them eventually). Students might even perceive this kind of activity as real work, while in fact they probably produced little or no work at all (despite the value that copied solutions might or might not have in their future learning).

Hence, the underlying analysis for this excerpt in relation to the assessment process can be stated as follows: design – creating a different moment in class by using a formative assessment task; implementation – showed students two exercises but gave them little time for them to do it on their own; results – the teacher mentored the solution, asking the class for input, but the majority of students only wrote the answers in their notebooks.

Then, in the second half of the class, the teacher intends to do something different. In E4 he suggests working in small groups:

Multimodal Narrative Excerpt 4: Teacher B (Resolution of Exercises in Small Groups and Interpreting and Applying the Information)

[...] announcing the study of the multiplication of matrices, I began by saying that I would propose "something different." I designed task 1 of the supporting text, indicating that, before knowing how to multiply matrices and without presenting any algorithm, the students would perform two subtasks, the final discussion of which would lead to the formalization of an algorithm. I also emphasized that the first subtask was not relevant to those who already knew the matrix multiplication algorithm. Suggesting that students organize themselves into groups of two or three members, I then distributed the task statement [...]. Observing the students' difficulty in beginning to solve the task, I tried to help them in the interpretation of the statement. I added that they had information about the multiplication of two matrices performed in Scilab and that the challenge was to interpret and apply the information provided in each subtask. I even stated that the multiplication of an element by the homologous element would not translate into the result presented. I also briefly reviewed the concept of internal products. Then I followed the work of the different groups, encouraging them and continuing to clarify students' doubts when requested. As time passed, questions about the statement began to give rise to questions about possible answers, which either I accepted, or I would point out the error, suggesting its reformulation. About 12 minutes after the task had started, I proposed a discussion of the results [...]. When I asked for an answer to subtask 1, the students remained silent. The impasse was unblocked when, at my request, Filipe proposed that each element of the first row of the first matrix should be multiplied with each element of the first column of the second matrix. I went on to describe the process, using the signage of the elements that I was considering, and the students began interacting more and more naturally.

For excerpt E4, the analysis is quite different: design – create a different moment in class promoting students' participation in collaborative work; implementation – students worked in groups with teacher's help, giving clues and clearing doubts; results – students made an effort to solve the proposed task and were able to get help with their doubts.

When comparing the excerpts E3 and E4:

- The design of E4 encouraged students' participation and involvement. E4 uses formative assessment (a strategy centered on the small-group exercises) differently from E3 (a strategy centered on the resolution of exercises in the classroom group).
- The implementation and results of E3 reveal that not all students had the opportunity, motivation and time to think about the question, but in E4 all students participated in small work groups with the teacher's help (formative assessment) before participating in the classroom group.

In fact, both pairs of MN excerpts show that in the same class the teachers took different approaches and consequently produced different outcomes. Probably their experience helped them generate (in time) a better solution to produce class engagement.

Assessment Tasks Observed With Multimodal Narratives

Teachers are aware that, in order for a task to fulfill its purpose, it should appear in students' learning path at specific moments. Tasks should also be aligned with learning objectives, foster knowledge mobilization and serve as a tool to broaden student competences to other contexts. Every experienced teacher knows that tasks should present a clear objective, should avoid being too directive (so they can foster students' development) and should be neither too simple nor too complex. The masterwork is to acknowledge students' capacities and needs, and design tasks that can foster their utmost learning.

Some of these characteristics can be identified in the same two MN examples.

Analyzing E1 with regard to this dimension (assessment task), it can be observed that, in terms of design, it provided an application to new contexts. In E2 and E4, some characteristics related to this dimension can be identified: the promotion of authentic work and the fostering of the mobilization of knowledge in an integrated way; this may be considered an achievable challenge as we referred above. In E4, the task can be identified as one that is aligned with learning objectives. In E3, the proposed task may have contributed to students' self-assessment.

Excerpt E5 (from the second MN) portrays an interesting task, where the teacher schematizes the class subject on the board, allowing students to situate themselves in the syllabus and in terms of what they will be developing, helping them to scaffold their knowledge.

Multimodal Narrative Excerpt 5: Teacher B (Construction of a Scheme on the Board to Help Scaffold Students' Knowledge)

[…] After 6 minutes, I finished the construction of the scheme […], saying that it would continue on the board and would be completed with the next concepts that would be discussed. […]

Closing this, I pointed to the scheme constructed at the beginning of the class and said that it was now possible to add other designations of matrices, which we would do later. […]

At the end, I completed the scheme constructed at the beginning of the class, purposely left on the board. Using a different color, I added the two new designations of matrices studied.

Analysis of E5 gives: design – foster the mobilization of knowledge in an integrated way; implementation – construction of a schematic summary on the board, which was completed during class; results – students got a clearer structured vision of what they were developing in class; even though the schema was constructed by the teacher, students were involved in the process.

In another excerpt from this class, E6, the teacher challenges the students to reach a conclusion on their own.

Multimodal Narrative Excerpt 6: Teacher B (Teacher Challenges Students to Reach a Conclusion on Their Own)

For the study of the concepts of a echelon form and row reduced echelon form of a matrix, I began by writing these designations as a title in the table. Next, I […] questioned the students about what they saw and what they thought of the colored spot in each of the matrices. Faced with the suggestions of being diagonal matrices or triangular matrices, I had to remind them that these would also have to be square matrices, which was not the case. Hugo even added the occurrence of zeros, before I, without giving more time, introduced the intuitive designation "on a ladder." Then, marking the case of matrices in the echelon form, I proceeded with the search for requirements for the identification of matrices of this nature. I started by asking whether any of the four matrices contained zero rows, and, after the students answered positively, I indicated the last two. However, the students remained silent when I questioned them about the zero rows' position. I then played with them by saying, between smiles, that I would have to enlarge the image and repeat the question. The students still hesitated to answer, until Hugo said they were "always on stairs" and Pedro said "last row." Therefore, I continued:

"May I say that they are at the bottom of the matrix? This is a requirement! To be a matrix in the echelon form, if there are rows of null elements ... all null ... they have to be ... they have to be the last rows of the matrix."

Despite teacher efforts to involve students in this development of knowledge in class, students seem to have difficulties following the teacher's thoughts. Some students make an effort to answer the teacher's questions, but they are unable to reach his objectives. Therefore, this task did not reach its full objective. Analyzing this excerpt (E6): design – propose a challenge aligned with the learning objectives; implementation – teacher asked students to identify some characteristics of matrices and if they could see a pattern; results – the majority of students were not following or understanding, even though the teacher tried to give hints and eventually the answer.

In excerpt E7 (from the first MN), the teacher idealized a laboratory activity in which the students should be able to perform a simple procedure and observe their own experiment as it changes color. The objective was for students to reflect on the conditions and try to explain the effect:

Multimodal Narrative Excerpt 7: Teacher A (Laboratory Work)

After the students had finished writing the answers on the board and in their notebooks, I handed out the necessary material for the execution of the second task, which consisted of the neutralization of an aqueous solution of sodium hydroxide through the action of the CO2 released by the students, as they spoke into the Erlenmeyer containing the solution. [...] I explained the task that they would carry out and exemplified it by speaking into the Erlenmeyer. During the execution of the task by the students, some confusion was generated due to the failure to obtain the desired results. The previously prepared solution was highly concentrated, and the students could not see the change of color despite successive attempts. As such, I had to dilute the solution and ask the students to repeat the procedure.

Unfortunately, the chemical solution was not of the right concentration, and the students became restless. The analysis of E7: design – promote authentic work fostering the mobilization of knowledge in an integrated way; implementation – each group was equipped with the lab material and had access to the procedure and the questions they should answer while the teacher monitored the task; results – students became frustrated with the lack of results, the teacher had to intervene, and the activity was repeated.

At the end of this class, the teacher decided to give the students the unaccomplished work as homework, as can be observed in excerpt E8:

Multimodal Narrative Excerpt 8: Teacher A (Teacher Challenges Students with Unaccomplished Work as Homework)

Finally, I showed the figure in the fourth task and recalled the pH values obtained in the previous solutions. I also mentioned the influence of these acids on the atmosphere, as regards the formation of acid rain. However, since we did not have any more time, I asked the students to complete task 4 at home [...].

The analysis of this excerpt: design – promote complementary work outside of class; implementation – due to the lack of time to finish the task in class, the teacher asked the students to complete it as homework; results – not observed.

Assessment Feedback Observed With Multimodal Narratives

Feedback observed with the MN focused on: (i) assessment feedback to the students and (ii) feedback for teachers.

Teacher mediation in the classroom is closely linked with the type of feedback, not only during class but also after class. Students should feel that this feedback is useful and important in completing the next phases of their work. This will potentially happen if the feedback is relevant and given assertively and in timely manner. Teachers also recognize its usefulness to help them manage the classroom curriculum (feedback for the teacher), to quickly address student doubts and misunderstandings, or to confirm that students are progressing satisfactorily.

The prime advantage of conducting formative assessment during class is the immediate feedback both students and teacher can get from the work performed and its matching with the learning objectives.

E2 and E4 are classic examples of promoting peer assessment successfully, providing timely feedback, contributing to students' reflection and providing aids when needed (and not before). This also helped teachers to acknowledge more clearly the students' progress.

In E6, the teacher also gets feedback from the students' knowledge development, mainly showing their difficulties in following the class development. This kind of feedback can help teachers perform any additional task that might help students overcome their difficulties.

In E7, the teacher received important feedback from students about the experiment. This feedback allowed her to act promptly and manage this curriculum activity in a better way, allowing students to proceed.

The following excerpt (E9) provides another example of how classroom feedback can be useful to the teacher:

Multimodal Narrative Excerpt 9: Teacher A (Classroom Feedback can be Useful for the Teacher)

I noticed that group III worked very well, even with one of the students with complicated behavior put there (Manuel). It is a group with a student (Joana) who shows great maturity for her age and manages to involve her colleagues to work with her, although she is a student who intervenes little in the extended discussions in class. They did not call me once, and, in the next phase (presentation and discussion), I realized that this was the group with the best answers [...]. I also noticed that she did not just dictate the answers to her colleagues, because she tried to talk to them and asked what they thought of their answers. [...] The students continued to work, but, in this final phase (about 3 minutes), there was already a greater buzz in the room, which indicated that some groups were already finishing, namely group III, which was the first one (about 4 minutes before the others). Group IV took the longest time. This group, although it seemed to be committed while I was in the room, did not work as much as it seemed at first sight. [...] It was only when they came to expose their answers on the board that I had the notion that they had worked less.

From excerpt E9, it becomes clear that with this feedback from students' work, the teacher becomes better able to understand who is working hard and who is not (regulatory feedback – centered on the task process, that is, on the strategies used to solve it). On the other hand, the teacher also identifies specific student characteristics, and this helps her understand who is able not only to evolve in their learning (self-regulation) but also help their colleagues, involving them in the process (peer feedback contributes to learning). So, for this analysis: design – promoting peer assessment and allowing feedback on student progress to the teacher; implementation – teacher walked around passing randomly through the groups, observing how students worked in their groups; results – the teacher became more conscious of the development and some characteristics of the students.

The collective feedback from students is received after their discussion in small groups, in the finalization of this task, transcribed in excerpt E10. This excerpt shows how the teacher finishes the task and how she decides to give feedback to students on its outcomes, calling a student from each group and gradually completing the answer on the board, with the help of the majority of the class.

Multimodal Narrative Excerpt 10: Teacher A (Collective Feedback From Students Received After Their Discussion in Small Groups)

As such, after everyone had answered the questions, one student from each group came to the board to answer a question. The students from the other groups made their contributions iteratively to improve and/or clarify the answer. In this way, the contributions of each group helped to enrich the final answer to each question. [...]

"Can anyone present any reason for it to be infertile?" I asked, addressing the class.

"Soil acidification," added Miguel.

I was questioning the other groups. In the end, I suggested:

"Your answer is fine. I would only add 'make soils infertile because of acidification.' Can you come and write it down, Manuel? You will only add in your answer: 'to make soils infertile because of acidification.'

This strategy offers several advantages. First, the students who are asked to stand out by giving the answers do not feel so exposed, since they were able to discuss them previously with their colleagues. Second, all students are able to see the achievements of their colleagues before the teacher's intervention – this can be very helpful in students' learning self-regulation. This peer collaboration is usually very profitable in terms of reaching all students, since it is given in their language in ways they understand (sometimes overcoming the same difficulties other students were also feeling). The kind of feedback students may receive from this kind of work can be richer than if it was unilaterally given (by the teacher). Third, the teacher also becomes aware of students' development and is able to clarify any common final doubts more accurately. Analyzing E10: design – students receive timely feedback from their peers and the teacher, and the teacher also gets a clearer idea of students' accomplishments; implementation – teacher called a group representative to write the answer on the board and asked for other groups' feedback; results – different groups constructed the final answer, mediated by the teacher, who finally validated it.

In the second MN, there is a classic example of a teacher giving immediate feedback during a class, validating students' remarks and using them to proceed (E11):

Multimodal Narrative Excerpt 11: Teacher B (Teacher's Feedback During Class)

Then Filipe asked whether "it could be just a zero on the right," alluding to the circumstance that in a line, only the element below the pivot was zero. This issue may have arisen from the fact that such a situation had not been explored. In this context, I could have used the first matrix of the example, but I decided to write another matrix in the table. I added a matrix with two rows, with the second pivot immediately to the right of the first one. [...] When identifying the pivots, Filipe returned to intervene:

"Because there is zero ... and because zero is below 2 ... if it was another number then it would not be...?"

I immediately validated the remark, recalling one of the requirements previously reported.

This excerpt portrays the events as they happened, plus the teacher's intention and the results for students' understanding. Analyzing E11: design – teacher fostered the students' reflection by giving pertinent clues and provided immediate feedback; implementation – teacher wrote an extra matrix on the board, addressing a student's question; results – this helped clarify the teacher's discourse; the student understood and was able to see the pattern; the teacher automatically gave him positive feedback (corrective feedback, focused on task performance).

DISCUSSION

The analyzed excerpts make it possible to identify some of the didactical aspects highlighted in the background section for each dimension (assessment process, assessment task and assessment feedback). Table 1 summarizes these identifications.

Regarding the dimension of assessment process, the most easily observable aspects are: (i) "create different assessment moments in class" and (ii) "use of formative assessment." The least present dimension in the MNs analyzed is "articulate formative and summative assessment" (e.g., summative assessments with formative purpose). The fact that this aspect was not observed does not mean that it does not exist; probably, it would be evident in a longitudinal analysis of several MN in the same discipline. Within these excerpts, the importance of formative assessment to involve students is clear, as it serves to encourage them to participate in their learning process. As Christoforou and Yigit (2008) state, these moments can make a difference to successful learning and can be achieved if students realize that their involvement will be assessed.

It was observed that similar assessment process designs could lead to different outcomes depending on the way the teacher implements them in the classroom. A few of the observed pertinent aspects include differences in the results observed in terms of student involvement, the possibility of autonomous thinking and responsibility in communicating.

Concerning the assessment tasks, the most observable aspect was "foster the mobilization of knowledge in an integrated way." In the examples given, the MN excerpts show how the teachers addressed the monitoring of student progress. By letting the students mobilize previous knowledge they feel pertinent to apply to new situations, teachers can help them scaffold their learning (Sabella & Redish, 2007). The

Table 1. Summary of didactical aspects identified in the 11 excerpts from MNs

Dimensions	Components Identified in Literature	Excerpts
Assessment process	Use of formative assessment	E1, E2, E3, E4
	Articulate formative and summative assessment (summative assessments with formative purpose)	Not directly observed in these 2 MN
	Foster student participation and responsibility	E1, E2, E4
	Diversify proposed activities (collaborative, individual)	E2, E4
	Provide several assessment moments during the course in the course curricula (and in classes in general)	Partially observed in both MN in the narrated class
	Create different assessment moments in class	E1, E2, E3, E4
	Allow the development of different competences, knowledge and attitudes	E2
Assessment tasks	Align tasks with learning objectives	E4, E6
	Promote authentic work	E2, E4, E7
	Complement class with out-of-class work	E8
	Foster the mobilization of knowledge in an integrated way	E2, E4, E5, E7
	Provide application to new contexts	E1
	Provide achievable challenges	E2, E4, E6
	Foster self-assessment	E3
Assessment feedback	Promote peer assessment and peer feedback	E2, E4, E9, E10
	Provide timely feedback	E2, E4, E10, E11
	Provide regular feedback	Observed in both MNs in the narrated class
	Provide relevant feedback: contribute to student reflection and progress (providing aids based on relevant and pertinent clues and only at key moments)	E2, E4, E11
	Provide feedback for the teacher	E2, E4, E6, E7, E9, E10

excerpts also illustrate the importance of the tasks being achievable challenges, aligned with the learning objectives, and promotors of authentic work. As Biggs (1999) and Lopes (2004) point out, attentiveness to these aspects facilitates a guiding line, helping students give meaning to new learnings. The aspects identified in the MN also show how teacher mediation can make tasks reveal different nuances. For instance, teachers could try to make tasks authentic until the end, so that it will be possible for students to think about them before discussion; also, teachers could turn tasks into collaborative and individual activities. This dependence between the teacher's mediation and the results obtained is highlighted by Lopes et al. (2010) as one of the crucial subjects in didactic research.

Different kinds of tasks were observed through the MN excerpts: e.g., lab work, the teacher working with the whole class to attempt the construction of knowledge, group work, homework, a recitation class with few attempts at interaction. Observing some of the outcomes through the MN shows that student involvement, responsibility and progress can be different depending on how teaching is conducted by the teacher.

With regard to assessment feedback, several excerpts (see Table I) focus on group work stimulating collaborative learning between peers (Lladó, 2014) and later peer feedback, showing that this is important for the regulation of learning (Nicol, 2008). Feedback efficiency is related to how teachers mediate learning (Bird & Yucel, 2015) and their teaching (Cruz, 2012). These examples show two important ways of implementing feedback mediation referred to in the literature review: (i) giving autonomy to students and (ii) providing aid based on relevant and pertinent clues, only at key times, which means, never before students have shown doubts (Lopes, 2004; Lopes et al., 2010; Viegas et al., 2015).

Peer assessment, peer feedback and feedback for the teachers are important aspects of assessment feedback. In some excerpts (see Table 1), the influence that teacher mediation can have on these processes, in terms of both teaching process management and learning process management, were here identified as crucial in class development. As Cruz (2012), Hattie and Timperley (2007) and Viegas, Lopes and Cravino (2010) point out, timely and relevant feedback will allow students to make better progress in their learning.

In addition, the analysis carried out with the MNs also shows that it is possible to study didactic aspects that concern teachers regardless of school grade or context. The presented MN excerpts covered two levels of education (secondary and university) and involved different subjects. It has been found that teachers' efforts and the results do not depend on these factors.

In the analyzed MN excerpts, it became clear that the teacher may act in a timely way, aiding students in their doubts and assisting class progress, if they receive feedback on students' learning. Table 1 highlights some of the identified aspects concerning assessment that were identified in the analyzed MN excerpts that can be directly linked to literature. The two analyzed MNs were not chosen in particular for being special in any way. As explained, they were used because they represented the aspects present in this book. This reinforces the importance any MN may have regarding any particular objective of the reader. In this chapter, the particular aspect was assessment. The MNs underlined not only different aspects of assessment but also the way the teacher conceived and designed them, the way they were mediated in the classroom and the outcomes of those decisions.

CONCLUSION

This chapter focused on the use of MN to visualize teaching practices regarding assessment and feedback, in terms of course curricula design versus its implementation in the classroom. The pursued objective was to better understand these different facets of a classroom through the MN. The analyzed dimensions were the assessment process, assessment tasks and assessment feedback (from teacher or peers and for the teacher). Excerpts from MNs showed how teacher mediation may affect students' involvement and learning in different contexts and at different levels. The importance of some particular aspects emphasizes key points found in the literature and here seen in practice, such as formulating an efficient task, using collaborative or individual learning, and giving students feedback on learning management.

Some aspects are more common or easy to identify than others; however, it was very helpful to understand them *in action*. The richness of this analysis using MN lies in showing teachers' efforts in terms of their intentions in the face of different assessment challenges (design). The MNs allow teachers to explain how they have implemented techniques in the classroom and lets readers observe (as if they were present in the classroom) what led the teachers to make some decisions, as well as how they adapted the design or, due to particular circumstances, chose to modify it, and how this affected the results, in terms of the students' attention, commitment and performance. Students' reactions, difficulties or breakthroughs underlined in the MN excerpts lead the reader through the teachers' aims and illustrates the outcomes.

The two analyzed MN were from different levels of teaching (one from high school, the other from university) and different topics (one from a science class, the other from mathematics); however, the same kinds of aspects emerged from them. This emphasizes the important role MN can play in showing how real teaching practices occur, in this case, with a special focus on assessment and feedback.

The fact that these identifications in teaching and learning processes emerge transversally supports the usefulness of MN for applications at other levels or contexts as objects of reflection, as illustrations or simply as inspirations for action to be taken by teachers, trainers or researchers on education. This reinforces the idea that teachers may use MNs as tools to reflect, improve or analyze their own practice or other teachers' practices, connecting didactical theories to observed practice. Teachers can use MNs to improve their teaching in several ways, namely, regarding assessment, and with different levels of commitment: by doing MN and going through the complete process of auto reflection toward the improvement of their own classes, as a process of research contributing to advances in this theoretical field, or simply by reading, comparing and understanding different practices and their outcomes and making personal adjustments or consolidations of their own practices. This shows the importance this tool can have in education by addressing parallel and complementary aspects of a classroom, making it a powerful tool in teaching.

REFERENCES

Arnold, I. (2016). Cheating online formative tests: Does it pay off? *The Internet and Higher Education*, *29*, 98–106. doi:10.1016/j.iheduc.2016.02.001

Biggs, J. (1999). *Teaching for quality learning at university*. Buckingham, UK: Society for Research into Higher Education – O. U. Press.

Bird, F., & Yucel, R. (2015). Feedback codes and action plans: Building the capacity of first-year students to apply feedback to a scientific report. *Assessment & Evaluation in Higher Education*, *40*(4), 508–527. doi:10.1080/02602938.2014.924476

Black, P., & William, D. (1998). Assessment and classroom learning. *Assessment in Education: Principles, Policy & Practice*, *5*(1), 7–64. doi:10.1080/0969595980050102

Black, P., & William, D. (2005). Lessons from around the world: How policies, politics and cultures constrain and afford assessment practices. *Curriculum Journal*, *16*(2), 249–261. doi:10.1080/09585170500136218

Boud, D., Cohen, R., & Sampson, J. (1999). Peer learning and assessment. *Assessment & Evaluation in Higher Education*, *24*(4), 413–426. doi:10.1080/0260293990240405

Christoforou, A., & Yigit, A. (2008). Improving teaching and learning in engineering education through a continous assessment process. *European Journal of Engineering Education*, *33*(1), 105–116. doi:10.1080/03043790701746405

Crisp, G. (2012). Integrative assessment: Reframing assessment practice for current and future learning. *Assessment & Evaluation in Higher Education*, *37*(1), 33–43. doi:10.1080/02602938.2010.494234

Crouch, C., & Mazur, E. (2001). Peer instruction: Ten years of experience and result. *American Journal of Physics*, *69*(9), 970–977. doi:10.1119/1.1374249

Cruz, E. (2012). *Da Avaliação do Impacte à Articulação da Investigação Práticas – O caso da Articulação na Formação Didáctica Pós-Graduada de Professores de Ciências e desafios futuros* [From impact evaluation to articulation of research practices – The case of articulation in post-graduate training of science teachers and future challenges] (Doctoral thesis). University of Aveiro, Portugal.

Dufresne, R., & Gerace, W. (2004). Assessing-to-learn: Formative assessment in physics instruction. *The Physics Teacher, 42*(7), 428–433. doi:10.1119/1.1804662

Earley, M. (2014). A synthesis of the literature on research methods education. *Teaching in Higher Education, 19*(3), 242–253. doi:10.1080/13562517.2013.860105

Etkina, E. (2000). Weekly reports: A two way feedback tool. *Science Education, 84*(5), 594–605. doi:10.1002/1098-237X(200009)84:5<594::AID-SCE3>3.0.CO;2-U

Evans, C. (2013). Making Sense of Assessment Feedback in Higher Education. *Review of Educational Research, 83*(1), 70–120. doi:10.3102/0034654312474350

Felder, R., Woods, D., Stice, J., & Rugarcia, A. (2000). The future of engineering education II. Teaching methods that work. *Chemical Engineering Education, 34*(1), 26–39.

Gibbs, G. (2003). Improving student learning through assessment [Editorial]. *Journal of Geography in Higher Education, 27*(2), 123–132.

Hattie, J., & Timperley, H. (2007). The power of feedback. *Review of Educational Research, 77*(1), 81–112. doi:10.3102/003465430298487

Laurillard, D. (2002). *Rethinking university teaching: a conversational framework for the effective use of learning technologies* (2nd ed.). London: Routledge Falmer. doi:10.4324/9780203160329

Lladó, A. (2014). Student perceptions of peer assessment: An interdisciplinary study. *Assessment & Evaluation in Higher Education, 39*(5), 592–610. doi:10.1080/02602938.2013.860077

Lopes, J. B. (2004). *Aprender e ensinar Física* [Learn and teach physics]. Lisbon: Fundação Calouste Gulbenkian.

Lopes, J. B., & Cunha, A. E. (2017). Self-directed professional development to improve effective teaching: Keys points for a model. *Teaching and Teacher Education, 68*, 262–274. doi:10.1016/j.tate.2017.09.009

Lopes, J. B., Silva, A., Cravino, J., Santos, C., Cunha, A., Pinto, A., ... Branco, J. (2014). Constructing and using multimodal narratives to research in science education: Contributions based on practical classroom. *Research in Science Education, 44*(3), 415–438. doi:10.100711165-013-9381-y

Lopes, J. B., Silva, A., Cravino, J., Viegas, C., Cunha, A., Saraiva, E., & Santos, C. (2010). *Investigação sobre a Mediação de professores de ciências físicas em sala de aula* [Research on science teachers' mediation in the classroom]. Vila Real, Portugal: Universidade de Trás-os-Montes e Alto Douro.

Mazur, E. (1997). *Peer instruction: a user's manual*. Prentice Hall.

McDonald, B., & Boud, D. (2003). The impact of self-assessment on achievement: The effects of self-assessment training on performance in external examinations. *Assessment in Education: Principles, Policy & Practice, 10*(2), 209–220. doi:10.1080/0969594032000121289

Mestre, J. (2001). Implications of research on learning for the education of prospective science and physics teachers. *Physics Education*, *36*(1), 44–51. doi:10.1088/0031-9120/36/1/308

Nicol, D. (2008). *Transforming assessment and feedback: enhancing integration and empowerment in the first year*. Quality Assurance Agency.

Nicol, D., & Macfarlane-Dick, D. (2004). *Rethinking formative assessment in HE: a theoretical model and seven principles of good feedback practice*. York, UK: HEA. Retrieved from http://www.enhancementthemes.ac.uk/pages/docdetail/docs/workshop/rethinking-formative-assessment-a-theoretical-model-and-seven-principles-of-good-feedback-practice-paper

Pedrosa-de-Jesus, H., & Watts, M. (2014). Managing affect in learners' questions in undergraduate Science. *Studies in Higher Education*, *39*(1), 102–116. doi:10.1080/03075079.2011.646983

Redish, E. (2003). *Teaching physics with the physics suite*. New York: John Wiley & Sons.

Rodrigues, F., & Oliveira, P. (2014). A system for formative assessment and monitoring students' progress. *Computers & Education*, *76*, 30–41. doi:10.1016/j.compedu.2014.03.001

Sabella, M., & Redish, E. (2007). Knowledge organization and activation in physics problem solving. *American Journal of Physics*, *75*(11), 1017–1029. doi:10.1119/1.2746359

Sadler, R. (1989). Formative assessment and the design of instructional systems. *Instructional Science*, *18*(2), 119–144. doi:10.1007/BF00117714

Sambell, K. (2016). Assessment and feedback in higher education: Considerable room for improvement? *Student Engagement in Higher Education Journal*, *1*(1), 2–14.

Sindelar, N. (2015). *Assessment powered teaching*. New York: Skyhorse Publishing.

Sondergaard, H., & Mulder, R. (2012). Collaborative learning through formative peer review: Pedagogy, programs and potential. *Computer Science Education*, *22*(4), 343–367. doi:10.1080/08993408.2012.728041

Viegas, C., Alves, G., & Lima, N. (2015). Formative Assessment Diversity to Foster Students' Engagement. In *Proceedings of ICL2015 – International Conference on Interactive Collaborative Learning* (pp. 929-935). Florence, Italy: Academic Press. 10.1109/ICL.2015.7318152

Viegas, C., Lopes, J. B., & Cravino, J. P. (2010). Learning through Assessment and Feedback In and Out of Class. In Innovations 2010 – World Innovations in Engineering Education and Research (pp. 33-46). Potomac, MD: iNEER.

Chapter 3
Dealing With Unexpected Situations in the Classroom:
Evidence From Multimodal Narratives of Teaching Practices

Ana Edite Cunha
Escola Secundária S. Pedro de Vila Real, Portugal

Cristina Marques
Universidade de Trás-os-Montes e Alto Douro, Portugal

ABSTRACT

This chapter examines teachers' practices in dealing with unexpected situations (USs) in the classroom and students' reactions to these practices. This study analyzes the multimodal narratives (MNs) of the classes of five teachers: three from basic education and two from higher education. The results show that the teachers identified USs and most of the time acted in a way to solve the problem by interacting with their students. This interaction took a variety of forms. The students reacted to these teachers' practices in different ways: clarifying their ideas and concepts, establishing links between ideas and/or concepts, answering or taking the initiative autonomously, correctly performing a task or solving a problem, and/or becoming more involved/active in their learning. In cases in which teachers did not act on the US, the students did not understand the task or took a passive attitude toward learning.

INTRODUCTION

This chapter presents an analysis of the performance of the classes of five teachers, three in basic education (students' ages from 13 to 15 years) and two in higher education, focusing on the decisions of the teachers when facing unexpected situations in the classroom and on the students' reactions to these decisions, in terms of learning.

The authors begin by presenting the reasoning of this study in terms of the classroom real-time decision making of teachers. The relevance of knowing and understanding teachers' decisions is discussed

DOI: 10.4018/978-1-5225-8570-1.ch003

to improve their pedagogical practice, especially when decisions are made when facing unexpected situations that occur in the classroom. It is argued that decisions must be taken after critical and systematic reflection on the students' thinking, their execution of tasks, and their learning, as well as the teaching practice. Although there is evidence that teachers and their actions condition the quality of students' learning, the characteristics of teachers' practices that determine improvements in teaching and learning are still unknown. Emphasis is placed on the relevance of teachers' data collection about their teaching and students' learning, so it can be analyzed, reflected on, and interpreted as a way of taking informed decisions, adapted to the reality of each classroom and of each student.

This study intends to identify the characteristics of teachers' decision making in face of unexpected situations in the classroom that contribute to improving students' learning.

BACKGROUND

The quality of the professional performance of teachers has been a concern among teachers themselves and investigators for a long time (e.g., Bishop, 1976; Datnow & Hubbard, 2016). In addition to the continuous updating of the knowledge to be taught, being a teacher requires constant adaptation to the classroom reality (Cunha, 2015; Santos, Lopes, & Cravino, 2011; Saraiva, Lopes, Cravino, & Santos, 2012). In their pedagogical practice, teachers have to be involved in a very complex process, which includes decision making (Lloyd, 2017). Teachers are often confronted with unexpected situations that occur in the classroom, that is, situations that they did not expect to happen, having to react and make decisions (Aho, Haverinen, Juuso, Laukka, & Sutinen, 2010; Bishop, 1976; Shavelson & Stern, 1981). That is, while teaching, teachers follow their "mental script," a process that is interrupted when unexpected events arise in the classroom. This stopping forces teachers to redirect their focus of attention and react to the new situation (Shavelson & Stern, 1981). According to Bishop (1976), taking decisions is a central activity of the teaching process, because the teacher has to know what to do, and how to do it, in the face of unforeseen situations that occur in the classroom. For example, when students do not understand or do not answer a question, when there are students who want to respond and others who do not, or when students make a mistake, the teacher has to make a decision, given a range of choices (Bishop, 1976). For this researcher, knowing how teachers make their decisions allows an understanding of the objectives and intentions of their teaching and the attitudes and scientific development of the students, making it possible to establish relationships between teaching theories and classroom reality. Borko, Roberts, and Shavelson (2008), after analyzing studies about classroom decision making, concluded that it emerged as a response to students' comments, incorrect responses or performance, lack of understanding, or even unruly behavior. Faced with these situations, teachers react in time, making teaching decisions such as questioning students, selecting students to respond, or giving appropriate examples. Understanding how decisions are made can help to improve the practice of decision making (Patel, Kaufman, & Arocha, 2002) and consequently make teaching more effective (Schoenfeld, 2015). That is, the articulation between thought and action that occurs during decision making facilitates better-quality learning (Lloyd, 2017).

However, it is not yet clear how knowledge about the desired practice is linked to the actual practice (Borko et al., 2008). Lloyd (2017) argued that critical reflection on their teaching practice helps teachers to improve their decision making. Other researchers have understood that decision making by teachers is one of the processes that generates reflection about what students think and about their learning, which promotes an improvement of teaching skills (Santagata & Yeh, 2016). That is, data related to the thinking and learning of students and their execution of tasks should be the object of reflection and interpretation by teachers in their decision making (Santagata & Yeh, 2016).

For good professional development to occur, teachers' ability and discipline to practice reflection and self-reflection about their professional performance cannot be neglected, as well as the exchange of experiences with their peers throughout their professional career (Cunha, 2015; Wright, Horn, & Sanders, 1997). The ability to reflect on their pedagogical practice, with the aim of improving it, implies that teachers are aware of the pedagogical decisions that they make in the classroom (Santos, 2017).

However, although there is increasing evidence that teacher performance influences student learning, the characteristics of teachers and of teaching that make it effective remain unclear (Santagata & Yeh, 2016). To these authors, it is clear that teaching in the classroom can be a powerful opportunity for teachers to improve their practices, resulting in planning and reflection on their teaching.

Furthermore, Hamilton et al. (2009) emphasized the importance of monitoring the learning of students, to be analyzed later to support the decision making, thus allowing the teacher to adapt the teaching to the individual characteristics of each student and to improve it. According to Dunn, Airola, Lo, and Garrison (2013), in data-based decision making, the teacher must have technological, statistical, and pedagogical knowledge to be able to analyze the data fully and improve student learning.

The detailed analyzis of the practice developed in the classroom, as well as the learning environment generated, is one of the well-known methods used over time to understand and improve teaching and learning. Following these ideas, the authors of this chapter believe that the need arises to describe with rigor what takes place inside the classroom. Since the classroom is a place where teachers and students interact, it is natural and evident that interpersonal relationships and different emotions and feelings happen in each environment of the teaching and learning processes. Therefore, an audio or video recording, and its transcription, may not be enough to describe and then study the relationships established within the classroom between teachers and students, as well as the quality of each teacher's practice. Therefore, it is essential to describe the true intentions of the teacher, what happened during the teacher's interaction with his or her students, the decision making, the reactions of the teacher him- or herself, and so on, which audio or video may not show (Cunha, Branco, Lopes, Cravino, & Santos, 2009; Lopes, Branco, & Jiménez-Aleixandre, 2011; Pinto et al., 2014). Therefore, to study the practices of a teacher in the classroom and to analyze how his or her performance, during a given task, influences the learning results obtained, it is very important to have data that capture and face the teaching and learning processes as a whole. Many of these goals are not fulfilled in their entirety using the available tools in education in the physical sciences, for instance "critical incidents" (Waight & Abd-El-Khalick, 2007), "diaries" (Hundhausen, 2002), and "noticing" (Mason, 2002).

More recently, an instrument has been developed that meets the need to record more rigorously the events taking place in the classroom without distorting the teaching and learning environment. This instrument is the Multimodal Narrative (MN) tool (Lopes et al., 2014), presented and described in detail in Chapter 1.

METHODOLOGY

The MN tool is an instrument that allows the recording of events that took place in the classroom, describing the contextualization of the class, the intentions of the teacher, the perceptions and reactions of the teacher and the students, the tasks proposed, and the work actually undertaken by the students. This description considers the class as a whole, preserving its complexity and reporting the phenomena in an integral way. Thus, it becomes possible to analyze and compare the narratives of various teachers (Larrivee, 2000; Seidel, Stürmer, Blomberg, Kobarg, & Schwindt, 2011).

This section describes specifically how teachers dealt with unexpected situations (USs) in the classroom in terms of decisions, based on the analyzis of each teacher's practice, and the students' reactions to these decisions in terms of learning. This analyzis was based on what happened in the classroom, taking as reflection material the MNs made by five teachers, three in basic education (students' ages from 13 to 15 years) and two in higher education (Cohen, Manion, & Morrison, 2011).

To perform the analyzis, codification of the teacher's practice was undertaken in each MN. Class excerpts were identified corresponding to the evidence of: (i) the detection/non-detection of a US; (ii) the utility of the US; (iii) no teacher reaction to the US; and (iv) the usefulness of reflections on teaching practices in the improvement and/or correction of these practices. The first attempt to codify the teacher's practice was made through evidence. A brief description of the different codifications was provided as well as their definition. Some categories of analyzis were inspired by the literature; however, most were obtained by systematic comparison with the MN excerpts, characterized by the same coding. After coding, using all the MNs, the settings were refined, and it was checked whether the selected excerpts were described well for each encoding. During this process, each encoding, as well as the definition, was not definitive, as some encodings can be grouped or divided. The teacher's practice was thus classified into three categories (micro-scale) of the unit of analysis Interaction (this designation refers to the interrelationships between teacher and students during the execution of the tasks by the students) (Lopes et al., 2014; Pinto et al., 2014; Santos, Lopes, Saraiva, & Cunha, 2015):

SU: USs are sometimes useful for teaching.
NR: The teacher passes by the US and does not react.
RUC: Reflection on teaching practices through mediation is useful for correcting practices a *posteriori*.

To undertake the codification of students' reactions, in each MN, class excerpts were identified corresponding to evidence of: (i) difficulties in understanding the task; (ii) a passive attitude toward learning: (iii) no learning; (iv) a response to the task according to what was expected; (v) an active attitude toward learning; vi) the establishment of links between ideas and/or concepts; and vii) a response or resolution of the task autonomously. The coding of students' reactions followed a process analogous to that used to encode teachers' practice. The students' reactions were thus classified into seven categories:

DU: The student has difficulty understanding the task.
CI: The student clarifies his/her ideas and concepts.
UR: The student understands the question/idea/task, responding/acting in the intended sense.
PA: The student has a passive attitude toward learning.

AA: The student has an active attitude toward learning.
EL: The student establishes links between ideas and/or concepts.
RSA: The student responds or solves the problem autonomously.

RESULTS AND DISCUSSION

After analyzing the narratives made by five teachers in basic education (three) and higher education (two), three categories of teachers' practice, concerning interaction as the unit of analyzis, and seven categories of students' reactions were identified (Tables 1, 2, 3, 4, and 5) (Böhm, 2004).

The results for teacher A presented in Table 1 allow the inference that:

- Teacher A, faced with the students' initial misunderstanding of the question raised, clarified the question by simplifying the language and reinforcing the main idea. This clarification allowed student A to understand the question and respond in the intended sense, establishing links between ideas and/or concepts. As the discussion was being developed in a large group (the class),

Table 1. Categorization of the teacher's practice and students' reactions in teacher A's class of higher education

MN Excerpts	US Description	Teacher's Practice	Students' Reactions
Teacher – "This idea of science as absolute, dogmatic truth is often constructed by us in a subliminal way. Can anyone tell us an example from everyday life that reveals this tendency to influence us?" Student A – "I did not understand the question." Teacher – "I was asking for a day-to-day situation that leads us to think that science is the absolute truth."	In view of the difficulty of student A in understanding the question, the teacher clarified it, simplifying it. Student B's response was not what the teacher intended. The teacher repeated the question but reinforced the main idea. Student A, who had initially said that he had not understood the question, responded according to the question posed, establishing links between ideas and/or		DU
Student B – "Sometimes it can even be the other way around. For example, when we hear that the universe has no end, we think that they are deceiving us and that it is not true."	concepts. The teacher took advantage of student A's response to establish a link between this response and day-	RUC	DU
Teacher – "Yes, but I was now thinking of situations that would lead us to think of science as truth." Student A – "When we hear a scientist say that pollution is raising the planet's average temperature, we believe because we have the proof."	to-day situations. From here, student A and student B were able to weave these and other considerations about what the teacher had intended with the initial question. During this conversation, the teacher was posing questions to		UR, EL
Teacher – The students were thoughtful, they looked like they were looking for an example, but no idea was coming up. "You have certainly watched a lot of ads on TV. Do you remember any of a scientist?" Student A – "I'm reminded of the advertising for Actimel, I think a scientist appears ...." Student B – "Yes, I also think so, it's to show that it's really good for health."	which the students were responding autonomously, establishing links between ideas and/or concepts.		EL EL, RSA

the teacher took advantage of student A's response to establish links between this response and everyday situations. Given this intervention of the teacher, student A and student B were able to weave some considerations about what the teacher intended with the initial question. From here, a discussion was established between student A, student B, and the teacher, in which the teacher was asking questions and the students were establishing links between ideas and/or concepts and responding autonomously, therefore tackling the question initially posed.

Analyzing the results of Table 2, relative to teacher B, it can be stated that:

- On several occasions, teacher B realized that he/she took teaching decisions that were not favorable to learning, such as presenting the knowledge already taught, giving the answer, or devaluing students' difficulties. At such times, the teacher chose to ignore that awareness and proceeded with his or her decisions. In these situations, the students did not assume a central role in their learning, leaving that role to the teacher, who was the transmitter of information or knowledge and they were the recipients. Therefore, the students had a passive attitude toward learning.

At other times, the students were not satisfied with the teacher's performance in the face of their difficulties. In these situations, the students insisted on receiving a response/clarification until the teacher tried to address the students' doubts. In one case, the teacher started questioning to try to understand the students' ideas, which provided discussion in the classroom and, consequently, the clarification of ideas and concepts, as well as the establishment of links between concepts by the students. In another case, the teacher failed to give the students clarification, ending the discussion by referring the clarification to a specialist. Therefore, the students' doubts remained.

From the critical analyzis of the MN of teacher C (Table 3), it is verified that:

- Teacher C experienced some USs during his classes, which were detected through student questions or procedures. On one occasion, the teacher took advantage of the US to talk about a relevant subject, something that he/she had not planned to do. Through the systematic questioning of the teacher, the students established links between concepts in an autonomous way, thus answering the questions posed. On another occasion, the US led the teacher to pose several hypotheses and test them along with the students. This decision allowed a student to take the initiative (autonomously) that led to the resolution of the situation.

The results of Table 4 for teacher D indicate that:

- Teacher D detected a US as a result of his students' passivity toward learning: the students did not interact with each other, were not sure about what they knew/did not know, did not take initiatives, and so on. The teacher acted, transforming individual doubts into collective questions, which resulted in a change in students' attitude toward learning from passive to active: the students began to participate in the class, expressing their difficulties, answering the questions that the teacher was asking, reviewing concepts already learned, establishing links between them, or performing simple tasks. In another situation involving a passive attitude among the students, the teacher

Table 2. Categorization of the teacher's practice and students' reactions in teacher B's class of higher education

<table>
<tr><th>MN Excerpts</th><th>US Description</th><th>Teacher's Practice</th><th>Students' Reactions</th></tr>
<tr><td>Teacher – Seeing most of the guys looking at me, and seeing no discussion, I said then: "I will remember what was a permanent regime ... or does someone want to do it for me?" Reminding me, already late, that it would have been much more interesting for them to do so, but having already said that I would do it, nobody wanted to move forward ... So, I had to go on.</td><td>After the decision that he, the teacher, would present knowledge that had already been taught, the teacher realized the inefficacy of his practice but decided to maintain his performance.</td><td>NR</td><td>PA</td></tr>
<tr><td>Teacher – After about 5 minutes, I ended up not being silent, because they were always asking for something ... or I was referring to something ... I noticed that they have many difficulties ... I questioned them about their progress and asked if anyone had managed to structure the reasoning.
Student A – "The flow will have a certain value ... at the beginning ... and T_i will have the same value ... I think ...."
Teacher – "I did not understand what it means ...."
Student A – "Q is equal ... so equals ...."</td><td>After the insistency of the students on receiving a response/clarification, the teacher started questioning to try to understand the students' ideas, but the students continued to have doubts.
After the teacher realized that the students were still having difficulties, he had to intervene until he realized what the students were saying and then they began to clarify and understand the concepts at stake and respond clearly to the questions that the teacher was posing.</td><td rowspan="2">RUC</td><td>DU
DU</td></tr>
<tr><td>Teacher – I continued without realizing it. So, I went to the blackboard to try to materialize: "Let's start from Q," I started, "and get the first layer: it will be the same as what? k? Is everything in SI?" I took the time to remind them.
Student A – "That's what I was saying, as the q is equal, the Ti 25 equals …."
Teacher – "… the same numerical value?" I asked incredulously, thinking that they were referring to match q = (Ti-25) ... There was some confusion among the students in explaining correctly and some slowness on my part, but I finally understood what they meant.
Student A – "This flow is equal to T_i for 25 °C ...."</td><td>Faced with the discussion that the teacher promoted, the students were able to build their reasoning and communicate their ideas, establishing links between ideas and/or concepts.</td><td>DU
CI, UR
EL</td></tr>
<tr><td>Teacher – "If you remember, in electrical circuits, we had electrical resistance between two points of?" I started. I ended up completing the answer, after 2 or 3 seconds with no one moving forward with the answer. I did not quite understand why, since the question was simple, and it was on the slide ... I thought maybe the students did not realize what I meant, and I kept going.</td><td>The teacher, who expected an answer from the students, chose to give them the answer.</td><td>NR</td><td>PA</td></tr>
<tr><td>Student P – "But the tooth also expands inward ...."
Teacher – "If it's the same, everything expands in the same way ...."
Teacher – "… in principle, yes ... I understand what you mean, but I believe the problem will be greater if they are different ... in any case, it's a question to put to the dentist the next time you're there ...."</td><td>Student P began a series of interventions, in which he insisted on an idea that did not agree with the teacher's idea.
The teacher snapped back, giving the explanation that was right and expected (yours). The exchange of arguments continued between the teacher and the student, without reaching agreement.
The professor finished the discussion, referring to clarification with other science professionals and ending the doubt.</td><td>NR</td><td>DU
DU
DU</td></tr>
</table>

Table 3. Categorization of the teacher's practice and students' reactions in teacher C's class of basic education

MN Excerpts	US Description	Teacher's Practice	Students' Reactions
Teacher – Student C asks if they all connect to the same electric extent. "Oh C, I should not say anything, but here is an important recommendation and we are going to do this survey now. You did well to ask that question, because it reminded me of 'what care should we have with the electric current?'" Student C – "We should not plug many appliances into the same electrical outlet." Teacher – "Otherwise, what happens?" Student A – "It can burn." Teacher – "Why?" Student D – "Because there is current overload."	It was the student's question that reminded the teacher of the need to talk about taking care with the electric current. Through systematic questioning of the teacher, several students were enumerating, and justifying, the care to be taken with the electric current.	RUC, SU	DU EL, RSA
Teacher – The teacher asked to connect the power supply to the extension and then to turn on the switch. Nothing happened. The teacher replaced one of the wires. The bulbs still did not light up. (...) They checked the power supply. The teacher asked student D to replace the link in the extension. The problem remained. The teacher looked for another power source. When he/she arrived, student D had the circuit running only with a light bulb.	There was a circuit that did not conduct the electric current. The teacher, in interaction with his students, made the decision to pose several hypotheses and to test them. A student took the initiative to make a change in the assembly of the circuit, which led to the resolution of the situation.	SU	EL, RSA

launched a challenge to the class in the form of a question–problem and included a worksheet with information about concepts and tasks. This decision led the students to interact, exchange ideas, perform tasks, establish relationships between concepts, and solve the initial problem. Therefore, they became more involved in their learning. Another teaching situation occurred, this time not unexpected, as it was the teacher himself who posed a question–problem with the purpose of causing cognitive conflict. Faced with this decision, the teacher needed to mediate the learning of the students. In particular, the decision helped the students to carry out the tasks using the resources available, mediating in a surgical and directed way but giving the students autonomy. Under these circumstances, the students manipulated resources (computational simulation) and established links between simulations and concepts to be learned and between concepts. Therefore, the students answered the initial question–problem autonomously.

- Teacher E detected a US (Table 5), but he/she did nothing immediately to solve it. In fact, the teacher only understood later that part of the students' difficulties in carrying out the proposed task was due to poor conception of the task. Therefore, before the teacher acted, the students did not understand the task, performing it badly, although the teacher interacted with the students, group by group, trying to mobilize their previous knowledge about acid-base indicators and to promote the establishment of links between concepts. Faced with the continuing difficulties of the students, the teacher corrected the task and continued the interactions with them. The students were gradually able to understand the proposed task and the knowledge involved. In another US, the teacher detected that the students had difficulties in carrying out the proposed tasks and he/she immediately started to mediate the students' work and learning. The teacher followed the work of the students group by group, referred to existing links between relevant concepts, but left the students work autonomously. As the students continued to experience difficulties for a long time, the

Table 4. Categorization of the teacher's practice and students' reactions in teacher D's class of basic education

MN Excerpts	US Description	Teacher's Practice	Students' Reactions
Teacher – (…) I noticed that they attached little relevance to the opinion of their co-workers. They preferred to listen to my opinion (…) To try to minimize the dependency on the teacher, I chose to transform the questions that they asked me into collective questions. One student thought that an outlet was an energy generator. I expressed that doubt as a question and put it to the class. Then they reviewed the production, transportation, and distribution of energy, which had already been taught. (…) Another question arose regarding a counter and an electric panel. Student M called a counter an electric panel. (…) This doubt was debated until all the students agreed to call it an electric panel.	The teacher identified that the students were not interacting with each other by discussing ideas. He acted by presenting the individual doubts/difficulties to the class, which provoked a discussion in the class. The students began to answer the questions and to put their doubts forward. Therefore, they were in a more active posture, establishing links between concepts that they had already learned.	**RUC, SU**	**PA** **AA, EL**
Teacher – I (…) noticed that, in general, they were worried about the history test and were somewhat "inactive" (they were starting the day). "I'm going to have to challenge them," I thought to myself, and then I asked the students: "With the equipment that is in the tray, can we measure the resistance of this conductor?" I showed them a copper wire wound in a propeller. The students seemed to be curious, so I passed around the conductor by hand. (…) Students A and B advanced the following hypothesis: Students A and B – "We set up an electrical circuit similar to the one in the previous class, but instead of the lamp we put the copper wire." Teacher – The students began to exchange opinions. (…) Students A and B asked me if they could set up the circuit. As the answer was yes, they got up and picked the equipment and set up the electrical circuit inside the tray. (…) The other students were aware of possible assembly errors. Student C said to student A: Student C – "The voltmeter is connected in parallel and the ammeter in series ...." Teacher – (…) Student A presented the circuit to the class, starting with the generator, reading the potential difference and the current intensity. (…)	Faced with some inactivity of the students, the teacher decided to challenge them, setting them a question–problem to answer. The attitude of the teacher promoted interaction between the students, causing them to exchange ideas among themselves and jointly to carry out an experimental set-up to respond to the problem.	**RUC, SU**	**PA** **AA, EL**
Teacher – "Shall we then calculate the resistance value of the conducting wire inserted into the circuit?" Student D – "How?" Teacher – The students registered the readings and calculated the resistance of the conductor. Student D – "So why does it always give the same value?" Student E – "Don't you see that the thread was always the same!?" Student F – "Ahh!! I have now realized ... this conductor has a resistance to the passage of electrons of 5.3 Ω …."	At a later stage of this experimental activity, the teacher relaunched the initial challenge, posing the initial question but reformulating it. Faced with the difficulties of the students, the teacher handed out a worksheet containing useful information about concepts and proposed tasks. In this way, the work triggered for students (discussion of ideas, establishment of relationships between concepts, etc.) by this strategy led them to solve the problem.	**RUC, SU**	**DU** **EL, RSA**
Teacher – I started by inviting students to read the challenge question: "What are the properties and applications of light?" There was some silence and nobody could answer. "Do not worry, because this is a 'big' question. It needs, therefore, a study of light that begins with the first challenge, proposed in tasks 1, 2, and 3, of the Learning Roadmap. Open the L1 and L2 digital features that will allow you to distinguish natural light sources from artificial light sources, luminous bodies from illuminated bodies (…)." Teacher – To construct the answer, the students opened without difficulty the hyperlink for the resources (...) and observed the trajectory of the light rays (…) I asked him (student A): "… so how does the light propagate?" And he answered promptly: Student A – "It propagates in all directions, in a straight line, in the form of electromagnetic waves." Student B – "It does not need material support to propagate."	Faced with the question–problem posed by the teacher, the students did not respond. This situation was not unexpected for the teacher, given his intention to provoke a cognitive conflict by asking the question. Under these conditions, the teacher had to mediate the students in solving the question. He set tasks for the students to complete in articulation with the use of the available resources. The execution of the tasks and the use of the resources by the students were guided by the teacher, in a directed, surgical way, and to allow students the maximum autonomy. Throughout this process, the students were discussing ideas with each other and with the teacher, performing tasks autonomously, mobilizing prior knowledge, and establishing connections between concepts. In this way, they answered the question initially posed.	**RUC, SU**	**DU** **EL, RSA**

Table 5. Categorization of the teacher's practice and students' reactions in teacher E's class of basic education

MN Excerpts	US Description	Teacher's Practice	Students' Reactions
Teacher – The students understood what they had to do. However, as the liquid solution was very concentrated, the expected color change upon breathing expiration did not happen quickly, and this generated much disturbance in the work of the groups.	The teacher detected the difficulty (high concentration of the liquid solution) but did nothing immediately to remedy it. At this stage, the teacher thought that the students understood what they had to do. However, the disturbance in the classroom was due not only to students' impatience to obtain results but also to the fact that the students thought that the intended effect was obtained with the voice and not with the expiration. This situation persisted for a long time, until the next US appeared (RUC).	**NR**	**DU** **DU**
Teacher – As I interacted with each group, I heard the other students trying to make their solutions change color. As the phrase written in step 2 states "order the solution to change color," some students shouted into the solution, which generated disruption in the lesson. I had not realized that, by setting the question in this way, the students would associate the color change with the voice and not with the CO_2 released into the solution as they spoke into the container. (….) I kept going through the groups to get them to realize what was at stake: "If this has this color, is it because we are in the presence of a … solution … what?" I asked the students in group II. (…) As the results were delayed, I advised the students to dilute the solution a little by adding water. Even so, the results were not immediately visible. (…) I chose to dilute each solution again. However, after 6 minutes of attempts, a student got the solution to change color.	The teacher detected the difficulty (poorly designed task) and interacted with the students, group by group, trying to mobilize their previous knowledge about acid-base indicators and promoting the establishment of links between concepts. At one point, he had to act in practical terms, diluting the initial basic solution. The conceptual difficulties of the students coexisted with a task that was poorly conceived by the teacher. These difficulties were solved with the persistent intervention of the teacher, who mediated the students almost individually.	**RUC**	**DU** **DU** **EL**
Teacher – After this exposure to the class, I indicated that they would now understand much better what was intended in task 4. (…) I explained that the scheme represented part of the water cycle and included the effects of air pollution. (…) "What we want is that you look at this figure and, based on what we have been talking about, answer those questions." Student A – "Oh teacher, check it out here." Teacher – I read what was written (…) and saw that (…) they had not been able to identify the acidic solutions that form in the atmosphere. "You can identify the acid that forms! Look here, this should provide some clue," I urged, pointing to the notebook, to the table that they had written with the name of the three acid solutions. Seeing that the students in student A's group were looking at the table and talking about it, I left them to work. Student B – "Oh teacher, here we have to do those things? Teacher – Student B was referring to the equations of words requested in c). (…) I noticed that she had written: Nitrogen Oxides + Sulphur Oxide -> Acid Rain. I pointed to each of the oxides and said: "This will give rise to one thing, and this to another. Both contribute to acid rain." (…) Teacher: I read what they had written (…) and questioned students C and D (…) "Did you remember the names sulfuric and nitric acid, out of nowhere?!" (…) Student D "No, it's because we saw them there!" said student D, pointing to table 3 of the task.	At the beginning of the class, the teacher reviewed the previous class to make sure that the students understood the relationship between the oxides responsible for atmospheric pollution and acid rain. The students demonstrated difficulties in carrying out the proposed tasks. The teacher, following the work of the students group by group, referred to the existing links between already-presented concepts, but left it to the students to try to give the desired answers autonomously. The difficulties remained with the students for a long time. The teacher insisted on the existence of links between concepts, but without giving the answers. This process was time consuming and persistent. Gradually, the students were establishing relationships between concepts, which allowed them to respond to tasks.	**RUC**	**DU** **DU** **EL, UR**

teacher had to accompany the students' learning almost individually and persistently, until they gradually began to establish relationships between concepts and respond to the tasks.

The aim of this study was to identify the characteristics of teachers' decision making, when faced with unexpected situations in the classroom, that contribute to improving students' learning. The teachers whose MNs were analyzed identified USs and most of the time acted to solve the problem by interacting with their students. This interaction took on the characteristics of mediation in the sense that Lopes, Cravino, Branco, Saraiva, and Silva (2008) referred to as "teacher action and language (verbal and not verbal) as a systematic answer to the students' learning demand in their specific development pathways to the intended curriculum learning outcomes (namely in terms of students' knowledge, competences and attitudes)" (p. 43). This mediation took a variety of forms, such as: i) clarifying questions previously posed (teacher A); ii) establishing links between ideas (teachers A and E); iii) asking the students successive questions and promoting argumentation with each answer (teachers A, B, C, and E); iv) providing relevant information and/or tasks to be executed (teacher D); v) formulating and testing hypotheses in an experimental problem (teacher C); vi) posing collective questions (teacher D); vii) posing a question–problem (teacher D); viii) accompanying the work of the students in each group or individually (teachers D and E); or ix) correcting deficient experimental planning (teacher E). Under these circumstances, students reacted to teachers' mediation in several ways: i) clarifying their ideas and concepts (CI); ii) understanding the question/idea/task and responding or acting in the intended sense (UR); iii) establishing links between ideas and/or concepts (EL); iv) responding, performing a task well, or solving a problem autonomously (RSA); and/or v) becoming more involved/active in their learning (AA).

However, there were times when teachers did not react to the USs. Therefore, they did not resolve them. Teacher B himself presented the knowledge that had already been taught, gave the answer, and devalued students' difficulties. Teacher E undertook nothing to correct the concentration of a solution in an experimental task. Faced with some of these teachers' practices, the students adopted a passive attitude toward learning, being the recipients of information or knowledge transmitted by the teacher (PA), or failed to understand the task, performing it badly (DU).

The students' reactions, in terms of learning, matched the various forms of mediation used by the teachers. Thus, when the teacher(s):

- *Clarified questions previously posed*
    - The students understood the question/idea/task, responding/acting in the intended sense (UR), *and* established links between ideas and/or concepts (EL).
- *Established links between ideas*
    - The students established links between ideas and/or concepts (EL);
    - The students established links between ideas and/or concepts (EL) and understood the question/idea/task, responding/acting in the intended sense (UR), but the teacher also had to use another form of mediation, namely *accompanying the work of the students in each group or individually.*
- *Asked the students successive questions and promoted argumentation with each answer*
    - The students established links between ideas and/or concepts (EL) and responded or Solved the problem autonomously (RSA);

- The students clarified their ideas and concepts (CI) and established links between ideas and/or concepts (EL);
- The students established links between ideas and/or concepts (EL), but the teacher also had to use other forms of mediation, namely *accompanying the work of the students in each group or individually* and *correcting deficient experimental planning.*

- *Formulated and tested hypotheses in an experimental problem*
    - The students established links between ideas and/or concepts (EL) and responded or solved the problem autonomously (RSA).
- *Posed collective questions*
    - The students had an active attitude toward learning (AA) and established links between ideas and/or concepts (EL).
- *Posed a question–problem*
    - The students adopted an active attitude toward learning (AA), established links between ideas and/or concepts (EL), and responded or solved the problem autonomously (RSA), but the teacher had to use one more form of mediation, at a certain moment, to help students to deal with some difficulties, namely *providing relevant information and/or tasks to be executed*;
    - The students established links between ideas and/or concepts (EL) and responded or solved the problem autonomously (RSA), but the teacher had to use other forms of mediation, namely *providing relevant information and/or tasks to be executed* and *accompanying the work of the students in each group or individually.*
- *Accompanied the work of the students in each group or individually, corrected deficient experimental planning*, or *provided relevant information and/or tasks to be executed*
    - The learning achieved by the students was the result of the concomitant use of other forms of mediation.

From the results obtained in this study, it can be stated that:

- When the teacher *posed a question–problem*, the learning promoted in the students reached higher quality levels (AA, EL, RSA). However, solving a question–problem is a task of great complexity, so the teacher also had to resort to other form(s) of mediation, namely *accompanying the work of the students in each group or individually* and/or *providing relevant information and/or tasks to be executed*. This result is in keeping with what other authors have argued, considering that teacher effort is an important attribute of mediation (Ditcher, 2001; Felder, Woods, Stice, & Rugarcia, 2000; Lopes, Cravino, & Silva, 2010). On the other hand, *accompanying the work of the students in each group or individually, providing relevant information and/or tasks to be executed*, and *correcting deficient experimental planning* are ways of mediating that the teachers did not use in isolation from other types of mediation in the MNs analyzed, so the learning achieved in these cases was due to an association of different types of mediation. This suggests that the forms of mediation mentioned last potentiated the *posing of the question–problem* in the development of the learning quality.

- When the teacher *clarified questions previously posed*, *asked collective questions*, *established links between ideas*, or *asked the students successive questions and promoted argumentation with each answer*, the students' learning reached levels of quality that were below those achieved when a *question–problem* was posed (in a general way, EL).
- It is relevant to note that there was one case in which the teacher mediated (teacher E), *asking the students successive questions and promoting or arguing with each answer* and *accompanying the work of the students in each group or individually*, but the students did not understand the task, performing it badly (DU), because the task suffered from deficient experimental planning.

Therefore, it is not enough to mediate using the forms that research shows to be able to promote learning. Being attentive to students' reactions, knowing how to interpret them, reflecting on the characteristics of teaching, in relation to both planning and implementation, and knowing when and how to mediate are fundamental for the good professional development of the teacher.

This conclusion reinforces the idea advocated by some researchers that it is necessary for teachers to improve their attitudes (Duschl & Wright, 1989) to influence the dynamics of the classroom, approaching the theory of practice (Millar & Osborne, 2009).

CONCLUSION

The MN tool can be considered an important formative tool because it describes teachers' action in more detail in the classroom. An MN allows the triggering of the reflective process, since it is more than a mere description of events. The elaboration and later reflection of MNs by each teacher enables them to analyze their practice at each moment of a class and throughout several classes. This process, performed in a systematic, deliberate, and conscious way, is extremely useful for the professional development of each teacher, since it allows deficiencies in the teacher's performance to be detected as well as enabling strengths to be found in the mediation conducted. This awareness makes it possible to correct, improve, and develop the teacher's practices.

By analyzing and comparing the MNs of several teachers, it is possible to characterize their practices, verify the relevance of some types of classroom interaction, and more easily identify the relevant mediation traits to deal with USs.

Therefore, MNs are potentially effective in relation to teachers' professional development, whether they are elaborated by the teacher himself (referring to his classes) or by another teacher. In this case, teachers, when reading the MNs of other colleagues, will be able to make comparisons with their classes, often revising them and being able to recognize aspects of their own practices that need evolution.

REFERENCES

Aho, E., Haverinen, H.-L., Juuso, H., Laukka, S. J., & Sutinen, A. (2010). Teachers' principles of decision-making and classroom management: A case study and a new observation method. *Procedia: Social and Behavioral Sciences*, *9*, 395–402. doi:10.1016/j.sbspro.2010.12.171

Bishop, A. J. (1976). Decision-making, the intervening variable. *Educational Studies in Mathematics*, *7*(1/2), 41–47. doi:10.1007/BF00144357

Böhm, A. (2004). Theoretical coding: Text analysis in grounded theory. In U. Flick, E. von Kardorff, & I. Steinke (Eds.), *A companion to qualitative research* (pp. 270–274). London: Sage.

Borko, H., Roberts, S. A., & Shavelson, R. (2008). Teachers' decision making: From Alan J. Bishop to today. In P. Clarkson & N. Presmeg (Eds.), *Critical issues in mathematics education* (pp. 37–67). Boston, MA: Springer. doi:10.1007/978-0-387-09673-5_4

Cohen, L., Manion, L., & Morrison, K. (2011). *Research methods in education* (7th ed.). London: Routledge.

Cunha, A. E. (2015). *Construção de práticas de referência no ensino da física para o ensino secundário. Tese de doutoramento.* Vila Real, Portugal: Universidade de Trás-os-Montes e Alto Douro.

Cunha, A. E., Branco, J., Lopes, J. B., Cravino, J. P., & Santos, C. (2009). Envolver os alunos de forma produtiva em aulas de ciências físicas: Evidências para uma proposta de uma ferramenta para ajudar os professores. In *Extra Nr. VIII Congresso Internacional sobre Investigación en Didáctica de las Ciencias* (pp. 1553–1558). Barcelona: Enseñanza de las Ciencias.

Datnow, A., & Hubbard, L. (2016). Teacher capacity for and beliefs about data-driven decision making: A literature review of international research. *Journal of Educational Change*, *17*(1), 7–28. doi:10.100710833-015-9264-2

Ditcher, A. K. (2001). Effective teaching and learning in higher education, with particular reference to the undergraduate education of professional engineers. *International Journal of Engineering Education*, *17*(1), 24–29.

Dunn, K. E., Airola, D. T., Lo, W.-J., & Garrison, M. (2013). Becoming data driven: The influence of teachers' sense of efficacy on concerns related to data-driven decision making. *Journal of Experimental Education*, *81*(2), 222–241. doi:10.1080/00220973.2012.699899

Duschl, R. A., & Wright, E. (1989). A case study of high school teachers' decision making models for planning and teaching science. *Journal of Research in Science Teaching*, *26*(6), 467–501. doi:10.1002/tea.3660260602

Felder, R. M., Woods, D. R., Stice, J. E., & Rugarcia, A. (2000). The future of engineering education II. Teaching methods that work. *Chemical Engineering Education*, *34*(1), 26–39.

Hamilton, L., Halverson, R., Jackson, S. S., Mandinach, E., Supovitz, J. A., Wayman, J. C., … Steele, J. L. (2009). *Using student achievement data to support instructional decision making*. United States Department of Education. Retrieved from http://repository.upenn.edu/gse_pubs/279

Hundhausen, C. D. (2002). Integrating algorithm visualization technology into an undergraduate algorithms course: Ethnographic studies of a social constructivist approach. *Computers & Education, 39*(3), 237–260. doi:10.1016/S0360-1315(02)00044-1

Larrivee, B. (2000). Transforming teaching practice: Becoming the critically reflective teacher. *Reflective Practice: International and Multidisciplinary Perspectives, 1*(3), 293–307. doi:10.1080/713693162

Lloyd, C. A. (2017). Exploring the real-world decision-making of novice and experienced teachers. *Journal of Further and Higher Education*, 1–17. doi:10.1080/0309877X.2017.1357070

Lopes, J. B., Branco, J., & Jiménez-Aleixandre, M. P. (2011). "Learning experience" provided by science teaching practice in a classroom and the development of students' competences. *Research in Science Education, 41*(5), 787–809. doi:10.100711165-010-9190-5

Lopes, J. B., Cravino, J. P., Branco, M. J., Saraiva, E., & Silva, A. (2008). Mediation of student learning: Dimensions and evidences in science teaching. *Problems of Education in the 21st Century, 9*, 42–52.

Lopes, J. B., Cravino, J. P., & Silva, A. A. (2010). *A model for effective teaching in science and technology (Metilost)*. New York: Nova Science Publishers, Inc.

Lopes, J. B., Silva, A. A., Cravino, J. P., Santos, C. A., Cunha, A., Pinto, A., ... Branco, M. J. (2014). Constructing and using multimodal narratives to research in science education: Contributions based on practical classroom. *Research in Science Education, 44*(3), 415–438. doi:10.100711165-013-9381-y

Mason, J. (2002). *Researching your own practice: Discipline of noticing*. New York: Routledge.

Millar, R., & Osborne, J. (2009). Research and practice: A complex relationship? In M. C. Shelley, L. D. Lore, & B. Hand (Eds.), *Quality research in literacy and science education: International perspectives and gold standards* (pp. 41–61). Dordrecht, The Netherlands: Springer. doi:10.1007/978-1-4020-8427-0_3

Patel, V. L., Kaufman, D. R., & Arocha, J. F. (2002). Emerging paradigms of cognition in medical decision-making. *Journal of Biomedical Informatics, 35*(1), 52–75. doi:10.1016/S1532-0464(02)00009-6 PMID:12415726

Pinto, A., Barbot, A., Viegas, C., Silva, A. A., Santos, C. A., & Lopes, J. B. (2014). Teaching science with experimental work and computer simulations in a primary teacher education course: What challenges to promote epistemic practices? *Procedia Technology, 13*, 86–96. doi:10.1016/j.protcy.2014.02.012

Santagata, R., & Yeh, C. (2016). The role of perception, interpretation, and decision making in the development of beginning teachers' competence. *ZDM Mathematics Education, 48*(1-2), 153–165. doi:10.100711858-015-0737-9

Santos, C. A. (2017). The role of teacher decisions in the classroom to develop students' epistemic activity. In J. B. Lopes, J. P. Cravino, E. S. Cruz, & A. Barbot (Eds.), *Teaching science: Contributions of research for planning, practice and professional development* (pp. 119–134). Nova Science Publishers.

Santos, C. A., Lopes, J. B., & Cravino, J. P. (2011). Patterns of didactic decisions made by teachers in physical sciences classrooms. *Problems of Education in the 21st Century, 36*, 76–89.

Santos, C. A., Lopes, J. B., Saraiva, E., & Cunha, A. E. (2015). Teacher decisions and didactical intentionality using computer simulations in physical science classrooms. In L. G. Chova, A. L. Martínez, & I. C. Torres (Eds.), *9th International Technology, Education and Development Conference* (pp. 5347–5356). Madrid: IATED Academy.

Saraiva, E., Lopes, J. B., Cravino, J. P., & Santos, C.A. (2012). How do teachers of physical sciences with different professional experiences use visual representations with epistemic functions in the classroom. *Problems of Education in the 21st Century, 42*, 97–114.

Schoenfeld, A. H. (2015). How we think: A theory of human decision-making, with a focus on teaching. In S. J. Cho (Ed.), *The Proceedings of the International Congress on Mathematical Education* (pp. 229–243). Seoul, South Korea: Springer International Publishing.

Seidel, T., Stürmer, K., Blomberg, G., Kobarg, M., & Schwindt, K. (2011). Teacher learning from analysis of videotaped classroom situations: Does it make a difference whether teachers observe their own teaching or that of others? *Teaching and Teacher Education, 27*(2), 259–267. doi:10.1016/j.tate.2010.08.009

Shavelson, R. J., & Stern, P. (1981). Research on teachers' pedagogical thoughts, judgments, decisions, and behavior. *Review of Educational Research, 51*(4), 455–498. Retrieved from http://rer.aera.net. doi:10.3102/00346543051004455

Waight, N., & Abd-El-Khalick, F. (2007). The impact of technology on the enactment of "inquiry" in a technology enthusiast's sixth grade science classroom. *Journal of Research in Science Teaching, 44*(1), 154–182. doi:10.1002/tea.20158

Wright, S. P., Horn, S. P., & Sanders, W. L. (1997). Teacher and classroom context effects on student's achievement: Implications for teacher evaluation. *Journal of Personnel Evaluation in Education, 11*(1), 57–67. doi:10.1023/A:1007999204543

Chapter 4
How a Specific Task Was Developed Through Multimodal Narratives

Domingos Kimpolo Nzau
Instituto de Educação em Ciência de Uíge (ISCED), Angola

ABSTRACT

Multimodal narratives (MNs) in the teaching of science and technology seek to report many aspects of a pedagogical nature in a complete and holistic manner. This chapter shows how six teachers of physics, from Angolan elementary education, without adequate pedagogical or academic training to teach physics, specifically the Newtonian concept of force, worked on a task during an intensive teacher training course. The MNs documents the professional development of the teachers in particular showing how they appropriated a tool (a specific didactic model) that was object of the training. The results achieved, with the help of multimodal narratives to register the occurrences in the seminar, once again reveal that MNs are as valuable a pedagogical tool in teachers' teaching practices as in their professional development (PD).

INTRODUCTION

Studies of teachers' PD, focusing on the improvement of teaching practices, have considered a register of activities undertaken in the classroom in educational science as a valuable, powerful, and usable instrument, because it reports the teaching practices in a complete and holistic way in the MN form. According to the scholars who developed them, MNs are able to portray the events occurring in the classroom in great detail (Lopes et al., 2014).

DOI: 10.4018/978-1-5225-8570-1.ch004

Researching mediation in the classroom requires the didactic action to be captured as a whole; in these terms, an MN is taken as a narrative of the events that occurred in the classroom when a specific task was performed.

The different works developed in the MNs that this work presents show that these tasks are of different natures and even point out that it is imperative to take into account their pedagogical aspects, as the intention is to improve not only the teaching practices of science and technology (S&T) in schools but also other pedagogical aspects, for instance teachers' training.

Central Idea of the Study

In the context of the continuous training of teachers, this chapter aims to show, in relation to a specific task, that MNs can also support the PD processes of physics teachers in Angolan general education, using the conceptual fields of Vergnaud (1987), in particular the Newtonian concept of force, considered as interactions between material systems.

This aspect is pointed as out one of the difficulties that students face in identifying and justifying the forces applied in material systems (Dumas-Carré & Goffard, 1997; Nzau & Costa, 1992; Viennot, 2002) in the teaching and learning process. Diagrams of objects–interactions (Dumas-Carré & Goffard, 1997), which were suggested as a didactic approach by these scholars to resolve this difficulty, were also taken into account in this study, which used MNs as a tool for data collection.

Issues

The study sought to qualify teachers for teaching practice, which many authors, such as Meirink, Meijer, Verloop, and Bergen (2009), Mushyikwa and Lubben (2009), and Starkey et al. (2009), among others, have defended as a way to overcome the weaknesses detected in teachers. It involved a group of six physics teachers in an exercise; each teacher was designated by the first letter of his or her name (M, A, B, C, P, and R), and they belonging to three schools of basic education in Cabinda Province (Angola) without adequate academic training or pedagogical principles to teach physics, namely Tando Zinze, Luvassa, and Barão Puna schools (Nzau, 2010).

The study was based on a specific didactic model of the conceptual field of Newtonian force (Nzau, 2013; Nzau, Lopes, & Costa, 2012; Pinto & Nzau, 2017) in the conceptual field models of Vergnaud (1987).

Study Limitations

Improving the quality of teaching practices, including evidence-based practices for science teaching, while ignoring the nature of professional and epistemological knowledge was questioned by Bradforth et al. (2015), because the conceptual nature and professional knowledge have other components, one of which is the ability to choose the research results that are appropriate for the situation (Winch, Oancea, & Orchard, 2015).

This didactic model was adopted in this program because it is considered to be constructivist, using conceptual teaching (Weil-Barais, 1995) in the conceptual field (Lopes, 2004), which Vygotsky, Leontíev, and Galperin (cited by Fosnot, 1996) advocated as playing an important role in the teaching and learning process of S&T. The associated seminar lasted five days (Appendix 1) and sought to respond to questions about the effectiveness of S&T teaching, principally physics in primary schools, particularly concerning the role that the conceptions of the pupils play in the process of their learning (Fosnot, 1996). It was designed to provide participants with increased training in physics teaching in general and the Newtonian concept of force in particular.

In fact, considering the small amount of time and a single concept, namely Newtonian force, although it is central to physics, little evidence has been presented about the PD of teachers in the teaching and learning process of physics, especially regarding a small number of teachers in seminars in comparison with the universe of existing teachers in Cabinda Province.

TEACHERS' PD IN CABINDA: CHALLENGES

There are several lines of research that have already focused on classroom interventions to investigate how they work in specific cases, such as task modification (Zangori, Forbes, & Schwarz, 2015), as well as the use of certain resources (Zacharia & De Jong, 2014) and theoretical tools (Berland et al., 2016).

Accordingly, the workshop sought to satisfy the participants' training needs in terms of the use of the concept of force as an interaction between material systems (Nzau, 2010, 2013) concerning the following:

1. Their ability to lead education better based on strategies involving constructivist didactic models that would allow them:
    a. To plan physics classes taking into account the previous knowledge of the students;
    b. To deal pedagogically with this knowledge in the context of the teaching and learning process;
2. Their preparation to know the evaluation mechanisms of the students' scientific knowledge, within the scope of this new educational perspective;
3. Their mobilization for training that tells them about the importance of the work of groups of teachers, as a space reserved for the discussion of the teaching materials to be used in each class, lesson planning, analysis of the teaching methods, and evaluation of the learning outcomes of the students, of the didactic materials, of the teaching methods used, and of the teaching and learning process of physics as a whole.

For this reason, the seminar had the following general objective:

To train the participants to rehearse the strategies of constructivist teaching proposed by the author to introduce the Newtonian concept of force as an interaction between material systems to the seventh-grade students in the schools of Cabinda Province (Angola), with the purpose of improving their learning in physics.

The general objective is aligned with the following specific objectives:

1. To contribute to the PD of teachers about the teaching of physics in schools of basic education;
2. To deepen their scientific knowledge about physics;
3. To know the structure of the situation of knowledge in the context of teaching and learning;
4. To know the epistemological specifications of teaching and learning environments;
5. To deal with the conceptions of students during the teaching and learning process of science;

6. To plan physical classes taking into account the students' conceptions;
7. To evaluate students' scientific knowledge, analyze it, and increase it;
8. To dominate the mediation mechanisms of education for the construction and development of scientific concepts during the teaching–learning process;
9. To develop teaching in which the pupils construct the scientific concept related to force, like interaction between material systems.

METHODOLOGY

The data collected in the MNs of this seminar were related to:

1. The concept of the class;
2. The concept of the didactic model;
3. The specific didactic model;
4. The structure of scientific knowledge;
5. The formative situation.

The data from the simulated class were collected not only through MNs but also through a written questionnaire developed for that purpose.

To evaluate the impact of the training acquired by teachers in this training action on the effectiveness of the didactic model used in teaching practices and on the effectiveness of the teaching practices resulting from teacher training in students' learning, some classes of these teachers were registered through MN and taken into account for the analysis of these aspects .From this study, we have as data in the MNs the teaching practices of the attending teachers, the students' perceptions about the classes in which they participated, and a report of the follow-up performed on the actions of each teacher.

Although the results to be presented on the performance of teachers in their schools are global, as an example, we will present only a few excerpts from the MNs of an attended classroom that a trained teacher has developed at his school.

DESCRIPTION OF THE TRAINING OF PHYSICS TEACHERS IN CABINDA

The development phase of the seminar, which aimed to address graduates' need for more knowledge about teaching practices, and the episodes of MNs identified in the development phase of the seminar are presented below.

Activities Undertaken on the First Day of the Seminar

The themes that were part of this day (Appendix 1) were presented as follows:

1. In the morning, in the classroom, after the usual greetings, the organizer of the seminar introduced the visitor to the participants. He was thanked for his turn, in the cordial manner and stated that he was willing to work all week. Organizing the participants and the teaching material into their

proper places, the organizer of the seminar started his intervention in the form of an exhibition with a slide projector to clarify the issues.

After the presentation, which lasted for about 1 hour, a cycle of debate began in which it was found that the themes discussed were not unfamiliar to anyone. However, the approach of the sub-themes caused the participants to become more aware of the challenge that must be faced in the process of teaching and learning physics regarding the aspects linked to the complex process of epistemic subjects.

2. The afternoon was dedicated to application exercises; the participants (Figure 1) conducted an analysis of a video-recorded class and were committed to identifying aspects related to those covered in the themes and sub-themes as finished products to be developed. The participants were thankful for what they learned in this activity.

Activities Undertaken on the Second Day of the Seminar

1. The second day was dedicated to the themes of the specific didactic model on the field of force to teach students the Newtonian concept of force as an interaction between material systems as well as the structure of the situation of scientific knowledge.

In the first theme, the participants gained knowledge about the fundamental aspects of this didactic model as an exchange of actions of two material systems that interact.

In the second theme, animated by a Portuguese invitee from the University of "Trás-Os-Montes e Alto Douro," who honored us with his visit and participation in the seminar, the teachers had the opportunity to learn about the structure of the situation of scientific knowledge in the scientific and educational contexts.

The two essential poles of the different epistemological approaches were highlighted, namely the object side and the subject side, which characterize traditional teaching. They were also informed of the

Figure 1. Work in groups of participants

main advantages of the new epistemological approaches that support the constructivist teaching that modern pedagogy suggests should be used in teaching practices.

In the approach taken to the structures of the situation of scientific knowledge in the two contexts, the eminent teacher clarified the actions that an epistemic subject can perform to know an epistemic object and to determine what it is necessary to undertake in the teaching and learning process for the student to become an epistemic subject (Figure 2).

2. The part devoted to the practical exercises, which took place that afternoon, began with a small scene oriented towards two participants connected by the hands and pulling each other, with their feet resting on sheets of paper. It was observed from this scene that one tended to be dragged by the other. The discussion about the spectacle led the participants not only to conclude that a description was different from an explanation of a fact or phenomenon but also to ask questions that would lead them to approach the interaction model in the sense of analyzing the actions carried out by the students.

The reading of part of the supporting text, accompanied by a debate that took place that afternoon, helped the participants to improve their understanding of the intervening material systems of the interaction model (Figure 3).

Activities Undertaken on the Third Day of the Seminar

The third day was dedicated to the analysis of a text designed to understand the concept of a specific teaching model about the conceptual field of force as a didactic tool in teaching.

1. In the morning, after the usual greetings among the participants present, the text prepared on the subject under discussion was distributed. This text, after being read and discussed by the participants, helped to clarify the nature of the Newtonian concept of force as interaction between material

Figure 2. Organization of the central planning aspects of teaching in a training situation
Source: Adapted from Lopes (2004).

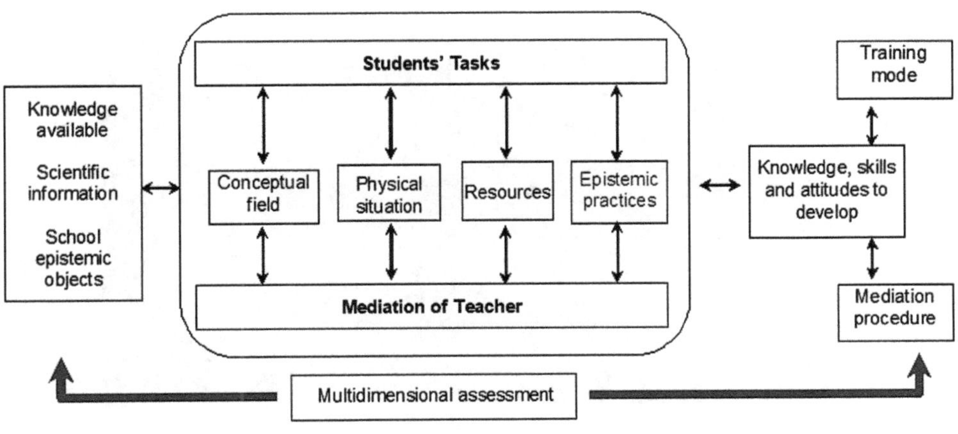

systems in contact or at a distance, particularly the latter, when planet Earth interacts with another material system located in the gravitational field.

The panels presented in Figure 4 (although the resolution of the picture is not very clear) reflect the information provided to the participants about the specific model of the conceptual field of force, revealing, in the first block of the frame, a balloon that falls, a situation for which the participants sought to identify, in groups, the material systems involved and the type of interactions observed; in the middle block, the groups sought to identify the type of applied forces and their corresponding senses when the balloon hit the ground; in the rightmost block, the physical situation, while changing the balloon falling into a child who kicks a ball, the discussion about the specific model in the conceptual field of force, in its essence, did not change.

The specific model of the conceptual field of force contributed even more to the participants' understanding of the difference between attractive and repulsive forces and therefore the identification of these forces (attractive and repulsive) from material objects. The approach in this communication led the participants to rethink their teaching practices, because it enriched their experiences.

Figure 3. Participants in the seminar with the invited Portuguese teacher

Figure 4. Example of some exercises performed by the participants

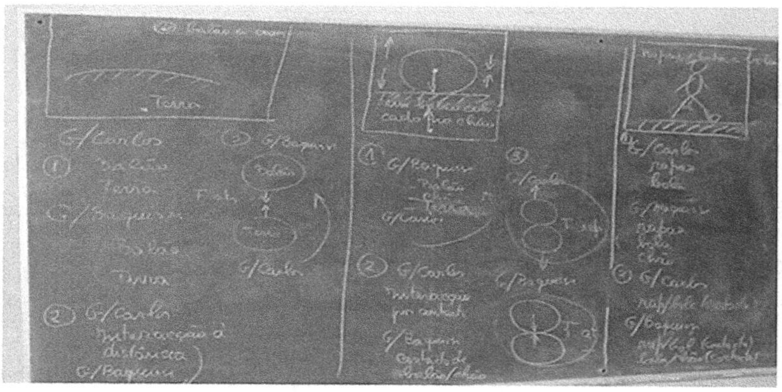

2. The afternoon of that day was devoted to the theme entitled "formative situations" as a tool to plan and manage the classes, slightly altering the initially established plan, in which the main speaker, once again, was the invited teacher.

How should the teaching be planned? The participants were informed, by the invited teacher, about the central aspects to be taken into account in planning the teaching and providing students with learning, the tasks to be performed by students, mediation by the teacher, the epistemic practices of production of knowledge, and the conceptual field of reference.

That afternoon, the practical exercises performed in a group aimed to reflect on the planning of a lesson linked to the theme entitled "types of forces (attractive and repulsive) and their effects (action at a distance and action by contact of material systems, respectively)." In the resolution of the exercises, which began with a discussion about the classification of the lesson plan to adopt, the participants agreed by consensus that each group would opt for any model of lesson planning. This work occupied the participants throughout the afternoon and continued during the afternoon of the following day.

Activities Undertaken on the Fourth Day of the Seminar

1. The morning of the fourth day was reserved for exercises, which aimed to solve problems related to the interaction model between material systems, constructing diagrams of interaction between material systems to identify their interactions, which characterize the types of force that work in each system. The participants initially encountered some difficulties, but with the help of the seminar leaders, the problems were solved.

During the debate, which served as an analysis of the work of the different groups, some concerns were raised about the ability of teachers of large groups to distribute, follow up, and discuss the tasks of the students, since this activity requires a considerable amount of time for the students to assimilate the different contents proposed. In response, the seminar leaders suggested that the time lost in completing these tasks is gained in other classes, which may involve exercising and applying the knowledge.

2. The afternoon continued with the resolution of exercises. These exercises ended in group work linked to the planning of classes, the subject of which, although it was in the interaction model, was the criterion of each group's work.

In the analysis of the work groups, it was observed that the general scheme of class planning did not differ among the different groups, so, to evaluate the commitment of the groups better, it was suggested that one of the participants should be selected to prepare and organize a simulated class. The choice was Teacher B.

Activities Undertaken on the Fifth Day of the Seminar

This was the day on which we examined the simulated class, which will be presented next.

RESULTS OF THE EXCERPTS FROM MNs COLLECTED IN THE PD PROGRAM OF PHYSICS TEACHERS IN CABINDA

We present in this section the various MNs registered in the classes attended in the phases of the simulated class and the independent performance of a teacher during the seminar.

In the Simulated Class

On the morning of the fifth day, dedicated to the simulated class, the chosen teacher conducted his class, considering his classmates as his students (Figure 5). He registered only one episode.

Single Episode

The teacher began the class with a greeting, followed by a brief conversation, which ended with the presentation of the content, initially writing on the board only the theme "strength and mass." Then he presented to the students a physical situation of a table being pushed, accompanied by the formulation of the following question:

"Why did the table move?"

The students' response in chorus was:

"It was because of the force that was applied to the table."

The teacher's comment was only to confirm what was said in the following terms:

Figure 5. Teacher B, while taking his class

"Yes, because it was the subject of the previous day's class."

He continued the lesson with the presentation of a new physical situation; this time, the teacher showed a drawing that represented a tree with falling leaves, asking the students the following question:

"Why do the leaves fall from the tree?"

After a short comment, the teacher concluded in the table the presentation of the summary by writing the sub-theme:

"Types of forces and their effects."

After providing the students with a brief commentary on some specific aspects of the Newtonian concept of force, he added, writing on the board the words "force of attraction, force of gravity, and repulsive force," that he was designating the existing types of forces.

Turning to the students, he asked them to identify the interactions and the type of forces acting on the leaves falling from the tree. The responses from the students showed that they had knowledge of the different forces acting on each leaf, during and after the fall; that is, during the leaf fall, the only force acting, disregarding the effects of air, was the attractive type force, and the interaction, in that case, was established between the leaf and planet Earth. After the fall due to the attractive force, the repulsive type of force was added in relation to the contact between the leaf and the ground. To seek to enrich the concept of force, as attractive or repulsive in nature, the teacher resorted to some examples about which he himself had doubts, such as the emergence of the tides of the sea being linked to the forces of attraction of the Moon and the Earth and confusing the fact that the force of gravity is not also an attractive force.

However, the example of the clash of two distracted people moving in a city street and immediately separating helped to clarify to students the concept of repulsive forces, since the immediate departure was compared with the effect of repulsion between material systems.

The lesson continued with some exercises involving imagining physical situations, sometimes by the students and sometimes by the teacher, identifying the interactions established between the different material systems and the corresponding forces that acted on them.

His lesson ended by distributing homework to the students.

The students reacted positively to these activities.

On the afternoon of that day, the work continued with the analysis of the class, during which the colleagues recognized the talent of Teacher B, despite some scientific and other pedagogical failures. This generated an interesting debate among the participants, in which each one made pertinent observations from the pedagogical point of view; for example, the pedagogical failure of starting to talk about a physical situation without having sufficiently explored the previous session, changing to another activity that consisted of the presentation of types of forces, and the failure from the scientific point of view regarding the aspects of the tides. The contribution of the invited teacher was particularly valuable in clarifying the connection of the Moon with the tides.

The intervention of the invited teacher focused more on the need to keep the class continually motivated from the beginning to the end, a fact verified at certain moments in the attended classroom. Another aspect on which the eminent academic insisted was the need to select physical situations that help the teacher to conduct a class as a mediator, since the tasks of the students should be foreseen at the time of the lesson planning.

The Independent Action of Teachers in Their Schools

MNs on How a Teacher Can Act Independently From His School

In the phase of independent acting of teachers, a teacher from the "Barão Puna" school, selected by chance, presented as episodes of MNs in his lesson the following:

Episode 1

After the usual greeting of the teacher to the students, he drew their attention to our presence in that room, because it was an accompaniment of a project that had been initiated with a formative action whereby he himself was the teacher of the class.

The visitors, after thanking the teacher for the cordial introduction, asked the students and the teacher not to worry about their presence in the room and to pretend that they were alone and that the cameraman was not there. Therefore, students should act as they normally do in class, asking questions, if necessary, and finally reacting as if they were accustomed to their teacher.

Then the teacher went to the board and wrote the date in the upper-right corner of the picture and in another corner the grade, discipline, and theme of the class that day.

Leaving the board, the teacher addressed the students and, to review the previous class, he said to them:

"Let's start talking about ...." Not concluding the sentence, he continued:

"We are on the subject of forces and mass. We will also take an approach to material systems' interaction."

"What does an interaction mean?" the teacher asked the students.

"What is interaction?" repeated the teacher.

"Interaction is the exchange of actions between two bodies," answered a student.

"The interaction is the exchange of ...," insisted the teacher without finishing the sentence.

"Of actions," the students concluded in chorus.

"Between two ...?" continued the teacher, leaving the sentence unfinished.

"Bodies," the students concluded.

"Between two bodies," confirmed the teacher, who continued by saying:

"When I speak of material systems, what is a material system?" the teacher asked the students, adding the following:

"We've talked about material systems, have we not?"

"Not yet," answered a student. A student, with her colleagues nodding, said in a loud voice,

"Teacher, not yet!"

"What?" asked the astonished teacher.

"Have a look at your notebooks. Feel free," the teacher said to the students.

One student got up and said,

"Material systems are the same as material bodies."

"Material systems are the same thing as bodies ...," the teacher, once again, left the sentence unfinished.

"Body materials," the students concluded.

"In physics, interactions between material systems are the same thing that exchanges actions between bodies ...?" The teacher asked the students to finish.

"Body materials," answered the students in chorus. "That's it," the teacher continued, asking the students the following: "The interaction is by direct contact, for example the ball on the floor, is not it? It is. Do the interactions determine the type of fo ...?" He left the students to finish, but he himself concluded: "Forces." He continued: "At this very moment, you are in a position to tell the kind of forces that each kind of interaction can produce. Any student can answer, here, what is force?" asked the teacher.

Two students rose at the same time, but the student behind gave up and left the classmate to answer the question. The student in front said, without noticing that another colleague was interested in the question:

"Force is the way two bodies are in interaction. Force is the way two bodies are in direct contact or ...?" the teacher left the incomplete sentence for the students to complete.

"Distance. "

"It is more or less that, the conclusion that we can reach," the teacher said to the students.

"The analysis of the interaction between material systems.That's very good. Let's continue with the review today."

Episode 2

Having that said, the teacher went to the board to complete the summary and ordered the students to write it in their notebooks. After the students finish the work, the teacher dropped a ball into the room that moved on the floor .

Addressing the students, he commented on the phenomenon in these terms: "Very well. I kicked the ball; the ball moved and stopped. I need a student to draw on the picture, more or less, a ball that moves on the floor, to make a scheme, a model of this drawing." A student stood up and went to the board to carry out the task, while the teacher asked the other students to draw it in their notebooks, in the following terms: "Let's all think. Each of you must make a drawing of a ball dropped to the ground. A drawing, do it all. A ball thrown to the floor moved and stopped. Let's go. Keep on thinking." While this happened, the student at the board drew without saying anything (blackboard reproduction 01). The teacher commented on the student's task.

"OK. This is our colleague's opinion," commented the teacher, pointing and gesturing with his right index finger the ball on the horizontal line, and added:

"This is the bottom, that's all." Then, the teacher went to the table to get the chalk and eraser. He cleaned first and designed (blackboard reproduction 02).

Figure 6. Blackboard reproduction 01

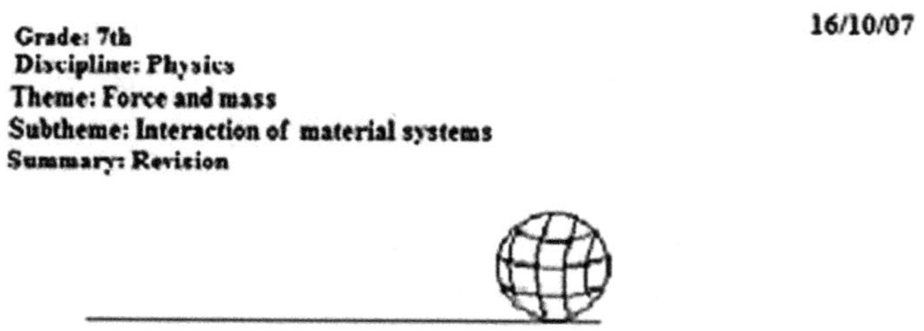

Figure 7. Blackboard reproduction 02

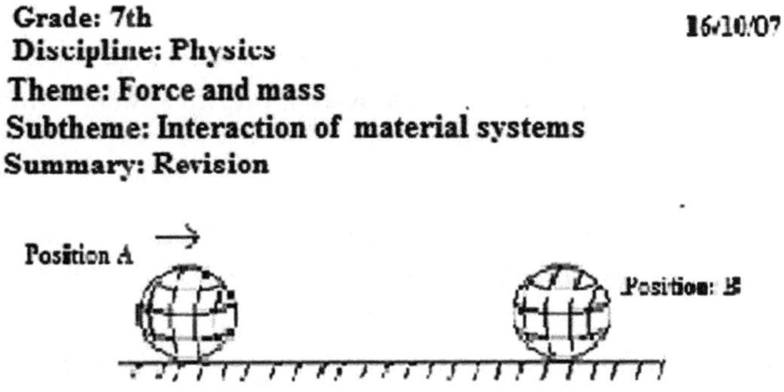

Returning to the class, he asked:

"What are the material systems, where the ball is in position, what are the material systems?"

A student stood up and answered, saying,

"The material systems of contact ..." he hesitated and the teacher insisted, in these terms:

"The material systems ...."

The student repeated:

"The material contact system ...." He did not continue the sentence and, pointing with his right index finger the drawing of the ball on the board, the teacher said:

"The material systems of this physical situation."

The student at the board replied:

"The ball and the Earth."

"Our colleague said it's the ball and the Earth: do you agree?" the teacher asked the students.

"Who has another idea?" the teacher asked the class again.

"What are the material systems that we will find in this situation? In this scenario, where is the ball in motion?" He pointed to the picture and continued:

"What are the material systems that we will find in this scenario?"

"In this first scenario and in this position A, what is it?" continued the teacher.

The slight silence of the class led the teacher to put the question in these terms:

"At position A and ball movement, what material systems are there? Or if you want to make another drawing or another scheme of this physical situation, what material systems can we find in this physical situation?"

The teacher momentarily stepped away from the board, addressing the class, but returned again to the board and, pointing to the drawing of the ball to the right of the board, insisted:

"Besides the ball and the floor, what other material systems are in this physical situation?"

The teacher himself resolved to answer the question by saying:

"We have Earth too," adding another drawing to the board (blackboard reproduction 03).

Pointing to them with his right index finger, the teacher said: "We have in this physical situation the ball, the ground, and the Earth."

Turning to position B of the ball on the board, the teacher told the students:

"If we consider position B, we are sure to find the same material systems. Now the ball has stopped; when the ball is stopped ...."

Drawing on the chart next to the previous scheme (blackboard reproduction 04), he added:

"Then it makes no difference." Pointing to the scheme with his right index finger, the teacher concluded:

"With the ball in motion, we will find as material systems the ball, the ground, and the Earth and also the ball at rest; let's find the ball, the ground, and the Earth."

Figure 8. Blackboard reproduction 03

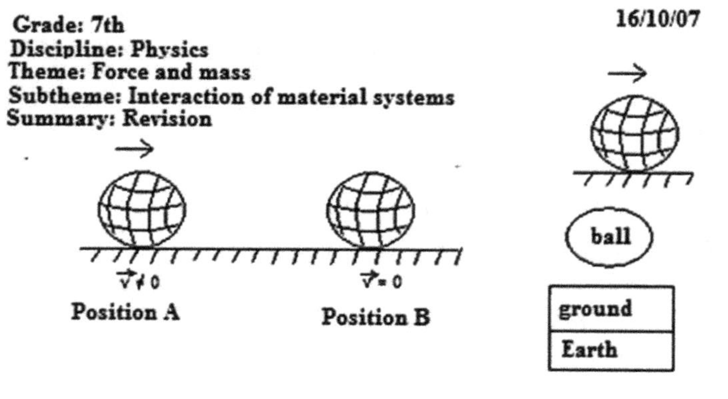

Episode 3

Continuing, the teacher said:

"So far we have already identified the material systems of a physical situation. Now let's identify the interactions. What are the interactions that we can identify in this physical situation?" the teacher asked the group.

Without waiting for the answer, he continued, using the diagram (blackboard reproduction 05), which he added to the board.

Drawing the same scheme, the teacher repeated his question, saying:

"Where is the stock exchange there?" He indicated the layout of the picture with his right index finger.

Figure 9. Blackboard reproduction 04

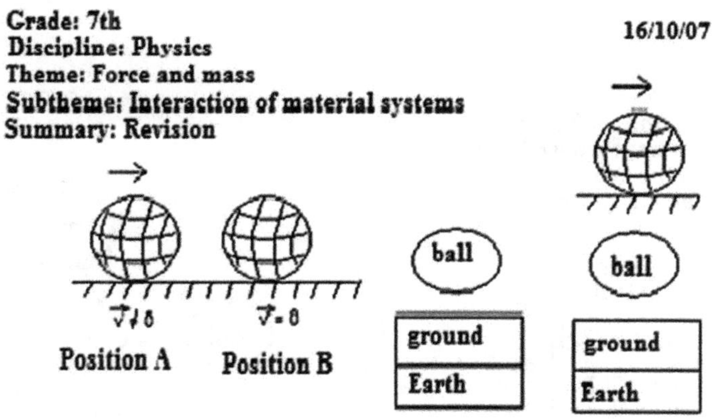

Figure 10. Blackboard reproduction 05

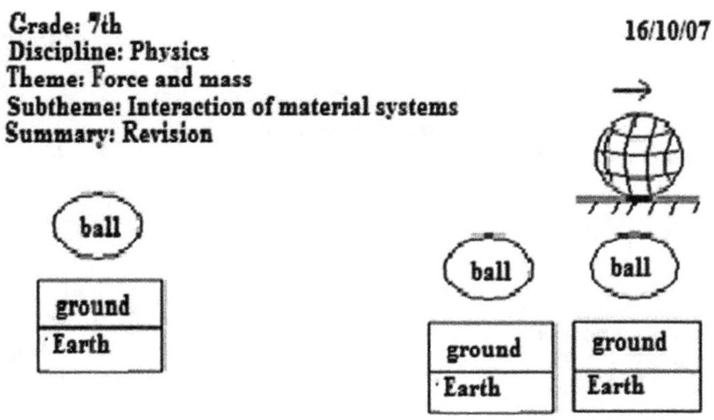

A volunteer student rose, went to the board, and, without comment, drew two broken lines with a double direction in the scheme (blackboard reproduction 06).

The lack of the arrow sign, at the origin of the broken line on the left side of the frame, caused the teacher to go to the board to complete the diagram (blackboard reproduction 07), saying:

"You forgot this sign here."

Pointing to the last scheme with his right index finger, the teacher recapitulated the student's task in the following terms:

"We have interaction on the ground and the Earth, do we not?"

"No," a student said.

The teacher, coming back to him, said:

Figure 11. Blackboard reproduction 06

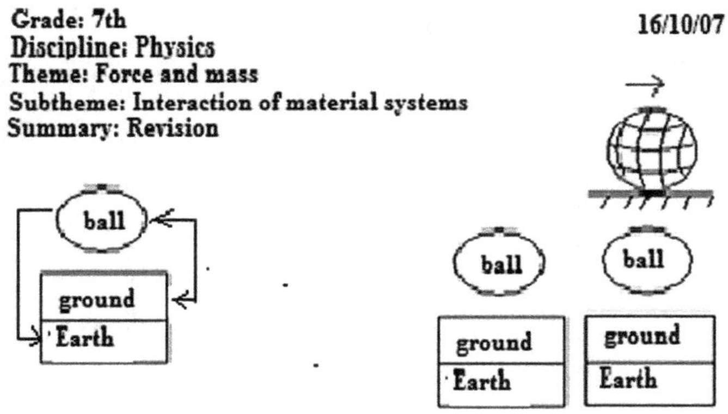

Figure 12. Blackboard reproduction 07

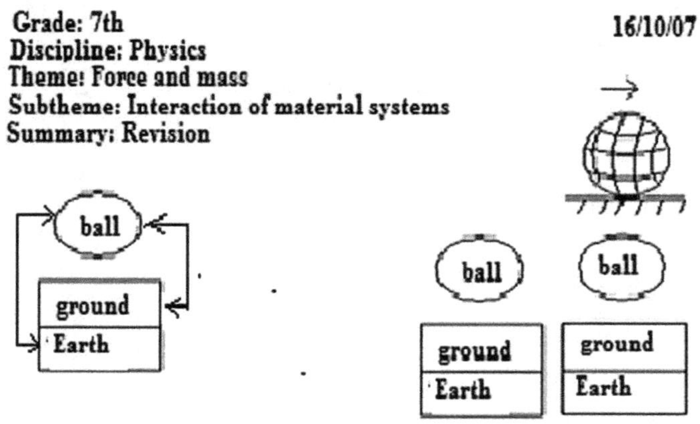

"It is not?" he asked.

The student said nothing, and the teacher continued:

"Now let's hear our colleague's opinion."

The student sustained his opinion as the following:

"The ball interacts with the ground."

The teacher reminded the student, pointing with his right index finger to the ball of the scheme, saying:

"We have the ball. So the ball is not in interaction with the Earth?"

"No," insisted the student.

"That is because it is only on the floor," concluded the student.

"There is no contact of the ball with the Earth. It is just the floor," argued the student.

The teacher, using the mercantile model of buying goods by order, reminded the student:

"You can buy goods by order, right?"

"There's no money," the student said to the teacher.

Then the teacher argued his idea in the following terms:

"The Earth has the ability to exchange actions at a distance. So we can see there's an interval there between the Earth and the ball. There is something that separates them, which we call the ground, yes or no?" "Yes, it is."

"Then the exchange of actions between the Earth and the ball is by contact or from a distance?" the teacher questioned.

"In the distance," the students replied in chorus.

"In the distance," the teacher confirmed and continued:

"Some are in contact, in fact, all the actions by contact are happening between the ball and the ..." asking with his right index finger for the word: "ground," the students concluded in chorus.

Still on the board, the teacher wrote only the word "Position" above the scheme under discussion, while saying: "Position A."

Pointing to the schematic with his right index finger, the teacher continued:

"We have already identified the interactions at position A. If we identify other interactions at position B, let us see if the same thing can happen. We have seen the ball in motion; now let us see the stopped ball. Is there any contradiction?" Moving his head and body from left to right and vice versa, he indicated a student to go to the board to identify such interactions, in the following terms:

"Go to the board to identify the interactions of the scheme of position B." Without comment, the student rose, went to the board, designed the scheme, and finished with the placement of the arrows as shown in blackboard reproduction 08.

The teacher, turning to the board, pointed to the down arrow and asked,

"Attention, where is this from? Where did it start?"

"Over here." The student showed the origin of the arrow by pointed to the black arrow (blackboard reproduction 09).

"Make it better," the teacher instructed.

The student, deleting the previous arrows, presented the scheme of the arrows in the figure that follows (blackboard reproduction 10).

"This kind of interaction is not near," the teacher commented and went to the board, cleared the problematic interaction, and corrected it by dividing the rectangle again by the diagonal, saying:

"Here." He pointed to the diagonal of the rectangle.

Figure 13. Blackboard reproduction 08

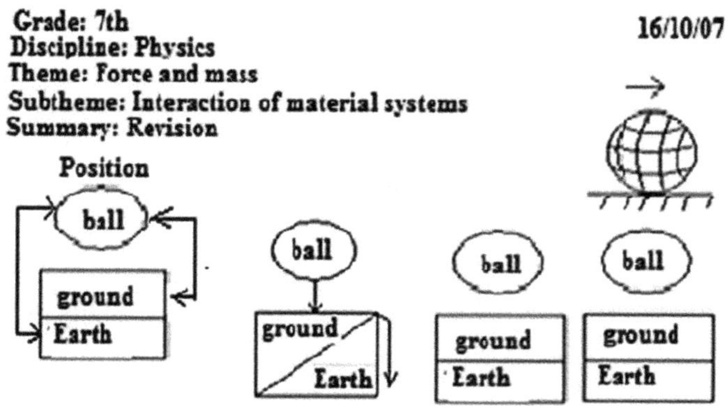

"I just wanted to vary, because it depends on ... I made a drawing here a little bit different, just to vary. Do not run away ... here we have the ground and the earth," he concluded.

Pointing to the scheme on the side where the rectangle is divided by a segment of line that touches the center of its widths, he added:

"Here we have the ground and the earth; I only changed the drawing, so there's no difference. So, it does not make a difference from the ground and the Earth," the teacher concluded.

Episode 4

To move onto another phase, the teacher began: "Let us see what kind of forces are realized between material systems. What kind of forces are they making between systems ...? "Materials," completed the students in chorus.

The teacher then went to the painting, erased the pictures, and continued by saying:

Figure 14. Blackboard reproduction 09

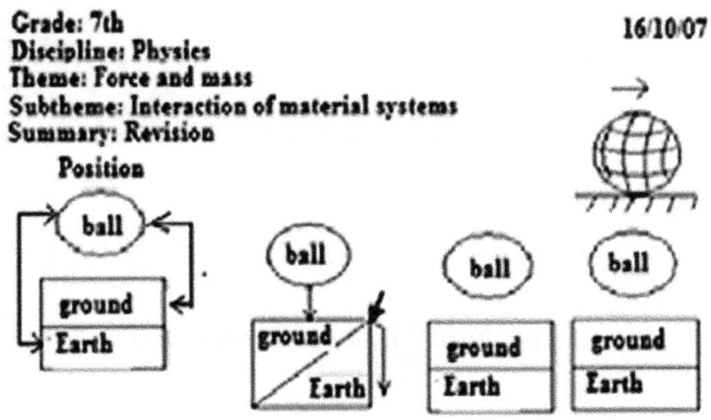

Figure 15. Blackboard reproduction 10

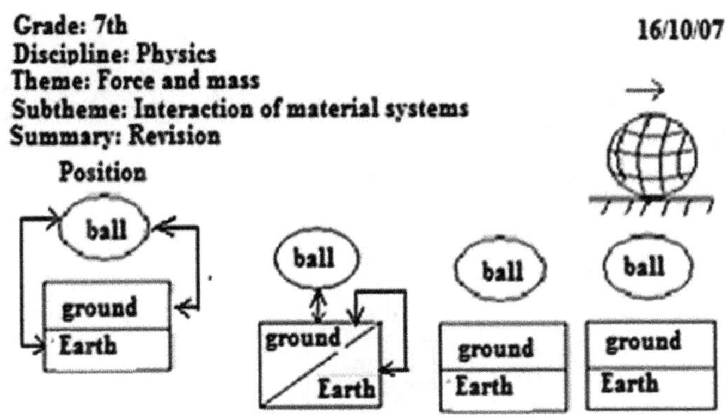

"Let us analyze what kind of forces will act in this situation. Come on ... We have to analyze. Remember that by contact the forces are of what nature?"

A student slowly spoke the word "repulsive," which the teacher did not understand well and asked him to speak loudly in a harsh tone, shouting,

"Speak louder. Rep. ... forces," the teacher himself concluded.

"And in the distance, what is the nature of the forces?" the teacher asked the students.

"Repulsive forces," a student replied.

"Forces ...?" the teacher insisted.

"Repulsive," the students answered in chorus.

"Repulsive forces," repeated the teacher, asking the students the following:

"Where do these repulsive forces act? Where are they going to act?"

Going to the board, the teacher said:

"The repulsive forces will act here." Placing an arrow over the ball pointing upward (blackboard reproduction 11), he continued:

"Acting where? On the surface of the ball," answered the teacher himself and continued, saying:

"What will act between the ball and the Earth? Is there an interaction at a distance, or not?"

"There is," the students answered in chorus.

"Then what forces?" the teacher asked the class.

Figure 16. Blackboard reproduction 11

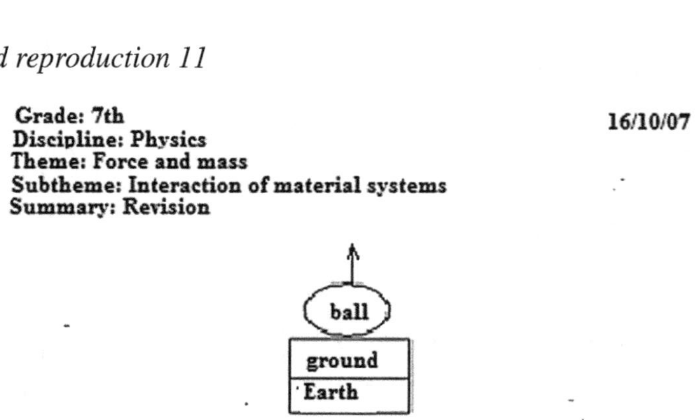

"Attractive," a student replied.

"What forces?" the teacher insisted.

"Attractive forces?" continued the teacher.

"Repulsive," another student said.

"Forces ...?" asked the teacher again.

"Repulsive," some students insisted.

"Attractive and repulsive forces," the teacher said; at the same time, he went to the board and placed the arrows, one from the center of the ball downward and another at the meeting point between the ball and the floor, but turned downward, and the other from the space reserved for the Earth but turned upward (blackboard reproduction 12).

Episode 5

Recapitulating the given matter, the teacher asked the class the following:

"What did we talk about?"

"Interactions," a student replied.

"About what ...?" the teacher insisted.

"Interactions," in a chorus, the students repeated.

"And we talked about what lately?" the teacher asked the students.

Figure 17. Blackboard reproduction 12

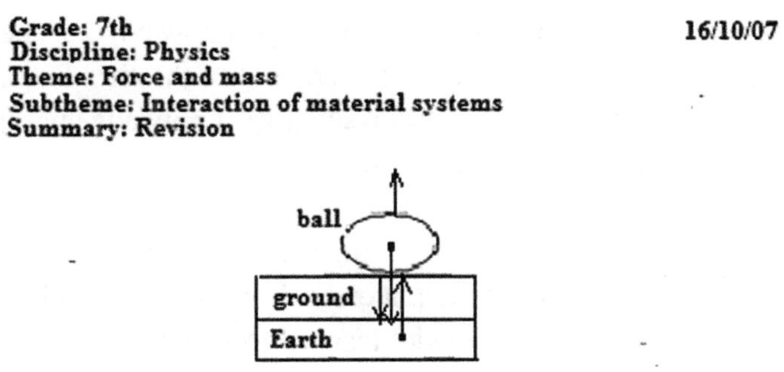

"Forces," a student said.

"About the forces that can act. OK?" asked the teacher.

"Yes," the students replied.

The teacher, on the board, wrote at the same time as saying the following:

"Forces ... ball force (blackboard reproduction 12). Let us see. This is the force (...)."

(The recording ended, but the narration continued.)

At the end of the class, we performed a quick five-minute review of the steps to be taken in building the concept of force, namely the physical situation, figurative representation, identification of material systems, identification of interactions between material systems, types of interactions, identification of forces, and types of forces.

At the end of the lesson, the team thanked all the students for the kindness that they had received in the class.

DISCUSSION OF THE RESULTS

The Seminar

Table 1 presents the overall results of the seminar on teacher education. The aspects considered in the analysis are the data obtained from the seminar report, the data obtained through the MNs of the classes watched, divided into episodes, the evaluation made by the participants of the seminar, and the data report of the teacher in charge of the training.

The table reveals that, during the seminar, the teachers were fully involved in understanding enough about the didactic model taught to introduce the Newtonian concept of force to their students, as the involvement of teachers in planned tasks in the seminar increased their meaningful learning. In addition, the participants were able:

1. To plan physics classes, taking into account the previous knowledge of the students;
2. To deal pedagogically with this knowledge in the context of the simulated class;
3. To know the evaluation mechanisms of students' scientific knowledge within the scope of this new educational perspective, which is constructivism.

The teachers acknowledged the importance of holding the training seminar, because many of the skills that were acquired were not in the domain of the majority. After the seminar, they sought to deepen the scientific and methodological aspects of the model learned to incorporate it into the planning of their classes, because they considered it to be fruitful and practical. The teachers sought:

Table 1. Data analysis of the teacher training seminar

Aspects Analyzed	Analysis Parameters Met or Not Met During the Seminar	Participating Schools					
		Tando Zinze	Luvassa		Barão Puna		
		M	A	B	C	P	R
Report of the seminar by the participants	Contact with the model of interaction between material systems for the introduction of the Newtonian concept of force through a text.	Y	Y	Y	Y	Y	Y
	Training through collaboration of scientific and methodological questions and even suggestions based on the personal experience of each teacher, including involvement in planned tasks.	Y	Y	Y	Y	Y	Y
Seminar evaluation by the participants	Negative aspects in the selection and organization of the seminar content.	N	N	N	N	N	N
	Negative aspects in conducting the seminar.	N	N	N	N	N	N
Report of the activities of the teachers after the seminar	Incorporation of the didactic model into the planning of the class.	Y	Y	Y	Y	Y	Y
	Manifestation of interest in deepening the didactic and scientific issues related to the didactic model studied.	Y	Y	Y	Y	Y	Y
	Schools struggle with administrative and organizational issues.	D	MD	MD	MD	MD	MD

Legend: Y – yes; N – no; D – with difficulties; MD – with many difficulties.

1. To overcome their difficulties of a pedagogical and scientific nature;
2. To incorporate the didactic model learned into the planning of their classes;
3. To express willingness to understand better the model of interaction of material systems and to put into practice progressive teaching for the construction of the Newtonian concept of force to their students.

From these considerations, we can say that the seminar achieved the objectives recommended, such as:

1. To form a group of active physics teachers, capable of dealing with the didactic model used in the teaching practices of their schools, thus contributing to the professional development of teachers on the teaching of physics in primary schools;
2. To have a positive perception of the influence of the didactic model used in the training action of these teachers, since they developed the scientific and pedagogical capacities necessary for teaching, in which the students (teachers) sought to construct the scientific concept related to force as an interaction between material systems.

Accordingly, we can infer that the seminar did indeed qualify teachers for the teaching exercise, which many authors, like Meirink et al. (2009), Mushyikwa and Lubben (2009), and Starkey et al. (2009), among others, have argued. This greatly contributed to overcoming the weaknesses detected in the diagnostic study of the same teachers. This inference partly indicates that the specific didactic model of the conceptual field of force has an influence on teacher training and as such can be used for this purpose.

In fact, this information was obtained thanks to the registration of the occurrences in the seminar through the MNs, which proved once again to be a valuable, powerful, and usable tool in educational science (Lopes et al., 2014), namely in the PD of teachers.

The Performance of a Teacher in the Simulated Class

The narratives of the simulated class provided the seminar participants with the following points of view:

1. The teacher had interesting positive aspects that need to be taken into account and kept, such as his presence in the classroom to organize teaching that resulted in true learning among the students;
2. The teacher needed to provide a new role of the physical situations in the process of teaching and learning, focusing his attention on classroom interventions to investigate how he works on specific aspects, such as the modification of the task, as defended by Zangori et al. (2015), the use of certain resources, as suggested by Zacharia and De Jong (2014), and the use of theoretical instruments, as argued by Berland et al. (2016).
3. Therefore, the selection of physical situations and use in his classroom should be harmonized, judiciously responding to purposes connected to: a) the imperative need to elaborate a model class so that it can serve as an example in the epistemic practices of the participants; and b) the need to emphasize the aspects of scientific theories of concepts, lacking in teachers, as we have seen with the problem of the tides.

In fact, this shows that the teacher in the simulated classroom performed well, thanks to the teaching narrations presented of his work.

The Independent Acting of a Teacher in a School

The analysis of the MNs of the class, using the dimensions related to the "traits of the teacher's mediation" and the "indicators of the use of the model" and its categories, without entering into detail, produced results that revealed that the teacher closely followed the didactic approach of the interaction model between material systems.

Always beginning the lesson with the presentation of examples that characterize physical situations of students' daily life and gradually moving to the more abstract aspects led the students to understand better the concept of force as the interaction between material systems.

Alongside the lesson, he engaged in considerable dialogic discourse, because it can be seen that, in almost all the episodes, the students were involved in learning activities, suggesting and discussing ideas projected from the physical situations presented.

Although using the same physical situation (moving the ball and then dropping it onto the ground), he managed to induce the students to perform the analysis of the surrounding material systems of the ball in motion and at rest and followed the approach of the concept suggested by Dumas-Carré and Goffard (1997), the system diagrams proposed by Viennot (2002), and the idea of forces as the interaction of material systems.

In these terms, it can be said that, in the teacher's mode of action, the following points were observed:

1. The analysis considers the teacher to have followed a pedagogical approach, not lacking difficulties, since some problems in managing the performance of the students of the class (directing questions to the class and indicating a student to respond) were apparent;

2. The teacher assumed the performance of the tasks alone, when the students remained silent when they were questioned. This action placed him in the position of a teacher and as sometimes authoritarian;
3. In spite of all this, we can note that the structural organization of his class can be considered as a tendency for constructivist didactics.

FUTURE RESEARCH DIRECTIONS

These findings show that, for the teachers, in addition to having a good technical and professional capacity, the training seminar may have been beneficial for them, thus agreeing with what they revealed at the end of the training seminar, that it actually contributed to responding to their minimum educational needs (learning to teach, teacher professionalization, and teacher training), in according with Montero (2001).

The MNs used in this analysis once again proved to be as valuable as pedagogical tools in teachers' teaching practices as in the task of professional teacher development (Lopes, 2017), due to their multimodal nature. Therefore, it is very important to develop them in future studies.

CONCLUSION

The results obtained in this study show that the continuing training of teachers from Cabinda, who presented some deficiencies of a scientific and even a pedagogical nature that did not allow them to develop teaching practices that would guarantee the students adequate scientific knowledge of physics, represents a method for the PD of the teachers. On the other hand, the option of including in their training program a constructivist teaching model, which served as a basis for introducing the Newtonian concept of force associated with its conceptual field, proved to be even more effective as a form of teachers' PD in the task of teaching and learning in Cabinda.

However, to analyze the organization, development, and description of the activities carried out in each of the phases, the MNs, which allowed the collection of data on specific teaching tasks and even of teachers' PD, proved to be an important pedagogical tool, because they sought to unveil in a complete and holistic way many aspects of a scientific, pedagogical, and didactic nature, such as those related to the construction of the concept taught as well as those related to the different tasks undertaken by the teachers in mediating the learning of their students.

Nevertheless, MNs have shown once again that they are important tools in the methodological organization, confirmed by the teaching practices carried out by these teachers, in this way proving that MNs are also powerful in supporting the PD processes of teachers, regardless of whether the research is dynamic or articulated with them.

However, although these results only covered a few Angolan secondary school teachers, they encouraged us to believe that, with exhaustive and comprehensive work, the teachers' PD in that Angolan province could be guaranteed.

REFERENCES

Berland, L. K., Schwarz, C. V., Krist, C., Kenyon, L., Lo, A. S., & Reiser, B. J. (2016). Epistemologies in practice: Making scientific practices meaningful for students. *Journal of Research in Science Teaching*, *53*(7), 1082–1112. doi:10.1002/tea.21257

Bradforth, S. E., Miller, E. R., Dichtel, W. R., Leibovich, A. L., Martin, J. D., Bjorkman, K. S., ... Smith, L. T. (2015). Improve undergraduate science education: It is time to use evidence-based teaching practices at all levels by providing incentives and effective evaluations. *Nature*, *523*(7560), 282–285. doi:10.1038/523282a PMID:26178951

Dumas-Carré, A., & Goffard, M. (1997). *Rénover les activités de résolution de problèmes en physique. Concepts et démarches*. Paris: Armand Colin.

Fosnot, C. (1996). *Construtivismo e Educação - Teoria, perspectivas e práticas. Maria João Batalha Reis (Trad.)*. Lisboa: Instituto Piaget.

Lopes, J. B. (2004). *Aprender e Ensinar Física*. Lisboa: Fundação Calouste Gulbenkian.

Lopes, J. B. (2017). Researching science teaching practices in the classroom over two decades: What matters? In J. B. Lopes, J. P. Cravino, E. Cruz, & A. Barbot (Eds.), *Teaching science: Contributions of research for planning, practice and professional development* (pp. 1–24). New York: Nova Science Publishers.

Lopes, J. B., Silva, A. A., Cravino, J. P., Santos, C. A., Cunha, A., Pinto, A., ... Branco, M. J. (2014). Constructing and using multimodal narratives to research in science education: Contributions based on practical classroom. *Research in Science Education*, *44*(3), 415–438. doi:10.100711165-013-9381-y

Meirink, J., Meijer, P., Verloop, N., & Bergen, T. C. M. (2009). Understanding teacher learning in secondary education: The reactions of teacher activities to changed beliefs about teaching and learning. *Teaching and Teacher Education*, *25*(1), 89–100. doi:10.1016/j.tate.2008.07.003

Montero, L. (2001). *A construção do conhecimento profissional docente. Armando Pereira da Silva (Trad.)*. Lisboa: Instituto Piaget.

Mushyikwa, E., & Lubben, F. (2009). Self-directed professional development – Hope for teachers working in deprived environments. *Teaching and Teacher Education*, *25*(3), 375–432. doi:10.1016/j.tate.2008.12.003

Nzau, D. K. (2010). *Das concepções dos slunos sobre força ao desenvolvimento de estratégias de ensino fundamentadas num modelo didáctico construtivista*. Tese de Doutoramento. Vila Real: Universidade de Trás-Os-Montes e Alto Douro.

Nzau, D. K. (2013). *Formação de professores de física na perspectiva construtivista. Proposta didáctica para a introdução do conceito da força*. Saarbrücken: Novas Edições Acadêmicas.

Nzau, D. K., & Costa, N. (1992). As concepções alternativas dos alunos sobre força como medida de interacção dos corpos. In 8ª Conferência Nacional de Física e 2º Encontro Ibérico para o Ensino de Física (pp. 570–571). Vila Real: Universidade de Trás-os-Montes e Alto Douro.

Nzau, D. K., Lopes, J. B., & Costa, N. (2012). Formação Continuada de Teacheres de Física, em Angola, com base num modelo didáctico para o campo conceptual de força. *Revista Brasileira de Ensino de Física, 34*(3), 3402. doi:10.1590/S1806-11172012000300014

Pinto, A., & Nzau, D. K. (2017). Training teachers how to plan for teaching science and technology. In J. B. Lopes, J. P. Cravino, E. Cruz, & A. Barbot (Eds.), *Teaching science: Contributions of research for planning, practice and professional development* (pp. 61–77). New York: Nova Science Publishers.

Starkey, L., Yates, A., Meyer, L. H., Hall, C., Taylor, M., Stevens, S., & Toia, R. (2009). Professional development design: Embedding educational reform in New Zealand. *Teaching and Teacher Education, 25*(1), 181–189. doi:10.1016/j.tate.2008.08.007

Vergnaud, D. (1987). In J. Piaget, P. Mounoud, & P. J. Brankart (Eds.), *Encyclopédie de la Pléiade Psychologie* (pp. 821–844). Paris: Gallimard.

Viennot, L. (2002). *Enseigner la physique*. Bruxelles: Éditions De Boeck Université.

Weil-Barais, A. (1995). La formation des connaissances en sciences expérimentales. In D. Gaonach & C. Golder (Eds.), *Manuel de Psychologie pour l'Enseignement* (pp. 410–437). Paris: Hachette Éducation.

Winch, C., Oancea, A., & Orchard, J. (2015). The contribution of educational research to teachers' professional learning: Philosophical understanding. *Oxford Review of Education, 41*(2), 202–216. doi:10.1080/03054985.2015.1017406

Zacharia, Z. T., & De Jong, T. (2014). The effects on students' conceptual understanding of electric circuits of introducing virtual manipulatives within a physical manipulatives-oriented curriculum. *Cognition and Instruction, 32*(2), 101–158. doi:10.1080/07370008.2014.887083

Zangori, L., Forbes, C. T., & Schwarz, C. V. (2015). Exploring the effect of embedded scaffolding within curriculum tasks on third-grade students' model-based explanations about hydrologic cycling. *Science & Education, 7*(24), 957–981. doi:10.100711191-015-9771-9

APPENDIX

Table 2. Plan of Teachers' Training Seminar

Period	Monday	Tuesday	Wednesday
Morning	1. Class on: a. Concept of a class; b. Components of a class; c. Epistemological viewpoints on the components of a class. 2. Application exercises in groups on: a. Analysis of descriptions of some lessons about the components of a class and epistemological viewpoints; b. Analysis of video-taped lessons; c. Collective discussion of the work of the groups.	1. Class on: a. Constructivist model. 2. Application exercises in groups on: a. Analysis of the components of the didactic model presented. 3. Class on: a. Structure of the scientific knowledge situation. 4. Application exercises in groups on: a. Analysis of the knowledge structures of a description or video recording.	1. Class on: a. Concept of a formative situation; b. Contents of physics that underlie it; c. Scientific relevance, social, environ-mental, and technological; d. Competencies to develop. 2. Application exercises in groups on: a. Tasks linked to the analysis of some physical situations with regard to the development of the learning and skills of students. b. Collective discussion of the work carried out in groups.
Afternoon	1. Class on: a. Didactic model concept; b. Different didactic models. 2. Application exercises in groups on: a. Examples of didactic models; b. Collective discussion of the work of the groups.	1. Application exercises in groups: a. Collective discussion on the work of the groups on the didactic model components presented and the structure of the scientific knowledge situation.	1. Class on: a. Model of interaction between material systems and a comparison with the didactic model studied. 2. Application exercises in groups on: a. The elements of the structure of the underlying scientific knowledge on the interaction between material systems; b. Collective discussion of the work done in groups.

Chapter 5
Using Multimodal Narratives to Study Relationships Between Concepts

António Barbot
Politécnico do Porto, Portugal

ABSTRACT

In conceptual teaching and learning, relationships between concepts play a key role. This chapter describes a case where two conceptual fields, water and energy, were treated in order to promote understanding of relationships between them. It deals with the importance of "potential relationships" and describes how multimodal narratives contributed to those elucidations and to these practices in the classroom and their interpretation, as well as to the professional development of teachers. The chapter elucidates concept use, relating concepts, ways of discerning relationships between concepts, loss of relationship opportunities that were created by students, and ways of verifying that certain relationships were achieved.

INTRODUCTION

This chapter deals with part of a classroom study about two specific conceptual fields, water and energy. It focuses on the problematic of the relationship between them, and not on details about the two conceptual fields. The chapter assumes two main objectives: to promote multimodal narratives as an instrument to study relationships between concepts, and to help teachers to clarify topics and relationships to be treated.

The following problem was identified: the need for studies and educational interventions in the classroom environment to promote understanding of relationships between water and energy in teaching. The relationships between water and energy (RWE) are multiple, as evidenced by the Proclamation of the United Nations Decade of Education for Sustainable Development (2002) and the Johannesburg Declaration on Sustainable Development (2002). Mechanical energy is used to produce electricity. The energy of the sun commands the hydrological cycle. Energy is used to produce potable water. From water, hydrogen can be extracted to be used as fuel. Food energy comes from life created with water and energy from the sun, received or stored on the planet. In education, water and energy are studied.

DOI: 10.4018/978-1-5225-8570-1.ch005

But relationships are scarce in teaching and research (Barbot, Lopes, & Soares, 2010, 2011; Cerqueira & Barbot, 2010). Water is the origin of life on Earth; it is the dominant element in the constitution of living beings, and it modulates the energy fluxes of the planet, namely, energy transfers (Allan, 2011). Energy is the base factor in the functioning and evolution of the world, of the planet, and of all human beings. The principle of energy conservation is crucial.

In education, when water and energy are studied, they typically are studied separately. Students learn that they are important for the sustainability of the planet and in particular for life, in which the human species is included.

THEORETICAL FRAMEWORK

The study referred to in this chapter (Barbot, 2014) assumes a constructivist view, which has solid and documented theoretical and practical bases, and currently constitutes the core of science education research. The study assumes that learning is always present and always interested, but this study does not focus on the qualities and or quantities of learning and its evaluation. It focuses on physical situations, tasks, models, resources, and physical and virtual models, and it makes suggestions tested in class as to its usability, feasibility, the students' motivation, time required and resources required. All this is intended to explore in class what is the central object of the study: the relations between water and energy, and to build on contributions to teachers, offering possibilities and elucidating the potentialities to treat such relationships. In terms of the teaching-learning dyad, the study focuses on teaching.

Novak and Gowin (1984), referring to Ausubel, report that, in meaningful learning, individuals relate new knowledge with relevant concepts and propositions that they already know. Bruner (1997) points to the importance of the culturally shaped notions by which people organize their conceptions of themselves, of others and of the world in which they live. Buty, Badreddine and Régnier (2012) point to the importance of how knowledge relates one thing to other. Therefore, a concept is much more than a word, using a concept is much more than just saying it, and learning meaningfully implies relating.

The study was carried out by a teacher-researcher, as described by Griffiths (1985), who studied his own practice in a natural classroom environment in an example of action research. It also had case study characteristics (Bogdan & Biklen, 1994; Stenhouse, 1988). The study had the following specifications: it involved, in a school year, classes of Initial Teacher Training courses in a school of education; it dealt with the relationship of two specific conceptual fields (water and energy), focusing on the teaching aspect as a specific instance of the teaching practice.

To relate is to understand better. Establishing links, identifying causes and effects, and building rich conceptual networks means knowing significantly more and being better prepared for the profession and for active citizenship (Barbot, 2017a). But a concept is not just a word. A concept, according to several authors, among them Toulmin (1977), has two aspects: representation and use. It is constituted by symbolic representations – natural language – and by other symbolisms, such as equations and other mathematical formalisms, graphical representations, taxonomy and computer programs. Simultaneously, it is also constituted by specifications of its modes of use, application procedures, and scopes and validity contexts.

The main problem was the need for studies and educational interventions in the classroom environment to promote the understanding of relationships between two concepts (here, water and energy) in teaching. In educational research, a problem is a need that exists in some context and that does not allow for something to function as well as desired. A problem may constitute a valid research problem only if the need that is claimed is well supported and if its formulation possesses a heuristic power that illuminates ways to tackle it. This may require extensive study and work. If this is achieved, the problem will constitute, simply by its formulation, a contribution made by the research study (Barbot, 2017b).

To solve a research problem consists of fulfilling, to some degree, the need that it expresses, and this will constitute the core contribution of that study. Usually, the need under consideration will not be completely fulfilled. Claims about having contributed to solving the problem should be accompanied by clear elucidations of what has been achieved and what remains to be done.

Multimodal narratives (MN) (Lopes et al., 2014) were of central importance in this study, as it was by means of this instrument that the collected data were organized.

RESEARCH METHODS

Data were collected with different instruments and methods. A survey was elaborated and applied to 116 teachers of the first and second cycles of Basic Education in the Porto and Vila Real districts of Portugal. The main objective was to know whether the respondents related or not, and how (in terms of topics, activities and resources), to the themes of water, energy and sustainability.

Three questionnaires were made: Questionnaire A (QA) centered on the theme of water; Questionnaire B (QB) focused on energy, and Questionnaire C (QC) focused on sustainability.

The application of these surveys was carried out as follows:

- With the collaboration of the directors of the departments of two groups of schools;
- With students from the School of Education P. Porto pursuing a master's degree;
- With teachers who cooperated in supervising the initiation to pedagogical practice of students of the School of Education of I. P. Porto.

Participants in the Classroom Activities

The field intervention involved the following students:

- Students of the first cycle of study in Basic Education, in the curricular unit "Physics for Education"
- Students of the first cycle of study in Basic Education, in the course "Chemistry for Education"
- Students of the second cycle of study in the first and second cycles of Basic Education, in the curricular unit "Didactics of the Natural Sciences in the First and Second Cycles of Basic Education"

This study focuses on classes of laboratory practice type.

It is, however, necessary to say something that is of key importance in characterizing the type of classes with which the study is concerned. There are important differences between activities based on the initiative of the students and activities that are based on the initiative of the teacher.

- As for the first type of activities, those based on the students' initiative are those that play the central role in the present authors' teaching practice, and they occupy a large part of their school time (about 10 lessons in 15).

- As for the second type of activities, those based on the teacher's initiative, they play a complementary role and occupy a small part of the school time (about 2 lessons in 15).

The activities on which this study focuses are of the type mentioned above in the second item: they are the initiative of the teacher. So it had to be, because it was the teacher who made the decision to study the theme of the relation between water and energy, and he selected the resources for that purpose. Therefore, this study does not deal with the formal evaluation and classification of students, a central aspect in teaching and learning. However, it must be emphasized that the classes with complementary activities (in this case, on water-energy relations) of the teacher's initiative are habitual in the present authors' teaching practice. They were neither created nor organized because of the specific need of the present study.

Audio recordings were made of the classes where the intervention of this study was carried out. The transcriptions of these audio recordings, as well as the collection of diverse documentation, constituted the material from which the multimodal narratives (MN) were constructed.

Multimodal narratives (Lopes et al., 2014) were developed with the aim of achieving a better understanding of teaching practices and their importance in science classes. They are based on independent data (audio or video recording, teacher documents, student documents, assignments, spatial organization, classroom facilities, photographs and used artifacts), and they incorporate extracts of such data. They also incorporate data on teacher descriptions, intentions and decisions, as well as descriptions of attitudes, silences and signs on the part of both the teacher and students. Multimodal narratives preserve the complex and holistic nature of teaching and learning, and describe these two processes in a multimodal way.

FIELD WORK: CONTEXTS, THEMES, RESOURCES AND ACTIVITIES

The field intervention was carried out in the initial teacher training for the first cycle of studies (Basic Education degree) and the second cycle of studies (Master in Education in the First and Second Cycles of Basic Education) that qualify students for teaching. The three intervention situations mentioned were diversified: they covered three different curricular units, four themes (the first part covers two themes, as indicated above), and undergraduate and master's degree students.

As will be described, the educational resources used were also diversified, including physical laboratory resources (models, material models) and virtual resources (computer simulators). A diversification was sought that, without creating dispersion, guaranteed divergence and adequate coverage, in order to strengthen the study. The following activities were carried out with these resources:

- Desalination with solar energy/urban water distribution
- Archimedes' law and gravitational potential energy
- Electric power production with fuel cells.

The following description is a summary of the tasks that were carried out. A script was used for each activity, and the general observations are derived from the MN analysis.

Desalination With Solar Energy/Urban Water Distribution

The materials used were a solar oven to desalinate water and a model of an urban water distribution system. Both resources were built with the collaboration of the students.

The solar desalinator consisted of a wooden box covered with an inclined glass. The box contained a tray where salt water was pumped in by an electric pump. When the desalinator was placed in the sun, the temperature inside the box increased, and the water evaporate. Upon contact with the glass, the water condensed, and it was drained and collected. Students handled the desalinator and discussed whether the glass should be hot or cold and what it would take for the water to evaporate and condense. They also observed that the inclination of the glass allowed the water to be collected, taking advantage of the gravitational force. The need to use energy to obtain unsalted water was discussed. There were debates on topics such as temperature, water evaporation, water condensation, energy (in various forms), water solubility, phase transitions, electrical circuits and energy cost savings. During the use of the urban water distribution model, there were debates about the need to elevate the water to reservoirs. Many students were unaware of the fact that this distribution is most often done by using gravitational force, after using electrical energy in its elevation. Several specific themes and conceptual fields were selected and studied, such as energy use in urban water distribution, pumps and filters, water treatment, gravitational field and water pressure in distribution pipelines (and their variation, namely between day and night). The activities with both models showed that it is possible to work with relationships between water and energy and with their relationships with sustainability.

Archimedes' Law and Gravitational Potential Energy

These classroom activities (described in Pinto et al., 2012) explored two computational simulations related to this theme. Work groups were formed, and an activity sheet was distributed to each group. Before the use of the simulation tools, the main physical conceptual topics were discussed with the students, namely: gravitational field, density, mass, weight and gravitational potential energy. The session began by exploring the simulation of the PhET project (Buoyancy, n.d.). The meaning of *equilibrium* was discussed, and students attempted to find as many equilibrium situations as they were able.

A computational simulation was then used in a Microsoft Excel spreadsheet (described in Silva, 1998). Some of the cases already handled when using the previous simulation were replicated, but now with the calculations made in the spreadsheet.

In a gravitational field, when a solid body that has been partially or totally immersed in a fluid is left with no external force beyond gravitational force, a static equilibrium situation will be reached if and when a situation of minimum total potential energy is reached. Students introduced data into spreadsheet cells and produced graphs of the potential energy variation of a block, of a fluid and of the system (block + fluid). They could predict and calculate the height of the immersed part of a block, depending on its density and on the density of the fluid.

Electric Power Production With Fuel Cells

These classes focused on the study of fuel cells using computer simulators and a fuel cell kit. The classes took place in a laboratory where computers with Internet access were available, where students accessed the computer simulations and did some research.

The class began with the formation of work groups. It started by screening a video that accompanied the kit. Then students opened the computational simulations. First, they explored the SEPUP (n.d. (a)) computational simulation related to water electrolysis; then they explored the SEPUP (n.d. (b)) simulation regarding the operation of a fuel cell. In each of the uses, students printed screens and wrote comments on each one. Students then proceeded to assemble the kit, which consisted of four main parts: (i) a photovoltaic panel where, through the photoelectric effect, the electric energy necessary for the electrolysis of water was obtained; (ii) a second part where the electrolysis of water was carried out, with storage of both resulting gases (oxygen and hydrogen); (iii) the fuel cell, where electric power supplied a small fan (iv).

The students performed several experiments: they turned off a part of the system, simulated a malfunction, covered the sun, put distilled water in the water container, turned off the gas supply pipes and tried to detect some variations of temperature, among other activities.

MULTIMODAL NARRATIVES ANALYSIS: CONCEPTS AND THEIR RELATIONSHIPS

Multimodal narratives were analyzed based on the creation of units and categories of analysis. The following tools were used: NVivo 10 software (from the company QSR) to make markings and SQL type queries, Statistica (from Statsoft) to create groupings and dendrograms, and the Microsoft Excel spreadsheet for calculation of correlations and their statistical significance.

A central aspect of MN analysis is the creation of units and categories of analysis. The first approach to MN analysis consisted of its re-reading, creation of preliminary categories of analysis and analysis with these categories. These were then refined until the final detailed categories of analysis were reached.

The scope, the framework, the problem and the research questions were created in relation to the theme and also with complementary guidelines of a preliminary and tentative nature. These aspects will be elucidated in the following. In dealing with the study of water, energy and water-energy relationships, three basic elements were established for the creation of categories. Because this study dealt with teaching in classroom environment, the basic elements were Teacher and Student.

The teaching (and learning) processes of this study were always related to specific resources, so this element was also chosen naturally and directly. The first analysis outlines indicated that in terms of resources, it was of interest to distinguish between the operational use of a resource and a reference to it, to better characterize the role played by the resource.

The Science, Technology, Society, Environment (STSE) element has been considered to better elucidate the nature of the references made to water, energy and, above all, their relationships: they may or may not have STSE characteristics, and it was provisionally considered that such a distinction could be of interest for this study. In fact, although it does not focus on this theme or on the related sustainability theme, these themes are necessarily present, given their transversality and comprehensiveness.

In this chapter, it is considered that learning is always present. But the chapter focuses on exploring what is the central object of study, relationships between concepts, and, in order to build contributions to teachers, offering possibilities and elucidating the potentialities to treat concepts and relationships between them.

The chapter reflects one of the aspects of the teaching and learning processes, which is one of the elements of the teaching-learning dyad: it focuses on teaching.

This process was rich in the reflection that emerged on the *concept of concept*, relationships between concepts and how these relationships can exist or not. The preliminary categories are presented in Table 1.

Explicit and Potential Relationships Between Concepts

Regarding Table 1, it is necessary to mention the explicit and potential elements when dealing with the relationships between water and energy.

It is important to begin by mentioning two aspects:

The reflections and options in this study about the explicit-potential pair had an intersection with the reflections related to the concept of concept. However, writing about this subject would be difficult, and it would certainly be confusing if the two reflections were presented together. Both reflections are here presented in an essentially sequential way: first the ones related to the explicit-potential pair and then the ones related to the concept of concept.

Before using the explicit-potential pair, the study used another "even more preliminary" notion: the pair "explicit-implicit." This happened naturally: if some relationships are considered as explicit, it emerges that others, which would be *not explicit*, should be considered as *implicit*. This seemed obvious. And yet such a seemingly natural and obvious choice concealed a major difficulty that only emerged after the MN analysis.

The difficulty with the "explicit-implicit" pair was as follows: was it possible to be certain that something was implicit in an action (verbal or non-verbal) of a student? And, of course, the answer was

Table 1. Preliminary analysis categories

Acronym	Full Description
S_Water	Student mentions the concept "water"
S_STSE	Student makes a reference to STSE
S_Energy	Student mentions the concept "energy"
S_ Explicit_RWE	Student establishes an explicit relationship between the concepts "water" and "energy"
Potential_RWE	Student performs verbal or non-verbal conduct that, from the point of view the teacher-researcher, has the potential to explore the theme of the relationship between water and energy
T_Water	Teacher mentions the concept "water"
T_STSE	Teacher makes a reference to STSE
T_Energy	Teacher mentions the concept "energy"
T_Explicit_RWE	Teacher establishes an explicit relationship between the concepts "water" and "energy"
S_Refers_Resource	Student concerns/performs something, referring to a resource
S_Uses_Resource	Student refers to something when operating or using a resource
T_Refers_Resource	Teacher concerns/performs something, referring to a resource
T_Uses_Resource	Teacher refers to something when operating or using a resource

no. In fact, it was not possible to guarantee that something not said or done was present in the mind, the intentionality of a student. One could wisely guess, reasonably suppose, or argue in depth, but one could not assure. To consider that there was something implicit was a subjective judgment about another, and in some specific cases in the MN analysis, there were situations where the present author felt the risk of treading the grounds of arbitrariness. And that could not be done. The concept of implicit itself contains something diffuse, difficult to parameterize, difficult to access by criteria. And that was not a good basis for categorization.

But would there be only explicit relationships, and "the others," not explicit ones? By analyzing the MN (and having in mind most of the classroom situations) it seemed that the answer was no. Among the non-explicit relationships, there were many differences. There were the purely unexplained, say, non-relationships, or, rather, there were passages from MNs in which there were no relationships and in which there were no anchor points for making relationships. There were also passages without relationships, but they contained potentialities for exploring relationships: the teacher just would have to detect those opportunities, start from there and, by just adding some information, soon make explicit relationships emerge.

It was to these situations and opportunities that the designation of "potential" was given. Consider that the potentiality of establishing a relationship does not constitute a subjective judgment about another: it is a didactic and scientific judgment on the part of a teacher. And this judgment can be grounded on educational and scientific terms. A teacher, a teacher-researcher in this study, can discern that something that occurs in a classroom environment can turn into an opportunity to establish a certain type of exploration and deepening. Grounding and sharing these opportunities can be a valuable contribution. It consists precisely in pointing out and describing with depth and substance something that may be useful to other professionals. In the case of this study, this type of contribution consists in providing events or situations with didactic potentialities for the deepening the understanding of the relationships between water and energy.

After clarifying these aspects and reviewing the MN analysis, it was found that there was an additional problem that had been felt in the analyses but that did not emerge sufficiently clearly: in the absence of the "potential" categorical element, some passages that did not fit the "implicit" element tended to be qualified with the "explicit" element. And this was a slightly forced and poorly sustained categorization, a kind of distortion. Of course, the intention was not to distort. It was an operative response to hesitant reflections like: "Well, you cannot guarantee that it will be implicit; but neither can it be said that nothing matters for the relationship; so, it is as if it were explicit." This, it must be emphasized, happened only in a few passages. But it did, and it caused discomfort and insecurity while affirming the conviction that something had to be done to improve categorization and analysis. The replacement of the element "implicit" with the element "potential" was a decisive contribution to such improvement.

The following excerpts are transcriptions of an MN in which an explicit relationship was previously considered to exist that later was considered to represent a potential relationship. Both were obtained from classes where resources were explored on desalination with solar energy/urban water distribution.

Multimodal Narrative Excerpt 1

Teacher: *(...) then how does water get to our homes?*
Student: *It must be because of the process itself ... of the pipes ...*
Teacher: *Is there a pipe from the water treatment plant for each house? (laughs)*

Student: *No ... there is a main pipe and then there are ramifications ...*
Student 2: *Well, there are scattered water tanks ...*
Teacher: *There are water reservoirs scattered ... we have talked about in reservoirs a bit, but before the reservoirs: water comes from the rivers, and then it goes to the water treatment plant, and then it goes to reservoirs.*
Student 3: *Go to the reservoirs.*
Teacher: *Exactly. But how do we put it in the reservoirs?*
Student: *Through plumbing ...*
Teacher: *Plumbing once again ... pumped.*
Student: *The energy that it takes ...*

Multimodal Narrative Excerpt 2

Students placed their hands on the glass and stated:
Student 1: *It's hot ...*
Student 2: *Because of the sun ...*
Teacher: *It's a little bit, but not much ... What do we want? Do we want it hot or cold? (There was some silence.)*
Student 2: *If it is hot, the water vapor would get there and it would not condense!*

In this last excerpt the word *water* appears, although in a fleeting way. The word *energy* does not appear, but it is obviously implicit. In the first analysis, this passage was considered to be an explicit relation ("it was as if it were explicit"). Then the possibility that there was a potential relationship was considered. And the justification for this *potentiality* is simple: if a student says that the sun heated the glass (*energy* transfer) and that the *water* vapor does not condense, then a teacher can explicitly explore relationships between water and energy (water evaporation, water condensation, energy in various forms, phase transitions).

And so, the pair of elements that came to be considered was explicit-potential. In the MN analysis, in each event it was considered that there was an explicit relation, a potential relation or neither.

The specification of the categories related to the explicit-potential pair was successively refined. Its final formulation is presented below; this is the one with which the final analyses were carried out:

- Regarding what is considered as potential relationships: cases in which students do not establish the relationship and do not show a conceptual understanding or capacity of use, but where they refer the core of the conceptual network of water and other parts of the conceptual network of energy, in a given verbal expression or a certain experimental action. This occurs in such a way that it becomes clear that the teacher has the opportunity to intervene to give a relational sense to the two aspects that the student is referring to.
- Regarding what is considered as explicit relationships: cases in which the student or teacher establishes a link between a component of one of the concepts (water and energy) and a component of the other. According to Toulmin (1977), this connection includes not only the symbolic representation component – words, equations, and so forth – but also the use: the context and/or the physical or virtual model, and/or the day-to-day situation, and/or a task being performed, and/or an application in another context.

In the next excerpt, students have established water-energy relations explicitly when they talk about water and refer to the resource. They refer to the resource and explain that it is possible to obtain an electric current and that the final product is water:

Multimodal Narrative Excerpt 3

Student 1: *[The fuel cell] separates the electrons from the protons (from hydrogen) ...*
Teacher: *Yes, and more?*
Student 1: *It produces electric current. This electric current ... the electrons and protons join with oxygen and form water again. [...]*
Student 2: *You can already make a water-moving car!*

This will be considered a contribution of this study because it gives other teachers indications of physical situations, models and tasks that can be used to deepen the relationship between water and energy, which is what this study is about. The students did not explicitly relate the concepts, but it was an opportunity for the teacher to advance their conceptual understanding by promoting the relationship that the teacher knows exists.

In the MN analysis, potential relationships are not considered when it comes to the teacher's actions. In this study, the teacher-researcher intends to promote relationships, and this is transversally present in all his academic and investigative activities.

Thus, in Table 1, the category S_RWE_Potential (student performs verbal or non-verbal action that, from the teacher-researcher's point of view, has the potential to explore the relationship between the concepts water and energy), doesn't have correspondence in the part concerning the teacher. The two respective rows are shaded in the table to highlight this asymmetry.

At some point in the study, it appeared that there was already a good elucidation of all the elements and categories in Table 1.

Referring to a Concept

As the preliminary analyses based on these categories continued, some questions arose that can be summarized as follows: What is to refer to the concept "water" or the concept "energy"? How can one specify/define/describe each of these concepts? Does the occurrence of one of the words *water* or *energy* necessarily mean that the respective concept is being evoked/invoked/referred to/used?

This originated other, and more complex, issues that were related to the ones stated before about the explicit and implicit elements: What are water and energy? What is the concept of relationship? What is evoking/invoking/referencing/using this concept?

That a concept is not just a word was known. And so, the occurrence of a word does not guarantee that the concept is sufficiently present (Toulmin, 1977) or that it is present with a degree of complexity and semantic completeness that allows one to infer that the student or the teacher is using the concept in a meaningful way. Or, that the individual is dealing with the concept with a level of understanding and intentional application such that it can be said that, from the point of view of teaching and/or learning, the concept is present and really being used. And even more: when the word is said, and even assuming that this is done with in-depth reflection and with a good understanding of its semantic field, is the concept being used or just being called by the concept's name? And, to go the limit: when the word is said, is the concept being used or only a word that has been torn from its meaning, for example "energy," when it might have been "force" or "heat" – without any relation to energy transfer?

Then there was another critical part, an intermediate process of long reflection. As will be seen, the description of this process consists, in part, in the explanation of a set of options that it was decided not to operationalize. But in research (and not only here), decisions about what not to consider are as important as decisions about what to consider. Both of these are closely related: focusing on one area of the field of study generally implies blurring or leaving out of sight adjacent areas. A set of reflections were presented that, for operational purposes, were partly abandoned, and a way of constructing categories of analysis was referred to that has not been used in its entirety. But to do so is important for two reasons: because they were those reflections, as well as the abandonment of a certain way of constructing categories that made it possible to find the alternative modes of operation that were used in this study, and because it is important to characterize how the study was developed, its story, and what it is and is not.

As described above, the fact that a concept is not just a word was known. A concept, according to several authors, among them Toulmin (1977), has two aspects: symbolic representation (natural language and others) and use. Toulmin writes that in order to do justice to the complexity of scientific concepts, three aspects, or elements, related to the use of such concepts must be distinguished: namely, (i) language, (ii) representation techniques and (iii) application of science. Thus, a concept is constituted by symbolic representations – natural language – and other symbolisms such as equations and other mathematical formalisms, graphical representations, taxonomies and computer programs – and simultaneously by specifying modes of use, application procedures, and scopes and contexts of validity. Thus, using a concept requires not only conceptual knowledge but also competences: knowledge at the level of linguistic representation and others, the ability to define scopes and limits of application, and usability, testing and experimentation. Toulmin proposes an epigram: a concept is an intellectual microinstitution, and the learning of a concept is comparable to initiation in a social institution.

Therefore, a concept is much more than a word, and using a concept is more than just saying it. And so the occurrence of a word in an MN by itself is far from necessarily signifying that a reference to a concept has been made and even further that it necessarily means that a concept has been used. However, in Table 2 there are categories such as "water" (student mentions the concept "water"), and no category makes explicit the use of a concept.

The next step in the process of creating the categories of analysis was to try to incorporate the two components of a concept (its symbolic representation and its use) for subsequent and new tentative analyses of MNs, considering this evolution. And before taking that step, it was necessary to prepare it: to specify the concepts of "water" and "energy," based on the aforementioned epistemological vision of Toulmin. That is, it was necessary to specify each one in terms of symbolic representation (natural language and other symbolism) and usage (e.g., application procedures, scopes, and contexts of validity).

Table 2 lists a set of illustrative descriptors of the two components of the two concepts "water" and "energy."

This would not be done exhaustively and completely, as this would be practically impossible. It would be necessary to do so with an adequate level of completeness, so that the specifications would be operationally useful in the analysis of MNs.

Partial MN analyses were performed by using this type of categorization. But some difficulties persisted: although the concepts of water and energy were clearer, what was meant by the use of a concept remained unclear. There were situations in which, from a semantic and educational point of view, one

Table 2. Examples of descriptors of each component of the concepts "water" and "energy"

Water		Energy	
Symbolic Representations	Uses	Symbolic Representations	Uses
H_2O Graphical and 3D representations of the molecule Equations of chemical reactions (hydrolysis, release of heat when H and O join, and so forth), thermal capacity, etc. Communication/sharing of meaningful expressions using symbolic representations, writing, drawing and so forth	A common substance, but with extraordinary properties (polar/ good solvent; it increases in volume when it passes from liquid to solid) Water distribution, transportation (flotation) Energy: hydro, nuclear, thermoelectric Human body, urine, kidneys, sweat and so forth	$\Delta E = W + Q + R$ $E = U \times I \times \Delta t$ Expressions about entropy, enthalpy in phase transition, etc. Principle of conservation. Various terms: energy, energy transfer, work, heat, radiation, potential, kinetic, free, enthalpy chemical, nuclear, electric and so forth Communication/sharing of meaningful expressions using symbolic representations, writing, drawing and so forth	Solar energy (radiation) that passes the electric energy; photovoltaic, wind and so forth Transport. Identification of new subatomic particles Food, health, and so forth Water distribution and others

could say that the concept was there, and that it was being used, but such use did not fit into the characteristics expressed by the descriptors in Table 2.

It became clear that the author was too attached to a certain epistemological specification of what a concept is, or that such a specification was being considered too rigidly and/or simplistically. But to continue the study by elucidating these epistemological and linguistic aspects (in addition to physical and technological ones) was a risk. It was necessary to reflect and assume pragmatically and clearly that this research is not about the concept of concept, but about teaching and learning concepts, and therefore about their use in teaching and learning contexts. This clarification was useful.

The following was clarified and assumed:

- It is important to establish what is meant by concept and in particular by specific concepts such as "water" and "energy." For the teacher-researcher, this is very important in order to organize ideas and the work.
- But the teacher-researcher must also bear in mind that when a concept is used, all its components do not have to be present, simultaneously and explicitly.
- The use of a word associated with a concept or another symbolic representation of this concept can be considered as a use of that concept – even if it is not an experimental, practical use, but a conceptual use, a unit of thought.
- It is important to continue to consider the types of use referred to in Table 2 that are "hands on" in different contexts: scientific, technological, experimental, and related to everyday life, but also to consider other types of use, such as "minds on"/propositional: the construction and use or sharing of symbolic expressions with meaning.
- It is important to consider that a use of propositional type has significance only if it gathers "hands on" uses or reference to such uses.
- It is not important to consider the occurrences of symbolic representation per se for analysis purposes. That is, the mere occurrence of a symbolic representation per se is not signaled in the analysis of an MN as corresponding to a conceptual occurrence marked in the corresponding category.
- It is not important to consider that there is a relationship between concepts if there is only correspondence between symbolic representations, e.g., words that name the concepts.

This resulted in a formulation that seemed appropriately profound and solid, which was the basis for the consequent reformulation of the categories and their use in further analyses of MN. The key aspects that led to the reformulation of the categories and the re-analysis of MNs are as follows:

- Not to consider for purposes of analysis the occurrences of symbolic representation per se, particularly not considering that there is a relationship between water and energy if in an event there is only coexistence of those words.
- Extending the concept of using a concept. A concept was then considered to be used if (and only if) an evocation/invocation/action (verbal or non-verbal) of a concept occurred, or if the student or teacher did any of the following in connection with a concept:
  - Used a resource.
  - Referred to a resource.
  - Engaged with STSE contexts.

In operational terms, the reformulation of the MN analysis in terms of the new categories was relatively simple. This was due to two factors. The first is that, although there have been substantive changes in the categories, the basic elements have remained essentially the same. The second factor has to do with the potential of the application NVivo 10 (from the company QSR), which was used for analysis of the MN. After the reformulation, the categories became those listed in Table 3.

Here is the description of the first three categories:

- Water_Student_STSE – Student refers to water with reference to STSE contexts;
- Water_Student_Ref_Resource – Student refers to water with reference to a resource;
- Water_Student_Uses_Resource – Student refers to water while using a resource.

This reformulation was carried out with queries on the database that had already been constituted in NVivo software. The intersections of categories have become the categories of the study that were used in subsequent analyses of the data.

The first category of Table 3 results from the intersection of the first two of Table 1.

In analyzing the narratives, it was considered that there was a change of event when a new category occurred or when a category ceased to occur. In this markup, NVivo software was extremely useful, since it makes it possible to have a more visual and clear view of the categories' occurrences by adding vertical bars at the right side of the text.

Main Results of the MN Analysis

Correlations Between Categories of Analysis

As said above, a Microsoft Excel spreadsheet was used for calculation of correlations between categories of analysis and their statistical significance.

In MN, regarding the desalination and distribution of water, in the STSE contexts, it is through the theme of energy that the students establish more explicit water-energy relationships, but when they use the resources, it is through the theme water that they more easily establish water-energy relationships. In the case of the teacher, it is through the theme of energy that more water-energy relationships are established, either in STSE contexts, or by *referring* to or *using* the resource.

Regarding the MN on Archimedes' law: (i) students establish explicit water-energy relationships during the use of the resource, both on water and energy; (ii) students do not establish explicit water-energy relationships with factors within the STSE scope. Making a water-energy relationship and, in

Table 3. Categories used in the analysis of the MN

Reformulated Categories – Acronyms
Water_Student_STSE
Water_Student_Refers_Resource
Water_Student_Uses_Resource
Water_Teacher_STSE
Water_Teacher_Refers_Resource
Water_Teacher_Uses_Resource
Energy_Student_STSE
Energy_Student_Refers_Resource
Energy_Student_Uses_Resource
Energy_Teacher_STSE
Energy_Teacher_Refers_Resource
Energy_Teacher_Uses_Resource
RWE_Explicit_Student_STSE
RWE_Explicit_Student_Ref_Resource
RWE_Explicit_Student_Uses_Resource
RWE_Explicit_Teacher_STSE
RWE_Explicit_Teacher_Ref_Resource
RWE_Explicit_Teacher_Uses_Resource
RWE_Potential_Student_STSE
RWE_Potential_Student_Ref_Resource
RWE_Potential_Student_Uses_Resource

addition, transferring that relationship to an STSE scope away from those activities, is a difficult step. These RWE only appear as potential, and about Energy. In addition, when the student uses the resource, there is a tendency to coexist regarding the themes Water and Energy, and to establish relationships between the two themes.

Regarding the MN on fuel cells, it is through the theme of energy that more conditions arise for the establishment of explicit water-energy relationships when students use or refer to the resource. For the teacher, whether referring to the resource or in an STSE context, it is the theme of water that establishes more water-energy relationships. It was the energy theme that most promoted the establishment of explicit water-energy relationships by students, which could be explained by the fact that, in debates and other interactions, energy played a central role, either because electricity was needed to produce hydrogen by electrolysis or by the possibility of producing electric energy from it. There were several opportunities for the teacher to explain water-energy relationships in an STSE context, promoting this type of establishment by the students.

Grouping Events in Relation to Water and Energy Relationships

In the MN on desalination and the urban water cycle, students' potential RWEs are almost always based on the reference to one of two topics: either water or energy. This indicates a way for the teacher to lead such potential relationships to a clarification. The importance of teacher intervention for students to establish explicit RWEs was evident.

In the MN on Archimedes' law, the importance of having considered the element of "potential RWE" was clear. The establishment of explicit RWEs on the part of the students happened mainly when there was an intervention on the part of the teacher. Exceptions were the events in which students commented on the graphs obtained in the computational simulation of the spreadsheet. In the MN on fuel cells, in general, students' explicit RWEs were established when they were using or referring to the resource. It is also possible to observe that there were no events with RWE that were teacher exclusive; that is, whenever the teacher established RWEs, so did the students. In addition, all student RWEs in an STSE context were potential.

CONCLUSION

Multimodal narratives played a key role in the data collection and in its analysis and interpretation.

In the analysis process, the category of potential relationships, related to students, played an important role. It was therefore relevant and fruitful to have considered the categorical element of potential relationships analysis. The MN were also essential here, because it was only through their analysis that the whole process of reflection could be carried out. In the case of the study to which this chapter refers, the concepts of water and energy were treated, used and represented.

However, the contributions of this chapter go beyond the scope of these specific concepts. In any other scientific area, in any other class context where it is intended to deal with concepts and their relations, this perspective can be taken into account about what a concept is, what is to relate concepts and what may be potential opportunities for those relationships to emerge. And this can exist only with MN analysis.

This perspective facilitates and potentially promotes additional studies that fill the need to relate some key concepts.

As for this chapter's main objectives, the data of the intervention were analyzed based on categories of analysis and with methods and instruments that may be useful for similar studies. Information about types of relationships between water and energy, as well as about their relationship with didactic resources and with STSE contexts, was constructed. Finally, some of these aspects and some of the aspects of teacher mediation were related.

This study promotes interdisciplinarity and transdisciplinarity. And this results because the theme of water, energy and their relationships is especially rich.

In fact, it can be said that the concept of energy is not only a transversal concept: it is *the* transverse concept, as many phenomena could be explained with the physics concept of energy. Water is a key element on the planet and in human activities and life itself, and the relationships between energy and water are important because of their importance per se and because of the importance of the relationships between them. The theme is unifying *par excellence*. This chapter (as well as the study to which it relates) is not nomothetic: it is not intended to specify what resources should be used, what activities should be performed, what tasks should be requested of the students, what relationships should be addressed or the types of mediation that should be used. And the mentioned themes can be interpreted for illustration purposes only.

It is intended, therefore, that this text be taken into account as a case study on theories and practices related to teaching and to teachers' professional development.

Some indications can be mentioned:

- Conducting research that is focused on in-depth content analyses of school manuals in order to: (i) elucidate how specific topics are related, (ii) draw inferences and suggestions to authors of manuals, teachers and researchers.
- Developing training actions and research, focusing on the relationship between topics, in the context of teacher training and other contexts of teacher professional development.

REFERENCES

Allan, R. (2011). The role of water vapour in Earth's energy flows. *Surveys in Geophysics*, 1–8. doi:10.100710712-011-9157-8

Barbot, A. (2014). *Água, Energia, Sustentabilidade e Educação Sustentada* [Water, Energy, Sustainability and grouded Education] (Unpublished doctoral dissertation). University of Trás-os-Montes e Alto Douro, Vila Real, Portugal.

Barbot, A. (2017a). Integrating the transversal theme of water-energy in initial teacher training. In Teaching science: contributions of research for planning, practice and professional development (pp. 173-188). New York: Nova Science.

Barbot, A. (2017b). Problems and Questions: Elucidation and Relevance for Research and Teaching. In Teaching science: contributions of research for planning, practice and professional development (pp. 325-336). New York: Nova Science.

Barbot, A., Lopes, J. B., & Soares, A. (2010). Água, Energia, Sustentabilidade e Educação Sustentada [Water, Energy, Sustainability and grouded Education]. In Resumos do 20º encontro Ibérico para o ensino da Física 2010 [Iberian meeting for the teaching of Physics 2010] (pp. 311–312). Universidade de Trás-os-Montes e Alto Douro.

Barbot, A., Lopes, J. B., & Soares, A. A. (2011). Relating Water, Energy and Sustainability to Improve Teaching and Learning in Physical and Environmental Sciences in Teacher Education for Classes of 6 – 12 Age Levels. In *ICERI2011 Proceedings* (pp. 1254–1261). ICERI.

Bogdan, R., & Biklen, S. (1994). *Investigação Qualitativa em Educação: uma introdução à teoria e aos métodos* [Qualitative research in education: an introduction to the theory and methods]. Porto: Porto Editora.

Bruner, J. (1997). *Acts of meaning: for a cultural psychology*. Lisbon: 70 Editions.

Buoyancy – project PhET. (n.d.). Retrieved from http://phet.colorado.edu/en/simulation/buoyancy

Buty, C., Badreddine, Z., & Régnier, J.-C. (2012). Didactique des sciences et interactions dans la classe: quelques lignes guidelines pour une analyze dynamique [Science teaching and interactions in the classroom: some guidelines for a dynamic analysis]. *Testing Research in Education in Science, 14*(1), 147-165.

Cerqueira, P., & Barbot, A. (2010). Ciclo Urbano da Água: Contributo para o seu tratamento no Ensino e na Educação [Urban Water Cycle: Contribution to its treatment in teaching and learning]. In Resumos do 20º encontro Ibérico para o ensino da Física 2010 [Iberian meeting for the teaching of Physics 2010] (pp. 339–340). Universidade de Trás-os-Montes e Alto Douro.

Griffiths, G. (1985). Doubts, dilemmas and diary keeping: some reflections on teacher-based research. In R. G. Burgess (Ed.), *Issues in educational research – qualitative methods. London*: Falmer.

Johannesburg Declaration on Sustainable Development. (2002, September). Retrieved from http://www.un.org/esa/sustdev/documents/WSSD_POI_PD/Spanish/WSSDsp_PD.htm

Lopes, J. B., Silva, A. A., Cravino, J., Viegas, C., Santos, C., Saraiva, E., ... White, J. (2014). Constructing and using multimodal narratives to research on science education: Contributions based on classroom practices. *Research in Science Education*, *44*(3), 415–438. doi:10.100711165-013-9381-y

Novak, J. D., & Gowin, D. B. (1984). *Learning how to learn*. Cambridge University Press. doi:10.1017/CBO9781139173469

Pinto, A., Barbot, A., Viegas, C., Silva, A. A., Santos, C. A., & Lopes, J. B. (2012). Teacher education using computer simulations – pre and in-service primary school teacher training to teach science. *10th International Conference on Computer Based Learning in Science (CBLIS)*.

SEPUP. (n.d.a). *Electrolysis Simulation*. Retrieved from http://www.sepuplhs.org/high/hydrogen/electrolysis_sim.html

SEPUP. (n.d.b). *Fuel Cell Animation*. Retrieved from http://www.sepuplhs.org/high/hydrogen/fuelcell_sim.html

Silva, A. A. (1998). Archimedes' law and potential energy: Modelling and simulation with a spreadsheet. *Physics Education*, *33*(2), 87–92. doi:10.1088/0031-9120/33/2/010

Stenhouse, L. (1988). Case study methods. In J. P. Keeves (Ed.), *Educational research, methodology and measurement: an international handbook* (pp. 49–53). Oxford, UK: Pergamon.

Toulmin, S. (1977). *Human understanding: the collective use and evolution of concepts*. Princeton, NJ: University Press.

Chapter 6
Looking at Linear Algebra Teaching Practices Through Multimodal Narratives

Cecília Costa
Universidade de Trás-os-Montes e Alto Douro, Portugal

Ricardo Gonçalves
Instituto Politécnico do Cávado e do Ave, Portugal

ABSTRACT

There is a lack of research studies on teaching practices in higher education, which does not contribute to a greater and better reflection on the failure in some disciplines, namely in linear algebra. Multimodal narrative (MN) is a facilitator of research in this field. This chapter describes how the use of MN of linear algebra classes was done and allowed reflection on and modification of the practices. Formative situations are presented: 1) planned and successful, 2) planned and less achieved, 3) unexpected and successful, and 4) unexpected and less achieved. Two transversal aspects to the teaching sequence are also presented: use of technology and geometric approach. This study made it possible to recognize the MN's potential as a tool to "observe" and analyze linear algebra teaching practices in diverse aspects. It also shows that MN analysis allowed linear algebra teachers to promote their professional development regardless of the teacher's experience.

INTRODUCTION

During the 1960s, introductory linear algebra courses appeared for undergraduates, primarily intended for math and engineering majors (Uhlig, 2003). According to some researchers (Hillel, 2000; Kleiner, 2007; Tucker, 1993), linear algebra is the first full-fledged mathematical theory with which students have contact. It is traditionally taught through an axiomatic approach, which means starting from the axiomatic definition of vector space, followed by deduction of associated concepts and results.

DOI: 10.4018/978-1-5225-8570-1.ch006

Vector, vector space, linear transformation, system of linear equations and *matrix* are the basic concepts of linear algebra. All are abstract concepts (although some have a degree of abstraction greater than others), since they are secondary concepts, that is, concepts resulting from the action of the subject on other concepts, as happens with most mathematical concepts. This characteristic justifies part of the complexity attributed to mathematics and some of the difficulties presented by students. Next, an excerpt of a multimodal narrative (MN) (Multimodal Narrative Excerpt 1) is presented that is illustrative of the concept abstraction level of linear algebra and the students' reaction to it.

Multimodal Narrative Excerpt 1: Teacher R (Subspace of a Vector Space)

Beginning by attending to the axiom on the null vector belonging to the set, soon some of the students' confusions were revealed.

"So what do you say?"

"It has to be all zero," Hélder suggested.

"Or ... who is this, Hélder, right now? This one here, is what?" I asked, pointing to the V in the expression $0v \in V$ *[v being a vector of V] that was still written on the board.*

"A vector," said Juliana, also entering the discussion.

"It's x, y, z, is not it?" replied Hélder.

"No!" I said in alarm. "This is the generic vector space V."

Over the decades, linear algebra teachers have found many recurring difficulties expressed by students as well as failure to complete the course. On the one hand, teachers' frustration with students' learning difficulties was recurrent, because for them the reasoning associated with the various concepts were quite obvious (Dorier, 2000; Dorier, Robert, Robinet, & Rogalski, 2000); on the other hand, many students had the "feeling of landing on a new planet and are not able to find their way in this new world" (Dorier, 1998, p. 142). Linear algebra didactic studies identified as conceptual difficulties in linear algebra learning the knowledge of prerequisites (mainly in logic, set theory and analytical geometry) and the use of formalism (students are not prepared for the level of abstraction that is required) (Dorier & Sierpinska, 2001).

The questioning and reflection about the way linear algebra was taught led to curricular reforms in several countries. In the United States, the Linear Algebra Curriculum Study Group (LACSG) played a major role in the change of linear algebra curriculum during the 1990s (Carlson, Johnson, Lay, & Porter, 1993), with influence in other countries. For example, the recommendation to opt for a matrix approach (Carlson et al., 1993), i.e., to explore the concepts from the matrices and based on concrete and practical examples in place of the axiomatic approach, is recognized in Portugal, as evidenced by the programs of linear algebra in various engineering courses.

In addition to the emphasis given to the matrix approach, the studies and reflection carried out by the LACSG (Carlson et al., 1993) and by other researchers (e.g., Dorier & Sierpinska, 2001; Harel,

1997; Trigueros & Possani, 2013) resulted in additional recommendations for teaching linear algebra. These included "use of technology; [...] exploration of tasks and linear algebra applications; knowledge of prerequisites; connections between the linear algebra concepts; scope of formalism; and attention in the languages and multiple representations of the concepts" (Gonçalves, 2018, p. 47).

In spite of these investigations and consequent recommendations, the above-mentioned problems in the teaching and learning of linear algebra remain, leading the present authors to believe that it is necessary and essential to observe teaching practices within the classroom. Unlike the situation in other levels of education, research studies on mathematics teaching practices in higher education remain scarce (Speer, Smith, & Horvath, 2010). This compromises the existence of specific didactic guidance for this level of education. Moreover, the difficulties known and identified in the learning of mathematics at the various levels of non-higher education, already widely covered in research in national and international mathematical education, have repercussions and become worse in higher education.

Given the relevance of linear algebra for the training of future professionals and researchers in the areas of science, technology, engineering and mathematics (STEM) and the higher degree of abstraction of the linear algebra notions, it is essential invest in research on teaching practices at the level of higher education, in particular with emphasis on teacher professional development.

Teaching practices are inherent to the teaching function and, therefore, reflecting on their own practices contributes to their professional development. It is a common practice of teachers – and those of higher education are no exception – to reflect on their own lessons. What is not so frequent is that this reflection be systematic and based on concrete facts and recorded; often, instead, it is reduced to more or less subjective, tacit and uncritical impressions (Cochran-Smith & Lytle, 1993). Lopes and Cunha (2017) argue that focusing on teaching practices is central to effective teacher professional development.

Ponte (2002) argues that the closer a teacher gets to academic research, the more the teacher has to gain from studying his or her own practice; this is close to the opinion of Alarcão (2001), who considers that every teacher has to be a researcher.

However, the authors' experience in studies focusing on the practices of Portuguese teachers has indicated that they are reluctant to open their classrooms, and this tendency is even more marked in higher education. Arguably, MN may constitute a key to the participation of more teachers in higher education in research on teaching practices, as it is not necessary for a stranger to intrude into the classroom to collect data, since (with advantage) teachers themselves can elaborate the MN.

In the following sections, based on the linear algebra teaching experience of the second author and his experience in constructing several MNs of the classes he taught, the authors reflected on the MN contribution to "observe" linear algebra practices and to promote the teacher's professional development. Notice that the most experienced teacher (first author) was able to "observe" these practices through MNs, although she did not attend any classes. The present study has the following objectives: to verify if MN allows the teacher and the researcher to reflect systematically on linear algebra teaching practices and, if so, to show that MN has the potential to allow others to "observe" and analyse these teaching practices. Supported by MNs, this Chapter also illustrates linear algebra formative situations in order to show the complexity of linear algebra contents referred to in the literature and to strengthen the relevance and need of teacher professional development.

MULTIMODAL NARRATIVES TO LOOK AT LINEAR ALGEBRA TEACHING PRACTICES

In a recent study, Gonçalves (2018) creates and implements a teaching sequence for linear algebra that has intentionally underpinned the recommendations referred to in the previous section, indicated by the literature on the teaching and learning of linear algebra. This study was implemented in a course in the first year and first semester of an electrical engineering course in a Portuguese public polytechnic institution. Essential to the accomplishment of this investigation was the observation of teaching practices for linear algebra. This is the reason why several classes were recorded and productions of the students, notes of the professor, etc. collected.

Investigating teaching practices requires knowing these practices and knowing them fully in all their complexity. MN brings together in a single document multimodal aspects that seek to approach the complexity typical of teaching practices, including what is not recordable by audio or video. This allows researchers to approach these practices even if they have not experienced them as teachers (according to Chapter 1) (Lopes et al., 2014).

Thus, the various data gathered to carry out the research were organized in a single instrument, the MN, and "for five of the sixteen classes, the teacher presented his (neutral) version of the 'history' about all the events that occurred in each class" (Gonçalves, 2018, p. 5). The teacher, who also assumes the role of researcher, constructed five MNs; two other researchers, taking into account the protocol (Lopes et al., 2014) detailed in Chapter 1, independently validated them. Table 1 presents the distribution of linear algebra contents addressed in the five classes that were the subject of MN construction, as well as the number of episodes of each.

The construction of the MN presupposes training by the teacher, in the sense of rendering the narrative as objective as possible, neutral and reliable with regard to the data. This learning implied that the teacher was distancing himself from the role he played as a teacher and could register his practice impartially. The process of building a MN is time consuming and is done in "layers." These layers add multimodal information and allow teachers (even during this process) to reflect on their practice.

The experience of the construction of MNs allowed the teacher to realize that the activity of reflecting on his practice was improving as he advanced in the writing process. The notes written at the end of the lesson, while permitting a reflexive act through which the teacher could modify his practice, were still very much based on the impression of the moment. When the teacher listened to the audio recordings and placed them in the MN, a qualitative leap was observed in this reflexive act, although still at the

Table 1. Multimodal narratives per class, content and number of episodes

Class	Multimodal Narrative	Content	Number of Episodes
2	MN_1	Matrix language. Matrix operations: addition and multiplication by a scalar; multiplication.	3
5	MN_2	Systems of linear equations: Gauss and Gauss-Jordan elimination methods.	4
7	MN_3	Systems of linear equations: problem solving.	3
9	MN_4	Inverse of an invertible matrix: calculation procedures.	5
14	MN_5	Linear combination of vectors. Subspace spanned.	4

level of common sense. At the end of the MN's construction, the teacher could already reflect in depth on his practice, generating a first analysis. However, only when it comes to MN content analysis (Bardin, 1997) does he gain a different perspective from the act of reflecting, now as researcher.

After being validated, the MNs are locked. From that moment on, they function as an inviolable document.

The MN, constituting a descriptive and multimodal account of these classes, allows a teacher to "revive" them as many times desired and with the distance and objectivity that allow systematic and deep reflection. The MN can be analyzed by other teachers, promoting debate and collaborative work. This is what happened in this case: the analysis and reflection on the practice of this teacher, made by both authors and supported in the MN, increased the level of scientific robustness and made it possible to recognize aspects such as those presented in the next section. It should also be noted that the focus of MN analysis might vary according to the aspects on which a teacher intends to reflect.

LOOKING AT LINEAR ALGEBRA TEACHING PRACTICES THROUGH MN

In this section, referring to excerpts from MNs, examples are presented of formative situations that occurred during the implementation of the referred teaching sequence (Gonçalves, 2018). Situations were selected such that the analysis of the MNs permitted classifying them as: i) planned and successful, ii) planned and less achieved, iii) unexpected and successful, or iv) unexpected and less achieved. In addition to these situations, examples are given of two transversal aspects of the teaching sequence: the use of technology and the geometric approach.

Formative Situations, Planned and Successful

Again taking into account the literature's recommendations for the teaching of linear algebra, an aspect that the teacher deliberately planned was the early introduction of the concept of linear combination of vectors (a concept of extreme importance in the study of vector spaces). Early introduction of key concepts of linear algebra is proposed by Lay (2012), where he establishes that "later generalizations of these concepts appear as natural extensions of familiar ideas" (p. x).

The strategy chosen began with the teacher relating the concept of linear combination (Multimodal Narrative Excerpt 2) to matrix multiplication and different ways of representing systems of linear equations (Multimodal Narrative Excerpt 3):

Multimodal Narrative Excerpt 2: Teacher R (Study of Vector Space Chapter)

Once we have explored the concepts of vector space and subspace of a vector space in the previous class, this lesson [...] is continued with the exploration of the concept of linear combination [...] and the approximation to the concept of subspace spanned.

Multimodal Narrative Excerpt 3: Teacher R (Connections With the Concept of Linear Combination)

In the first place, I recalled that this designation [linear combination of vectors] had already appeared in the supporting text, and I have shown the two cases where this had happened: in the context of an exercise on matrix multiplication, and in the context of the approach to the study of systems of linear equations. I recapitulated each of the cases, trying to inform the students about the expression associated to the concept of linear combination.

For example, in an exercise (1.13), a linear combination was established as a product of a matrix A by a vector X, that is, AX as a linear combination of the columns of A, with the coordinates of vector X as scalars. The teacher hoped that students would develop intuition about the concept of linear combination of vectors and when it came time to formalize the concept took advantage of what had already been studied.

The teacher seeks to show how the concept of linear combination can function as a tool when it comes to multiplying matrices, as the following excerpt (Multimodal Narrative Excerpt 4) from the MN illustrates:

Multimodal Narrative Excerpt 4: Teacher R (Connections With the Concept of Linear Combination)

Reviewing the cases, I began by saying:

"You even did this exercise in class, which was 1.13, where I gave you here ... look, here is the concept of linear combination. At that time I asked you ... try to adapt this information in the multiplication of matrices."

In the course of the class, the students demonstrated the recognition of matrix addition and multiplication of a matrix by a scalar, as well as systems of linear equations, with the concept of linear combination of vectors (Multimodal Narrative Excerpt 5):

Multimodal Narrative Excerpt 5: Teacher R (Connections With the Concept of Linear Combination)

"But this must be with some system," Fernando said now.

"Develop," I said.

"I don't know. As a matrix ... making an augmented matrix with the result ... we might be able to find out the a, b, and c," Fernando added.

"That smells of something we know, doesn't it?" I said.

Fernando had just mentioned the point where I intended to reach. He said that we would continue with the writing of linear equations.

Given that one form chosen to represent systems of linear equations was using the linear combination of the vector of independent terms as a function of the remaining vectors of unknowns, when studying the linear combination of vectors in the vector space chapter, students easily associated the search for the scalars to solve systems of linear equations before the teacher indicated it.

The detailed analysis of the MN led to the conclusion that the strategy of introducing early the concept of linear combination of vectors made it possible to counter the students' tendency not to see the possible relations between the different concepts of linear algebra suggested by the researchers (Lapp, Nyman, & Berry, 2010). The system of categories resulting from the content analysis of MNs also yielded information about two relevant aspects: 1) The teacher insisted on the idea that, in the relation between the concepts, some function as tools for others (the linear combination concept allows for a different way of multiplying matrices, and representing a system of linear equations or solving systems of linear equations becomes a part of the procedure to investigate the possibility of writing a vector as a linear combination of others); 2) the nature of the relationships established by the students between these concepts was close to those established by the teacher, which points to the impact and dependence of teacher work in this area.

Formative Situations, Planned and Less Achieved

For the introduction of matrix multiplication, the teacher elaborated a task, taking into account two recommendations of the literature: to provide the exploration of tasks and to connect with students' previous knowledge (prerequisites). In particular, the teacher also considered the non-standard views of the matrix product AB proposed by Carlson (1993). The task called "algorithm for matrix multiplication" consisted of two subtasks that led to two different ways of multiplying matrices.

The teacher began by presenting the task to the students (Multimodal Narrative Excerpt 6):

Multimodal Narrative Excerpt 6: Teacher R (Matrix Multiplication)

Already in the second half of the class, announcing the study of matrix multiplication, I began by saying that I would propose "something different." I showed task 1 [...], indicating that before they knew how to multiply matrices and without presenting any algorithm, the students would perform two subtasks, and final discussion would lead to the formalization of an algorithm.

Given the usual tendency to generalize procedures to add matrices and multiply a matrix by a scalar, the teacher warned about this fact and gave more explanations on how to perform the task (Multimodal Narrative Excerpt 7):

Multimodal Narrative Excerpt 7: Teacher R (Matrix Multiplication)

I continued to rule out the possibility of multiplying only homologous elements. I insisted that the challenge was to use the statement information of each subtask to obtain the initial product matrix.

The first subtask suggested the use of the dot product of two vectors, already known from secondary school, to multiply matrices. Nevertheless, the teacher recalled the notion of dot product of two vectors (Multimodal Narrative Excerpt 8):

Multimodal Narrative Excerpt 8: Teacher R (Matrix Multiplication)

[...] in the particular case of the first subtask, I reviewed the dot product of two vectors according to their coordinates. After this explanation, I waited for the students to finally try to answer the task, and I started to circulate around the room, trying to find out about the work done in each group.

The idea was interesting, but it was demanding. The students expressed difficulty in their achievement, as the following excerpts from the MN show (Multimodal Narrative Excerpts 9 and 10):

Multimodal Narrative Excerpt 9: Teacher R (Matrix Multiplication)

When I asked for an answer to subtask 1, the students remained silent. The deadlock was unblocked when Fernando, at my request, proposed to multiply each element of the first row of the first matrix by each element of the first column of the second matrix. Then I went on to describe the process, using the signage of the elements I was considering, and the students were interacting more and more naturally. At this point, I took the opportunity to address the basic requirement for the possibility of multiplying two matrices in light of the size of the matrices.

Multimodal Narrative Excerpt 10: Teacher R (Matrix Multiplication)

"It's multiplying," Fernando began hesitantly.

"Say ... is multiply who by whom? Say! Don't be ashamed," I said, trying to keep the student going.

"u1 by v1" [u1 and v1 being the first components of vectors u and v, respectively, that were intended to multiply], continued Fernando.

Here the student was speaking generically, according to the definition presented in the statement for dot product, so I asked him to concretize:

"The 3 by 2 ..." Fernando said, looking at the matrices that I had transcribed on the board [...].

The MN analysis suggests that the (unusual) strategy used to teach matrix multiplication using an operation with vectors already known to the students – the dot product of vectors – was not achieved as expected, and its use is quite questionable for the understanding of the mathematical concept under study. The students' performance in solving the task points to the fragility of the idea.

Still, the excerpts presented evidence that parts of the task were accomplished. The teacher-led discussion led some students to discover relevant aspects of matrix multiplication, namely, the criteria for the operation to be well defined (eliminating the possible error of multiplying matrices by the homologous elements) and the possible relation of this operation with other operations with matrices. As a mitigating factor, the greater difficulty in responding to subtask 2 was already anticipated by Dubinsky (1997), when speaking of the symbolic complexity associated with the ways of multiplying matrices proposed by Carlson (1993).

It is also worth mentioning the positive fact of the task to propose alternative procedures, which allowed the students to develop critical sense, as shown below (Multimodal Narrative Excerpt 11). The students, confronted by the teacher, infer the algorithm for matrix multiplication conceptually, which is more significant and efficient in terms of time.

Multimodal Narrative Excerpt 11: Teacher R (Matrix Multiplication)

"Ah ... I'm going to ask you which is the best algorithm? ... which is the most time-saving?"

"The first," said some students.

"The first or the second?" I insisted.

"The second," I heard someone say.

"Professor, this is almost impossible," said Carla, referring to the second algorithm.

"Of course it's the second," said Fernando, contradicting himself.

"The second? ... This one is faster?" I asked Fernando, for having changed his mind.

"But at least it's safer," Eduardo said, not knowing which one he was referring to.

"The first! ... The first one is much faster!" defended Carla, turning to her colleagues who were against her.

"You have freedom ... choose whatever you want!" I said, closing the discussion.

Formative Situation, Unexpected and Successful

Since students attended the electrical engineering course, and in order to include linear algebra applications involving real situations and students' daily lives (Carlson et al., 1993), the teacher prepared a task called "Wheatstone Bridge." This presupposes an application of systems of linear equations to electrical circuits by using Kirchhoff's laws. Here is an excerpt from an MN (Multimodal Narrative Excerpt 12) where the teacher describes this task and the didactic intentions underlying it:

Multimodal Narrative Excerpt 12: Teacher R (Introduction to Solving Systems of Linear Equations)

This task has three subtasks. It was designed to work as an approximation to new concepts in concrete situations and in the students' formation context. The first subtask is to be performed in the present class and presupposes the writing of associated linear equations of Kirchhoff's laws in an electric circuit, called the "Wheatstone Bridge." The second and third subtasks, to be carried out in subsequent classes, will involve the writing of the linear system in matrix language and the resolution of it using the Gauss elimination method, respectively.

The teacher introduced the task and, in particular, the first subtask in the class of the mode reported in the MN (Multimodal Narrative Excerpt 13):

Multimodal Narrative Excerpt 13: Teacher R (Introduction to Solving Systems of Linear Equations)

I began by referring to a task that students would have to develop, in about five minutes, as an approximation to the introduction of the Gauss elimination method. I showed the task and asked the students about the circuit presented:

"On a circuit that has a particular designation, that's Wheatstone Bridge, have you heard?"

"Wheatstone Bridge? Yes," said Fernando.

"Is the name familiar to you?"

"No!" answered Juliana.

"It has four resistors," continued Fernando, and Eduardo, who was beginning to remember, corroborated.

"But in this case we have five resistors!" I clarified.

As the image of the projected circuit was somewhat blurred, I enlarged the image and continued:

"In this case, do we have how many resistors? Five resistors and one ... one ..." [...]

"A power supply," Eduardo said.

"A power supply," I repeated.

Given the silence of the remaining students, I deduced that the knowledge about the circuit was unfamiliar.

The reading of the MN indicates that the teacher's prediction that the context of the task was within the students' knowledge would not have been very correct. Despite the great applicability of linear algebra in several STEM courses, the use of these applications in the teaching of these courses is difficult because linear algebra is usually taught in the first year and the first semester, and given the students' lack of training in the specific area of the course at that time.

The MN analysis shows that only two students seemed to know the electric circuit presented (Multimodal Narrative Excerpt 13), which the teacher did not consider relevant for the follow-up of the lesson plan. However, knowing Kirchhoff's laws was fundamental, so he explained them through an example (Multimodal Narrative Excerpt 14):

Multimodal Narrative Excerpt 14: Teacher R (Introduction to Solving Systems of Linear Equations)

As this aspect was irrelevant to the development of the task, I decided to circumvent the situation and speak in terms of the application of Kirchhoff's laws in an electric circuit. To ensure that the students knew and correctly applied Kirchhoff's laws, I showed an image of an example situation.

This adjustment also implied the option for a more interventional action (Multimodal Narrative Excerpt 15):

Multimodal Narrative Excerpt 15: Teacher R (Introduction to Solving Systems of Linear Equations)

My intention was to give full autonomy, corroborated by the initial revision of Kirchhoff's laws in one example. However, the students soon began to ask questions about what was really being asked: "But is it to do for all nodes?"; "Professor, what do you want to do with this?" Looking at the groups, I became aware of some expressions of hesitation, anxiety and discouragement. I interpreted the situation as gaps in understanding about what was actually requested. I then decided to follow the groups' work more closely, to clarify the statement and to safeguard progress in the resolution of the subtask.

The reading of episode 2 of the MN, corresponding to this situation and transcribed below (Multimodal Narrative Excerpt 16), shows more detail of what happened in that part of the lesson.

Multimodal Narrative Excerpt 16: Teacher R (Introduction to Solving Systems of Linear Equations)

Pointing to the image at the left, I related it to the application of the Kirchhoff's current law. Then, concentrating on the image at the right, I tried to identify the signal of the current and to advance with the equation of the circuit. I then stated that this was the scope of the subtask: to write the equations that they identified in the circuit designated as Wheatstone Bridge.

This is an example where the teacher's reflection about his decision to change strategy is essential. On the one hand, this decision was the result of an impression during the lesson, and, on the other, it did not allow the full achievement of the objectives established for this task, since it was planned to be carried out autonomously by the students and ended up having the teacher's direct collaboration. In spite of this, the later analysis of the MN makes it possible to verify that, unlike the situation reported in the previous subsection, in this case, the modifications made by the teacher due to the emergence of unexpected situations did not jeopardize the students' understanding of the mathematical concept under study.

Nevertheless, through MN it was found that the time spent (about 19 minutes) was higher than the teacher had initially estimated (about 5 minutes) to address the introductory part of this task.

Formative Situation, Unexpected and Less Achieved

Gauss and Gauss-Jordan elimination methods were presented in the same class, one after the other, as two possibilities for solving systems of linear equations. The teacher said that the choice of one of them should be careful. He advised Gauss's elimination method when the system of linear equations was to be solved by hand and the Gauss-Jordan method when the students used Scilab, since it is the method underlying the solution given by this computer algebra system (CAS). After this explanation, the teacher often used the Gauss elimination method for solving systems of linear equations, as evidenced by the following excerpt of MN (Multimodal Narrative Excerpt 17), not giving the same importance to the Gauss-Jordan elimination method.

Multimodal Narrative Excerpt 17: Teacher R (Gauss and Gauss-Jordan Elimination Methods)

Reformulating the speech, I asked now:

"What do you need to solve systems of linear equations?"

"Gauss ... Gauss's method." I listened.

The teacher used the Gauss-Jordan elimination method mainly as a procedure for the determination of the inverse of an invertible matrix, which happened in a later class.

The teacher introduced the procedure for calculating the inverse of an invertible matrix starting from the definition of inverse matrix and doing the resolution of several systems of linear equations simultaneously (Williams, 2001), using the Gauss-Jordan elimination method (Multimodal Narrative Excerpt 18):

Multimodal Narrative Excerpt 18: Teacher R (Gauss and Gauss-Jordan Elimination Methods)

[...] I started by remembering the procedure to find the inverse of an invertible matrix, according to the Gauss-Jordan elimination method introduced in the last lecture.

The senior teacher suggested this strategy and, at the time, it did not raise difficulties for the students. However, the analysis of the MN, which allowed comparing two events from different classes, made it possible to perceive in more depth something that some students had misunderstood.

The following excerpt (Multimodal Narrative Excerpt 19) explains the confusion between the Gauss-Jordan elimination method for solving a system of linear equations and the procedure for calculating the inverse of an invertible matrix:

Multimodal Narrative Excerpt 19: Teacher R (Gauss and Gauss-Jordan Elimination Methods)

When I added an arrow between the systems of linear equations and Gauss to the scheme and the symbol [A|B], I heard some "ahs" of surprise. [...]

Here Eduardo posed a surprising question, which made me see how far the students' confusion can go, about the concepts and procedures I taught:

"Why do not you put A bar I?"

After realizing that the student intended to say [A|In], I told him that this had to do with the most recent procedure they learned for computing the inverse of an invertible matrix, related to the Gauss-Jordan elimination method.

An explanation for the identified confusion may be that the mathematical representation used for the augmented matrix of a system of linear equations, to which Gauss-Jordan elimination method is applied, [A|B], is similar to the representation of the matrix to apply the algorithm for computing the inverse matrix, [A|In]. However, this first justification did not convince the authors and led them to a new reading and a deeper analysis of the MN. This led them to question the most appropriate time to introduce each of the methods of solving systems of linear equations: simultaneously (as was done) or at different moments, giving time for the consolidation of each method, and only then to apply the Gauss-Jordan elimination method to find the inverse of an invertible matrix.

However, in the specific case of the use of Gauss and Gauss-Jordan elimination methods for solving systems of linear equations, it was found that the students could develop a critical attitude regarding the most appropriate method to consider when solving systems of linear equations, as exemplified by the following excerpts from MNs (Multimodal Narrative Excerpts 20 and 21):

Multimodal Narrative Excerpt 20: Teacher R (Gauss and Gauss-Jordan Elimination Methods)

When Fernando asked me if he had to put it in the "echelon form and reduced by rows," I warned that it did not exist. It was just one form, the echelon form, or the other, row reduced.

"What do you think you need?" I asked.

"Row reduced."

"To solve a system? Look at our schema, over there," I said, pointing to the scheme built on the board, which I had not erased yet.

"It's also on the step."

"Ready! Enough!" I said.

"But if I put it in the row reduced echelon form I would know what was the x ..."

Multimodal Narrative Excerpt 21: Teacher R (Gauss and Gauss-Jordan elimination methods)

I suggested that Manuel consider how to solve a system of linear equations. I had noticed that the student had used the Gauss elimination method in the first question by writing the augmented matrix in echelon form. As the system had two equations, but only one unknown, I hinted that there was a faster option.

Transversal Aspects of the Teaching Sequence

Teaching linear algebra through the use of technology is a didactic recommendation long known in the literature. The teaching sequence in Gonçalves (2018) included this as a transversal aspect, in particular using CAS and dynamic geometry software (DGS). Another aspect referred to in the literature is the geometric approach. This is considered relevant, but without constituting itself as a central aspect or leading students to become "trapped" in the geometric world. In the teaching sequence, this language was adopted (with students developing the geometric intuition of concepts, such as vector subspace or linear combination of vectors), and students have been able to relate it to algebraic language.

On the Use of Technology

The MNs showed that the teacher used the technology with different functions, namely: to discover new concepts, to clarify theoretical aspects (here, to clarify means to consolidate or exemplify), to calculate and to anticipate. Discovering and clarifying concepts with the help of CAS were functions previously suggested by researchers (Carlson et al., 1993; Hillel, 2001), to which the teacher also associated the DGS to explore, where possible, the visual nature of concepts. These functions refer, respectively, to moments before and after the concepts are formalized. The categories of calculations and the anticipation of some concepts emerged from the data, although the teacher anticipated that the exploration of more significant examples (Harel, 1997) would naturally imply calculations that are more complex.

For instance, and relating with the example given for a formative situation that was planned and successful, the teacher used GeoGebra and Mathematica software as DGS to introduce the concept of the linear combination of vectors. In GeoGebra, he built a figure with vectors with two coordinates, representing the coefficients of the unknowns and the constant terms, and based on that figure he explained the following, as registered in a MN (Multimodal Narrative Excerpt 22):

Multimodal Narrative Excerpt 22: Teacher R (Connections With the Concept of Linear Combination)

"Are you seeing, for now, what is the format of a linear combination? It is to have a vector that can be obtained at the expense of others, provided that ... what? If you multiply the others by any scales, which can be determined to give that vector," I concluded.

In the example presented as a formative situation that was planned and less achieved, the teacher proposed to students to solve an initial task for matrix multiplication by using Scilab. From the result obtained in Scilab of the multiplication of two matrices, the teacher expected that students would deduce the row-column multiplication method, based on the concept of dot product of vectors (Multimodal Narrative Excerpt 23).

Multimodal Narrative Excerpt 23: Teacher R (Matrix Multiplication)

"This is what we know ... Scilab did the multiplication and gave this," I said, pointing to the result of the multiplication shown on the screen.

In relation to formative situations that were unexpected, successful and less achieved, both involved systems of linear equations, which was a topic that involved great use of technology (GeoGebra, Mathematica and Scilab) in their different functions, shown in the following excerpts from MNs (Multimodal Narrative Excerpts 24, 25 and 26).

Multimodal Narrative Excerpt 24: Teacher R (Introduction to Solving Systems of Linear Equations)

I showed the [...] geometric representation in the context of an example of the supporting text:

I began by asking the number of equations:

"Five," the students replied.

"Believe it!" Juliana reacted immediately. [...]

I then circled the figure in Mathematica, and the students said that the planes intersected one line and that the system of linear equations would be consistent and with many solutions. [...]

After saying the command for solving the system of linear equations (A\B), the students were saying the solutions obtained. [...] I established that in these cases, we were only finding particular solutions [...].

Figure 1.

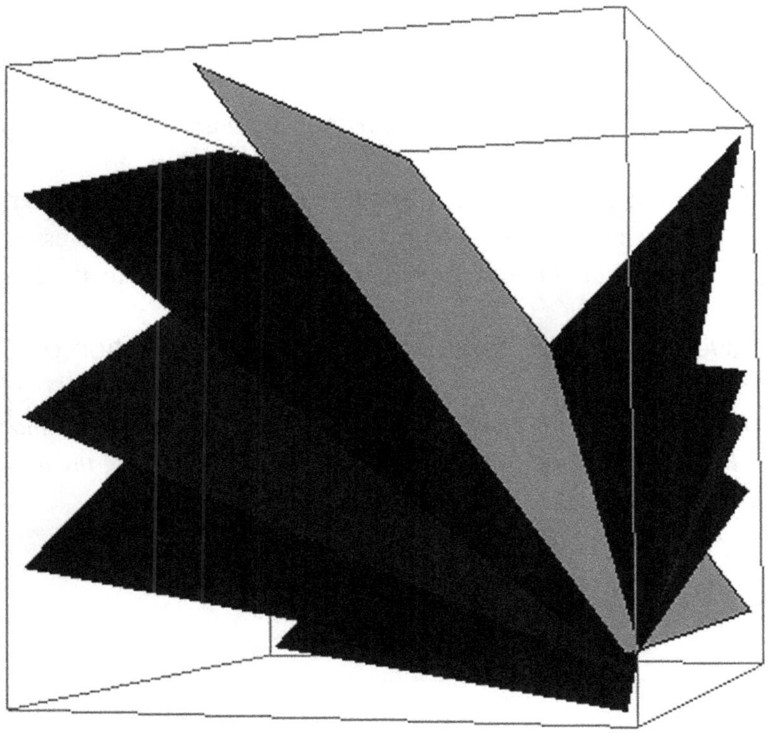

Multimodal Narrative Excerpt 25: Teacher R (Gauss and Gauss-Jordan Elimination Methods)

Returning to Scilab, I presented the final result, highlighting the command used, rref, and I mentioned that the row reduced echelon form found was unique.

Multimodal Narrative Excerpt 26: Teacher R (Gauss and Gauss-Jordan Elimination Methods)

[...] When Juliana asked me if they had to solve the system of linear equations by hand, I said I would use Scilab to be faster.

The use of technology also facilitated the achievement of the geometric approach to certain concepts, as mentioned in the next subsection.

On the Geometric Approach

The analysis of MNs made it possible to verify that the geometric approach worked as a complement and never as a central element of the teaching practices. The teacher associated the DGS with exploring, where possible, the visual nature of concepts, with positive reactions from students (Multimodal Narrative Excerpt 27):

Multimodal Narrative Excerpt 27: Teacher R (Subspace of a Vector Space)

I began by showing the geometric representation of the set.

"Go through zero, over there!" said Hélder immediately, after I pointed out the origin of the referential.

Still in relation to the example presented in the first formative situation, it was verified that the interpretation of the concept presented by the geometric approach was one of the essential aspects for the students conceptualize the concept of linear combination of vectors (Multimodal Narrative Excerpt 28).

Multimodal Narrative Excerpt 28: Teacher R (Linear Combination of Vectors)

"It is impossible with operations with these vectors to find who? w! Or, put it more formally, it is ... w is not written as a linear combination of v_1*,* v_2 *and* v_3*," I said, presenting the answer to the question under consideration. [...]*

Figure 2.

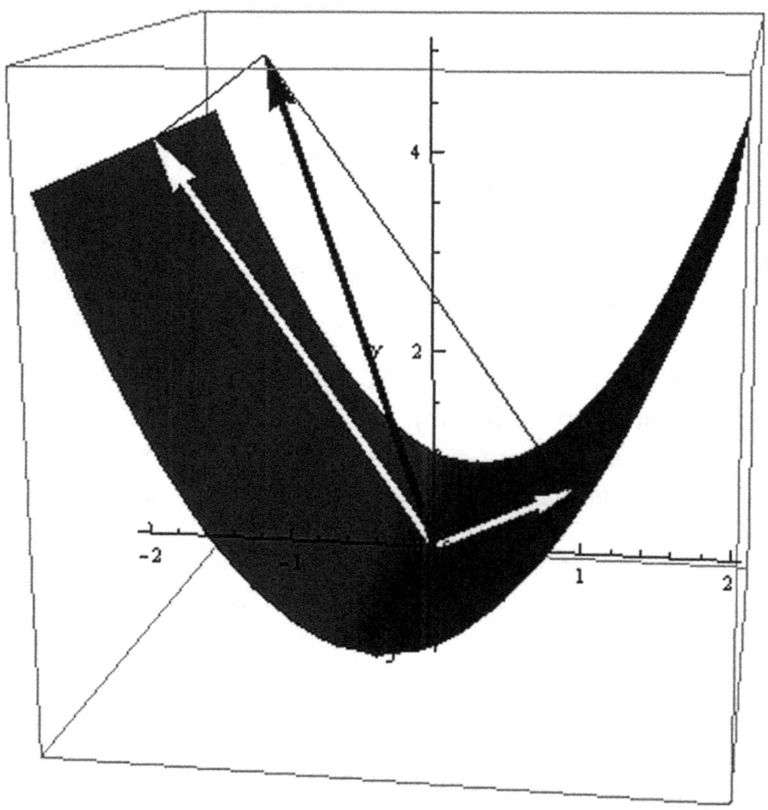

As for the geometric interpretation, we established that the vector w would not be in the same plane as the others, showing it in Mathematica. By rotating the figure conveniently, we found that the vector w was almost perpendicular to the other vectors.

Also in the case of solving and classifying systems of linear equations, the teacher used the geometric approach (Multimodal Narrative Excerpt 29):

Multimodal Narrative Excerpt 29: Teacher R (Introduction to Solving Systems of Linear Equations)

Analogously, I extended the discussion of systems of linear equations with three equations and three unknowns to the relative position of planes in space. In this context, I used intuitive geometric designs, using different representations of plans, the walls of the room, the floor, the ceiling, the tabletops and sheets of paper.

Figure 3.

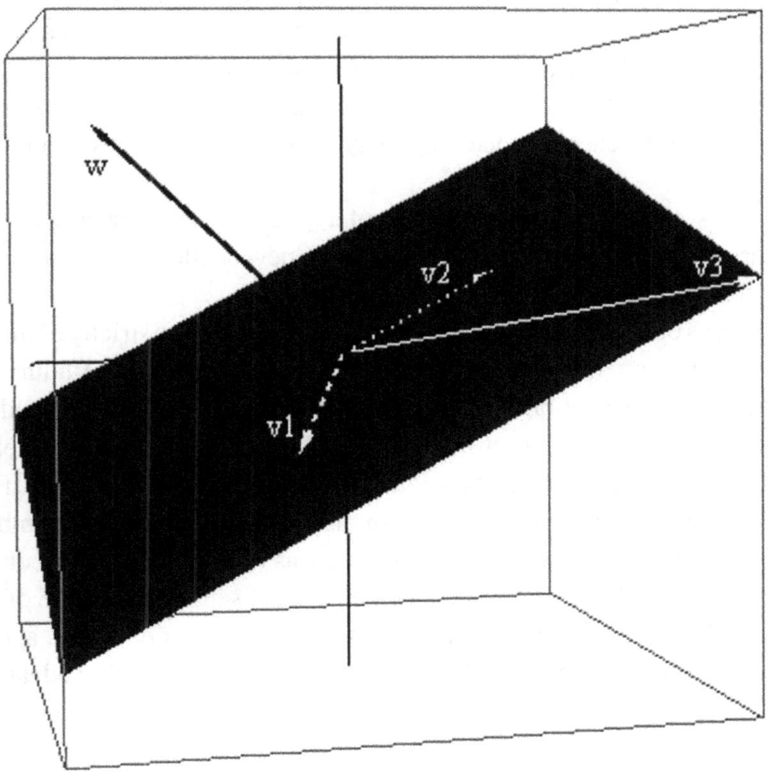

CONCLUSION

This study confirmed that MNs allow the teacher and the researcher a systematic reflection based on concrete and recorded facts (Lopes et al., 2014). It also makes it possible to go beyond the teacher's memory, in the sense that these registers allow teachers and researchers approach events that occurred in different classes and whose analysis can reveal relationships between them. This is what happened in the example presented to illustrate an unexpected and less achieved formative situation.

The reflection presented about the teacher's experience in the construction of MNs is in line with the one advocated by Alarcão (2001) and Ponte (2002). The closer the teacher's reflection on his or her own practice is to the research reflection, the better it is.

Although it was anticipated that the use of MNs would promote the professional development of the teacher teaching linear algebra for the first time, it was verified that the senior teacher also became enriched professionally due to the analysis of MNs and collaborative discussions with a colleague. This confirms the experience described by Luke and Rogers (2015) and that established by Lopes and Cunha (2017) on the relevance of articulating the processes of collaborative work with researchers and peers. In the present case, both teachers had the roles of linear algebra teachers and researchers.

This study showed the MN's potential as a tool to "observe" and analyze linear algebra teaching practices. The MNs were adequate to the content taught and allowed the refinement of analyses, which is a benefit for the study of mathematical teaching practices. By bringing together multimodal registers and approaching the complexity of a lesson, teachers and researchers alike can analyze problems from a variety of perspectives, which together make for deeper answers.

The examples presented for planned or unexpected formative situations that were more or less achieved serve to alert linear algebra teachers to the complexity of teaching practices and their implications for professional development.

For example, task resolution and the exploration of applications associated with linear algebra concepts create new learning opportunities, but they also pose new challenges to the teacher, as in the case of the task "Wheatstone Bridge."

Although several ways of interconnecting the concepts, not only those strictly of linear algebra but also those related to prerequisites, are well studied in the literature, more difficult is finding the most proficient forms (as was found in the connection between the Gauss elimination method and the Gauss-Jordan elimination method, and the connection of matrix multiplication to the prerequisites). In addition, the moment chosen for the introduction of a certain concept requires the teacher's scientific and pedagogical reflection for learning to be achieved, as happened in the example of the linear combination of vectors.

In the examples of less achieved formative situations, it was verified that the introduction of alternative procedures translated into the development of students' critical sense. In the case of matrix multiplication, it led them to establish the row-by-column algorithm as the most effective, and in the case of the Gauss and Gauss-Jordan elimination methods, they decided when it was more advantageous to use one or the other.

The teacher's use of different software was relevant, and it was justified by helping in the interpretation and understanding of processes (numerical or visual exploration) and not in obtaining products (only performing calculations). The geometric approach allowed students to relate different languages and representations of the concepts.

Hence, the work of reflection by teachers who opt for didactic innovations, in collaboration with the work of researchers in this field, is essential. The knowledge that researchers construct from MNs can be returned to the teachers, in collaborative cycles and with results in two aspects: in the production of studies on teaching practices and in teachers' professional development.

REFERENCES

Alarcão, I. (2001). Professor-investigador: Que sentido? Que função? [Teacher-researcher: What sense? What function?]. In B. P. Campos (Ed.), *Formação profissional de professores no ensino superior* [Professional training of teachers in higher education] (Vol. 1, pp. 21–31). Oporto, Portugal: Porto Editora.

Bardin, L. (1997). *L'analyse de contenu* [Content analysis]. Paris: Presses Universitaires de France.

Carlson, D. (1993). Teaching linear algebra: Must the fog always roll in? *The College Mathematics Journal*, *24*(1), 29–40. doi:10.1080/07468342.1993.11973503

Carlson, D., Johnson, C. R., Lay, D. C., & Porter, A. D. (1993). The Linear Algebra Curriculum Study Group recommendations for the first course in linear algebra. *The College Mathematics Journal*, *24*(1), 41–46. doi:10.1080/07468342.1993.11973504

Cochran-Smith, M., & Lytle, S. L. (1993). *Inside-outside: Teacher research and knowledge*. New York, NY: Teacher's College Press.

Dorier, J.-L. (1998). The role of formalism in the teaching of the theory of vector spaces. *Linear Algebra and Its Applications*, *275/276*, 141–160. doi:10.1016/S0024-3795(97)10061-1

Dorier, J.-L. (2000). Epistemological analysis of the genesis of the theory of vector spaces. In J.-L. Dorier (Ed.), *On the teaching of linear algebra* (pp. 191–207). Dordrecht: Kluwer Academic.

Dorier, J.-L., Robert, A., Robinet, J., & Rogalski, M. (2000). The obstacle of formalism in linear algebra. In J.-L. Dorier (Ed.), *On the teaching of linear algebra* (pp. 85–124). Dordrecht: Kluwer Academic.

Dorier, J.-L., & Sierpinska, A. (2001). Research into the teaching and learning of linear algebra. In D. Holton (Ed.), *The teaching and learning of mathematics at university level: an ICMI study* (pp. 255–273). Dordrecht: Kluwer Academic.

Dubinsky, E. (1997). Some thoughts on a first course in linear algebra at the college level. In D. Carlson, C. R. Johnson, D. Lay, A. D. Porter, A. E. Watkins, & W. Watkins (Eds.), *Resources for teaching linear algebra* (Vol. 42, pp. 85–105). Washington, DC: Mathematical Association of America.

Gonçalves, R. (2018). *A investigação em didática da álgebra linear: Uma perspetiva do seu contributo para o ensino por meio da reflexão sobre a prática* [Research on the teaching and learning of linear algebra: A perspective of its contribution to teaching through reflection on practice] (Unpublished doctoral dissertation). UTAD, Portugal.

Harel, G. (1997). The Linear Algebra Curriculum Study Group recommendations: moving beyond concept definition. In D. Carlson, C. R. Johnson, D. Lay, A. D. Porter, A. E. Watkins, & W. Watkins (Eds.), *Resources for teaching linear algebra* (Vol. 42, pp. 107–126). Washington, DC: Mathematical Association of America.

Hillel, J. (2000). Modes of description and the problem of representation in linear algebra. In J.-L. Dorier (Ed.), *On the teaching of linear algebra* (pp. 191–207). Dordrecht: Kluwer Academic.

Hillel, J. (2001). Computer Algebra Systems in the learning and teaching of linear algebra: Some examples. In D. Holton (Ed.), *The teaching and learning of mathematics at university level: an ICMI Study* (Vol. 7, pp. 371–380). Dordrecht: Kluwer Academic. doi:10.1007/0-306-47231-7_33

Kleiner, I. (2007). *A history of abstract algebra*. Boston, MA: Birkhäuser. doi:10.1007/978-0-8176-4685-1

Lapp, D. A., Nyman, M. A., & Berry, J. S. (2010). Student connections of linear algebra concepts: An analysis of concept maps. *International Journal of Mathematical Education in Science and Technology, 41*(1), 1–18. doi:10.1080/00207390903236665

Lay, D. C. (2012). *Linear algebra and its applications* (4th ed.). Boston, MA: Addison Wesley.

Lopes, J. B., & Cunha, A. E. (2017). Self-directed professional development to improve effective teaching: Key points for a model. *Teaching and Teacher Education, 68*, 262–274. doi:10.1016/j.tate.2017.09.009

Lopes, J. B., Silva, A. A., Cravino, J. P., Santos, C. A., Cunha, A., Pinto, A., ... Branco, M. J. (2014). Constructing and using multimodal narratives to research in science education: Contributions based on practical classroom. *Research in Science Education, 44*(3), 415–438. doi:10.100711165-013-9381-y

Luke, N., & Rogers, C. (2015). Responding to uncertainty: Teacher educator professional development through co-teaching and collaborative reflection. *Learning Landscapes, 8*, 245–259.

Ponte, J. P. (2002). Investigar a nossa própria prática [Investigate our own practice]. In Grupo de Trabalho sobre Investigação (Ed.), Reflectir e investigar sobre a prática profissional [Reflect and investigate on professional practice] (pp. 5-28). Lisbon: Associação de Professores de Matemática.

Speer, N., Smith, J. III, & Horvath, A. (2010). Collegiate mathematics teaching: An unexamined practice. *The Journal of Mathematical Behavior, 29*(2), 99–114. doi:10.1016/j.jmathb.2010.02.001

Trigueros, M., & Possani, E. (2013). Using an economics model for teaching linear algebra. *Linear Algebra and Its Applications, 438*(4), 1779–1792. doi:10.1016/j.laa.2011.04.009

Tucker, A. (1993). The growing importance of linear algebra in undergraduate mathematics. *The College Mathematics Journal, 24*(1), 3–9. doi:10.1080/07468342.1993.11973500

Uhlig, F. (2003). A new unified, balanced and conceptual approach to teaching linear algebra. *Linear Algebra and Its Applications, 361*, 147–159. doi:10.1016/S0024-3795(02)00319-1

Williams, G. (2001). *Linear algebra with applications* (4th ed.). Sudbury, MA: Jones and Bartlett.

Section 3

Professional Development of Science and Technology Teachers Using Multimodal Narratives

In this section of the book, the use of multimodal narrative (MN) in the practice of Science and Technology teachers is contemplated including a view to their professional development. Throughout the chapters, situations in which action and object centered reflection potentiates professional knowledge and sharing of experiences are presented. Despite these cases being referent to specific themes, since MNs focus on classroom actions and practices, they may prove to be rich in strategies and approaches. They are thus useful in any context, particularly in Science and Technology teaching. The reading of this section can be done from two perspectives: first, from the perspective of teachers interested in performing MN of their classes as a way to better understand their practice to reflect on their work or even to use them for didactic purposes; second, by the reading of MN performed by other teachers (even in different realities) to identify strengths and weaknesses of classroom practices and to better understand the outcome of certain strategies. Regardless of the perspective, it will be possible to use the MN to enrich the practice of teaching.

Chapter 7
Using Multimodal Narratives in Professional Development to Support Collaborative Work:
Improving Teaching Practices to Enhance Learning

J. Bernardino Lopes
Universidade de Trás-os-Montes e Alto Douro, Portugal

ABSTRACT

In this chapter, the authors discuss professional development (PD), self- and hetero-directed based on cases of physics education teachers in secondary and higher education and the role that multimodal narratives have in both processes. They adopt the PD perspective, which assumes that PD should be evaluated by its impact on teaching practices improvement. In this context, the PD aims to improve, incrementally, teaching practices. This process is difficult and complex and involves both teachers themselves and other professionals. Multimodal narratives (MN) are a tool that makes it possible to make teaching practices public and thus open to analysis by the teacher, other professionals, and researchers. The reported PD uses MNs as a tool that supports the teacher's collaborative work with peers and researchers. This collaborative work results in important incremental improvements in teaching practices.

INTRODUCTION

This chapter focuses on professional development (PD) centered on improving the quality of teaching practices with impact on student learning in line with the current definition of PD (Soebari & Aldridge, 2015). The PD of teachers involves multiple dimensions, one of which is the improvement of the quality of teaching practices and students' achievement (Desimone, 2009; Fischer et al., 2018). There may, however, be PD without there being improvement of teaching practices in classroom lessons (Bakkenes, Vermunt & Wubbels, 2010; Vescio, Ross, & Adams, 2008). The most recent conceptualization of PD

DOI: 10.4018/978-1-5225-8570-1.ch007

emphasizes that improving teaching practices in classroom lessons with effects on the quality of student learning is an essential component of the definition of PD (Desimone & Garet, 2015; Wei, Darling-Hammond, Andree, Richardson, & Orphanos, 2009). It is necessary, therefore, that teaching practice in the classroom be central to the PD itself (Vescio, Ross, & Adams, 2008; Wei et al., 2009). As teaching practices are difficult to change (Bakkenes et al., 2010; Lopes & Cunha, 2017; Morris & Hiebert, 2011; Osborne, 2014), it is necessary to consider PD as incremental: a PD that allows successive iterations of the teaching practice and verifying in what conditions a certain aspect of teaching practices works and is effective to students' learning. Thus, the PD, centered on the improvement of teaching practices, must be based on successive cycles of action research, each of them (Lopes & Cunha, 2017) with: (a) design of a proposal for teaching based on intentions, explicit didactics, past experience and research results; (b) implementation of the teaching proposal and data collection; and (c) analysis of data collected to determine the effectiveness of teaching practices and seek new lines of action in teaching practices or refinements of them.

The central idea of this chapter is to show how a tool – Multimodal Narratives (Lopes et al., 2014) – that helps to collect, organize and aggregate data about classroom teaching practices can be used in PD (self- or hetero-directed) centered on teaching practices. The PD is based on collaborative processes of a teacher with peers or with researchers (Avalos, 2011). In these processes, Multimodal Narratives (MN) of lessons allow that: a) teachers collaborate with each other using the MN as a tool to identify objectively what promotes or inhibits students' learning; and b) researchers study the dynamics and processes of teaching and point out new directions. Thus, the main objectives of this chapter are:

Objective One: Conceptualize PD articulating teaching practices and research evidences.

Objective Two: Show how PD centered on teaching practices can incrementally improve the quality of teaching practices and students' learning.

Objective Three: Present the different roles that MN can play in promoting PD focused on teaching practices.

A FRAMEWORK FOR PROFESSIONAL DEVELOPMENT ARTICULATING TEACHING PRACTICES AND RESEARCH EVIDENCES

Teaching practice in classroom lessons is an essential component of a teacher's professional practice in any curricular area. The teaching practice as a professional practice has a tacit dimension that articulates with the explicit dimension in a way that is different according to the theoretical perspective of the authors (Baird, 2004; Cook & Brown, 1999; Nonaka & Takeuchi, 1995; Winch, Oancea, & Orchard, 2015). Understanding the nature of teaching practice is important because only then is it possible to conceive and frame the theoretical and practical problems that its improvement involves.

Cook and Brown (1999) and Nonaka and Takeuchi (1995) conceive professional knowledge as a combination of four areas of professional development. Two are epistemological in nature: (a) explicit individual knowledge (which refers to the conceptual knowledge appropriated by individuals) and (b) tacit individual knowledge (which refers to the know-how an individual reveals only in situations and actions). The remaining two are ontological in nature: (a) explicit knowledge of the group (which refers to shared knowledge among individuals in a group or institution and manifests itself in the shared culture and stories of the group) and (b) tacit knowledge of the group (which refers to knowledge tacitly shared between individuals, revealed in the modus operandi of the group in situations and in its organization).

This conception of professional knowledge, which finds echoes in other authors (e.g., Lustick, 2009; McIntyre, 2005; Tardif, 2000), shows that professional knowledge has a different nature from the knowledge produced by research. Thus, it is understood why research in Science Education does not have a direct impact on the teaching practices of science teachers due to theoretical and practical problems related to the nature of professional knowledge (Winch, Oancea, & Orchard, 2015).

Research into science teaching practices studies how teachers really act and how they transform conceptual knowledge into know-how (and vice versa) and identifies the actions and discourse interactions that can be part of each teaching practices pattern (with characteristics of good or less good practices). It also explores what characteristics are likely to improve, in a sustainable way, the quality of teaching practices and students' learning (Lopes, 2017).

Why Is It Important to Articulate Research and the Quality of Teaching Practices?

Given the nature of the teacher's professional knowledge, the articulation between research evidences and teaching practices pose several practical and theoretical problems (McIntyre, 2005; Winch, Oancea, & Orchard, 2015). Even so, some authors argue that research evidences should be massively used in science teaching practices (e.g., Bradforth et al., 2015). However, to add to the theoretical studies referred to above presenting difficulties in the massive use of research evidences, there are empirical studies that show the difficulty of appropriating research results that influence change in teaching practices (Vescio, Ross, & Adams, 2008) or even in the resources that professional communities provide (Jagger & Yore, 2012). Despite this, Bradforth et al. (2015) are right when they say that there are consolidated results and so there is no good reason not to incorporate them into teaching practices. However, it is necessary to have a more complete notion of teachers' professional knowledge, which allows a constructive relationship between teachers and research knowledge (Winch, Oancea, & Orchard, 2015). In particular, it is necessary to be aware that the research can contribute only to teachers' technical knowledge (Winch, Oancea, & Orchard, 2015), and professional knowledge has other components and needs (Lopes, 2017; Masuda, Ebersole, & Barrett, 2013). Therefore, how to perform PD with impact on teaching practices remains a challenge because there are several constraints (Fischer et al., 2018; Lopes, 2017; Lopes & Cunha, 2017).

On the one hand, improving teaching practices consistently during the teacher's career needs considerable effort and favorable circumstances (Bakkenes et al., 2010; Fischer et al., 2018; Morris & Hiebert, 2011; Stigler & Hiebert, 2009; Hiebert & Morris, 2012; Richter, Kunter, Klusmann, Lüdtke, & Baumert, 2014). On the other hand, there are evidences showing that in order to change teaching practices, it is not enough to change teachers' attitudes, beliefs and perceptions (Guskey, 2002). The current PD models focused on teaching practices can: (a) use little evidence of research perspective (Desimone & Garet, 2015; Fischer et al., 2018) and (b) point only to evidence-based teaching practices perspectives (Bradforth et al., 2015). To face the weaknesses of these two current PD models, a new perspective on PD is necessary, centered in teaching practices and articulating the teaching practices from the perspective of the teacher's knowledge, with research evidences of teaching practices to improve its quality and impact on students' learning (Lopes, 2017; Lopes & Cunha, 2017). This new perspective on PD implies improving the teaching practices of a given teacher throughout his or her career (Fischer et al., 2018; Lopes, 2017; Lopes & Cunha, 2017).

Several authors have drawn attention to the need for effective articulation between teaching practice and research (Korthagen, Loughran, & Russell, 2006; Lustick, 2009), although Avalos (2011) has shown, in a meta-analysis study, that professional development (PD) has not been based on the articulation of these two strands. Even the PD model centered on teaching practices proposed by L. Desimone (Desimone, 2009; Desimone & Garet, 2015), and widely accepted, does not incorporate research to improve teaching practices.

It is in this context of few alternatives to make possible that articulation that Lopes (2017) proposed another kind of approach to articulate research evidences and teaching practices, having as background the collaboration between teachers and researchers in the scope of the PD of teachers. This proposal is in line with empirical studies that show that teaching practices can be at the center of PD, both in its practical and investigative dimensions (Lopes & Cunha, 2017; Vanassche & Kelchtermans, 2016). It is also in agreement with the recent reconceptualization of PD that affirms that the improvement of teaching practices is one of the key aspects of its definition (Desimone, 2009; Desimone & Garet, 2015; Soebari & Aldridge, 2015; Wei et al., 2009).

As a consequence, teacher training and PD must be reconceptualized. Lopes and Cunha (2017) propose a model of incremental PD throughout the career of a teacher that integrates the knowledge of the community of teachers with the knowledge of the community of researchers, by using collaboration with both communities and the use of different tools, in particular Multimodal Narratives (MN).

The following sections we will develop two central ideas based on the research of our research group (Branco, 2018; Cunha, 2015; Lopes & Cunha, 2017; Lopes, Viegas, & Cravino, 2010) that are crucial for PD focused on teaching practices in the classroom:

- Teaching practices can be the central context of effective PD, making it possible to articulate the knowledge of professional practice with research evidences; and
- PD through teaching practices can be done only if appropriate tools are used to that effect to help teachers to stop referring only to the perception of their own teaching practices.

Key Concepts

In the articulation between research and teaching practices, the research needs to consider two central aspects: (a) helping teachers to develop a research focus on teaching practices (Hopkins, 2014) and (b) recognizing the epistemological status of teaching practices to understand the nature of this knowledge as well as the role they play both in the teaching work process and in the professional identity of teachers (Tardif, 2000).

The characteristics of teachers' teaching practices depend on how teachers integrate several types of knowledge in their professional practice (Shulman, 1986; Jones & Cowie, 2011; Fischler, 2011). For example, the experience of participating, as researcher, in science education research projects allows the teacher to be attentive to some important research results in the teacher's mediation actions and creates willingness to explore ways to integrate them into their professional practice (Lopes, 2017). This integration is difficult and better done if it is in the context of PD (Corrigan, Dillon, & Gunstone, 2011; Fischler, 2011; Jones & Cowie, 2011; Luft & Hewson, 2014). Thus, professional development is the scope in which evidences from research and the professional knowledge of teachers can be articulated to build new knowing or refine the existing knowing (Lopes, 2017).

To address the question of the articulation between research and teaching practices, it is necessary to remember that PD can be triggered by external pressure (directed by PD programs) (Fischer et al., 2018) and, in this case, their effects can end when the program has been completed (Desimone, Garet, Birman, Porter, & Yoon, 2003). This kind of PD is here called hetero-directed PD. PD can also be triggered by the teacher's own decision and will. This is called in the literature self-directed PD. Self-directed PD is defined as the professional development arising from a teacher's own initiative; i.e., the PD process is determined and initiated by the teacher (Van Eekelen, Vermunt, & Boshuizen, 2006). Hence, self-directed PD can provide an internal locus of control that ensures that teachers continue to build, through their career, new understanding of teaching and learning (Mushayikwa & Lubben, 2009).

Avalos (2011) found, in the above-referenced meta-analysis study, that PD needs two main learning processes to stimulate teachers to alter or reinforce teaching practices: collaborative work with researchers and collaborative work with peers. These collaborative processes can be structured or semi-structured (such as partnerships, collaborative networks) or informal contexts (such as workplace interactions).

The first collaborative process (with researchers) can take place in the scope of school-university or teacher-researcher partnerships (Avalos, 2011) and aims at finding evidences about teaching practices in order to implement evidence-based practices (Burbank & Kauchak, 2003) or creating conditions for joint work and joint contributions (Avalos, 2011). The second collaborative process (with peers) can take place in the scope of teacher networks and teams, communities of practice and communities of learning (Avalos, 2011); it concerns how teachers work together and share practices to generate reflection and teaching proposals (Luke & Rogers, 2015; Vescio, Ross, & Adams, 2008).

Professional Development Articulating Teaching Practices and Research Evidences

This section, based on Lopes (2017), proposes a model of articulation between teaching practices and research evidences to enhance students' learning using PD as scope in which occur transformations of professional knowledge. First, it is necessary to remember that articulating research and teaching practices is recognized as necessary (McIntyre, 2005). It is a path attempted by several research teams (Luft & Hewson, 2014; Luke & Rogers, 2015), but it remains a challenge to put into practice with the expected impact (Avalos, 2011; Jagger & Yore, 2012).

As professional knowledge is a combination of individual and community-shared knowledge, each of them with implicit (tacit) and explicit components (Baird, 2004; Cook & Brown, 1999), science teachers need to develop specific knowing, shareable with their community of professionals (Fischer et al., 2018). This knowing can be developed only within the framework of Knowledge-as-Action (Jones & Cowie, 2011; Fischler, 2011), i.e., in a PD framework. Lopes (2017) points out that as a result of this essential characteristic, any collaborative strategy between science education researchers and science teachers needs to be clear that when a teacher collaborates with researchers within the context of research work, he or she learns from the professional community of researchers and learners some of their skills and practices. In the same way, a researcher who collaborates with teachers in the context of teaching work in the professional community of teachers also learns some of their skills and practices.

Therefore, Lopes (2017) proposes that "professional development is the scope in which the research evidences produced by the research of sciences teaching practices is transformed into knowing that progressively is tuning until it becomes operative and effective, that is, that improves in fact the quality of the students' learnings" (p. 8). This approach to professional development can take advantage of the possible articulation between the collaborative work with colleagues and with researchers, both centered in teaching practices. The articulation between these two types of collaborative work can take place in action-research cycles of short to medium duration (Lopes, 2017). In some phases of the action-research cycle the work with colleagues predominates, and in other phases the work with researchers predominates (Lopes, 2017). This approach was used in several cases and during several action-research cycles (Branco, 2018; Cunha, 2015; Lopes & Cunha, 2017; Lopes, Viegas, & Cravino, 2010).

Figure 1 illustrates the main aspects to taken into account in the problem of improving science teaching practices and how they are interconnected. The PD framework (Lopes, 2017) is used as the scope in which the transformation and articulation of the two knowledges (professional knowledge from teaching and research evidences from research) occur to produce new knowing and improve teaching practices, with impact on students' learning outcomes.

To give substance to professional development that articulates science teaching practices with research evidences to enhance students' learning, the following are necessary:

- *Working collaboratively with a group of teachers.* The group of teachers should be available to develop professionally in a systematic and prolonged way (Lopes & Cunha, 2017). The work of the group of teachers should benefit from the accumulated knowledge regarding the functioning of communities of practice (Desimone, 2009; Desimone & Garet, 2015; Fischer et al., 2018).
- *Working collaboratively with a research team.* The fundamental criterion for researchers is to choose teachers based only on the teachers' interest in improving their teaching practices. The work should be centered on the teaching practices themselves and not looking for practices of "good" teachers and describing them to propose them as a model, as proposed by some authors (e.g., Alsop, Bencze, & Pedretti, 2004; Loughran, 2012). The starting point is to adopt the teachers' interest in improving their teaching practices; that is, the first and decisive step for the col-

Figure 1. Framework articulating science teaching practices and research evidences to enhance students learning
Source: Lopes, 2017.

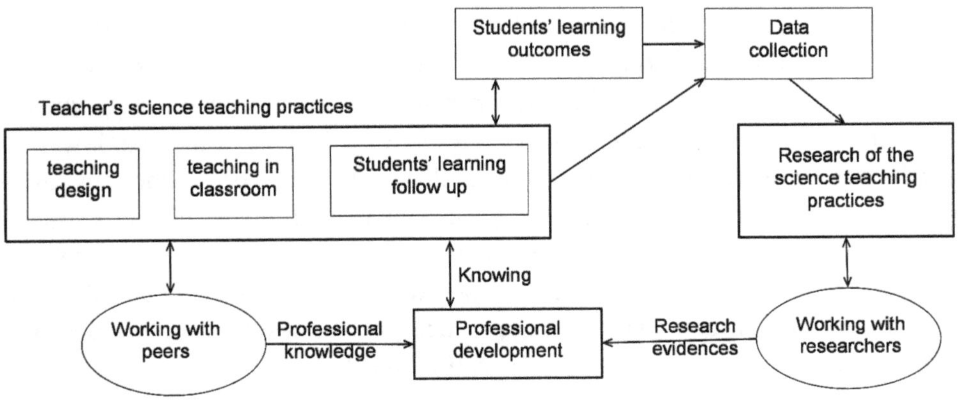

laborative work of teachers with researchers is to identify what is important for the teachers to improve in their practices (Lopes, 2017). Then, it is necessary to start the negotiation process in order to develop two interconnected components: a) develop a research focus (Hopkins, 2014) and b) choose and develop a theoretical perspective that makes it possible to see the problem in other ways and to generate an object of study (Lopes & Cunha, 2017). Developing a research focus implies adopting the concerns of the teachers and transforming them into research problem in order to formulate and refine research questions until they can be answered in an empirical way (Lopes, 2017).

- *Adopting tools that allow collaborative and productive work in each community and in the articulation between the two.* Collaborative processes are facilitated and become productive when they are based on tools for: (a) planning science teaching (Lopes, Cravino, & Silva, 2010); (b) aggregating and transforming data from teaching practices in a public and sharable description, i.e., a multimodal narrative tool (Lopes et al., 2014) used by both teachers and researchers; (c) analyzing data in a research perspective; and (d) teachers' monitoring of their own practices, in a given dimension, confronting these teaching practices with recommendations (Lopes et al., 2012). These tools can be used in both collaborative environments; however, in each one the tools are used in different ways and with different functions (Cunha, 2015; Lopes, 2017; Lopes & Cunha, 2017).

The Role of Multimodal Narratives for the Refinement of Teaching Practices

One of the first obstacles to overcome in PD through the refinement of teaching practices is the teacher's self-perception about their lessons. The research we have developed (Lopes, Cravino, Branco, Saraiva, & Silva, 2008) makes it clear that teachers realize that their initial perception of a lesson does not correspond entirely to what actually happens. This awareness is made possible by the use of Multimodal Narratives (MN) of what happens in a classroom (actions and speech), in a descriptive and chronological way, done by the teacher and based on collected data, following a protocol described in Lopes et al. (2014). This protocol makes it possible to organize data collection, aggregate all data collected and place it in a readable and condensed form to be easily used and handled (Lopes et al., 2014; see chapter 1 for more details). A MN is an account written by the teacher, with chronological, descriptive and multimodal characteristics, of the actions and speech of the teacher and students, based on data collected and following the referred protocol (Lopes et al., 2014). A MN is a self-contained and concise document that may later be analyzed, avoiding the difficulty of handling multiple data sources (Lopes et al., 2014, p. 416). It incorporates excerpts from data collected as well as the teacher's description of her or his intentions, decisions, attitudes, silences or gestures. All MN used in the present research, according to the protocol followed, have the same structure and focus. Each MN was revised so that it was self-contained and in accordance with the data collected. Finally, a member of the research team checked whether each MN was self-contained, was in accordance with data and was readable. Once validated, it was not changed. The MN can be made public (available in on online collection) and can be used for any research or professional purpose. As a tool, it has two functions: (a) it is the main document through which research in teaching practices is performed, and (b) it offers excerpts of real teaching practices, making it possible to discuss them and obtain insights from them. In addition, as teachers elaborate MNs, they become aware of the characteristics of their own teaching practices and allow others (colleagues or researchers) to help them improve their teaching practices in specific aspects.

It is known that reflection on professional practice allows for learning and stimulates PD, and this has been advocated by several authors (e.g., Alarcão, 1996). Thus, the availability of MN of one or more lessons as a reference permits teachers' reflection on their professional practice and facilitates the awareness of aspects related to the practice about which teachers may never have reflected before and that need improvement, taking into account the development of student learning in the classroom (Lopes et al., 2014).

The tools shared by professionals of a given profession support collaborative processes where reflection and sharing of experiences is of decisive importance to PD (Morris & Hiebert, 2011). Thus, MNs, in the context of teacher PD centered in teaching practices, are an important tool insofar as they (Lopes & Cunha, 2017):

- allow teachers to reflect, on their own and together with peers, based on real situations of teaching practices;
- allow researchers and teachers to develop an in-depth analysis of teaching practices in order to identify key aspects of teaching practices that are effective in improving the quality of student learning.

SELF-DIRECTED PROFESSIONAL DEVELOPMENT USING MULTIMODAL NARRATIVES

Self-directed professional development (PD), in order to be possible and improve teacher teaching practices, requires that (Lopes & Cunha, 2017):

- Certain requirements are met (personal will to improve teaching practices, more than two of action-research cycles, clear didactic intention);
- That the focus of PD be on teaching practices themselves and that this focus evolves to specific aspects of teaching practice;
- That collaborative processes of the teacher with colleagues and researchers be developed based on shared tools, one of which is fundamental: Multimodal Narratives (MN) of the lessons that are studied.

Our research team has developed several cases of self-directed PD in which the two collaborative processes (with peers and with researchers) articulate each other (Lopes & Cunha, 2017). In all of them, MNs have played a central role in the progressive improvement of teaching practices. The following paragraphs describe two cases.

A Case of Self-Directed Professional Development in Engineering Education

In this case, described in Lopes, Viegas and Cravino (2010), the teacher-researcher aimed to increase students' interest in the Physics classes involved in engineering courses and to improve students' skills in solving problems. The study focused on improving the quality of a teacher-researcher's and a colleague's teaching practices and had two action-research cycles. Each action-research cycle had three phases: a) diagnosis and planning of teaching, b) implementation and data collection, and c) elaboration of MNs and their analysis to enhance teaching practices in the next action-research cycle. The involved teachers (teacher-researcher and a colleague) jointly developed materials and tasks to be used in the Physics classes of engineering courses (first phase of each action-research cycle).

According to MN protocol (Lopes et al., 2014) the data collected were of two types: independent data and teacher-dependent data. The first group of data includes audio recordings; written documents produced by the teacher; photographs of the blackboard and of the students' layout in the room; documents produced by the students; resources used; and characteristics of the school, the students and the teacher. The second group, dependent data, includes the teacher's intentions, actions (which may or may not be on record), reactions, attitudes, silences, gestures and decisions. The teacher-researcher produced the MNs of his own lessons and of his colleague's lessons (first intervention) based on the data collected and following the procedures indicated in the protocol to elaborate a MN (Lopes et al., 2014). Each MN has two parts (Lopes et al., 2014): a) contextual information regarding the lesson (students' characteristics, space organization, environment, reference to the episodes) and b) a multimodal description of each episode (beginning with the presentation of the task and ending with the closure of the task or the teacher's decision to end it), focusing on the teacher's and students' discourse, interactions and actions. Finally, the MNs were validated (Lopes et al., 2014), taking into account: a) the protocol for their elaboration (to ensure the characteristics and structure of an MN), b) the data collected (to ensure its adequacy to the data) and c) its readability.

Analysis of the MNs of several lessons by the two teachers involved identifying specific aspects of the teaching practices of each one. First, each classroom teacher's approach was, as expected, different, even though they used the same materials, which they had designed together. It was possible to identify that while the teacher-researcher tried to use the questions to show the pertinence of the subjects, often before their exploration in the classroom, generating moments of discussion and asking the students to talk to each other before intervening, the other teacher used these questions as a complement after the subjects had been presented, conducting a discussion with the whole class. In this way, the interventions of this teacher were centered on the answers of the faster and more participatory. Although the time factor was a real and legitimate constraint to this option, the teacher believed that in this way he fulfilled the curriculum and saved time. The teacher-researcher – in addition to benefiting himself by having elaborated the MN of his own classes because he was aware at the time of various aspects of his own teaching practices – was also able to reflect on his teaching practices and to improve specific aspects of planning and teacher mediation in the classroom.

Multimodal Narrative Excerpt 1, from one of the lessons in the second action-research cycle, illustrates how three distinctive features of teacher-researcher action took shape:

- Gaining the students' attention before starting a new subject and trying to show its pertinence;
- Ensuring that all student contributions are considered and, in particular, ensuring that all the arguments presented by the students are discussed, even if at a later time;
- Allowing students time to work autonomously on the proposed task.

Multimodal Narrative Excerpt 1: Lesson of Teacher-Researcher

This was followed by an example that would allow us to see if the students were actually following up. An alloy of two materials: copper and steel. We knew their conductivities and they appeared on the slide (figure below). The first question was very simple (the only one that is on the slide at this point). I said ... which is the material that is the best conductor? (Figure 2)

Without difficulties they identified the highest k.

I went on to the next question (showing the second question on the slide, figure above):

"I invented it: now we make a 'sandwich' with these two materials and what we want to find out is what material will we feel a greater decrease in temperature."

From my previous experience in recent years, I knew this was not an obvious issue for most students. The trend turns out to be the connection of the best conductor to the higher ΔT.

That is why I considered the question very important and emphasized that they would have to think before answering.

I asked them to think individually, and then they tried to argue with their colleague on the side, and I request to they do not respond loudly...

They thought and argued for about 40 seconds and when the discussion subsided I asked for the answer.

Almost all respond that it was the copper.

I repeated: "The copper will feel a greater temperature decrease ... this is what you think." I tried to perceive by their faces if this was really what they wanted to say. "Let me try to understand," I continued

Figure 2.

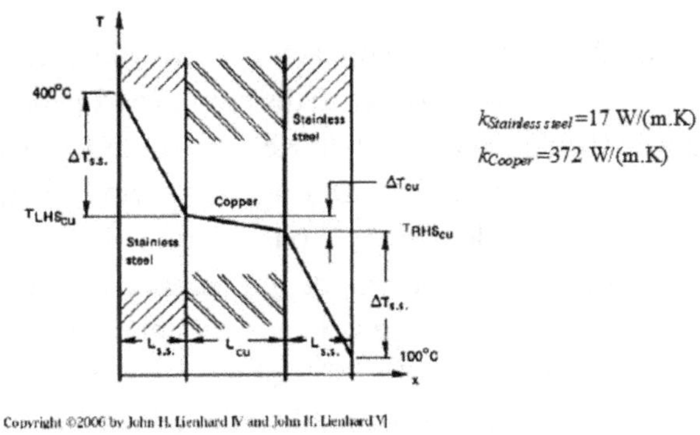

Student P: "…the question is not very clear..."

Teacher: "That's what's worse ..."

I then tried to clarify and remind them what the temperature slope would depend on, using what I already had on the board.

I explained, "The question is, a material in which there will be a greater variation, that is to say a greater slope, in which this temperature is more different from the one." I pointed. "…OK?" I asked student P more specifically to see if he had noticed, that he was affirmative...

"Anyone want to change their minds?" I continued.

Some students changed their answers, but it seemed to me that most of them still did not understand it. I chose to explain that if there was a layer that offered more resistance to the passage of heat – now that they already had knowledge of electricity ... I reminded them – would that imply that such material would be a worse conductor or a better conductor?

"Worse," some said.

I said, "Of course, worse conductor!" – reinforcing that idea – and continued, "And the temperatures here (pointing to the extremes) will become more alike or more different? Think about this..."

There was silence. […]

The analysis of MNs of both teachers after the second implementation made it possible to identify some specific aspects of their teaching practices (in particular, the three referred to above and present in the MN excerpt) that predominate in the teacher-researcher (Lopes, Viegas, & Cravino, 2010). In this study, it was shown that these differences in teaching practices played a decisive role in increasing students' interest in physics and in developing their problem-solving skills.

A Case of Self-Directed Professional Development in Secondary Education

The study developed by Cunha (2015) reports the teacher's own PD over 10 years (2004–2014). The focus of the teacher's teaching practices was always to develop experimental work focusing on improving students' epistemic practices (Berland et al., 2016; Kelly, 2014; Sandoval & Reiser, 2004) and their productive engagement (Engle & Conant, 2002). During that period, three complete action-research cycles occurred. Each action-research cycle had three phases: (a) collaborative work with the teacher's peers to develop a teaching proposal (exploration guide, preparation of experimental material, teaching mediation to promote students' productive engagement and epistemic practices), (b) implementation of the teaching proposal and data collection to make MNs of the respective lessons, and (c) collaborative work with researchers to analyze the MNs for the publication of papers in professional and academic forums and the identification of aspects of teaching practices that work well, are not to be used or could be adapted in the next research-action cycle.

The teacher's intervention in the first cycle took place in 2006, the second in 2008, and the third in 2012. Each intervention was centered on an experimental task that unrolled over several lessons (three or four lessons in a total of 315 to 360 minutes). The task structure was always the same: initial challenge (contextualized), planning to respond to the challenge, implementation of the experimental activity, and interpretation of the results and answer to the problem. In 2006, the challenge was about the difference in sound quality emitted by AM or FM radio stations; in 2008, about why there are several repeater antennas in communications (mobile or satellite); in 2012, on the background radiation in the universe.

MNs of lessons were constructed by using data collected from the same type as those mentioned in the previous case, using the same protocol, having the same structure and characteristics, and validated in the same way.

The MNs were analyzed in four dimensions according to the teacher's and research purposes and the respective theoretical framework: a) actions of the teacher to promote the students' productive engagement, b) actions of the teacher to promote students' epistemic practices, c) indicators of the students' productive engagement and d) students' epistemic practices that actually took place in the classroom.

During the research, the MNs were used in a systematic way, and, therefore, after the analysis was completed, in each MN the excerpts related to a given category of a given dimension of analysis were identified. Due to their characteristics and to the characterization of several excerpts with the analysis framework used, the MNs made it possible for teachers to quickly identify aspects of teaching that needed to be considered from their own point of view, during the collaborative work with peers in the following research-action cycle. That is, the MNs worked as a tool during collaborative work with researchers for research purposes and migrated to collaborative work with colleagues also as a tool, but with other purposes and uses.

Below is a brief account of how the research phase of the second cycle of research was triggered and how it influenced the planning phase of the third action-research cycle, as well as the role MNs played in both, with regard to MN excerpts 2 to 5.

How the MNs Allow the Articulation of the Research Phase of the Second Action-Research Cycle With the Planning Phase of the Third Action-Research Cycle

As soon as the classes of the second implementation (2008) were finished, teacher A developed an MN of each lesson using the data collected. Each MN was reformulated until it was consistent with the data, complete, self-contained, legible and multimodal. Finally, a researcher validated each MN.

When the MNs of the second implementation were concluded, a period of collaboration between the teacher and the researchers began in which they analyzed the MNs. Based on the MN analysis, focusing on the productive work of the students and how the teacher induced it, the teacher proposed, elaborated and presented two communications to international conferences.

After this work, the researcher challenged teacher A to make a deeper analysis of the students' productive engagement, using the same empirical material, and to make a study of the evolution of the students' productive engagement from the first to the second implementation, using the same process of analysis but refining the categories (from a codification of MN) and doing statistical analysis of the quantitative results to be published in an international journal with peer review (Cunha, Lopes, Cravino, & Santos, 2012). One of the aspects identified as important to engaging students is presenting the task as a challenge (see MN excerpts 2 and 3). In these excerpts, it is possible to identify that the task presented in 2008, although it also refers to a real situation (like the 2006 task), has more contextual information, and the challenge is more open and clearer than in the 2006 task.

Multimodal Narrative Excerpt 2: Teacher A (Main Task From Lessons of 2006)

Some broadcast radio stations broadcast in AM and FM. Is there any difference in the sound emitted by one type of emission or the other? Formulate a response hypothesis.

Multimodal Narrative Excerpt 3: Teacher A (Main Task From Lessons of 2008)

In communications for mobile and satellite, certain frequency bands of microwave are used. In large cities, high towers are built that support a set of parabolic antennas in order to allow point-to-point propagation of microwaves above the tops of buildings.

How can we model this physical situation in the laboratory?

Why are there multiple repeater antennas? Is there a limit to the distance between them?

Why do antennas lie at appreciable levels at ground level?

Formulate a hypothesis for answers for each one of the questions.

Meanwhile, in December 2011, the first meetings with schoolmates took place to outline the planning for the use of computational simulations in Physics teaching in a context of experimental work. In the meetings of teachers of the same discipline at the school (in which teacher A participated), the implementation of the experimental work was discussed through the presentation of a exploration guide proposed by this teacher, as well as its implementation. The group also discussed how to engage the students, mentioning the categories identified in the article referred to and using the related MN excerpts.

The teachers studied the MN excerpts with regard to how to pose the task as a challenge and how to keep it as a challenge until the end of the lesson, using excerpts from the MN. Also, they discussed how to give the students some autonomy in performing the tasks, an important aspect to promote the students' epistemic practices, again using excerpts from the MNs and taking into account the referred research paper. MN excerpt 4 illustrates two modes (of five identified in the paper) of keeping the task as a challenge: indicating/manipulating resources for students to be able to work autonomously and asking students to clarify the challenge they are addressing.

Multimodal Narrative Excerpt 4: Teacher A (Keeping the Main Task of 2008 as Challenge)

[...] I just listen and encourage them to respond for themselves by suggesting that they consult the textbook and discuss it with colleagues. [...]

[...] Now read what you wrote and think a little ... Why do we have several antennas in certain areas? Why do we have several along the way? How can we model this physical situation in the laboratory? [...]

As a result of this reflection and work with MN, the experimental work was modified in the presentation of the physical situation, in terms of how the problem is presented and how the teacher requests the comparison between the conclusions obtained with the initial response given to the problem. In mediation planning, the five ways of keeping the tasks as a challenge (one of the most important results of the paper) and ways of granting authority and autonomy to the students (a condition for good epistemic practices) were used (see MN excerpt 4). Planning for the third implementation ended in March 2012.

At this stage of planning, a key aspect to consider was how to present the task as a challenge. Although all the tasks include a reference to a real situation that one wanted to study from the physical point of view, there are aspects that have changed from year to year: the initial challenge was more open and clear in 2008 and 2012 than in 2006, and the physics context was more explicit (see MN excerpts 2, 3, and 5). MNs were important not only for comparing the formulation of tasks but especially for analyzing how students became involved in each, and the efforts that teachers made to keep each task as a challenge until it was completed.

Multimodal Narrative Excerpt 5: Teacher A (Main Task From Lessons of 2012)

In 1965, the background radiation was detected, which occurred 13.7 billion years ago, when the universe began to expand and, at the same time, began its consequent cooling. This radiation was interpreted as a residue of the Big Bang. The wavelength of the background radiation, which corresponds to the maximum power radiated by the universe, is 0.107 cm.

a) Is it possible to know the current temperature of the universe? What is your value?

b) Justify the statement, "Due to the expansion of the universe and its consequent cooling, the original electromagnetic radiation has shifted to the red."

Task: Propose a course of research, using the simulation on the radiation of a black body that is provided to you, to obtain a reasoned answer to the problems presented.

Suggestions: Write down the ideas you want to check; explain the "experiences" that you intend to do to verify your ideas and respond to the problem.

The third implementation started in April 2012. Each teacher used the joint planning as understood. In the classroom implementation, teacher A was particularly attentive to the aspects identified in the article referred to above. In addition, he made sure that the students: (a) correctly understood the task, (b) had time to give their answer and (c) discussed their answers. When a difficulty arose, the teacher did not immediately give the answer, but instead raised a new question so that the students could reflect on the subject and be able to proceed autonomously. He followed what everyone was doing through direct observation or mid-run questions to verify whether they understood what they were doing or not.

Teacher A received a response from the editor of Revista Electrónica de Enseñanza de las Ciencias in July 2012, and the article, after some changes, was finally published.

In summary, MNs were used during the research phase as a tool to find evidence that supported a research result relevant to the academic community and to teachers. The same MNs were used as tool to construct new teaching proposals that overcame the characteristics of the teaching practices identified previously, in particular identifying the excerpts that were worked in the research context.

This research demonstrates the importance of the collaborative processes of the teacher with peers and with researchers for his PD and the decisive role of the MN. Specifically,

- The teacher becomes aware of aspects that he had never thought about,
- The teaching practices are researched in order to identify specific aspects that have made them effective in student learning,
- The MN may be taken as a reference of teaching practices that were objects of reflection, sharing and insights in the collaboration with peers.

HETERO-DIRECTED PROFESSIONAL DEVELOPMENT USING MULTIMODAL NARRATIVES

This section briefly reports two cases of hetero-directed PD, described in Branco (2018), centered on the use of MN made by a researcher of lessons by two teachers of Physical-Chemical Sciences who taught at different schools in the Portuguese education system (teacher B and teacher C) and their students (13–14 years old). In both cases, the PD was based on: (a) data collection to elaborate MNs from some lessons and (b) a process of attempting to find solutions for teaching practices considering, as a starting point, the analysis and reflection on MNs. The work was carried out over three years, in which there were two complete action-research cycles. The first action-research cycle had three phases.

Phase One – Evaluation Phase Based on Data Collection: In this phase was collected data of teachers' teaching practices in the classroom for two lessons per teacher (audio recording, class characterization, student list, classroom plan, students' notebooks, resources made available by the teacher, presentations made by the students), followed by researcher interaction with the teachers as they described their intentions, decisions and perspectives. The researcher elaborated MNs from two lessons from each teacher. After each MN was completed, the teacher was asked to verify whether anything needed to be corrected. Finally, the MNs were validated as described in Lopes et al. (2014). The researcher analyzed the MNs of the two teachers in a research register. The process was therefore guided by a theoretical framework, research questions, and reproducible and recognizable analysis procedures. In particular, in this phase of the study the central question was: What characteristics of the teaching practices of teachers are promoters of epistemic practices in students?

From the MN analysis, the researcher identified patterns of teacher mediation, and the most frequent was an inhibiting pattern of teaching practices on students' epistemic practices, characterized by Branco (2018): (a) the teacher requests an immediate response to a question, often without presentation of any physical situation or contextualization; (b) the teacher presents a physical situation, using the evocation or images and/or information contained in worksheet without contextualization; (c) the teacher use teaching practices whose traits do not promote the occurrence of epistemic practices, such as addressing many situations without deepening any, which shortens the work of students; (d) students manifest few epistemic practices, and these are fairly elementary, since students have no opportunity to think or perform any task. (See MN excerpt 6.)

Multimodal Narrative Excerpt 6: Lesson of Teacher C (Teacher Mediation Pattern Inhibiting of Epistemic Practices)

The teacher asks, "Oh, Beatrice, I am going to raise another question: Is all light visible?" as he writes the question on the board.

"No," John answered immediately.

"Beatrice, your colleague John said no; do you not agree?" The teacher is still trying to draw attention to this new question.

"I agree," she says.

"Then write in your notebook 'No'." The teacher asks, "So what is the light we can see?" (1h; 06min; 32s)

"That's like the sound. We hear from 20 to 20000 Hz," Daniel replies.

"Very well, Daniel ... 'We only see a part of the radiation that reaches us ...'," says the teacher to Beatrice, writing on the board, taking advantage of Daniel's response. "Can I ask you something else now?"

Phase Two – Intervention of Researcher in Order to Direct the Professional Development of the Teachers Involved: The MNs were presented to teachers, after the researcher analyzed each one. Then the researcher and teachers made a joint analysis and discussion of the MNs. The teachers expressed some surprise with certain episodes that occurred in their classes. They began to become aware of certain aspects that went unnoticed and began to ask for suggestions to go beyond what seemed to them a less appropriate practice. For example, teacher B found that there was a student who dominated the interventions, not giving the opportunity for colleagues to participate, and teacher C found that does not give an opportunity for students to expose and follow up on their ideas. The results of the MN analysis were presented and discussed with the participating teachers (see MN excerpt 6), who, in addition to agreeing with the analysis, found it very interesting and enriching in order to get them to reflect and improve their practices. In spite of this, teachers B and C expressed some fear about: (a) not doing well, (b) being disoriented if the students asked many questions and (c) that students would not get enough information to carry out the proposed tasks,

Figure 3.

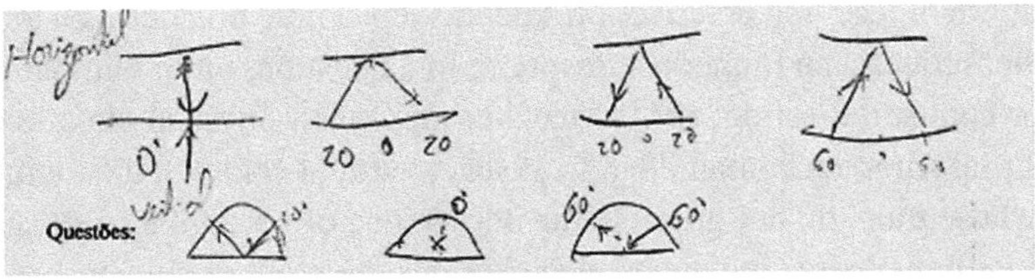

without their constant intervention. During this PD program, teachers B and C often questioned the researcher on how to act when students proposed something unforeseen.

Phase Three – New Implementation of Teaching Practices (With Data Collection Similar to Phase One) With the Support of the Researcher to Discuss or Clarify Any Questions: Progressively, teacher B stopped using the textbook exclusively. He began to prepare worksheets and to suggest activities that he had to read aloud so that everyone could understand what was asked for. He adopted greater flexibility in the planning of his classes and in the management of the classroom. He was very happy to participate in this PD program and revealed that his students also shared this feeling, because at the end of a class two students expressed: "We like to work with the model because we understand the subject better." Teacher C began by suggesting that students think of materials to bring to class in an attempt to stimulate their curiosity and creativity. He asked them to present their work results orally. He also began to make the planning of his classes more flexible. In one class, because of this flexibility, he forgot to correct a response given by the student, which he kept on the board. Almost at the end of the class he returned to the question after having to think about how he was going to approach the subject again, without leaving the context. He also said, "I am not fully aware if I am on the right path." Despite these doubts and uncertainties on the part of the teachers, the teacher mediation patterns that promote epistemic practices became more frequent in the classes of both the second intervention (with researcher follow-up) and the third intervention (without researcher follow-up).

The second action-research cycle also had three phases.

Phase One (Second Cycle) – Evaluation Phase Based on Data Collection: Based on data collected, the researcher elaborated MNs of six lessons per teacher by using the same procedures indicated above. The MN analysis, in a researcher register, was guided by the same analysis framework and research questions. The teaching practice that most promoted students' epistemic practices (see MN excerpt 7) is characterized by Branco (2018): (a) the teacher requests work that requires the elaboration of a response, with a physical situation presented and contextualized through objects or equipment; b) the teacher always asks for additional aspects to the students' responses in order to clarify the epistemic practices that occur; (c) at certain moments, the teacher asks students to consider new perspectives in the approach to the task, values the students' work epistemically or reinforces their understanding through a synthesis or additional information; and (d) students manifest various types of epistemic practices, some of them elaborated.

Multimodal Narrative Excerpt 7: Lesson of Teacher C (Teacher Mediation Pattern Promoting of Epistemic Practices)

Meanwhile, the students in groups 3, 4 and 5 perform activity A. The teacher circulates through the groups. He notices that the students in group 4 are having some difficulty: Stephanie does not understand how to place the slot to obtain the experimental results. The teacher says for the student to place the slot very close to the lamp. He then suggests varying the distance between the slot and the lamp. After a few tries the students execute this correctly. They then answer the questions (Figure 3).

Q1: *The angle of the incident light is always symmetrical to the angle of the reflected light.*

Q2: *The plane is vertical. The normal to the surface at the point of incidence and the reflected light are in the same plane." "The conclusion is the same when the plane is horizontal.*

Laws of Reflection:

1. The incident ray (i), the normal to the surface at the point of incidence and the reflected ray (r) are in the same plane.

2. The angle of incidence (α) and the angle of reflection (β, are equal.

Phase Two (Second Cycle) – Researcher Feedback and New Implementation of Teaching Practices: In this phase, the researcher gave feedback to the teachers about the MN analysis of lessons from the second teaching implementation. However, the researcher did not discuss with teachers how to conduct the third teaching implementation, nor did the researcher give support to them during the teaching implementation.

Phase Three (Second Cycle) – Evaluation Phase Based on Data Collection: Based on data collected, the researcher elaborated MNs of three lessons per teacher by using the same procedures indicated above. The MN analysis, in a researcher register, was guided by the same analysis framework and research questions.

The main result of this research (Branco, 2018) was: The professional development of teachers occurs by changing patterns of teaching practices. In particular, changing certain features of mediation allowed teachers' teaching practices to evolve to promote more and better epistemic practices among students. The change observed in teachers' practices was supported by a process of reflection based on Multimodal Narratives, where the interaction with the researcher was determinant.

CONTRIBUTIONS AND DISCUSSION

In all cases, there was articulation between the teaching practices and the research of the teaching practices themselves, and this articulation made it possible: (a) to produce research results on characteristics of teaching practices in the sciences that promote the quality of student learning in specific aspects and (b) to induce incremental improvements in the teaching practices of the teachers involved.

The cases presented and their analysis show the following key contributions:

- *The PD model articulating teaching practices and research evidences of the teachers practices themselves is effective in improving the quality of teaching practices.* It is possible to articulate the collaborative process among teachers with the collaborative process between teachers and researchers in the scope of PD of teachers that improves their teaching practices. This result confirms that these collaborative processes are fundamental in PD (Avalos, 2011). It also confirms that, regardless of the differences between professional knowledge and research knowledge (Winch, Oancea, & Orchard, 2015), they can enrich each other. It also shows that although the most disseminated PD models do not incorporate research results (Desimone, 2009; Desimone & Garet, 2015; Fischer et al., 2018), there are PD models, in particular what the present authors have proposed and have been using, that are equally effective in improving teaching practices when they incorporate research results (Lopes, 2017; Lopes & Cunha, 2017) and are perhaps more effective in the long term. This PD model needs future research in three important strands: (a) how it can be scaled to a context of greater dissemination of its use, (b) what is the long-term effectiveness after the end of formal collaboration with the research team, and (c) how, more broadly, to help teachers to develop their own MNs by using the proposed protocol to do that.
- *The improvement in teaching practices is incremental.* Improvements in teaching practices are incremental and promote student learning in specific aspects. This improvement stems from the incorporation of research evidences of the teaching practices of the teachers involved. This result is compatible with the theoretical analysis that points to the difficulty of incorporating research results into teaching practices (Winch, Oancea, & Orchard, 2015)) and the empirical results confirming this difficulty (Avalos, 2011; Jagger & Yore, 2012), despite the proposal of Bradforth et al. (2015) of the massive use of research evidences in teaching practices. A future research direction is to study the effect of incorporating research evidences, in the context of PD, that come from other studies.
- *The teaching practices patterns change by the incorporation of new traits.* It is possible to make safe progress in increasing the quality of teaching practices by characterizing, in a research context, specific traits of teaching and their effects on specific aspects of learning. If these traits are altered with the incorporation of research evidences and appropriated by professional knowledge, in the context of PD, the new traits progressively change the patterns of teaching practices (Branco, 2018; Cunha, 2015; Lopes & Cunha, 2017; Lopes, Viegas, & Cravino, 2010). A future research direction is exploring the multiple dimensions and aspects of science learning and studying how teaching practices can develop them.

- *Multimodal Narratives is fundamental tool for effective PD.* Multimodal Narratives, as a tool, plays a decisive role in PD as:
    - They heighten teachers' awareness of specific aspects of their own teaching practices as they elaborated the MNs (or collaborated closely with a researcher to elaborate them).
    - They allow teachers, individually or collaboratively, to analyze critically or simply use excerpts to pinpoint traits that should be reinforced, improved or avoided according to their effects on student learning.
    - They allow researchers to study, within the scope of research work with academic integrity, characteristics, contexts and teaching patterns that are decisive for student learning.
    - They also serve as a support for collaborative work between teachers and between teachers and researchers, as they provide a tool that is understood and workable by stakeholders.

A future research direction regarding the use of Multimodal Narratives, as a tool, in the context of PD is a) to identify ways to help teachers to develop their own MNs by using the proposed protocol for this and b) to explore strategies of PD using MNs that involve a larger number of teachers.

CONCLUSION

This chapter presents a model for professional development (PD) according to its most recent conceptualization focusing on the incremental and sustained improvement of teaching practices. It reports two cases of self-directed PD (two studies) and two cases of hetero-directed PD (one study) in which the PD model used articulates the teaching practices and research evidences of the teaching practices themselves.

The mentioned studies make it possible to identify important contributions:

- The PD model articulating teaching practices and research evidences of the teachers' practices themselves is effective in improving the quality of teaching practices.
- The improvement in teaching practices is incremental.
- The patterns of teaching practices change through the incorporation of new traits.
- The patterns of teaching practices change through the incorporation of new traits.

The paper elucidates how the use of Multimodal Narrations (MN) is central to PD processes that impact on the incremental improvement of teaching practices in the context of a PD model that articulates collaborative work with peers and collaborative work with researchers. In particular, the paper shows how MNs can support the collaborative work of the teacher with peers and/or researchers in order to change and improve teaching practices. In short, MN are tools to support PD for the following reasons:

- They allow teaching practices to be kept in a public register that can be easily consulted by teachers, researchers and other professionals.
- They allow professional practices to be studied by the teacher or by others.
- They allow reference cases of teaching practices to be established, which can be the objects of reflection, scholarly study, illustration or insight.
- They support collaborative work between teachers and between teachers and researchers as they provide a common tool.

ACKNOWLEDGMENT

This work is funded by the ERDF e European Regional Development

Fund through the COMPETE Programme (operational programme for competitiveness) and by National Funds through the FCT e Fundaç~ao para a Ci^encia e a Tecnologia (Portuguese Foundation for Science and Technology) within project [PTDC/CPE/112303/2009]

REFERENCES

Alarcão, I. (1996). Reflexão crítica sobre o pensamento de D. Schön e os programas de formação de professores [Critical reflection on D. Schön's thinking and teacher training programs]. In *Formação Reflexiva de Professores: Estratégias de Supervisão* [Reflective Teacher Training: Supervision Strategies] (pp. 9–40). Porto, Portugal: Porto Editora.

Alsop, S., Bencze, L., & Pedretti, E. (2004). *Analysing exemplary science teaching*. Maidenhead, UK: McGraw-Hill Education (UK).

Avalos, B. (2011). Teacher professional development in teaching and teacher education over ten years. *Teaching and Teacher Education*, *27*(1), 10–20. doi:10.1016/j.tate.2010.08.007

Baird, D. (2004). *Thing knowledge: A philosophy of scientific instruments*. Berkeley, CA: University of California Press.

Bakkenes, I., Vermunt, J. D., & Wubbels, T. (2010). Teacher learning in the context of educational innovation: Learning activities and learning outcomes of experienced teachers. *Learning and Instruction*, *20*(6), 533–548. doi:10.1016/j.learninstruc.2009.09.001

Berland, L. K., Schwarz, C. V., Krist, C., Kenyon, L., Lo, A. S., & Reiser, B. J. (2016). Epistemologies in practice: Making scientific practices meaningful for students. *Journal of Research in Science Teaching*, *53*(7), 1082–1112. doi:10.1002/tea.21257

Bradforth, S. E., Miller, E. R., Dichtel, W. R., Leibovich, A. K., Feig, A. L., Martin, D., ... Smith, T. L. (2015). Improve undergraduate science education: It is time to use evidence-based teaching practices at all levels by providing incentives and effective evaluations. *Nature*, *523*(7560), 282–285. doi:10.1038/523282a PMID:26178951

Branco, M. J. (2018). *Desenvolvimento das práticas de mediação dos professores em contexto de ensino de Ciências Físicas - O papel de ferramentas de ajuda* [Development of the teachers' mediation practices in the context of Physics teaching: The role of tools for help] (Unpublished doctoral dissertation). Universidade de Trás-os-Montes e Alto Douro, Vila Real, Portugal.

Burbank, M. D., & Kauchak, D. (2003). An alternative model for professional development: Investigations into effective collaboration. *Teaching and Teacher Education*, *19*(5), 499–514. doi:10.1016/S0742-051X(03)00048-9

Cook, S., & Brown, J. (1999). Bridging epistemologies: The generative dance between organizational knowledge and organizational knowing. *Organization Science*, *10*(4), 381–400. doi:10.1287/orsc.10.4.381

Corrigan, D., Dillon, J., & Gunstone, R. (Eds.). (2011). *The professional knowledge base of science teaching*. Dordrecht, The Netherlands: Springer. doi:10.1007/978-90-481-3927-9

Cunha, A. E. (2015). *Construção de práticas de referência no Ensino da Física para o Ensino Secundário* [Construction of Physics Teaching practices of reference for Secondary Education] (Unpublished doctoral dissertation). Universidade de Trás-os-Montes e Alto Douro, Vila Real, Portugal.

Cunha, A. E., Lopes, J. B., Cravino, J. P., & Santos, C. A. (2012). Envolver os alunos na realização de trabalho experimental de forma produtiva: O caso de um professor experiente em busca de boas práticas [Engage students in the performance of experimental work in a productive way: the case of an experienced teacher in search of good practices]. *Revista Electrónica de Enseñanza de las Ciencias, 11*(3), 635–659.

Desimone, L., Garet, M., Birman, B., Porter, A., & Yoon, K. (2003). Improving teachers' in-service professional development in mathematics and science: The role of post-secondary institutions. *Educational Policy, 17*(5), 613–649. doi:10.1177/0895904803256791

Desimone, L. M. (2009). Improving impact studies of teachers' professional development: Toward better conceptualizations and measures. *Educational Researcher, 38*(3), 181–199. doi:10.3102/0013189X08331140

Desimone, L. M., & Garet, M. S. (2015). Best practices in teachers' professional development in the United States. *Psychology. Social Education, 7*(3), 252–263.

Engle, R. A., & Conant, F. R. (2002). Guiding principles for fostering productive disciplinary engagement: Explaining an emergent argument in a community of learners classroom. *Cognition and Instruction, 20*(4), 399–483. doi:10.1207/S1532690XCI2004_1

Fischer, C., Fishman, B., Dede, C., Eisenkraft, A., Frumin, K., Foster, B., ... McCoy, A. (2018). Investigating relationships between school context, teacher professional development, teaching practices, and student achievement in response to a nationwide science reform. *Teaching and Teacher Education, 72*, 107–121. doi:10.1016/j.tate.2018.02.011

Fischler, H. (2011). Didaktik: An appropriate framework for the professional work of science teachers? In D. Corrigan, J. Dillon, & R. Gunstone (Eds.), *The professional knowledge base of science teaching* (pp. 31–50). Dordrecht, The Netherlands: Springer. doi:10.1007/978-90-481-3927-9_3

Guskey, T. R. (2002). Professional development and teacher change. *Teachers and Teaching, 8*(3), 381–391. doi:10.1080/135406002100000512

Hiebert, J., & Morris, A. K. (2012). Teaching, rather than teachers, as a path toward improving classroom instruction. *Journal of Teacher Education, 63*(2), 92–102. doi:10.1177/0022487111428328

Hopkins, D. (2014). *A teacher's guide to classroom research*. Maidenhead, UK: McGraw-Hill Education.

Jagger, S. L., & Yore, L. D. (2012). Mind the gap: Looking for evidence-based practice of science literacy for all in science teaching journals. *Journal of Science Teacher Education, 23*(6), 559–577. doi:10.100710972-012-9271-6

Jones, A., & Cowie, B. (2011). Moving beyond deconstruction and reconstruction: Teacher knowledge-as-action? In D. Corrigan, J. Dillon, & R. Gunstone (Eds.), *The professional knowledge base of science teaching* (pp. 51–63). Dordrecht, The Netherlands: Springer. doi:10.1007/978-90-481-3927-9_4

Kelly, G. J. (2014). Inquiry teaching and learning: Philosophical considerations. In M. R. Matthews (Ed.), *Handbook of Historical and Philosophical Studies in Science Education* (pp. 1363–1380). Springer.

Korthagen, F., Loughran, J., & Russell, T. (2006). Developing fundamental principles for teacher education programs and practices. *Teaching and Teacher Education, 22*(8), 1020–1041. doi:10.1016/j.tate.2006.04.022

Lopes, J. B. (2017). Researching Science Teaching Practices in the Classroom over Two Decades: What Matters? In J. B. Lopes, J. P. Cravino, E. S. Cruz, & A. Barbot (Eds.), *Teacher Science: Contributions of Research for Planning, Practice and Professional Development* (pp. 1–26). New York: Nova Science Publishers.

Lopes, J. B., Cravino, J. P., Branco, M., Saraiva, E., & Silva, A. A. (2008). Mediation of student learning: Dimensions and evidences in science teaching. *Problems of Education in the 21st Century, 9*(9), 42-52.

Lopes, J. B., Cravino, J. P., & Silva, A. A. (2010). *Effective Teaching for Intended Learning Outcomes in Science and Technology (Metilost)*. New York: Nova Science Publishers.

Lopes, J. B., & Cunha, A. E. (2017). Self-directed professional development to improve effective teaching: Keys points for a model. *Teaching and Teacher Education, 68*, 262–274. doi:10.1016/j.tate.2017.09.009

Lopes, J. B., Silva, A. A., Cravino, J. P., Santos, C. A., Cunha, A., Pinto, A., ... Branco, M. J. (2014). Constructing and using multimodal narratives to research in science education: Contributions based on practical classroom. *Research in Science Education, 44*(3), 415–438. doi:10.100711165-013-9381-y

Lopes, J. B., Silva, A. A., Cravino, J. P., Viegas, C., Cunha, A. E., Saraiva, E., ... Santos, C. A. (2012). Instrumentos de ajuda à mediação do professor para promover a aprendizagem de alunos e o desenvolvimento profissional dos professores [Tools to help teacher mediation to promote student learning and teacher professional development]. *Sensos, 2*(1), 125–171.

Lopes, J. B., Viegas, C., & Cravino, J. P. (2010). Improving the Learning of Physics and Development of Competences in Engineering Students. *International Journal of Engineering Education, 26*(3), 612–627.

Loughran, J. (2012). *What expert teachers do: Enhancing professional knowledge for classroom practice*. New York: Routledge. doi:10.4324/9780203851470

Luft, J. A., & Hewson, P. W. (2014). Research on teacher professional development programs in science. Handbook of Research on Science Education, 2, 889-909.

Luke, N., & Rogers, C. (2015). Responding to uncertainty: Teacher educator professional development through co-teaching and collaborative reflection. *Teacher Research: Integrating Action, Observation, and Reflection, 8*(2), 245–260.

Lustick, D. (2009). The Failure of Inquiry: Preparing Science Teachers with an Authentic Investigation. *Journal of Science Teacher Education, 20*(6), 583–604. doi:10.100710972-009-9149-4

Masuda, A. M., Ebersole, M. M., & Barrett, D. (2013). A Qualitative Inquiry: Teachers' attitudes and willingness to engage in professional development experiences at different career stages. *International Journal of Progressive Education, 79*(2), 6–14.

McIntyre, D. (2005). Bridging the gap between research and practice. *Cambridge Journal of Education, 35*(3), 357–382. doi:10.1080/03057640500319065

Morris, A. K., & Hiebert, J. (2011). Creating shared instructional products: An alternative approach to improving teaching. *Educational Researcher, 40*(1), 5–14. doi:10.3102/0013189X10393501

Mushayikwa, E., & Lubben, F. (2009). Self-directed professional development: Hope for teachers working in deprived environments? *Teaching and Teacher Education, 25*(3), 375–382. doi:10.1016/j.tate.2008.12.003

Nonaka, I., & Takeuchi, H. (1995). *The knowledge-creating company: How Japanese companies create the dynamics of innovation*. New York: Oxford University Press.

Osborne, J. (2014). Teaching scientific practices: Meeting the challenge of change. *Journal of Science Teacher Education, 25*(2), 177–196. doi:10.100710972-014-9384-1

Richter, D., Kunter, M., Klusmann, U., Lüdtke, O., & Baumert, J. (2014). Professional development across the teaching career. In S. Krolak-Schwerdt, S. Glock, & M. Böhmer (Eds.), *Teachers' Professional Development. The Future of Education Research* (pp. 97–121). Rotterdam: Sense Publishers. doi:10.1007/978-94-6209-536-6_7

Sandoval, W. A., & Reiser, B. J. (2004). Explanation-driven inquiry: Integrating conceptual and epistemic scaffolds for scientific inquiry. *Science Education, 88*(3), 345–372. doi:10.1002ce.10130

Shulman, L. (1986). Those who understand: Knowledge growth in teaching. *Educational Researcher, 15*(2), 4–14. doi:10.3102/0013189X015002004

Soebari, T. S., & Aldridge, J. M. (2015). Using student perceptions of the learning environment to evaluate the effectiveness of a teacher professional development programme. *Learning Environments Research, 18*(2), 163–178. doi:10.100710984-015-9175-4

Stigler, J. W., & Hiebert, J. (2009). *The teaching gap: Best ideas from the world's teachers for improving education in the classroom*. New York: Simon and Schuster.

Tardif, M. (2000). Os saberes profissionais dos professores e conhecimentos universitários [The professional knowledge of teachers and university knowledge]. *Revista Brasileira de Educação, 13*, 5–24.

Van Eekelen, I. M., Vermunt, J. D., & Boshuizen, H. P. A. (2006). Exploring teachers' will to learn. *Teaching and Teacher Education, 22*(4), 408–423. doi:10.1016/j.tate.2005.12.001

Vanassche, E., & Kelchtermans, G. (2016). Facilitating self-study of teacher education practices: Toward a pedagogy of teacher educator professional development. *Professional Development in Education*, *42*(1), 100–122. doi:10.1080/19415257.2014.986813

Vescio, V., Ross, D., & Adams, A. (2008). A review of research on the impact of professional learning communities on teaching practice and student learning. *Teaching and Teacher Education*, *24*(1), 80–91. doi:10.1016/j.tate.2007.01.004

Wei, R. C., Darling-Hammond, L., Andree, A., Richardson, N., & Orphanos, S. (2009). *Professional learning in the learning profession: A status report on teacher development in the United States and abroad*. Dallas, TX: National Staff Development Council.

Winch, C., Oancea, A., & Orchard, J. (2015). The contribution of educational research to teachers' professional learning: Philosophical understandings. *Oxford Review of Education*, *41*(2), 202–216. doi :10.1080/03054985.2015.1017406

Chapter 8
Using Multimodal Narratives in Teacher Education

José Alexandre Pinto
Politécnico do Porto, Portugal

ABSTRACT

Teacher training processes must incorporate a reflective dimension as a strategy for professional development. The pursuit of a professional identity and the need to give personal meaning to theoretical principles grounds the emergence of young teachers' reflection. In this chapter, multimodal narrative (MN) is presented as a tool to support reflexive approaches in the development of teachers' professional knowledge. The data collected about the perspectives of student teachers and supervisors who experienced the use of MN show the interest of these actors about the tool and about the processes of its use. This chapter presents and discusses constraints identified throughout the pilot study of using a MN in teacher training that the authors developed. It also presents a proposal for the use of MN in the context of initiation of the professional practice that includes an adapted version of the MN tool and a phased process of its use.

INTRODUCTION

As future education professionals, teachers during their training need to develop a set of skills that enable them to establish a theory-practice dialectic. In this sense, the adoption of a reflexive attitude proves extremely important for their professional future. It is up to the teacher to take a proactive attitude toward the research premise and view it as a training strategy.

This chapter attempts to contribute to ensure an enriching experience for young teachers during their professional practice initiation. The present work try to value the perspective of those who have undergone this challenge but also those who, with more experience, take responsibility to follow them in order to become better professionals. Thus, the focus of this work is on the day-to-day work of teacher trainers and student teachers, particularly in supervising and internship tasks. The importance of the vocational training dimension in teacher education courses is well identified if one considers its preponderance

DOI: 10.4018/978-1-5225-8570-1.ch008

in the curricular background for teaching qualification in Portugal. This principle is in harmony with guidelines that emphasize training within the teaching profession (Nóvoa, 2009; Perrenoud, 1998).

The multimodal narrative (MN), as a document, is a written story of a class, with a timeline, and describes all events that take place in the classroom. The MN is focused on actions and languages adopted by the teacher and students during the accomplishment of a task. Its characteristics allow the preservation of the holistic and complex nature of teaching practices.

MN characteristics have aroused the present authors' attention by their potential to incite reflection and to nourish it by the richness and objectivity of the classroom objects and the events that they provide. Next, this chapter presents the authors' work, explaining the context and its actors, which refers to an attempt to value the processes of initial teacher training, in a professional context, incorporating the use of MNs in the training practices.

The authors sought clues, based on the perspective of the student teachers and supervisors, to establish a functional proposal, recognized by the individuals involved and with potential to increase value in the supervising process, throughout a reflective approach.

The pathway was oriented toward the promotion and deepening of reflection, in an attempt to interpret the reality of the classroom. The work was based on collected data, namely, on information, doubts and feelings of the student teachers. The first aim was to provide student teachers with a space for reflection that would challenge them to think about the dynamics of their classes. In this way, the authors intend student teachers to identify what they consider appropriate in view of building their autonomy and professional knowledge. In this sense, the main objectives are defined as: to design a way to operationalize the integration of MN in the process of internship supervision and to evaluate the perspective of student teachers and in-service teachers regarding their interest in the MN in the internship context.

SUPERVISION AND REFLECTIVE TEACHER TRAINING

From the perspective of many authors, teacher training must be structured around reflection as a strategy for valuing pedagogical practice and the development of professional autonomy. This perspective of teachers' education presents itself in an opposite direction to a vision of a prescriptive and external matrix in which teachers fundamentally seek the application and propagation of normative pedagogy perceived as effective (e.g., Alarcão & Roldão, 2010; Sá-Chaves, 2000; Schön, 1987; Zeichner & Liston, 2014).

Assuming the purpose of a reflective teachers' education, it is expected that, in contrast to assuming a role of users of knowledge developed by the academies, teachers can themselves assume a posture of reflective practice by throwing their critical point of view on such knowledge, contextualizing it in their practice (Farrell & Ives, 2015; Wallace, 1991). A significant part of the life experiences of all young teachers is related to their experiences as students. These experiences support their convictions, beliefs, attitudes and values about the meaning of teaching and learning. It is essential to consider such prior and personal knowledge, as it can hinder change and innovation. In this context, it is crucial to support changing teachers' beliefs (Brownlee, Ferguson, & Ryan, 2017). It is in this perspective that Handal and Lauvås (1987) consider that one of the main functions of supervision is to develop the teachers' practical theories, making them sensitive to change.

Theory and practice are two central concepts in teacher education. As Korthagen (2012) warns, it is easy to pull the training process to one of these extremes. On the one hand, adopting a theory-in-practice approach, traditionally prevalent, one attempts to teach the most important knowledge in the expectation that the student teacher will apply it directly in practice. Or, on the other hand, promoting a school-based approach, one hopes that by experiencing the daily action of experienced professionals, the student teachers build their knowledge. This approach, close to trial and error, also entails great weaknesses. Korthagen (2012) states that in both approaches there seems to be a gap between theory and practice. In order to overcome this, Korthagen proposes a realistic approach to teacher training that is very explicit in the example that the author presents, which is an analogy to the captain's training process; this serves in the present authors' eyes as the scenario sought for their approach. Korthagen (2010) presents a five phase model of reflective practice in teacher education designed with the principles of a realistic approach, the ALACT model: action, looking back on the action, awareness of essential aspects, creating alternative methods of action and trial. This model is an example where the process starts with students' own experiences (Korthagen, 2010, p. 414):

1. Starting with the concrete practical;
2. Promoting systematic reflection on student teachers' own feeling, thinking and acting; learners' feeling, thinking and acting; the role of context; and the relationships between these aspects;
3. Building on interaction among students and between students and teacher educators;
4. Using a three-level model (gestalt, schema and theory); and
5. Integrating theory and practice as well as several disciplines.

Among the many definitions of reflection that may be found, there are many differences, but there are also many unities. Thinking about the most important similarities, one notes the importance of emotions (Zeichner & Liston, 2014) as well as critical agency (Hickson, 2011; Zeichner & Liston, 2014). These are aspects that are found repeatedly in the work of several authors on reflective practice (e.g., Jamil & Hamre, 2018; Dennis, Gelfuso, & Sweeney, 2018).

In this sense, Zeichner, Payne and Brayko (2015) argue that "the creation of new hybrid spaces (…) where academic, school-based, and community-based knowledge come together in less hierarchical and haphazard ways to support teacher learning" is important.

According to Vieira (1993), the reflective practice in teacher training, which is the purpose of the present paper, should follow the principles of i) focus on the subject and processes, ii) knowledge problematization, iii) theory and practice integration, and iv) metacognitive introspection, which is framed in the following assumptions: a) practice generates theory; b) the good professional is a reflexive subject; and c) a professional with autonomy train autonomous professionals. With regard to this last point, the author remembers the relevance of the supervisor and the style that the supervisor adopts. The present authors do not see themselves reflected in a model of normative supervisor, who assumes all control, granting a low degree of freedom and autonomy to the student teacher. On the contrary, they have the perspective of those who understand the supervisor as a promoter of emancipatory communicative relations, a promoter of democratic relations between supervisor and trainee (Sá-Chaves, 2000). In a scenario of diversity in supervisory styles, Glickmann, Gordon, and Ross-Gordon (2001), characterizing different supervisory styles, emphasize the difficulty in adopting one style, bearing in mind the multiplicity and complexity of professional situations. Alarcão and Sá-Chaves (1994) propose an ecological approach for the professional development of teachers, considering the multifaceted context in which supervisor and supervised subject face the whole process in a permanent dynamism. The present authors intend that the supervisor's style also inspires the reflective dimension of teacher training and believe that this is important to ensure. They recall the contribution of Nóvoa (2009), in the same bearing as Gardiner and Lorch (2015), when he identifies the need for a teacher training built within the career, and the present authors agree with him when says that he prefers a "deliberative transformation, insofar as the teaching work does not translate into simple transposition, because it presupposes a transformation of knowledge, and requires a deliberation, that is, a response to personal, social and cultural dilemmas" (Nóvoa, 2009, p. 210).

INTERNSHIP SUPERVISION WITH INTEGRATION OF THE MN TOOL

In all that has been mentioned, reflection emerges as a central strategy in teacher training: reflection fueled by interaction between subjects (supervisors and trainees) and reflection centered on theory and practice, establishing bridges between the two in a personal construction of knowledge. The challenge of knowing how to build a formative course that welcomes and encourages reflection and through it harmonizes theory and practice is identified. The search for concrete objects and focus in this scenario, which deepens teachers' practices, was the reason that led the present authors to look to MN as a rich and potential tool.

Participants and Supervision Practices

This study took place in three academic years and involved working with three groups of students (2014/2015 group A, 2015/2016 group B and 2016/2017 group C) that corresponded to the classes of an initial primary teacher training course. All the groups, the first two with 24 students and the other with 22, were made up mostly of female students, aged between 21 and 23 years. The work was based on the articulation between two curricular units of the course, one focused on science education (Didactics of Natural Sciences) and another on teaching practice.

Emphasizing the collaborative dimension as a training strategy, internship groups are formed of four elements each seeking to emphasize collaborative work and peer reflection. The groups are distributed by the training schools according to the availability of the schools and the preference of the student teachers. This task is always the responsibility of the students. The four elements have in common the school, the classes and the supervisor. The group is subdivided into dyads, and to each one is distributed a class. At the end of the first semester, the students exchange classes. All the internships were distributed by four internship schools based in the area of the formative institution. The organization of all internship activities was always one of the first tasks and permanently involved the student teachers and supervisors. This task allowed the structuring of a work plan that made explicit: presence in the class, times of weekly meetings with supervisors and types of intervention in the class (observation, cooperation or regency). An attempt was made to incorporate into the sequence of intervention of the student teachers cycles of observation, of cooperation and of regency, in order to potentiate bases for a sustained reflection. Both the student teachers of a dyad and the in-service teacher responsible for the class are present in all classroom activities. The supervision process intertwined two dynamics of interaction between the supervisor and the student teachers. The first dynamic was marked by more informal work, face to face or by distance communication, usually related to reflection around events in the training schools, ideas about strategies or resources used or to be developed and implemented, or difficulties related to management of the curriculum or the classroom, among other needs and challenges that students felt and wanted to share with the supervisor. The second dynamic, previously scheduled and always fulfilling the observation cycle (Vieira & Moreira, 2011), included pre-observation meetings, observation, and post-observation meetings. The moments of reflection were mainly supported in notes elaborated by the in-service teacher, supervisor and student teachers. Sometimes records have personal characteristics and structure, while at other times the records are based on instruments focused on a certain dimension often found in and adapted from the literature.

MN Integration in Supervision of Internships: First Approach

The first attempt in the present research was to transfer to teacher training work with multimodal narratives that the researchers had developed in the university-based research group. This first approach was based a) on the belief that the reflection around concrete experiences or events makes it possible to relate theory and practice and, in this way, to construct a professional identity and b) the authors' experience of seeing preserved in MN classroom dynamics with great potential for reflection, which they would like to see deepened in the supervising process of student teachers. At first it was decided to act in two curricular spaces. In the curricular unit Didactics of Natural Sciences (first semester), the results of the investigation have gained new interest. In addition to the strict thematic that inspired them, the authors also began to value the methodological options of research expressed in them, particularly those concerning the elaboration and use of MN. In addition to raising awareness and training for research in education, the present authors promote, with their students, training aimed at knowledge about the purpose and structure of MN. Its nature and structure, ways of elaborating and validation strategies were explored. The authors also recommended the exploration of MN examples carried out within their research team. On the other hand, in the other curricular unit, Supervised Teaching Practice (annual), it was decided to encourage student teachers to choose one lesson that they had implemented in training schools and to construct a MN of it. This MN would be validated by the pair (the other student teacher of the dyad) that was present during all activity in the classroom. The MN should be analyzed in the meeting between student teachers and their supervisors. It was always explicit that this was an individual option, with no direct impact on the final classification and with freedom to change decisions along the way. The authors were aware that this was the first approach.

About 37% of the students in group A and 25% of those in groups B and C accepted the challenge. In the first exercises of MN construction, signs of consolidated aspects were observed that were not considered to constitute difficulties for student teachers. One the other hand, signs of pronounced fragilities were detected that would certainly have an impact on future work. One important fragility is visible in the next excerpt of MN. There was a pronounced tendency to avoid detailing the teacher's action. For example, the excerpt does not provide details about how the student teacher challenged the class to present and discuss their ideas about the video presented. It is also impossible to know how the teacher's action was developed to try to involve the students in the answer to the questions written on the blackboard, nor about how the student teacher acted in order to improve the students' participation in the task.

Multimodal Narrative Excerpt 1: Teacher A (Embryonic Development in Humans)

Begin at 9:25 a.m.

[...]

The teacher began by informing the students that they were going to see a video on the development of the embryo. So she put the video on, turned off the lights in the room, and as the students visualized it, she moved to the side of the room and it was necessary to ask some students to be attentive and not talk. When the video ended, and after a dialogue, the teacher informed the students that they should transcribe into their daily notebooks the questions she would write on the board, and respond to them.

Afterward she wrote the following questions on the board:

Where does the embryo develop?

What organ controls the nutrients provided to the new being?

What is the name of the internal communication channel established between the mother and the new being?

When does the embryo begin to be called a fetus?

How long is the gestational period?

[...]

When the teacher tried to begin correcting the questions, the students were impatient, and it was necessary to adjust the comportment of the students several times. Then, the questions began to be answered:

[...]

The accentuated occurrence of these difficulties in the first MN produced by one of the student teachers, combined with the non-identification of them in the validation process made by the pair, alerted the researchers immediately to work on this. Another very visible situation in the first approach was related to the structure of the MN produced. In general, student teachers interpreted the general structure of an MN well. All of them integrated the planned sections, namely, the classroom setting; a description of the space where it took place, including a room plan; the development of the MN by episodes; and the use of multimodal elements. However, the difficulty in interpreting the role of multimodal elements in that construction was very visible. Images with little potential are often found because they are repetitive or little related to the written text and, in several cases, grouped at the end of the MN as an attachment. Thus, the multimodal elements lost the potential to show the dynamics in the classroom. Another aspect that also relates to the structure is the concept of episode.

The episode corresponds to the work developed during a task. It begins with the presentation of a task and ends with its conclusion. The episode can be terminated prematurely on the teacher's initiative. It should correspond to a medium-sized unit and not to an immediate occurrence.

As this is a structuring concept of MN, it has often been assumed to refer to any activity, often with a very short development.

The integration of MN in the routines of an internship requires a strong perspective of its added value, shared among all the intervening subjects. Many reasons can influence its use and dictate the success with which it is made. However, none will be more decisive than, with recognition of its formative value, its acceptance as a tool that can genuinely help in the professional development of student teachers.

The researchers, conscious that they were in an initial search for ways of integrating MNs, always tried to be attentive and close to everyone.

The MN examples obtained from the first approaches were a rich source of data and resources that founded the improvement of the proposal and guided the work with the following groups. A very relevant aspect was to have "imperfect" examples to show and with which to work. Until then, students had explored examples of completed and validated MNs. The researchers' attempts to integrate alerts for difficulties that were anticipated seem not to have been sufficient at first.

From the first to the second and third groups, the researchers maintained the proposal in the previous terms. However, in an effort to add value to the intervention, the researchers integrated examples of acquired experiences and sought to include in-service teachers in the approach. Since it was not possible to have systematic training for all, the decision was made to work closer with two in-service school teachers who had previously worked with the present authors and who knew the MNs. They were given particular relevance in the task of validating the MNs. The joint effort allowed an evolution in the quality of the MNs, as is explicit in the following excerpt. This evolution produced more detailed MNs that better fulfilled the structure. This evolution of MNs provided encouragement and enabled the researchers to anticipate the potential impact of MNs on reflective processes.

Multimodal Narrative Excerpt 2: Teacher A (Invasive Species)

Episode 1: invasive species

Class time: 120 minutes

Begin at 10:45 a.m.

Contextual information:

This is a group of students between the ages of 8 and 10 who attend the 3rd year of schooling. The group consists of 18 children: 6 females and 12 males. One of the students is repeating the 3rd year, 3 are retained in the 2nd year, and 1 attends the 3rd but is enrolled in the 4th year. They have an individual educational plan.

The school does not have a laboratory; however, there is a small washbasin in the classroom, and there is also a balcony that facilitates more practical activities. The organization of classroom space varies

Figure 1. Classroom Plan

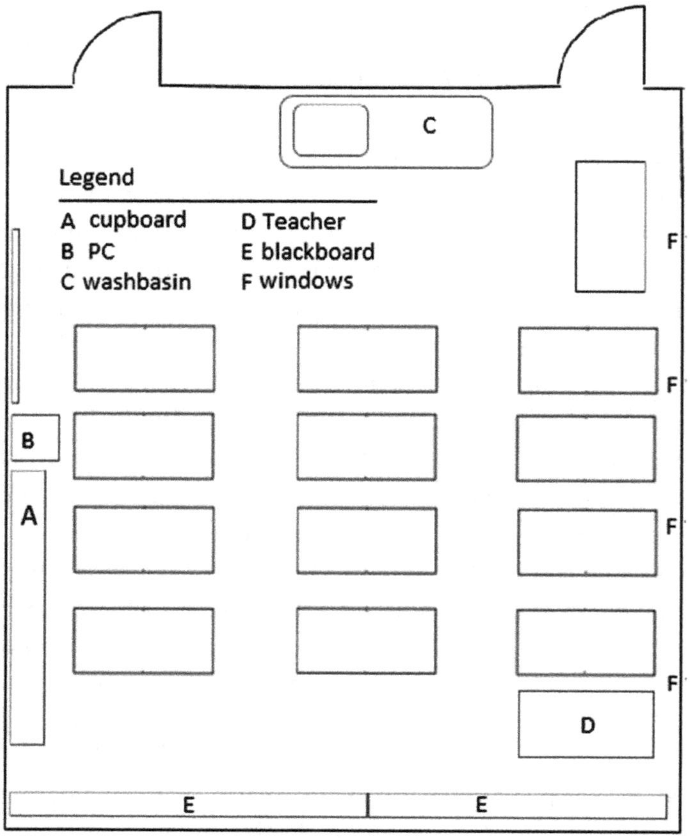

throughout the school year for several reasons, and there is freedom to change it whenever the activities developed require.

Episode 1

I started the lesson by observing and reflecting on the results obtained in the experimental activity (regarding invasive species), questioning the students about the problem that we had defined previously, so that, orally, I recalled the work that we were developing. Since the tray in which the seeds were planted was always in the classroom, I chose to hold it and stood by the table, inclining it to make it easier to see.

I asked the students to answer, given the observation made over the course of the day, what would be the solution to our problem question, "What is the invader?" Marco very quickly said, "Species 1." As everyone agreed with Marco, I asked them to explain what happened.

Teacher: "What happened?"

"Species 1 grew on the fire," said Marco.

"The burn has grown and the unburnt one has not grown," added Diana.

Teacher: "And with species 2 what happened?"

"It did not grow on both sides," said Thomas.

Teacher: "Why do you think nothing happened?"

"Because it's not invasive," Joao said.

Teacher: "But normally, even if it was not invasive, it should have grown when not burned, right? A plant when it is not burnt usually grows. Why did this happen?"

"It did not have time," David said.

"Perhaps you have not had time yet. It's like the pines, they take a lot of time to grow," Marco added.

Teacher: "Yes, you may not have had enough time. So, in the species they claim to be the invader, why did it not grow up when not burned?"

"Because it's invasive," Ricardo said.

Teacher: "Is it invasive and what is the characteristic?"

"It only grows when it is burned, otherwise it will not grow," said Thomas.

Teacher: "Is it always like this?"

"Depends, it's because this plant is fireproof," Filipe said.

"When you have a fire, when the fire is over, it's not ruined, it's still alive," Marco said.

Teacher: "Yes, but as we have seen, the seed is still 'alive' and can grow."

"Teacher, if there is a fire, does not the plant burn?" asked Miguel.

Teacher: "The plant can burn, but the seeds are able to germinate because it resistant."

As is clearly visible in the excerpt presented above, the MN began to incorporate more detailed information regarding both the intentions assumed by the student teacher and the interactions that occurred throughout the work developed. It was verified that there was care in initiating the MN, including sufficient contextual information so that the environment where the class was developed was better known. This care was extended to the first paragraph of the episode, where detailed information was added on how the lesson theme was started and how it was done to guarantee a full observation of the material that was intended to be shown. Acknowledging the relevance of the mentioned dimensions' incorporation in the MN corresponds to an important gain in the sense of using them as a support for reflective dimensions during the initial teacher training process.

Multimodal Narrative Excerpt 3: Teacher A (Invasive Species – Excerpt 2 Continuation)

Then I presented the students with the photos we took during the experimental activity, starting with the first day, in order to recall the initial global aspect. I presented the photos that were taken a second time, about a week apart, where changes were already taking place, and I asked them to tell me what was possible at this stage.

[...]

Figure 2.

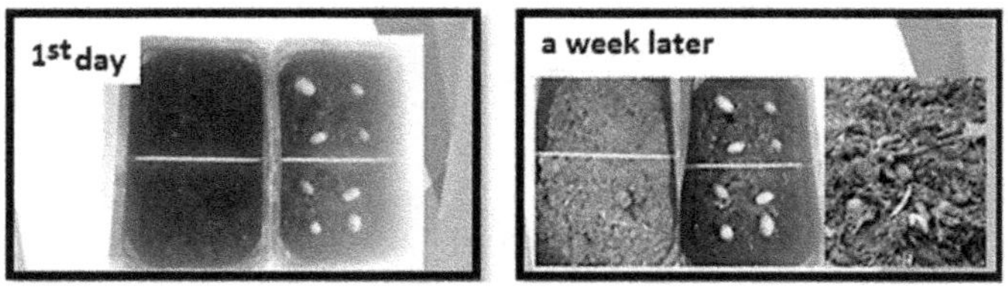

Figure 3.

Figure 4.

Teacher: "On the third day, what happened?"

"It grew a little longer," said Miguel.

Teacher: "And how much did it grow?"

"A centimeter and a half," said Thomas.

Teacher: "Exactly; we made that record in our table.

And after a week, what did we already have?"

"It grew up, looks like a pine tree," said Marco.

"It already had leaves," David added.

Teacher: "And this part here?" I pointed to the photograph.

"It's another one to grow," Miguel said.

"And what happened to species 2?"

"It did not grow up," she told Diana.

Teacher: "Exactly, nothing had happened yet. And how were you yesterday?"

"I was getting big leaves," Marco said.

"I was a little longer..."

Teacher, is this a bush?" asked Thomas.

Teacher: "No, it's a tree, but it grows that way first."

"Teacher, after growing so much, it was like that," said Marco, with his hands representing the pitch the plant had acquired.

Teacher: "It was a little tart, was not it?"

"Is it a shrub?" asked Marco again.

"Is it a pine tree?" asked Miguel.

"In this case it will be a tree, but only when it starts growing" I explained again, trying to be more explicit. I nodded and said it would not be a pine. "What kind do you think it is?" I asked the question to the whole class and at the same time they answered:

"It's the burned kind," said Marco.

"It's the invader," Thomas said.

"It is species 1," said Miguel.

Teacher: "But what will the name of the plant be?"

"A pine," said Marco again.

Teacher: "A pine tree? Will it be?"

"Sure!" said Miguel.

"It could have been the bucks we saw here that were invasive, but…in this case they are."

"A bush," Marco insisted again.

Teacher: "Which bush? I asked for a name..."

"A strawberry," said Thomas, and many of his colleagues looked reprobate.

"Will it be?"

"A tree," he said to Diana.

"Yes, it's a tree, but that's what I said, I want to know the name..."

Since there were no more ideas, we said we'd see.

"So our invasive species is the one that's here, is it?"

"Black Acacia," read several students.

"And do you understand what it is?"

Several students made a few attempts to read what was written.

Teacher: "And what is this (Acacia decurrens)?"

"It's its scientific name," said Ricardo.

"Exactly, the scientific name of this plant. And now we'll see more features. What is this here? Can you see it?"

"Teacher, the picture from below looks like bacteria," said Diana. (...)'

In the sequence mentioned previously, the MN excerpt presented illustrates the caution taken by the student teacher in providing information that best characterizes the interactions that occurred throughout the developed work. This characteristic is visible by the presence of expressions that detail such interactions, such as, *"I asked the question to the whole class"* or *"Marco insisted again."*

In this excerpt, it can also be seen how the student teacher was cautious in providing information about the reasons that led him to a certain initiative. This is visible by the presence of expressions such as, *"Since there were no more ideas"* or *"Several students made a few attempts to read what was written."*

As is clearly visible in the MN excerpt presented above, the MN began to integrate images with potential to better explain what happened and the classroom environment.

The incorporation of all these characteristics in the MN developed by the student teachers constitutes a determinant milestone, as much to the relevance of its use as to a support of reflective dimensions in the initial teacher training process.

Figure 5.

Perspectives of In-Service Teachers and Student Teachers

Student Teachers' Perspective

To support an interest in evaluating the viability and importance of MN use, data were collected from the participants involved. At first, data were collected informally but with profound and meaningful results. A focus group was promoted to collect information about how to improve the practice of MN use in internships. This was conducted with the group of student teachers who agreed to develop MNs on a regular basis and with the in-service teachers who accompanied this process, taking advantage of routine conversations. Student teachers' written records were also frequently obtained and used in construction of the Internship Report.

Although the number of students who decided to accept the challenge of building MNs remained low and decreased from the first to the second and third groups, the quality obtained fueled the researchers' expectations about the potential of their use. In the effort to understand the reasons why few have accepted the task of elaborating and working with MNs, one reason arises immediately – the ever-present complaint: volume of work. The following excerpts of some written reflections are representative of the general consensus regarding the interest and value expressed by student teachers about the use of MNs.

The multimodal narrative presents a moment of a supervised teaching practice class. The execution of the same made me come back to the moment and reflect on what happened, about the dialogues that existed between the students and me and the strategies used. This is very helpful. (Student teacher, group A)

Many aspects of the class seemed to be non-existent but were reality. By building the MN it gave to become more aware of how it really was. I can look better to myself as a teacher and I know better what I can improve. (Student teacher, group B)

I have no doubt that we all find the MNs very valuable and should do them, but the truth is that they give a lot of work, and for everyone. Often the colleague cannot validate or the professor takes time, and we do not have it at the right time.

In short, the multimodal narrative is an instrument that makes the students return to thinking about the lesson, mainly, in the dialogues elapsed and in the decisions taken. (Student teachers, group C)

Difficulties were also identified. In this sense, the researchers were attentive to the testimonies of those who had not decided to elaborate MNs. In written testimonies or informally, the student teachers showed little confidence regarding their use, namely, regarding time spent, structure and maintaining neutrality, among other aspects.

In-Service Teachers' Perspective

In-service teachers involved in internship also recognized the value and interest in the use of MNs. They recognized that their elaboration requires care, the planning of tasks and a significant investment of time. The following comments particularly emphasize their testimony on aspects that are valued.

When we discuss the lesson with a MN, it is easier to say what we want. On the one hand, we have the examples, and it seems that we remember things. For example, as I have interns for a long time, I know that many times during class I see something I want to say and even take note, but then with time and other ideas I sometimes end up not knowing well what I wanted at that moment. MN seems to shed light on those moments. [...]

Sometimes it seems that student teachers do not even believe that what we are saying happened. I do not want to say they are dishonest, but the idea remains ... sometimes they even make me stay in doubt...

The truth is that MN helps to become aware of things... (In-service teachers, groups B and C)

The testimonies show consensus about gains due to the use of MNs, especially regarding: a) awareness and objectification of practices, b) preservation of information for present and future memory, c) incentive for reflection on the practices and d) fostering collaborative relationships between student teachers, in-service teachers and supervisors.

In summary, based on data and experiences during the work, the researchers recognize MNs as a valuable tool in the context of initial teacher training, in particular due to their contribution in reflection dynamics. Therefore, MNs should be strategically integrated into the supervision routines of internships. The feasibility of the use of MNs in the context of internships is recognized, but it needs to be planned.

MN'S INTEGRATION IN SUPERVISION OF INTERNSHIPS: SOLUTIONS AND RECOMMENDATIONS

After identification of the challenges – namely, time constraints, difficulty with structure, difficulty in fulfilling the nature of the data, difficulty maintaining neutrality, difficulties in timely interaction and the task of encouraging the engagement of student teachers in the MN process – a proposal was created for the integration of MN in the context of initial teacher training. This proposal is based on two pillars:

1. *To value its use in curricular units (theoretical and practical) developed during the course in the superior institution.* MNs have the potential to make an interesting contribution in curricular units of a more theoretical nature. They allow students to explore, discuss and share their views around concepts, combining them with classroom events in close contexts (schools they know, with students and/or teachers they know), presented clearly and authentically. This dynamic has an interest in promoting motivation for reflection, bringing about situations related to the subjects and their surroundings, and facilitating the appropriation of theoretical orientations, facilitating their confrontation with real examples. Another advantage is that, being familiar with the work around MN, students acquire knowledge about this tool, which facilitates its later use to produce their own MNs in the context of internships, in classes conducted by themselves or their colleagues.
2. *To adjust MNs to the internship conditions.* The need for time management involved is the basis for the proposal of a shorter MN structure. MNs should integrate the initial contextualization and the development of a single episode. The responsibility to decide on which episode to consider is shared between student teachers, in-service teachers and the supervisor. The adapted version of MN comprises:

Table 1. Proposal for the integration of MN in the context of initial teacher training

Time	Tasks	Actors
Pre-observation	Lesson plan elaboration Sharing options and deciding which episode to consider in MN	Student teacher regent initiative All (student teachers, in-service teacher and supervisor)
Class observation	Classroom management Data collection for MN Collection of data in general	Student teacher regent Student teacher (pedagogical pair) In-service teacher
Post-observation	Elaboration of MN Validation of MN MN sharing for everyone and meeting scheduling Classroom reflection meeting	Student teacher (pedagogical pair) in interaction with student teacher regent In-service teacher Student teacher (pedagogical pair) All (student teachers, in-service teacher and supervisor)

a. **Contextual Elements:** Specific time, space and circumstances of the classroom; teacher's and students' characteristics; classroom plan and predominant spatial position of teacher and students. Being that student teachers characterize in detail all context at the beginning of the stage, in the adapted version of MN, student teachers are asked to focus only on particular aspects that are different in the class reported.
b. **Description of the Storyline:** These units around the tasks are called episodes, and they chronologically mark the time of the class. In the adapted version of the MN tool, student teachers are asked to identify the different episodes within the class.
c. **Descriptive Part:** Each episode is narrated from its beginning until its end. In the adapted version of the MN tool, student teachers are asked, in accordance with an in-service teacher, to select one episode. The student teacher must present the reasons for selecting the particular episode and narrate it.

The MN should be done by the pedagogical pair in articulation with the student teacher regent. Validation will be ensured by the in-service teacher in articulation with student teachers. Once this course is complete, there is a post-class meeting schedule with participation of all, including the supervisor. In a next phase, the functions are exchanged between the pedagogical pair and the process repeated. The following table summarizes the tasks.

CONCLUSION

This chapter reports a proposal to incorporate the MN tool in the supervising process in the context of initial teacher training. MNs were seen by teachers and student teachers as a viable tool with great potential for promoting professional skills and scaffolding a pathway to immerse in reflection processes. In particular, this chapter emphasizes the contributions of MN to awareness of practices, the objectification of class moments, preservation of information for present and future memory, encouragement of reflection on practices through the retrospective look that it requires, and the fostering of collaborative relationships between student teachers, in-service teachers and supervisors involved in internships. The data collected support the idea of great potential in reinforcing and reaching the theoretical principles that have been outlined and presented, of which the following serve as highlights: motivation for reflection, valuing reflection as a professional development strategy, and construction of professional identity. Challenges that should be considered were identified, in particular: (i) managing time and workload and ii) engagement of student teachers in the MN process. The chapter presented a proposal for integrating MN at the stage that incorporates the received criticisms and overcomes the obstacles identified.

The described pathway and the achieved results make it possible to move forward in the integration of MN in the internship supervision process. The present authors are currently implementing MN integration under the conditions described in a training course for primary teachers and collecting data on the whole process, with an interest in evaluating the dimensions of reflective work and its depth. This collaborative construction, achieved in the interaction between the different actors, is still a starting point. It remains necessary to continue a systematic approach to validate all convictions regarding the gains from the use of MN, which is a task for future work.

REFERENCES

Alarcão, I., & Roldão, M. C. (2010). Supervisão: Um contexto de desenvolvimento profissional dos professores [Supervision: A context of professional development of teachers] (2nd ed.). Odivelas: Edições Pedago.

Alarcão, I., & Sá-Chaves, I. (1994). Supervisão de professores e desenvolvimento humano: Uma perspectiva ecológica [Teacher supervision and human development: an ecological perspective]. In J. Tavares (Ed.), Para intervir em educação: Contributos dos colóquios (pp. 201-232). Porto: Edições CIDInE.

Brownlee, J., Ferguson, L., & Ryan, M. (2017). Changing teachers' epistemic cognition: A new conceptual framework for epistemic reflexivity. *Educational Psychologist*, *52*(4), 242–252. doi:10.1080/00461520.2017.1333430

Dennis, D. V., Gelfuso, A., & Sweeney, S. (2018). Trying and reflecting: Two novice teacher educators' first attempts at guiding pre-service teachers' reflection on literacy field experiences. *Education 3-13*, *46*(4), 456-468.

Farrell, T., & Ives, J. (2015). Exploring teacher beliefs and classroom practices through reflective practice: A case study. *Language Teaching Research*, *19*(5), 594–610. doi:10.1177/1362168814541722

Gardiner, W., & Lorch, J. (2015). From "outsider" to "bridge": The changing role of university supervision in an urban teacher residency program. *Action in Teacher Education*, *37*(2), 172–189. doi:10.1080/01626620.2015.1004601

Glickman, C. D., Gordon, S. P., & Ross-Gordon, J. M. (2001). *Supervision and instructional leadership: A developmental approach*. Boston: Pearson.

Handal, G., & Lauvås, P. (1987). *Promoting reflective teaching: Supervision in action*. Milton Keynes, UK: Open University Press.

Hickson, H. (2011). Critical reflection: Reflecting on learning to be reflective. *Reflective Practice: International and Multidisciplinary Perspectives*, *12*(6), 829–839. doi:10.1080/14623943.2011.616687

Jamil, F. M., & Hamre, B. K. (2018). Teacher reflection in the context of an online professional development course: Applying principles of cognitive science to promote teacher learning. *Action in Teacher Education*, 1–17.

Korthagen, F. (2010). How teacher education can make a difference. *Journal of Education for Teaching*, *36*(4), 407–423. doi:10.1080/02607476.2010.513854

Korthagen, F. (2012). A prática, a teoria e a pessoa na formação de professores. *Educação, Sociedade & Culturas*, *36,* 141-158.

Nóvoa, A. (2009). *Para uma formação de professores construída dentro da profissão* [For a teacher training built within the profession]. Retrieved from http://www.revistaeducacion.educacion.es/re350/re350_09por.pdf

Perrenoud, P. (1998). La transposition didactique à partir de pratiques: Des savoirs aux compétences [Didactic transposition from practices: From knowledge to skills]. *Revue des Sciences de l'Education*, *24*(3), 487–514. doi:10.7202/031969ar

Sá-Chaves, I. (2000). *Formação, conhecimento e supervisão: Contributos nas áreas de formação de professores e de outros profissionais* [Training, knowledge and supervision: Contributions in the areas of teacher training and other professionals]. Aveiro: Universidade de Aveiro.

Schön, D. (1987). *Educating the reflective practitioner*. San Francisco: Jossey-Bass Publishers.

Vieira, F. (1993). Observação e supervisão de professores [Observation and supervision of teachers]. In F. Sequeira (Ed.), *Dimensões da Educação em Língua Estrangeira* [Dimensions of Foreign Language Education] (pp. 69–90). Braga: Universidade do Minho.

Vieira, F., & Moreira, M. A. (2011). *Supervisão e avaliação do desempenho docente: Para uma abordagem de orientação transformadora* [Supervision and evaluation of teacher performance: Towards a transformative orientation approach.]. Lisbon: Ministério da Educação. Retrieved from http://www.ccap.min-edu.pt/docs/Caderno_CCAP_1-Supervisao.pdf

Wallace, M. J. (1991). *Training foreign language teachers*. Cambridge, UK: CUP.

Zeichner, K., & Liston, D. (2014). *Reflective teaching: An introduction* (2nd ed.). New York: Routledge.

Zeichner, K., Payne, K., & Brayko, K. (2015). Democratizing teacher education. *Journal of Teacher Education*, *66*(2), 122–135. doi:10.1177/0022487114560908

Chapter 9
Multimodal Narratives as a Tool for In-Service Teachers in an Online Professional Development Course

Daniela Pedrosa
Universidade de Trás-os-Montes e Alto Douro, Portugal

Gonçalo Cruz
Universidade de Trás-os-Montes e Alto Douro, Portugal & Universidade Aberta, Portugal

Leonel Morgado
Universidade Aberta, Portugal & LE@D, Portugal

ABSTRACT

This chapter presents how multimodal narratives were employed as a self-reflection tool within an online professional development program for in-service teacher training at Universidade Aberta, Portugal during two editions of a pedagogic practice course. The chapter includes the aspects that raised issues and those that trainees performed correctly. This is done in three stages: beforehand, upon initial contact with multimodal narratives, and after providing feedback to trainees. The most relevant issues were in aspects directly required to enrich the narrative. Aspects related to multimodal narrative structure and features were completed successfully. It is recommended that future attempts to employ multimodal narratives in this context adapt learning resources and pedagogic support practices by employing formative feedback and continual support during the trainees' process of exploring and exploiting multimodal narratives.

INTRODUCTION

The "Pedagogic Practices Seminar" course took place at Universidade Aberta (UAb), Portugal, in the second semester of the 2015/2016 and 2016/2017 school years, as part of an online training program for the in-service professional development of teachers. This was a formal, non-graduation program.

DOI: 10.4018/978-1-5225-8570-1.ch009

The trainees were active non–higher education teachers who aimed to qualify for recruitment group 550 (Computing) of the Portuguese Ministry of Education.

The goal of this course is for students to develop scientific and pedagogic skills for creating and developing pedagogic projects in computing education. They must identify, reflect upon and solve learning issues, combining pedagogic and research components to improve their practices and their professional development. Hence, it is important to provide trainees with knowledge of research techniques that enable them to develop those competences, one of them being multimodal narratives (MNs). MNs are typically documents that describe chronologically, in a self-contained and multimodal way, what a teacher and students do and say in a given context of teaching and learning (Lopes et al., 2014), incorporating multimodal data. They are drawn up in accordance with the guidelines of a protocol defining three stages of implementation: data collection, preparation of the MN and validation. In this way, a MN is a self-contained and concise document that can be analyzed later, avoiding the difficulty of dealing with multiple data sources (*ibid.*). MNs are indeed tools in support of teaching and professional development, since they help teachers understand their decision-making process and the intent of their actions (Lopes et al., 2010). They enable teachers to observe the actual events within their classes (sometimes distinct from their initial perceptions or recollections), and they facilitate the identification of which pedagogic strategies were most effective (*ibid.*). They hold the potential to help improve teachers' professional development (Lopes et al., 2014).

There were several learning activities in the Pedagogic Practices Seminar course. The contact of trainees with MNs took place in the third, fourth and sixth topics of the course, which involved the trainees in reading about their use and then creating several MNs related to episodes in their current professional practice.

The initial contact with the MN technique was in the third topic.

This chapter identifies which aspects raised the most significant issues upon first contact as well as those that were correctly performed by the trainees, over two editions of this course (2015/2016 and 2016/2017). The same analysis was done after providing formative feedback (which was provided only in the 2016/2017 edition).

In both editions of the course, the trainees' initial contact with MNs was challenging, both regarding its initial use and its improvement.

Four categories were analyzed, referring to essential aspects of a MN: Category 1 – *Presents the essential aspects of a MN*; Category 2 – *Presents a well-structured MN*; Category 3 – *Narrating an episode with its essential aspects* and Category 4 – *Enriching the MN*. Most issues occurred in categories 3 and 4, i.e., the aspects that detail and enrich a MN.

In categories 1 (*Presents the essential aspects of a MN)* and 2 (*Presents a well-structured MN*), which are related to MN structure and features, even though several issues emerged, these were infrequent and/or solved after providing feedback (2016/2017 edition only). Therefore, the results suggest that in both editions (2015/2016 and 2016/2017), the students understood the importance of the MN and satisfactorily identified the main features of a MN.

It is also possible to conclude that formative feedback is essential for students to identify, reflect upon and improve their processes for creating a MN. This arises from observing that trainees of the 2016/2017 edition, who had the opportunity to improve their assignments after receiving feedback, managed to overcome and understand their shortcomings in executing a MN.

It is recommended that future attempts to employ multimodal narratives in this context adapt the instructions and models provided for creating a MN, and also that deadlines be adjusted in order to provide formative feedback on trainees' assignments.

The remainder of this chapter is structured as follows: the following section, "Background," summarizes the main cornerstones of UAb's pedagogic model and presents the theoretical underpinning of the present authors' practice and research; then, the "Context" section presents the course where this research took place, detailing its structure and goals, its public and how it evolved over the two editions under study. The methodology is described in the subsequent section, including data procedures. The penultimate section details the results, and the chapter concludes with a final reflections section, which highlights the major conclusions and their links to the background literature.

BACKGROUND

The Universidade Aberta adopted a specific virtual pedagogical model (Pereira, Mendes, Morgado, Amante, & Bidarra, 2008), based on four cornerstones:

1. **Student-Centered Learning:** The student is an active agent and builder of knowledge, able to manage and devote time/effort to the learning process within a learning community that relies on collaborative learning through dialogue and peer interaction.
2. **Flexibility:** This aspect is based on the historical origins of distance learning (DL), namely, learning anywhere, at any moment. The teaching-learning process is asynchronous and continuous, occurring independently of the time-space dimension of students' and teachers' locations, facilitating students' management of learning according to their professional and personal lives.
3. **Interaction:** The model focuses on including student-student interaction through the creation of spaces for discussion in the virtual environment, not just student-content and student-teacher interaction. This requires adequate instructional design and adoption of strategies that stimulate the engagement of students, as well as their commitment and management to achieve learning objectives.
4. **Digital Inclusion:** The adoption of information and communication technologies to support learning activities contributes to diminishing the societal divide between info-included and info-excluded citizens, by empowering them to use these essential modern media of communication and networking.

The development of the Pedagogic Practices Seminar for in-service teachers was thus inspired by these four institutional pedagogic model cornerstones.

Teachers' ability for self-reflection holds the potential for transforming their pedagogic practices, by enabling the development of new actions and improvements. Reflective practice is seen as a facilitating mechanism for change in people, in organizations, in processes and in contexts (Feucht, Brownlee, & Schraw, 2017).

Reflective practice provides teachers with opportunities to develop their professional skills, enabling their awareness and reflection on how to transform their classrooms (Oliveira & Serrazina, 2002). However, developing adequate reflection practices can be a complex process. Current literature proposes that teachers move beyond reflection and exploit epistemic reflexivity. That is, they should employ the available information to rethink their own conceptions and cognitive-epistemic beliefs: an internal reflection that may lead to a new conception of the self and, through that, lead to new ways to act (Feucht, Brownlee, & Schraw, 2017).

Thus, the present authors employ training strategies that ask trainees to reflect upon their practices (Schön, 2000), by combining small-scale interventions with action research features (Cortesão & Stoer, 1997; Coutinho et al., 2009), and articulate this with Bruner's spiral curriculum principles (1960). The rationale is that a reflective approach, involving action-research projects, supports teachers' increase in knowledge about their professional practice and about tacit knowledge. That is, when teachers (course trainees) reflect in action and about their action, they are involved in a research process that enables them to understand their own processes and improve their teaching practices (Oliveira & Serrazina, 2002).

MNs appear in this context due to their potential as a tool to support the identification and reflection upon educational problems, enabling the course trainees to get in touch with the hands-on use of methods of high epistemic rigor.

MNs support teachers' practice and professional development (Lopes et al., 2014) by helping teachers understand their decision-making process and the intent of their actions (Lopes et al., 2010). They enable the teacher to observe the events in their classes, which are sometimes different from their initial perceptions. And they facilitate reflection about which pedagogic strategies were the most effective ones (*ibid.*). Further, MNs enable teachers and researchers to analyze students' learning in depth (Lopes & Cunha, 2017).

Since building MNs is a complex process, it is recommended that one follow the Lopes et al. protocol (2014) in order to create them adequately. The protocol is important because it has established guidelines for data collection and a process by which the MN is built and validated, incorporating the aspects and structure that define a MN.

Therefore, it is fundamental to provide support for professionals/researchers in their first contact with MNs, so that they are created adequately.

This was a challenge due to the context of learning: distance learning (DL), a format where hitherto there had been no prior case of lecturing about the use of MNs or including their use as assignments. DL requires that learners (in this case, trainers) develop certain learning and self-learning skills (Cazan, 2014), namely, in structuring their learning, engaging with course content and managing their study time effectively (Broadbent, 2017).

It is also important to understand how students can use self-regulation strategies (Zimmerman, 2013) appropriately to improve their learning in DL (Broadbent, 2017) and to build their personal learning environment (Dabbagh & Kitsantas, 2012). For this, as Järvelä et al. (2015) explain, not all students need the same kind of support, and there is a need to clarify what type of support is most appropriate for each student to adjust his/her cognitive, motivational and emotional strategies.

Formative assessment is an approach that resorts to a set of evaluations to aid teaching and learning (Shute & Kim, 2014). It employs students' results to adjust instruction, in order to support effective learning. Formative assessment takes into account each individual and his/her characteristics, and it is interactive throughout the development of learning and reflection tasks. It also provides teachers with information that enables them to revise their instruction and improve their teaching-learning process (Shute & Kim, 2014; Reinholz, 2016).

According to Santos (2016), formative assessment aims to assess, but with the goal of helping people learn. The educational goal is to help students become proactive in their learning processes (Black & Wiliam, 2009; Reinholz, 2016).

Adopting formative assessment is more than just providing teachers with tools to help them implement different types of assessment, combine tests, analyze data and adjust instruction (Shute & Kim, 2014): it is also essential to provide students with tools that help them improve their motivation and problem-solving skills (*ibid.*). Formative feedback helps students' self-regulation, motivation and self-awareness about the progress in their work (Hattie & Timperley, 2007; Shute & Kim, 2014), because it provides them with guidance on how to improve their learning processes and encourages their reflection capabilities (Morris & Chikwa, 2016; Nicol & Macfarlane-Dick, 2006).

CONTEXT

Pedagogic Practice Seminar

This course was part of a fully online professional development program for in-service teachers. It was a formal non-graduation program provided by the Universidade Aberta. Its target was active teachers in non–higher education who aimed to qualify for recruitment group 550 (Computing) of the Portuguese Ministry of Education.

The goal of this course was to enable trainees to apply computing didactics knowledge to their pedagogic practice, to be able to identify educational issues, and to reflect and find solutions for these issues. From that reflection, students would draft an intervention proposal, focusing on which strategies to use and providing literature support for the proposed pedagogic practices.

The overarching framework for the intended pedagogic practice was the use of computing as a meta-tool for knowledge, enabling problem analysis and solving, as well as enabling enhanced communication and understanding of ideas (diSessa, 2001).

The course adopted a project-based pedagogic approach, combining it with Bruner's spiral curriculum principles (1960). Throughout the semester, each trainee would design and develop a computing didactics project, refining and improving it across several iterations, by including new topics and techniques, or by resorting to field trials.

The course started in May and ended in July, in both editions. It was organized with six learning topics, where trainees had to complete several tasks, with provided support materials (readings, sample videos, instructions). Online fora were used for debating and sharing doubts. The course structure was provided to the students as a learning contract, which was the subject of discussion and acted as a mediator between the learning requirements and students' needs and interests, as foreseen by UAb's pedagogic model (Pereira et al., 2008, p. 29).

Among the various tasks required of trainees, MNs were created as part of topics 3, 4 and 6. The first contact with MNs, therefore, was in topic 3.

Topic 1 was devoted to presenting the course, the teaching team and the trainees; negotiating the learning contract (e.g., clearing doubts and tuning assessment goals); and identifying themes for each trainee's project (length: one week).

Topic 2 addressed computing as a metatool, both from a theory perspective and from a hands-on perspective by trying out tools for computer programming and game creation targeted at youngsters. The trainees drafted a short report of their use of the tools and a first sketch of their project, where they applied theory to their theme (length: two weeks).

Topic 3 was devoted to developing the trainees' contact with action research methodology (AR), applying it to the teaching of computing. The trainees' main tasks were: clarifying the timeline for development of their project (the one sketched in topic 2), in order to define how to approach it as AR, and a first attempt at creating a MN of a class/activity, as a data collection method supporting the reflection stages of AR (length: two weeks).

Topic 4 explored techniques for the participatory creation of computing and multimedia artefacts using low-tech models. Trainees had to conduct a prototyping activity with their students or other volunteers, where they would apply the computing didactics knowledge they were studying in the course. Trainees would also report that activity as a MN (length: two weeks).

In Topic 5, trainees increased their knowledge on the state of the art of computing didactics and rendered their study in concrete form via their ongoing didactics projects, by drafting a chapter where they would explain how the main concepts from the state of the art in the field were impacting their projects (length: two weeks).

The final topic was devoted to producing the final version of their projects, under supervision of the course teaching staff. Trainees' tasks were: revising and improving the didactics projects under drafting from early in the semester; testing those projects by conducting three trial sessions, registered as synthetic MNs; and drafting the final version of the projects (written report), based on the trial results and including a final reflection (length: one month).

Differences in Pedagogic Procedures for Tutoring and Feedback

2015/2016 Edition

Each learning topic had a forum for clearing doubts associated with its activities. These fora were the main tools for providing teaching support to trainees over the course of the activities. Trainees mostly used these fora by posing questions on the requested procedures for each task and on the assessment criteria; they also used the fora to present completed assignments. The provided feedback was never used to enable students to improve their assignments and submit new versions. The provided feedback was used only as a final assessment of each task, alongside a grade.

2016/2017 Edition

As in the previous edition, each learning topic had a forum for clearing doubts associated with its activities. These fora were the main tools for providing teaching support to trainees over the course of the activities.

The main difference regarding the previous edition was the creation of a process for formative feedback and tutoring. From topic 2 onward, individual feedback was provided for each task, enabling the trainees to improve their projects as they developed them.

The tutoring and feedback process was based on Nicol and Macfarlane-Dick's best practices model (2006), which involves the active tutoring of learning processes with the goal of enabling self-regulation of trainees' learning toward accomplishing the proposed aims.

The formative feedback provided specific comments on mistakes, along with improvement suggestions (Shute & Kim, 2014). The aim was that trainees would reflect about their own learning (Zimmerman & Moylan, 2009) so that they could assess their own success or failure in each assignment and proceed with necessary corrections (Panadero, Jonsson, & Strijbos, 2016).

Regarding trainees' initial contact with MNs, the request in this regard during topic 3 was developed across five stages (2016/2017 edition only):

1. Clarifying early doubts trainees expressed about the assignment.
2. Trainees would send a first version of the MN to the teaching staff, who would analyze it and identify which MN aspects were correctly developed and which had issues.
3. Providing formative feedback on this first MN version. The teaching staff drafted trainee-specific comments, so that each one could improve his/her MNs.
4. Clarifying trainees' doubts after the feedback (and about it).
5. Trainees would submit a post-feedback version of the MN. The teaching staff would analyze it and report on which aspects were correctly developed and which had issues.

Trainees' Characteristics

2015/2016 Edition

A total of 36 trainees (26 male, 10 female) were enrolled, of which 29 completed the course (21 male, 8 female). Four trainees (3 male, 1 female) were completely inactive throughout the course.

All trainees who completed the course held a bachelor's degree (*licenciatura*) in one of the following fields: informatics, electric engineering, information technology, computer networks or information systems.

Most trainees were active teachers and/or vocational trainers, with a few being specialized technical staff (computing services staff at educational organizations).

With regard to location, most were based in different regions of the Portuguese mainland (north, center, south); three came from the islands of Madeira and one from the islands of the Azores.

2016/2017 Edition

A total of 19 trainees (14 male, 5 female) were enrolled, of which 14 completed the course (10 male, 4 female). Of the remaining 5 students, 2 (male) completed the early activities but dropped out and 3 (2 male, 1 female) were completely inactive throughout the course.

All trainees who completed the course held a bachelor's degree, and 6 held postgraduate degrees (1 attended a postgraduate program, 4 held master's degrees, and 1 held a doctoral degree). The bachelor's degrees were in the following fields: informatics, electric engineering, information technology, computer networks and information systems.

Most trainees were active teachers and/or vocational trainers, with a few being specialized technical staff (computing services staff at educational organizations). One was a businessman.

Regarding location, they were based in different regions of the Portuguese mainland (north, center, south), except one, who came from the islands of Madeira.

RESEARCH METHODOLOGY

Procedures for Data Analysis and Concept Definitions

The process of constructing a MN (Lopes et al., 2014) takes place in three essential phases, established through protocol, with indications for: (a) the collection of data from the classroom, (b) the narrator's construction of the MN, making a single document, and (c) validation.

In fact, MNs must be analyzed by other researchers who independently attest to their accuracy, reliability and legibility. Each MN must present the set of characteristics and qualities defined for MNs as well as coherence in its internal structure. In this way, the validation process of a MN occurs when each researcher proceeds to read the MN and simultaneously consults the data that gave rise to it. In case of doubt or gaps, it is requested that the MN be reformulated and subsequently re-evaluated until these aspects and their reliability are guaranteed (Lopes et al., 2014).

For this work, the present authors investigated four analysis categories, based on the items required for correctly developing MNs, namely:

Category 1: Presents the essential aspects of a MN;
Category 2: Presents a well-structured MN;
Category 3: Narrating an episode with its essential aspects; and
Category 4: Enriching the MN.

The purpose of this analysis was to establish: which aspects revealed issues on the part of trainees when creating their MNs (both initially and – for the second edition – also after feedback), which aspects were correctly executed, and which issues were solved after feedback.

The analysis categories are organized according to the characteristics of MNs (Table 1) and developed from literature on MNs' essential features regarding their structure and creation (Lopes et al., 2010, pp. 18–25) and based on the MN protocol by Lopes et al. (2014). They were also based on the instructions and templates provided by the teaching staff to the trainees.

The results are presented in tabular format, showing how many trainees correctly developed their MNs, how many presented issues while developing their first version and how many presented issues in the post-feedback version (the latter only for the 2016/2017 edition). The column reporting post-feedback results also includes the number of trainees that managed to overcome the issues.

RESULTS

2015/2016 Edition

Regarding Category 1 – *Presents the essential aspects of a MN* (Table 2), the main issues that students encountered while creating their MNs were verifiability (n = 28), multimodal aspects (n = 18), modular presentation (n = 3) and MN goal (n = 3). Other aspects were mostly correctly executed.

Regarding Category 2 – *Presents a well-structured MN* (Table 3), in the first phase, the main issues for students while creating their MNs were at the context level: class context description (n = 20); details

Table 1. Defining the categories and aspects under analysis.

Categories	Aspects	Definition
Category 1: Presents the essential aspects of a MN	MN goal	Whether it is aligned with the goals of a MN
	Narrative/description	Whether its format is descriptive
	Unbiased perception	Presents an objective and unbiased narrative
	Multimodal	Includes multimodal data
	Modular	Presents the various episodes
	Verifiable	Includes data for checking its truthfulness
	Comparable	Enables comparisons with similar situations
	Demonstrates teacher's intent	Describes the teacher's ultimate goals
	Usefulness for professional development and research	Presents aspects that enable a teacher to reflect upon his/her professional practice
Category 2: Presents a well-structured MN	Classroom description	Detailed presentation of contextual data regarding each of these aspects
	Students description	
	Class context description (topic, references, etc.)	
	Details about spatial organization and location of teacher and students	
	Details on number of students, total class duration and episode length, class goals, etc.	
	Synthetic narrative of the entire class, with the sequence of episodes	Synthetic report of the episodes composing the narrative
	Episode(s) narrative(s)	Report of each episode, with any required multimodal events
Category 3: Narrating an episode with its essential aspects	Total task duration (including pauses and partial times)	Presentation of data regarding each organizational aspect
	Description of resources provided by the teacher and other resources	
	Identification of assigned tasks	
	Teacher's and students' actions, regardless of their capture modality and status	Description of all collected teacher and students' actions
	Teacher's perceptions	Report of teacher's perceptions
	Data on teacher's intents and decisions	Presentation of data about the decision made by the teacher
	Dialogues between the teacher and the students, and among students	Dialogue excerpts
	Data on final products achieved by the teacher and the students	Presentation of data on results produced by the teacher and the students
Category 4: Enriching the MN	Required multimodal data (e.g., gestures, silence, schematics, photos, document snippets, time recordings)	Enriching the MN with various kinds of multimodal data
	Required pieces/summary of dialogues	
	Explanations of reactions, silences, stances, etc. by students and the teacher	
	Teleological data (e.g., intents, decisions)	

Table 2. Category 1: Presents the essential aspects of a MN (2015/2016 edition).

Aspects	Initial Execution	
	Correctly executed (n= 28)	Execution issues (n= 28)
MN goal	25	3
Narrative/description	28	0
Unbiased perception	28	0
Multimodal	10	18
Modular	25	3
Verifiable	0	28
Comparable	28	0
Demonstrates teacher's intent	28	0
Usefulness for professional development and research	28	0

Table 3. Category 2: Presents a well-structured MN (2015/2016 edition).

Aspects	Initial execution	
	Correctly executed (n= 28)	Execution issues (n= 28)
Classroom description	28	0
Students description	28	0
Class context description (topic, references, etc.)	8	20
Details about spatial organization and location of teacher and students	16	12
Details on number of students, total class duration and episode length, class goals, etc.	14	14
Synthetic narrative of the entire class, with the sequence of episodes	24	4
Episode(s) narrative(s)	26	2

on number of students, class duration, etc. (n = 14); details about spatial organization and location of teacher and students (n = 12); and synthetic narrative of episodes (n = 4).

In Category 3 – *Narrating an episode with its essential aspects* (Table 4), in the initial phase, most trainees presented issues on all aspects, with the exception of aspects related to the actions of the teacher and the students, regardless of whether they were recorded as audio or in some other way (n = 0), and aside from aspects of dialogue between teacher and students and among students (n = 2), rendering this an issue-heavy category.

Most trainees had issues with total task duration (including pauses and partial times) (n = 22) and also with other aspects, such as data on final products produced by the teacher and the students (n = 13), on data on teacher intents and decisions (n = 9), and on teacher perceptions (n = 7). It is clear that in this initial phase, the trainees were unable to correctly execute these MN aspects.

Table 4. Category 3: Narrating an episode with its essential aspects.

Aspects	Initial execution	
	Correctly executed (n= 28)	Execution issues (n= 28)
Total task duration (including pauses and partial times)	6	22
Description of resources provided by the teacher and other resources	22	6
Identification of assigned tasks	22	6
Teacher and students' actions, regardless of their capture modality and status	28	0
Teacher perceptions	21	7
Data on teacher intents and decisions	19	9
Dialogues between the teacher and the students, and among students	26	2
Data on final products achieved by the teacher and the students	15	13

Lastly, Category 4 – *Enriching the MN* (Table 5), in the initial phase, most students presented issues on all aspects, apart from the aspect on pieces/summaries of dialogues (n = 2).

Along with Category 3, Category 4 is also an issue-heavy category. Figure 1 is an example that demonstrates the issues identified in this category, in which there is no presentation of the required multimodal data (e.g., gestures, silences, time recordings). Furthermore, there are no descriptions or explanations, for instance, regarding the reactions and stances by students and the teacher during the use of Dreamweaver software. There are no summaries of dialogues or even teleological data that could enrich the narrative (e.g., intents of the students and teacher behaviors when using Dreamweaver).

Table 5. Category 4: Enriching the MN.

Aspects	Initial execution	
	Correctly executed (n= 28)	Execution issues (n= 28)
Required multimodal data (e.g., gestures, silence, schematics, photos, document snippets, time recordings)	5	23
Required pieces/summary of dialogues	26	2
Explanations for reactions, silences, stances, etc. by students and the teacher	6	22
Teleological data (e.g., intents, decisions)	20	8

Figure 1. Excerpt from part of a MN with issues in its enrichment.

2. Narrative of one episode
Schedule: 2 p.m. to 5:20 p.m. - 20 minutes of break
I did a review from the last class in which we had done na introduction to Dreamweaver. That is, I taught students how to format the background, create the structure with one table and the insertion of text and images, and also how to resize images in Photoshop.
V. started to open the program, he is a student that has some multimedia creation knowledge.
V. changed places with A., she asked to start and nobody opposed her.
A. asked how to start, C. answered: by opening the HTML option. A. said what background colour should we choose, or do you prefer and image as background? A2. said, colour is better, it's more beautiful and clean. V2. said, choose the light grey, B. said, light blue, A. experimented both colours, and with almost complete consensus, chose light blue.
A. with my support inserted the table with 3 lines.

Issues

- There are no required multimodal data (e.g., gestures, silence, schematics, photos, document snippets, time recordings)
- There are no explanations for reactions, silences, stances, etc. by students and the teacher
- There are no required pieces/ summaries of dialogues
- There are no teleological data (e.g., intents, decisions)

2016/2017 Edition

For Category 1 – *Presents the essential aspects of a MN* (Table 6), in the initial phase the main issues of trainees while executing MNs were verifiability (n = 10), multimodal aspects (n = 8) and modular presentation (n = 5). The remaining aspects were mostly executed correctly by trainees. After feedback, the issue distribution was impacting these same aspects.

Regarding verifiability, only 1 of the 10 trainees corrected it. Four trainees performed no corrections, and the other trainees didn't submit a new MN after feedback. Regarding multimodal aspects, only 2 of the 8 trainees corrected them. For 3, the issue remained, and the others didn't submit a new MN after

Table 6. Category 1: Presents the essential aspects of a MN.

Aspects	Initial execution		Post-feedback execution	
	Correctly executed (n= 14)	Execution issues (n= 14)	Correctly executed (n= 10)	Execution issues (n= 10)
MN goal	11	3	10 (3 corrected)	0
Narrative/description	13	1	10 (1 corrected)	0
Unbiased perception	11	3	10 (2 corrected)	0
Multimodal	6	8	7 (2 corrected)	3
Modular	9	5	9 (3 corrected)	1
Verifiable	4	10	6 (1 corrected)	4
Comparable	10	4	9 (1 corrected)	1
Demonstrates teacher's intent	14	0	10	0
Usefulness for professional development and research	12	2	10 (1 corrected)	0

feedback. Lastly, the modular presentation was the most corrected aspect (3 of 5 trainees). One trainee didn't correct it, and another didn't submit an updated MN after feedback.

In Category 2 – *Presents a well-structured MN* (Table 7), the main issues were at the level of context: of spatial organization and location (n = 9); of synthetic narrative of episodes (n = 8); of classroom context (n = 6); and of details, such as number of students, total class duration and episode length, class goals, etc. (n = 6). After feedback, there were improvements, and the aspects where issues remained were: spatial organization and location (n = 3), and synthetic narrative of episodes (n = 2).

In Category 3 – *Narrating an episode with its essential aspects* (Table 8), most aspects exhibited issues in trainees' MNs, aside from aspects describing actions of the teacher and the students (n = 1) and describing resources (n = 3), rendering this an issue-heavy category.

Table 7. Category 2: Presents a well-structured MN.

Aspects	Initial execution		Post-feedback execution	
	Correctly executed (n= 14)	Execution issues (n= 14)	Correctly executed (n= 10)	Execution issues (n= 10)
Classroom description	12	2	9	1
Students description	13	1	9	1
Class context description (topic, references, etc.)	8	6	10 (2 corrected)	0
Details about spatial organization and location of teacher and students	5	9	7 (5 corrected)	3
Details on number of students, total class duration and episode length, class goals, etc.	8	6	9 (3 corrected)	1
Synthetic narrative of the entire class, with the sequence of episodes	6	8	8 (5 corrected)	2
Episode(s) narrative(s)	10	4	9 (3 corrected)	1

Table 8. Category 3: Narrating an episode with its essential aspects.

Aspects	Initial execution		Post-feedback execution	
	Correctly executed (n= 14)	Execution issues (n= 14)	Correctly executed (n= 10)	Execution issues (n= 10)
Total task duration (including pauses and partial times)	5	9	7 (3 corrected)	3
Description of resources provided by the teacher and other resources	11	3	9 (2 corrected)	1
Identification of assigned tasks	7	7	10 (6 corrected)	0
Teacher and students' actions, regardless of their capture modality and status	13	1	9	1
Teacher perceptions	5	9	8 (3 corrected)	2
Data on teacher intents and decisions	5	9	11 (3 corrected)	3
Dialogues between the teacher and the students, and among students	6	8	9 (5 corrected)	1
Data on final products achieved by the teacher and the students	0	14	10 (10 corrected)	0

All trainees exhibited issues in the narratives of final products (n = 14). Other aspects, such as describing the task organization with partial times (n = 9), teacher perceptions (n = 9) and data on teacher intents and decisions (n = 9), highlight the difficulties trainees experienced in this first phase.

After feedback, there was significant improvement in virtually all aspects, but most significantly in the one with the greatest number of issues (final products): 10 out of 14 trainees improved it, correctly executing this aspect.

Below is a sample excerpt from a MN, where the trainee improved her MN after feedback by adding a paragraph about final products. Before feedback:

Before the break, the first goal in the job proposal was reached: inserting cliparts into the text document, resizing them and placing them on a page. We concluded this 3rd episode when I asked them if it would be the ideal time to pause and ask them to keep the work done so far. [The trainer. [Trainee C left this sentence hanging here in the initial version.] (Trainee C, April 2017, part of the improvement of her first MN)

After feedback, the trainee included more information:

[…] it matters mentioning that I had foreseen as specific goals of this that the students should be able to insert graphic elements in word-processing documents; and also, able to communicate on-line by sending e-mail messages (including attaching the final document, the result of completing the proposed assignment). All students were skilled in demonstrating their learning. (Trainee C, April 2017, part of the improvement of her first MN)

Most trainees corrected the aspect of identification of assigned tasks: 6 out of 7 trainees with execution issues corrected this aspect. One trainee didn't submit an improved version. Similarly, the aspect regarding dialogues between the teacher and the students exhibited many improvements: 5 out of 8 trainees corrected it, 1 was unable to improve it, and 2 didn't submit an improved MN.

Lastly, in Category 4 – *Enriching the MN* (Table 9), aside from the aspect of pieces/summaries of dialogues (n = 2), most MNs had issues. Along with Category 3, Category 4 was also issue-heavy.

Multimodal data (n = 9), teleological data (n = 7) and explanations for behaviors (n = 6) were the aspects that most trainees failed to execute correctly.

Table 9. Category 4: Enriching the MN.

Aspects	Initial execution		Post-feedback execution	
	Correctly executed (n= 14)	Execution issues (n= 14)	Correctly executed (n= 10)	Execution issues (n= 10)
Required multimodal data (e.g., gestures, silence, schematics, photos, document snippets, time recordings)	5	9	7 (3 corrected)	3
Required pieces/summary of dialogues	12	2	9	1
Explanations for reactions, silences, stances, etc. by students and the teacher	8	6	7 (1 corrected)	3
Teleological data (e.g., intents, decisions)	7	7	9 (4 corrected)	1

After feedback, some issues remained; however, these were fewer than initially (e.g., Figure 2). The aspect of behavioral explanations was the least improved issue: 1 out of 6 trainees executed it correctly, 3 trainees exhibited the same issue, and 1 didn't deliver an improved MN. Regarding multimodal aspects, 3 out of 9 trainees corrected them, 3 exhibited the same issue, and the others didn't submit an improved version. Lastly, regarding teleological data, 4 out of 7 trainees corrected this aspect, 1 exhibited the same issue, and the others didn't submit an improved version.

Figures 2 and 3 are examples that demonstrate the issues identified in this category, in which there are no multimodal data or teleological data, nor explanation of behaviors, gestures or other forms of enrichment.

These results will be discussed and linked with the background literature in the following section.

Figure 2. Multimodal narrative sample. In this case, the trainee incorrectly split the two modalities (text and image) over two sections of the document, instead of keeping them in the same section.

"Contextualization
(...)
Characterization of the sessions: (...) Each training session (and whenever else it was necessary) included work proposals (see in Annex IV an example of the activity for 04/11/2017), designed by the trainer and totally directed towards the training needs of the group, since this is a tailor-made training course for individual trainees."

PROPOSTA DE TRABALHO Nº 7

CURSO:	Iniciação à Informática					AÇÃO:	Iniciação à Informática_1		
UNIDADE/MÓDULO:						LOCAL:	Mirandela		
MODALIDADE:	Formação à medida								
HOMOLOGAÇÃO:	Não se aplica					PROJETO Nº:	Formação Não Financiada		
FORMADOR:	[redacted]					DATA DE REALIZAÇÃO:			
FORMANDO:									
TEMÁTICA:	Processador de Texto								
MATERIAL A UTILIZAR:	Computador, *MS Office Word*								
PROPOSTA DE TRABALHO:	Individual	x	Em grupo		Nº de elementos		DURAÇÃO:	120 Min	Sessão nº: 9

Figure 3. Excerpt from part of a MN that, after feedback, still presents issues in its enrichment.

	Issues
2. Narrative of one episode Before the beginning of the class, I and my colleague prepared the area and the circuit with black adhesive tape for the robot to follow the line. We created a set of long curves, but also some tighter ones and even some 90° ones. The students came shortly after 2 p.m., maybe because it was a more informal class. We asked one of the students to describe the robot. The student said that just as with Programmable Controllers (PC), we have digital inputs and outputs. - But contrary to the PCs that have predefined inputs and outputs, here we have a processor that has configurable pins which we can define individually whether each pin is for input or for output... - I said. - Yes, we have pins that control the engines through a bridge -H, that is this piece here... - I inter--rupted Tomás' thought and asked "What is that of bridge-H?" - The bridge-H allows us to provide electrical current to rotate the engine and enables it to rotate forward and backwards - said the student. - Because we have pins assigned to control the direction, and also one pin of PWM that controls the engine speed. We have two engines, one for each wheel. We can move forward, backwards, turn slowly one engine with the other one stopped or quickly turn, by putting one engine moving forward and the other moving backward.	There are no required multimodal data (e.g., gestures, silence, schematics, photos, document snippets, time recordings) There are no explanations for reactions, silences, stances, etc. by students and the teacher There are no teleological data (e.g., intents, decisions)

FINAL REFLECTIONS

This effort is the first one reporting processes and outcomes about MN distance education didactics. The outcome of this use of formative feedback is in line with Kim and Shute's (2014) report of the impact of formative assessment on the effectiveness of teaching and learning. The present authors' revision of their own instruction and teaching process across sessions was directly influenced by the outcome of formative assessment. This result is in line with the conclusions of the present authors as well as Reinholz's (2016) report about impacts on the improvement of the teaching process. However, the present study did not collect direct data supporting Black and Wiliam's (2009) expectation of formative assessment contributing to making students more proactive in their learning processes.

In both editions of the course (2015/2016 and 2016/2017), the high number of execution issues reported in the tables of the results section led to the conclusion that the initial contact of trainees with the MN technique was challenging regarding its initial execution. In the second edition (2016/2017), although the trainees were able to improve their MNs with feedback, the process (execution of a MN) remained challenging. In fact, building a MN is a complex process (Lopes et al., 2014), and creating them adequately is challenging.

Of the four analysis categories and their aspects, it is visible that the highest prevalence of issues were in Category 3 – *Narrating an episode with its essential aspects* and in Category 4 – *Enriching the MN*. That is, the most issues occurred on the very aspects that detail and enrich the MNs (see Tables 4 and 5 of the 2015/2016 edition and Tables 8 and 9 of the 2016/2017 edition).

In both editions of the course, it was visible in Category 3 that, during the initial phase, the trainees were unable to execute aspects regarding attained results (final products) and description of partial times, perceptions and decision-making data. Category 4 also exhibits several issues on aspects of enriching the MN beyond the aspects of Category 3. The difference between the two editions was that in cases where the trainees obtained feedback (in the second edition) and made corrections, they were able to improve the aspects that had presented flaws. This again confirms the importance of specific feedback, because it helps toward students' self-regulation, motivation and self-awareness about the progress of their work (Hattie & Timperley, 2007; Shute & Kim, 2014).

However, the basic aspects, part of Category 1 – *Presents the essential aspects of a MN*, were mostly correctly executed in both editions of the course, although some trainees presented some execution issues. In the second edition (2016/2017), due to the feedback, the issues that presented problems were solved. The MNs were properly executed post-feedback.

It is thus feasible to conclude that the trainees understood the purpose, aspects and usefulness of the MN technique, in both editions, and that in cases where a problem arises they can easily correct these problems with specific feedback (as occurred in the second edition). In Category 1, the aspects with more issues were verifiability, multimodal aspects and modular presentation. The former two may be due to unsuccessful understanding by trainees of the need to include data in support of the MN (audio-visual recordings, didactic materials, etc.).

To avoid similar situations, a proposal for future implementations of the course is to create a set of instructions that are straightforward and unequivocal, asking trainees to include this kind of support material.

Furthermore, since the MN template that was provided to trainees (and this was emphasized in the assignment instructions) was limited to two pages (although it was possible to send extra material as appendices), this may have limited trainees' initiative for exploring and enriching their MNs with diverse types of data. Thus, it is proposed that templates and instructions for MNs should be improved.

Still regarding Category 1 and the aspect of modular presentation (presenting/summarizing the various MN episodes), even though there were specific instructions for executing this aspect, issues occurred in both editions of the course. A hypothesis for this is that the fact that the instructions required only the actual narrative of a single episode may have led students to focus on that aspect and disregard the synthesis of the remaining episodes. Should this hypothesis be confirmed on further research, it would strengthen the need for improving the assignment instructions.

Regarding Category 2 – *Presents a well-structured MN,* both editions of the course confirmed that trainees didn't execute several of its aspects correctly, particularly those regarding greater details and enriching the MN. The main issues were at the level of describing the spatial organization and location, the class context, and the synthetic narrative of episodes.

In the 2015/2016 edition, it should be recalled that trainees received feedback only on their final MNs (for assessment purposes) and had no opportunity to improve their assignments after the teaching staff identified the issues and provided feedback. This is seen as a methodological limitation of this edition.

In the 2016/2017 edition, which included a tutoring and feedback process, most trainees submitted a post-feedback improved MN, where they managed to improve some of the issues affecting their MNs. However, some trainees still exhibited issues in their post-feedback MNs. One must also consider that 4 out of 14 trainees in this second edition did not submit an improved post-feedback MN, which limits the feedback impact assessment.

Another aspect that may be analyzed in subsequent research is: why did only a few trainees requested clarifications on the provided feedback? Only 4 out of 14 trainees in the 2016/2017 session debated the feedback they received.

The authors acknowledge the importance of tutoring and feedback to enable trainees to overcome their difficulties in executing MNs at an initial phase. For future editions, it is necessary to provide feedback on the aspects where most issues arose. This should include not only final feedback but also formative feedback.

Another aspect that can be improved is the definition of task duration, since MN creation is time consuming, requiring preparation, data collection, data processing and actual execution (narration). There were regular requests by trainees to postpone task delivery, and for future editions it is proposed that more time be allocated to the creation of MNs.

It is advisable to provide tutoring and support by researchers experienced in using MNs, providing trainees with accounts (e.g., video testimonials), feedback and samples in support of the professionals and researchers who are in the initial phase of using MNs.

The overall recommendation by the authors of this chapter is that MNs should be used in educational/ training contexts, since the trainees of both editions appear to have improved their reflective skills by the use of MNs, having reported that MNs provided a tool that helped them identify educational problems. While this is not validated in this chapter, the authors raise it as an enticing possibility for further research and leave a final testimony as a sample:

This attempt at using multimodal narratives was important as an instrument for recording information during the sessions. If this kind of records did not exist, it would be practically impossible to memorize all episodes, and subsequently difficult to reflect upon them and understand if the pre-established goals were successfully reached. As a trainer, I found the experience interesting, and have the intent and will to repeat it more often, not just for an episode, but for all episodes in a session. The records within the multimodal narrative enable me to have a better perception, a real perception of the classes and improve my performance in professional practice and teaching. (Trainee N, April 2017, reflection upon his first contact with MNs)

ACKNOWLEDGMENT

The authors would like to thank all trainees of the course, in both editions, and in particular those who kindly allowed their MNs to be disseminated as examples: Cecília Gomes, Nuno Peixoto and Ricardo Pereira.

REFERENCES

Black, P., & Wiliam, D. (2009). Developing the theory of formative assessment. *Educational Assessment, Evaluation and Accountability*, *21*(1), 5–31. doi:10.100711092-008-9068-5

Broadbent, J. (2017). Comparing online and blended learner's self-regulated learning strategies and academic performance. *The Internet and Higher Education*, *33*, 24–32. doi:10.1016/j.iheduc.2017.01.004

Bruner, J. S. (1960). *The process of education*. Cambridge, MA: Harvard University Press.

Cazan, A. M. (2014, July). Self-regulated learning and academic achievement in the context of online learning environments. *The International Scientific Conference Elearning and Software for Education*, 90-95.

Cortesão, L., & Stoer, S. R. (1997). Investigação-acção e a produção de conhecimento no âmbito de uma formação de professores para a educação inter/multicultural [Action research and the production of knowledge within a teacher training course for inter/multicultural education]. *Educação Sociedade & Culturas*, *7*, 7–28.

Coutinho, C. P. S., Dias, A., Bessa, A., Ferreira, F., Vieira, M. J. R. C., & Regina, S. (2009). Investigação-acção: Metodologia preferencial nas práticas educativas [Action research: the preferred methodology within educational practice]. *Revista Psicologia. Education and Culture*, *13*(2), 355–379.

Dabbagh, N., & Kitsantas, A. (2012). Personal learning environments, social media, and self-regulated learning: A natural formula for connecting formal and informal learning. *The Internet and Higher Education*, *15*(1), 3–8. doi:10.1016/j.iheduc.2011.06.002

diSessa, A. (2001). *Changing minds: Computers, learning, and literacy*. Cambridge, MA: MIT Press.

Feucht, F. C., Brownlee, J. L., & Schraw, G. (2017). Moving beyond reflection: Reflexivity and epistemic cognition in teaching and teacher education. *Educational Psychologist*, *52*(4), 234–241. doi:10.1080/00461520.2017.1350180

Hattie, J., & Timperley, H. (2007). The power of feedback. *Review of Educational Research, 77*(1), 81–112. doi:10.3102/003465430298487

Järvelä, S., Kirschner, P., Panadero, E., Malmberg, J., Phielix, C., Jaspers, J., ... Järvenoja, H. (2015). Enhancing socially shared regulation in collaborative learning groups: Designing for CSCL regulation tools. *Educational Technology Research and Development, 63*(1), 125–142. doi:10.100711423-014-9358-1

Lopes, J., Silva, A., Cravino, J., Santos, C., Cunha, A., Pinto, A., ... Branco, M. (2014). Constructing and using multimodal narratives to research in science education: Contributions based on practical classroom. *Research in Science Education, 44*(3), 415–438. doi:10.100711165-013-9381-y

Lopes, J. B., & Cunha, A. E. (2017). Self-directed professional development to improve effective teaching: Keys points for a model. *Teaching and Teacher Education, 68*, 262–274. doi:10.1016/j.tate.2017.09.009

Lopes, J. B., Silva, A. A., Cravino, J. P., Viegas, C., Cunha, A. E., Saraiva, E., ... Santos, A. (2010). *Investigação sobre a mediação de professores de ciências físicas em sala de aula* [Research on physical sciences teacher mediation within classrooms]. Vila Real: UTAD.

Morris, C., & Chikwa, G. (2016). Audio versus written feedback: Exploring learners' preference and the impact of feedback format on students' academic performance. *Active Learning in Higher Education, 17*(2), 125–137. doi:10.1177/1469787416637482

Nicol, D., & Macfarlane-Dick, D. (2006). Formative assessment and self-regulated learning: A model and seven principles of good feedback practice. *Studies in Higher Education, 31*(2), 199–218. doi:10.1080/03075070600572090

Oliveira, I., & Serrazina, L. (2002). A reflexão e o professor como investigador [Reflection and teacher as a researcher]. In APM-GTI (Ed.), Reflectir e investigar sobre a prática profissional (pp. 29-42). Lisbon: APM.

Panadero, E., Jonsson, A., & Strijbos, J. W. (2016). Scaffolding self-regulated learning through self-assessment and peer assessment: Guidelines for classroom implementation. In *Assessment for learning: Meeting the challenge of implementation* (pp. 311–326). Cham, Switzerland: Springer International Publishing.

Pereira, A., Mendes, A. Q., Morgado, L., Amante, L., & Bidarra, J. (2008). *Universidade Aberta's pedagogical model for distance education: A university for the future*. Lisbon: Universidade Aberta.

Reinholz, D. (2016). The assessment cycle: A model for learning through peer assessment. *Assessment & Evaluation in Higher Education, 41*(2), 301–315. doi:10.1080/02602938.2015.1008982

Santos, L. (2016). A articulação entre a avaliação sumativa e a formativa, na prática pedagógica: Uma impossibilidade ou um desafio? [Combining summative and formative assessment in pedagogical practice. An impossibility or a challenge?]. *Avaliação e Políticas Públicas em Educação, 24*(92), 637–669.

Schön, D. (2000). *Educando o profissional reflexivo: Um novo design para o ensino e a aprendizagem* [Educating the reflective professional: a new design for teaching and learning]. Porto Alegre, Brazil: ARTMED.

Shute, V. J., & Kim, Y. J. (2014). Formative and stealth assessment. In *Handbook of research on educational communications and technology* (pp. 311–321). New York: Springer. doi:10.1007/978-1-4614-3185-5_25

Zimmerman, B. J. (2013). From cognitive modeling to self-regulation: A social cognitive career path. *Educational Psychologist*, *48*(3), 135–147. doi:10.1080/00461520.2013.794676

Zimmerman, B. J., & Moylan, A. R. (2009). Self-regulation: Where metacognition and motivation intersect. In D. J. Hacker, J. Dunlosky, & A. C. Graesser (Eds.), *Handbook of metacognition in education* (pp. 299–315). New York: Routledge.

Chapter 10
Improving Visual Representations' Impact in Mathematics Teaching With Multimodal Narratives

Paula Montenegro
Universidade de Trás-os-Montes e Alto Douro, Portugal

Cecília Costa
Universidade de Trás-os-Montes e Alto Douro, Portugal

J. Bernardino Lopes
Universidade de Trás-os-Montes e Alto Douro, Portugal

ABSTRACT

This chapter focuses on the impact that multimodal narratives (MN), elaborated to a first intervention (MN1) and to a succession of interventions (MN2), had in the professional development of a teacher to increase the impact of visual representations in the teaching and learning of mathematical content. The study follows an interpretative paradigm, with design science research and case study investigation methods. The participants are two groups of students (mean age 10.7 years and 11.3 years) and their Mathematics teacher. The data were collected from the students' written records, lesson plans, and teacher's notes. The data collected were organized in a MN. The results show that the MN is a tool that facilitates the process of professional development at the self-reflection level and in combination with researchers. The authors conclude that MN is an efficient and useful tool for professional development that aims to improve teaching practices and student learning.

INTRODUCTION

This chapter refers to the intermediate phase of a research study with the general objective of exploring ways to increase the impact of the use of visual representation as an epistemic tool in teaching and learning different mathematical contents of basic education. In the initial phase of the study, specifici-

DOI: 10.4018/978-1-5225-8570-1.ch010

ties of the use of visual representations in teaching were identified that were not evident either to the teacher (first author) or to the students; this led to the need for a professional development phase to better understand the identified specificities and explore ways to improve the impact of the use of visual representations (Montenegro, Costa, & Lopes, 2018). During this stage of professional development, the teacher used a multimodal narrative (MN1) (already done in the first phase of the research study) to start her professional development process; she further elaborated and analyzed a multimodal narrative (MN2) to record her reflections and organize the data she collected about her attempts to enhance the impact of the use of visual representation as an epistemic tool in classroom activities and what she considered relevant for this purpose.

The objective of this chapter is to evaluate the importance of multimodal narratives in the teacher's self-reflection and to develop her knowledge about the use of visual representations in teaching and learning, exploring ways to increase the impact of their use in the classroom.

The constant adaptation of mathematics curricula, together with the changes in competencies evidenced by students resulting from curricular changes, bring teachers to a continuous need for professional development.

According to Ponte (1994), the need for growth and various acquisitions is recognized in professional development, and the role of the fundamental subject is attributed to the teacher himself. Thus, the teacher is in constant development and it is up to him/her to decide when, how, and what s/he wants to study, or to get involved in a given project. However, the training provided is not always adequate to the needs of certain teachers' professional development. This was the case for the teacher of our study, which is why she opted to become involved in a well-structured professional development based on research on her own practices, developed in collaboration with two researchers in a PhD scope.

Through her experience as both a teacher and as a learner, the teacher was convinced that visual representations played an important role in teaching and learning. She received positive feedback from students when using visual representations in teaching, but she did not understand the reasons why students were reluctant to use them.

That is why we look for an answer to the following problem: what actions should the teacher take to improve her teaching practice using visual representations? In more detail, we seek to know how to use a visual representation to represent a mathematical concept, how to do a demonstration using a visual representation, how to explain something through a visual representation, or how to use a visual representation to get to other types of more abstract representation—for example, algebraic. In this chapter, we focus on identifying the role of MNs and how they were used in the professional development process of the teacher to enhance teaching practice using visual representations.

THEORETICAL BACKGROUND

Although tolerated, the use of visual representations in mathematics teaching and learning is not accepted by teachers and students with the same status as other more widely used representations (e.g., arithmetic and algebraic) (Montenegro, Costa, & Lopes, 2018). Hence, a phase of professional development is needed to explore ways of using visual representations in teaching. A professional development model focused on the improvement of teaching practices in line with the most recent conceptualization (Soebari & Aldridge, 2015) is adopted, based on the articulation with research on teachers' own teaching practices (Fischer et al., 2018; Lopes & Cunha, 2017).

Thus, during the professional development phase, it is necessary to collect large amounts of data from various sources. To organize these data, multimodal narratives (MNs) have been elaborated. This is a tool that can help teachers' self-reflection to identify directions for their own professional development, leading them to perceive their intentions and the processes that led to their decision-making, and helping to understand effective ways of promoting learning (Lopes et al., 2014). The teacher, when confronted with the narration of his class, can initiate a whole process of reflection (individual or shared) that allows him to become aware of important aspects of his professional practice (Lopes et al., 2014). The elaboration of an MN combines writing and reflection, allowing a retrospective and deep analysis of the classroom events, actions, and decisions, even in a research process in collaboration with other researchers, it being important for teachers to improve their teaching practices in a context of a professional development.

Visual Representations

Mathematical objects are abstractions that do not exist in the real world, it being possible to access them only through representations. Goldin (2008) defines a representation as any entity that can replace another. Each of the systems of verbal, symbolic, and visual representations possess specific codes and rules. A mathematical activity requires cognitive operations that are carried out using different systems of representation (Ainsworth, 2006), as well as the transformation of a representation to other in the same or in another representation system (e.g., Duval, 2006). Symbolic (e.g., Montenegro, Campos, & Aires, 2017) and verbal representations (e.g., Morgan et al., 2014) are accepted and recognized in mathematical activity; however, interest in visual representations is controversial (Goldin & Kaput, 1996) mostly due to the difficulties involved in their use. The most common visual representations in the mathematics class are graphs, figures, tables, and schemes. A visual representation is "a collection of graphical symbols that visually encode causal, functional, structural, and semantic properties and relationships of a represented world—either abstract or concrete" (Sedig & Sumner, 2006, p. 2).

Thus, a visual representation can include verbal, numerical, and other types of symbol as far as they are connected to mathematical understanding and respect all the requirements of the given definition.

Visual representations are used in mathematics classes mainly as representations of a given mathematical concept. For instance, it is common to represent a function through an algebraic expression (symbolic representation) and by a graph or a table (both visual representations), or to represent a fraction through symbolic, verbal, and visual representations (Table 1).

We can consider that the three representations are static in the sense that they represent the same mathematical entity in different ways. We highlight the fact that the visual system allows different representations—in this case of visual representations in Table 1 in discrete unities (above) and in continuous unities (underneath). Going from one to the other, even in a different system, does not require more than

Table 1. Representation of a rational number in the three systems of representation.

System of representation		
Verbal	Symbolic (numerical)	Visual
Two thirds	$\frac{2}{3}$	
		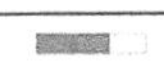

the knowledge and mastery of certain procedures. However, the characteristics of visual representations allow a dynamic utilization if its usage allows different ways to explore these and other mathematical contents (Montenegro, Costa, & Lopes, 2018). The ability to think and communicate information can be identified by visualization and defined as

> *…the ability, the process and the product of creation, interpretation, use of and reflection upon pictures, images, diagrams, in our minds, on paper or with technological tools, with the purpose of depicting and communicating information, thinking about and developing previously unknown ideas and advancing understandings (Arcavi, 2003, p. 217).*

The importance of visualization in mathematics learning seems today to be recognized as a key component of reasoning in problem-solving and demonstrations. Nevertheless, in mathematical education, visualization is a source of many difficulties. Dreyfus (1991) refers, among others, to cognitive difficulties in the utilization of visualizations. These refer to the bigger or smaller difficulty of what is visual, as well as the lack of routines and procedures that visual reasoning requires, or the necessity of achieving a flexible and competent transformation between visual and analytical (symbolic) representations of the same situation—which Duval (2006) defines as conversion—which is the core of mathematical understanding and the cause of many difficulties (Arcavi, 2003; Duval, 2006). Dreyfus (1991) also refers to sociological difficulties, including the idea that analytical representations of a sequential nature seem to be more pedagogically appropriate and efficient than visual representations.

Due to the low value attributed to visual aspects in the classroom, there is still reluctance in the usage of visualization in mathematics (Dreyfus, 1991; Goldin & Kaput, 1996). The National Council of Teachers of Mathematics (NCTM, 2008) adds to this the need for teachers to provide students with situations with diverse visual representations and to present them with new forms of representation useful in the solving of certain problems. The role of visual representations in knowledge is studied in several areas because the dynamic use of visual representations in the classroom is complex.

The individual skills involved in the flexible use of representations are complex constructs; hence research studies with different perspectives are needed in different areas to understand these skills and their acquisition (Heinze, Star, & Verschaffel, 2009).

In addition, the teachers ascribe limited relevance to the use of schemes (an example of visual representation) in classroom practice; however, supporting understanding of those representations could increase their relevance (Norton & McCloskey, 2008) and consequently allow more efficient use. In this sense, studies that explore a more dynamic use of visual representations are important and relevant. In the case in which the study is developed by a teacher, it makes sense to frame the professional development.

Professional Development

The way we teach (which includes resources, with an emphasis on technologies, diversification of methodologies and teaching strategies, etc.), whose impact is reflected in improved outcomes (with students learning more and better), is the current concern in education. The specialization of a teacher includes three main domains: (i) construct the professional identity; (ii) be a member of a professional community; and (iii) improve the teacher's professional practice skills (Shagrir, 2010). Teacher professional development refers to those processes that improve the knowledge, skills/abilities, or attitudes related to teachers' work (Sparks & Loucks-Horsley, 1989). This professional development must be continuous

(Hargreaves, 2007) to go along with the evolution of society. There are several models of professional development of teachers. Sparks and Loucks-Horsley (1989) indicate:

- Autonomous professional development: the teacher creates and executes certain activities that he believes develop his knowledge.
- Professional development based on reflection: the teacher knows, analyzes, evaluates, and questions his own teaching practice—that is, the teacher reflects on his own practice.
- Professional development through curricular and organizational development: the teacher undertakes a certain project seeking to improve the education quality in the solution of his own problems.
- Professional development through short training courses.
- Professional development based on research: the teacher improves his practices considering the results of action research, identifying and diagnosing problems in his own practice, often in collaboration with other researchers (Lopes & Cunha, 2017).

Use of the trajectories of teachers' learning to structure professional development allows the development of several aspects of knowledge (Bargagliotti & Anderson, 2017). Three characteristics of professional development that teachers value are the long duration, the interactions with peers to compare student performances, and the opportunity to think and learn about a specific topic (Norton & McCloskey, 2008). The teachers also consider professional development during work, and not only before assuming the position, important, allowing connections between theory and practice (Shagrir, 2010). One of these indicators is that the teacher is involved in proving and promoting critical thinking and problem solving between teachers and/or future teachers and exploring theoretical and practical problems in teaching and learning (Shagrir, 2010).

As mentioned, one mode of professional development can be research on one's own practice, through self-reflection, whose purpose is to contribute to clarifying the problems of practice by seeking solutions and reinforcing the teacher's professional competence by adding research in dealing with problems (Ponte, 2004). Lopes and Cunha (2017) argue that professional development based on teaching practices integrating the results of action research cycles is efficient in improving the quality of teaching practices and students' learning. Visual representations are innovative teaching tools with a positive impact on the development of epistemic competencies (Sandoval & Reiser, 2004; Saraiva, 2017). Their use requires the development of professional knowledge that allows the teacher to comprehend both their utility and relevance of their support (Sandoval & Reiser, 2004; Saraiva, 2017).

According to Ponte (2004), the fundamental condition in the research on own teaching practice is the identification of a relevant problem (theoretical or practical) for which a convincing answer is sought methodically that must be disclosed to the community concerned, discussed, and validated publicly. However, one of the biggest problems in the research on own teaching practice, beyond the nearness of the researcher and the participants, is the nearness between the researcher and the object of study. To overcome this problem, Ponte (2004) suggests three routes for the researcher:

1. Resort to the theory—experience accumulated by ancestors.
2. Take advantage of belonging to a group—examine one's perspectives directly with critical peers, creating distance from one's perceptions and prejudices.
3. Take advantage of the debate outside the group, helping to relativize one's own perspectives.

During the research on own teaching practices, data will be collected. One way to guarantee the truthfulness of the study is to use different data sources. The teacher has the possibility of collecting the data through class recording (audio and video), written individual records, photos of what teachers and students write on the black/white/interactive board, and diverse documents—namely class plans, tasks, records of meetings between pedagogical peers, etc. The use of these diverse sources of data will allow the triangulation of data, bringing feasibility to the study. The elaboration of MN (Lopes et al., 2014) allows organizing all the collected data in a single document.

Multimodal Narrative

An MN is a detailed description of a class, having as its base certain documents (Lopes et al., 2014). These documents can be, for instance, audio and video recording of a class, the class plan, tasks, teacher records, student records, etc., that will be the base for its construction. This variety of data enrich the narrative, but at the same time allows its validation.

An MN is an instrument that refers to a process and a product. There is a protocol for an MN process that proposes: i) a set of indications about the collection of classroom data, the context, and the use of multimodal registers; ii) a set of steps for the narrator to make a single document describing the course of action and discourse among teacher and students in the classroom, which aggregates and transforms the collected data and incorporates the teacher's point of view in a descriptive way; and iii) a set of steps to validate an MN that include verification of the essential characteristics of an MN and its internal structure (Lopes, Viegas, & Pinto, 2018).

The product is a document prepared with the support of an independent data set with a multimodal, self-contained, validated, public, and shareable description of a teaching practice (Lopes, Viegas, & Pinto, 2018).

Usually, the narrative is written by the teacher himself to include his intentions and perceptions and is built in two essential parts:

1. A summary, including the presentation and contextualization, identification of the subject, the objectives, space and temporal descriptions, and other information considered key.
2. The narrative of episodes, with a beginning and an end that can constitute several classes, one class, or part of one class.

The teacher can narrate the direct or indirect discourse, including excerpts considered important to the reader's understanding of the class or to justify actions and decisions. Finally, external experts validate the MN based on the available data that, if needed, can be accessed directly.

The elaboration of an MN allows organization of the data and facilitates the analysis. In carrying out the self-reflection inherent in the analysis, teachers identify guidelines for their professional development, identifying the actions necessary for their improvement (Lopes et al., 2014). The MNs, as shown by Lopes and Cunha (2017), are an important tool to facilitate the work of articulation between teachers and researchers in a context of scholarly research production.

METHODOLOGY

The professional development chosen by the teacher was carried out through research with reflection on her own teaching practice. The teacher identified and diagnosed problems in her practice and, in collaboration with researchers, sought to find a solution, reviewed literature on the area of research, organized her action with other researchers, and put the results found to peers for their consideration.

This study is inserted in a broader study with a first intervention, where the referred problem (about exploring ways to increase the impact of the use of visual representations as an epistemic tool in teaching and learning in different mathematical contents of basic education) is identified and where the viability of visual representations as a tool to create knowledge by the students is evaluated. A set of intermediary interventions is followed to mature this idea and prepare a second intervention that leads to a possible solution to the problem. In this chapter, we refer specifically to the impact that the elaboration of multimodal narratives (MN1 and MN2), elaborated to the first intervention (MN1) and the succession of interventions (MN2), had on the professional development of the teacher and on the organization of the broader study.

This study follows an interpretative paradigm, with design science research (Hevner, 2007) and case study (Yin, 2010) investigation methods. The methods are complementary: the former supported the intervention in a problem identified, followed by the development of a possible solution through the usage of an artefact—visual representation; the latter allows for a detailed analysis of the research situation.

We consider the unit of study (Ponte, 1994) constituted by the teacher, her peers, and the students over time. The case study made important contributions as it helped to clarify details and, at certain points in the study, allowed the teacher to guide the direction of the broader study. The case study also complemented the teacher's knowledge by allowing a detailed analysis, describing and explaining each of the phases of the study.

The application of this method is based mainly on the elaboration of the MN that allowed a detailed and descriptive analysis of the situation under study.

To find an answer to the referred problem, there was the need to design a succession of interventions implemented in the class from which data of a diverse nature were collected and MN2 constructed. The data are of a descriptive nature, reported by images or words and wrapped in the meanings that the subject gives them, as well as in the characteristics of the content that sustains them (Bogdan & Biklen, 1994).

The Participants

The participants were a class of 18 students aged 10–13 years and their mathematics teacher, who worked together throughout the period of study, and her colleagues. The selection of the students was based on their potential benefit from the creativity in the solution of the task, as usually they did not use visual representations in mathematics classes, and on the likely variety of difficulties shown during its solution. The teacher was predisposed to explore visual representations and trusted in her capacity to respond to each of the difficulties shown by her students in using visual representations.

The First Intervention and the Task

The first intervention consisted of an algebra task exploration that led the students to explore the generalization of a figurative pattern (sequences and regularities) which represented a numerical sequence (3, 5, 7, 9, 11, ...).

The selection of this content was intentional because its exploration easily allows the utilization of three systems of representation: verbal, symbolic (numerical and algebraic), and visual.

The students worked in groups (four/five students each). During the solution of the task, the teacher gave her students freedom to choose their strategy and to use whatever representations they wanted to. They used numerical representations and visual ones (schemas and table). She intervened when she considered it pertinent. She appealed every time it was possible and pertinent to visual representations, exploring these forms of utilization. At some point in time the task was solved; the teacher had to help the students (two of the four groups) to deviate from an inappropriate strategy (direct proportion) or difficulties in calculation procedures relating the numerical values in the table (columns and rows). She showed these two groups a strategy based on a visual treatment (color the corner square grey; see the figure on the right, in Multimodal Narrative Excerpt 2, p. 10) that resulted in the students using the representations in the three systems of representation and the realization of several conversions between them (see the figure in the Multimodal Narrative Excerpt 4, p. 11). This intervention was described in detail in MN1 and was the object of self-reflection by the teacher. In the results section, we identify the evidences of this work that the teacher wanted to add to her professional development. Through this intervention, the teacher identified the realization of visual treatments as the action needed to increase the impact of the use of visual representations.

Succession of Interventions

The succession of interventions occurred over one and a half years, a period when the research teacher reflected on her own practice concerning the use of representations and the three systems of representation (verbal, symbolic, and visual), particularly visual representations, during mathematics classes. These interventions were instrumental in the professional development of the teacher to deal more effectively with multiple representations, visual representations, and visual treatments.

With more focus on the previous academic year, the teacher planned her work in such a way as to provide diverse learning situations using the three systems of representation in varied mathematical contexts of all the curricular topics.

The teacher initiated the succession of interventions with a prior analysis of the tasks. She analyzed the tasks that students were to perform in terms of representation diversity. She classified them according to the system(s) of representation used in the problem statement and, more importantly, identified the systems of representation she expected students to use. As the teacher wanted to use the three types of representation as often as possible, if the tasks would not allow it, she would have to adapt them. The teacher conducted an analysis of all the tasks in the 'Areas of plan figures' chapter of the schoolbook adopted to characterize the representations used in the problem statement of the tasks and the suggested and expected representations in the students' solutions. The teacher noted that most of the tasks in the adopted mathematics schoolbook that students had to solve resorted to a system of representations, usually verbal, presenting visual or symbolic representations as a complement (Montenegro, Campos, &

Aires, 2017). This work gave her some flexibility in the knowledge of the characteristics of tasks about representations. In the following school year, she started by exploring the content 'Natural numbers – multiples and dividers' (39 lessons), resorting as often as possible to visual representations, having as her objectives that the students were familiar with the three types of systems of representation and could understand them as equivalent representations (Montenegro, Costa & Lopes, 2017a). The teacher used a Venn diagram to represent visually the divisors of a natural number and the greatest common divisor (gcd) of two or more integers, and a numerical line to represent visually the multiples of a natural number, or least common multiple (lcm) of two numbers, including some treatments. During the exploration of content with multiple representations, the teacher, whenever possible, used as a teaching approach the realization of visual treatments in any type of representation, looking for a better understanding of the effects of their use and making them familiar to students. To finish this study of the content, the teacher planned a task that consisted of identification of the representations used in the solutions of four problems (about divisors, multiples, gcd, and lcm) to be solved by her students in groups (Montenegro, Costa, & Lopes, 2017c), and a test to be solved by her students and those of another class whose activity did not value the multiple representations nor the visual ones (Montenegro, Costa, & Lopes, 2017b).

This was the content that the teacher explored in more detail, and from which she got most of her professional knowledge, being described in the results section. In the content she taught later, she sought specific knowledge. After this, she also made interventions in geometric content (angles and triangles). She explored a routine exercise applied to consolidate the classification of triangles in terms of the sides and the angles, and how to enrich it from the point of view of the representations used (Montenegro, Costa, & Lopes, 2017d).

The Data

During the first intervention, data were collected from the class audio record, each of the students' written registers, the teacher's class plan, and her notes.

During the successive interventions, data were collected from the teacher's weekly plans, students' daily notebooks, photographic records of the blackboard, and the teacher's notes.

Data Analysis

With the collected data, the teacher wrote two multimodal narratives: one on the first intervention (MN1) and the other on the succession of interventions (MN2). In both, the teacher added actions and decisions, and some justifications, to increase reader understanding; at the same time, she promoted her self-reflection. The writing of narratives followed the protocol described in Lopes et al. (2014).

The writing of multimodal narratives made possible the agglutination of data in a single document, which was used to make the data analysis. In the case of MN1, the narrative consisted of a detailed description of what happened in the class (actions and language). The elaboration of this multimodal narrative was also based on the description and justification of the teacher's intentions and decisions in different phases of the class. It consisted of the two parts already mentioned above, and was validated by two external researchers, who affirmed the collected data to confirm their conformity to the narrative and verify that it was self-contained and legible.

At the beginning, we identified the strategies chosen by the students to solve the task and the representations (which were different within the same group). We also identified the moments of difficulty felt by students during the solution of the task, and which situations were overcome through visual representations. In a second phase, we conducted the data analysis, paying special attention to those situations in which the teacher realized that the use of visual representation influenced learning, attempting to answer the research problem. After the MN1 analysis, the teacher began a set of interventions from which she elaborated MN2, based on the collected data and following the same protocol (Lopes et al., 2014). This consisted of the two same parts: the presentation and context, and episodes' description.

RESULTS

Findings From Multimodal Narrative of Intervention 1

In this first intervention, we found the viability of using visual representations with the same status as symbolic and verbal representations and found that they could be adapted to other situations, satisfying a dynamic utilization of visual representations. Visual representations proved to be something that can be used in a knowledge-building situation during teaching and learning (Montenegro, Costa, & Lopes, 2018).

The first relevant evidence for the use of visual representations that the teacher found was the ease with which students could act on representations from any representation system, highlighting relevant information (Excerpt 1 of Multimodal Narrative 1: teacher's collection of data on students' work).

The elaboration of MN1 facilitated the identification of these situations.

Multimodal Narrative Excerpt 1: Teacher C (Patterns)

Later, I realized that students who wrote the resemblance of the sequence's figures, signalizing some elements, highlighting them, [...] used this schema [...] to verify other figures of the sequence and answer other questions of the task.

The teacher concluded that the students could use the visual representations to sign/highlight some characteristics (Multimodal Narrative Excerpt 2) in different ways. On the left side of the figure of Multimodal Narrative Excerpt 2, they used small lines to highlight some information; on the right side, the students used color with the same objective.

Figure 1.

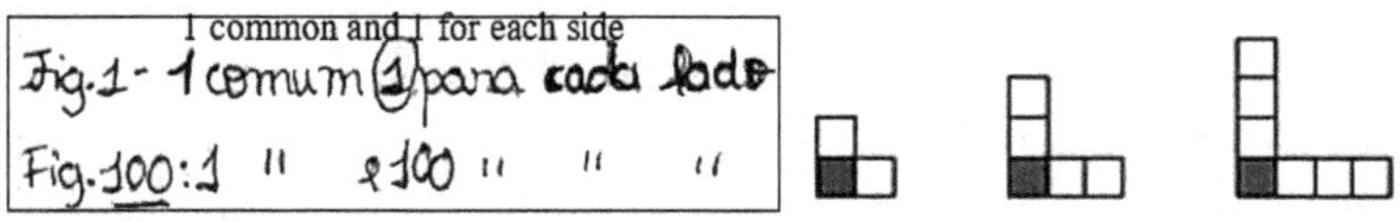

Multimodal Narrative Excerpt 2: Teacher C (Patterns: Two Representations - Verbal and Visual: Figure 1)

Elaboration of the multimodal narrative allowed the teacher to identify the conditions wherein the groups of students used the three representation systems simultaneously. The next excerpt of the MN1 identifies the teacher's intentions and her decision on the way to address the analysis of the characteristics of the sequence (Multimodal Narrative Excerpt 3). The elaboration of the multimodal narrative, articulating the collected data (written registers of students, audio record of the class, intentions of the teacher) facilitated this analysis.

The second evidence the teacher found was of the use of visual representations to justify reasoning. In the specific case, a visual treatment of the visual representation provided through the use of color (see Multimodal Narrative Excerpt 2, on the right) allowed students to show the inadequacy of the strategy (Multimodal Narrative Excerpt 3) and guided them to another that was more appropriate, promoting the use of multiple representations and conversations between them (Multimodal Narrative Excerpt 4).

Multimodal Narrative Excerpt 3: Teacher C (Patterns)

I verified with the students the impossibility of designing the 100th term, and then the necessity to find other way to find it. I suggest to students to look very carefully, so they could see the regularities in the figures of the sequence. I directed their attention, so they saw that all figures had a similar design: they had a square in the corner that was common to all figures and the same number of squares on each side.

All the students: Aaaahhh!

They gave the answer to the 110th term and verified that their first solution was wrong.

The teacher concluded that she could use representations in the three representation systems simultaneously as complements and verification of reasoning. Multimodal Narrative Excerpt 4 shows one of these examples, in which the aim was to determine which is the number of the sequence figure that will have 81 squares, using a representation of the same situation in the three representation systems.

The third evidence the teacher found was the ease of converting representations through the recreation of representations through visual treatments (Multimodal Narrative Excerpt 4 and Multimodal Narrative Excerpt 5).

Multimodal Narrative Excerpt 4: Teacher C (Patterns)

To solve the question, the student fixed his attention on the rule for growing the sequence, in natural language (verbal representation system, in the Figure 2). He made his drawing (visual representation system, in the Figure, center) and removed the corner square from the total of squares. Then he divided by two (symbolic representation system, right).

The elaboration of the MN1 allowed the teacher to identify the situations in which students resorted to less conventional uses of visual representations, as we see in Multimodal Narrative Excerpt 5.

Figure 2.

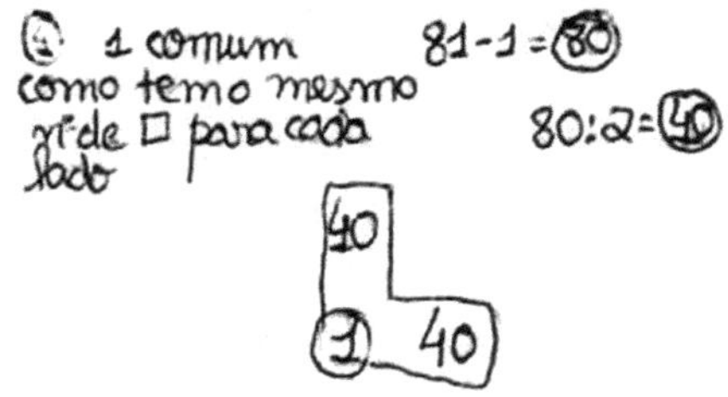

Multimodal Narrative Excerpt 5: Teacher C (Patterns)

Later, I realized that Group I used too small signs in their representation as a schema...

The fourth evidence found by the teacher was the use of visual treatments to translate the reasoning performed during the task resolution (Multimodal Narrative Excerpt 6).

The teacher found a use of visual representations (arrows) to represent the order or the relation/operation between the different sequence terms (Multimodal Narrative Excerpt 6) in a numerical representation that allowed the students to somehow convey their reasoning.

Multimodal Narrative Excerpt 6: Teacher C (Patterns)

The students used small arrows to indicate an order to the right to indicate when they should have added to obtain the following term of the sequence (Figure 3).

One of the results that the teacher found was the need to continue her professional development. If, on the one hand, during the first intervention, she found one dynamic way to use visual representations, on the other hand, that dynamic form was still limited to the generalization of figurative patterns and, consequently, to algebra. For that reason, it was necessary to find ways to give a similar use to visual representations in other mathematical contexts. The lack of similarity of this mathematical content (patterns) with other mathematical content impeded their extrapolation to new situations. For this reason, she felt the need to continue investigating—that is, the teacher, in her role as the fundamental subject in her own professional development, recognized the need to develop and acquire new tools that would allow her to use and manipulate visual representations, promoting her professional development, on the one hand and, on the other hand, improving her performance in their use. The teacher had to develop

Figure 3.

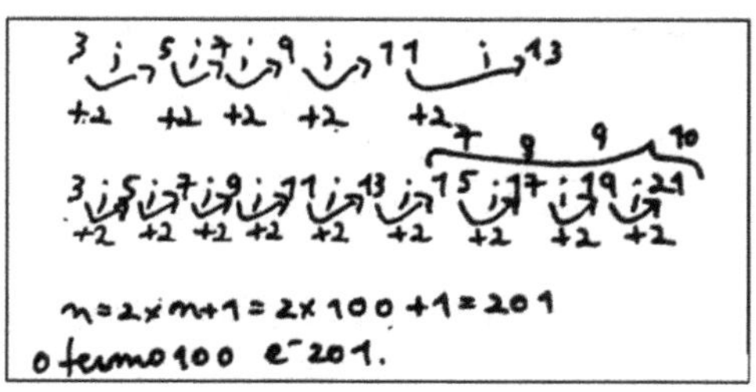

Figure 4.

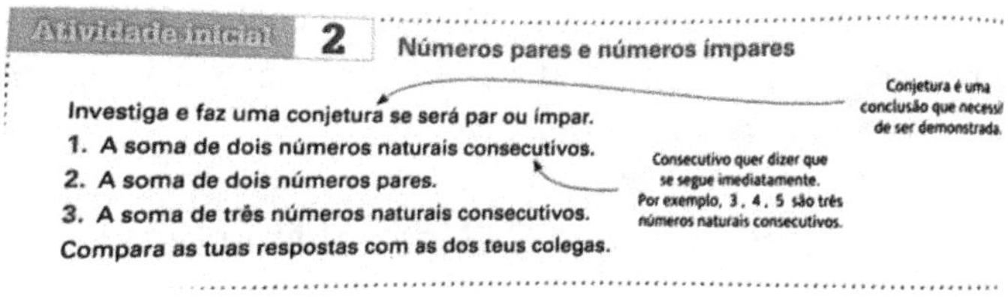

Atividade inicial 2 Números pares e números ímpares

Investiga e faz uma conjetura se será par ou ímpar.

1. A soma de dois números naturais consecutivos.
2. A soma de dois números pares.
3. A soma de três números naturais consecutivos.

Compara as tuas respostas com as dos teus colegas.

Conjetura é uma conclusão que necessi de ser demonstrada.

Consecutivo quer dizer que se segue imediatamente. Por exemplo, 3, 4, 5 são três números naturais consecutivos.

Table 2. Translation of Figure 4

Translation:
Investigate and make a conjecture when it will be even or odd. 1. The sum of two consecutive natural numbers. 2. The sum of two even numbers. 3. The sum of three consecutive natural numbers. 4. Compare your answers with those of your colleagues.

this dexterity to get her students to use visual representations. For that, the teacher made a succession of interventions following the first one. She had many questions and few answers. She sought to use the three representation systems in all classes, with conversions between representations. She also sought to know which visual representation she should use to represent the mathematical object of a specific content—namely, the divisors or the multiples of a natural number, and how to get students to use these representations, or how to solve the task visually. How would students react? Would they use these representations? Would they present other difficulties inherent in the design of these representations? Would they distinguish different representations for the same object? Would they realize the equivalence of representations? The teacher planned her activities over one and half years, having as a research objective to answer as many questions as possible.

Findings from Multimodal Narrative of the Succession of Interventions

The teacher, basing her reflection on the evidences identified in the first intervention, planned a succession of interventions for the following school year. The elaboration of the MN2 allowed the teacher to identify the class preparation as essential, with special emphasis on the characterization of the tasks to explore in terms of representations. The first need the teacher felt was to know the tasks in terms of transformations of representations that may be essential to a good task exploration (Multimodal Narrative Excerpt 7).

Multimodal Narrative Excerpt 7: Teacher C (Natural Numbers)

Activity Explored in Whole Class (Figure 4): *'Even and odd numbers', p.12. Its statement is written in a verbal representation. It is expected that students use numerical representations in their answers. The range of answers is limited.*

The elaboration of MN2 allowed the teacher to identify some situations where it is possible to enrich a learning process with other transformations if she makes some changes to the task, as we see in Excerpt 8 MN2. The teacher, in her practice, can enrich the tasks she selects, adding other types of representation. This was the first change in her teaching practice.

Multimodal Narrative Excerpt 8: Teacher C (Natural Numbers)

After completing these tasks, during the week indicated, the teacher realized that during the lesson she must manage the representations to be used and the transformations to be made. This can take advantage of the students' suggestions regarding the representations to be used, knowing which students use most easily.

The second change in the teacher's practice was the constant concern to vary the types of representation and to provide diverse transformations of representations during the classes. The elaboration of the MN2 allowed identification of the teacher's concern related to the use and exploration of representations during the classes and identification of which representations the students used and what types of representation transformations they made. Excerpt 9 of MN2 shows this concern.

Multimodal Narrative Excerpt 9: Teacher C (Natural Numbers)

Application of the distributive property of multiplication in relation to the addition of natural numbers in the calculation of areas of plane figures.

I sought to provide the widest possible range of representations in different systems of representation (Figure 5). The students used representations in three systems of representation: discursive in natural

Figure 5.

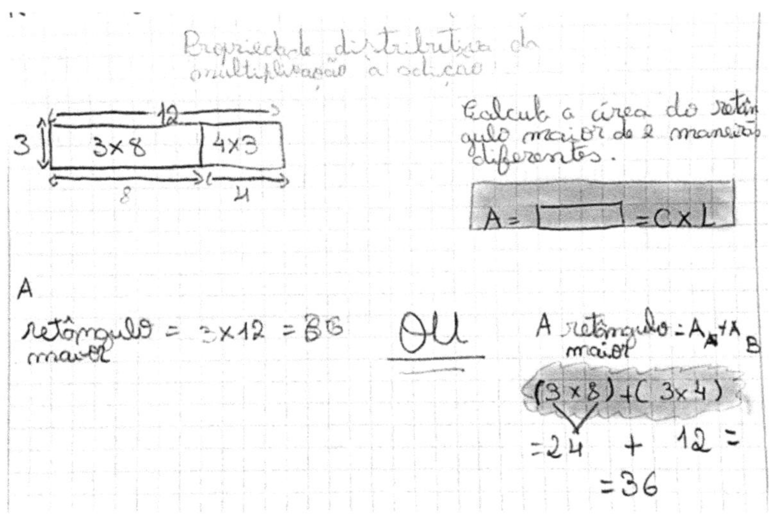

language, discursive symbolic (numerical), and non-discursive iconic (geometrical figure) The students made treatments in symbolic and iconic registers and they made conversions among the three registers.

Regarding her professional development, the teacher began to be able to better classify the tasks regarding the representations and transformations of representations that allowed her to identify more quickly the tasks where it was more difficult to provide a greater diversity of representations (Excerpt 10 of the Multimodal Narrative 2).

Multimodal Narrative Excerpt 10: Teacher C (Natural Numbers)

Consolidation tasks are the most difficult to vary representations. Numerical representations most often require numerical representations.

The teacher also realized the possibility of varying the representation systems used in the answers and some situations that were propitious to the use of visual representations (Excerpt 11 MN2).

Multimodal Narrative Excerpt 11: Teacher C (Natural Numbers)

It is possible to solve a problem in the three representation systems. The syntheses allow the use of visual representations.

The teacher realized that the use of multiple representations must be frequent since students take time to feel comfortable with their use. However, mechanization cannot be allowed, as there is a risk of diminishing its impact (Excerpt 12 of MN2).

Multimodal Narrative Excerpt 12: Teacher C (Natural Numbers)

Students need time to exercise different representations. However, there is the danger of mechanizing if the work is repetitive. It is necessary to balance their use.

The teacher confirmed the performance of visual treatments through color, which enabled the diversification of representations (Excerpt 13 of MN2).

Multimodal Narrative Excerpt 13: Teacher C (Natural Numbers)

The introduction of color can be used in representations to highlight relevant information and better establish correspondence between representations. Diversifying the way you present the resolution increases the types of representation.

The teacher concluded that facilitating the use of multiple representations should give her students the opportunity to choose the representation that was easier for them to work with, and to go slowly, demanding resolutions in another type of representation (Excerpt 14 of MN2).

Multimodal Narrative Excerpt 14: Teacher C (Natural Numbers)

Searching for the representation that fits best can be difficult. Each student should be able to choose the representation with which he wants to work at an early stage. A combination of representation systems can facilitate the reading, writing, and organization of information. Representations with elements of representation systems may be of added value to the teacher.

The teacher concluded that the use of visual representations is characteristic of the theme and that changing the subject requires a new work of analysis and self-reflection (Excerpt 15 of MN2).

Multimodal Narrative Excerpt 15: Teacher C (Natural Numbers)

The use of visual representations is not evident; it allows a great variety and the decision to choose the best or the most appropriate. The more you use it, the more you want to use it, at least in an initial phase of understanding something, which can be content (as a sign) or problem solving (with visual treatments).

After this work of reflection and analysis of her own teaching practice, the teacher felt more secure in using different representations and knowing ways to extract more value from the visual representations through the accomplishment of visual treatments. At the end of this phase, the research team reoriented the broad study based on the results obtained during this stage of the professional development.

DISCUSSION AND FUTURE RESEARCH DIRECTIONS

The MN1 and MN2 that the teacher constructed facilitated the data analysis, condensing the data in a single document, and promoted her self-reflection, which she did in a structured way. From this reflection, the teacher found a starting point to explore other situations in which to use visual representations in relation to other mathematical content.

The self-reflection done through this multimodal narrative served as a device for the design of a succession of interventions, with the objective of the teacher's professional development. The teacher, based on knowledge of the research carried out in the area, found a way to operationalize the analysis of the characteristics of the representations through the work of Duval (2006). This operation allowed her to reflect on the professional knowledge acquired (see Excerpts 10 to 15 of MN2).

On the other hand, the MN allowed the opportunity to work together with other researchers, initially to validate the multimodal narratives and later to produce knowledge recognized by the community (Montenegro, Campos, & Aires, 2017; Montenegro, Costa, & Lopes, 2017a, 2017b, 2017c, 2017d, 2018).

Indeed, the succession of interventions was a privileged environment for the teacher's professional development, in the sense that the execution followed by the self-reflection gave important information through the analysis of the way that students worked and thought about: (i) their knowledge of the characteristics of the tasks in terms of representations and their transformations; and (ii) the form, as the teacher included visual representations in the teaching. This development increased the teacher's knowledge about teaching and learning (Collins, Joseph & Bielaczyc, 2004; Lopes & Cunha, 2017; Morgan et al., 2014; Norton & McCloskey, 2008) using visual representations by promoting her knowledge about the characteristics of those interventions and the processes of conception and development.

From this succession of interventions and analysis of MN2 resulted the publication of some papers. The most common representation system the teachers demand in students' answers is the symbolic system (Montenegro, Campos, & Aires, 2017). However, the teacher can contradict this trend. For example, Montenegro, Costa, and Lopes (2017d) refer to a routine exercise to which they added a visual representation (table) which increased the range of transformations of representations, namely conversions. A similar situation is described in this chapter, in Multimodal Narrative Excerpt 8, in which the answer organized in a Carroll diagram increased the range of representations. The teaching of mathematical content can appeal to the joint exploration of representations of different representation systems. In this respect, Montenegro, Costa, and Lopes (2017b, 2017b) analyzed two learning situations which differed in the use (joint or not) of different representation systems in the teaching of divisors of a natural number. That way, giving a more dynamic character to the visual representations—i.e., using them for an overview of other types of representation, for example, verbal (Multimodal Narrative Excerpt 2, left) or symbolic representations (Multimodal Narrative Excerpt 3)—the static representation is transformed into a dynamic one, making possible the exploration of other ways of exploring the tasks. When someone analyzes his own practice, therefore, we recommend the elaboration of MNs, because they allow a structured analysis.

The MN proved to be an effective document in a situation of professional development, based on the self-reflection of a teacher who investigates her own practice with the collaboration of other researchers. However, it would be relevant to study the use of MNs on each of the other models of professional development considered by Sparks and Loucks-Horsley (1989)—i.e., autonomous professional development, curricular and organizational professional development, and professional development through short training courses.

Finally, a brief note about the pedagogical implications of the teacher's work. The teacher must characterize the representations in the statement of the tasks and the representations that are expected in the students' answers. However, as there are no clear indications in the curricula, the exploration and reinvention of representations is based on the teacher's pedagogical decision, according to her beliefs.

CONCLUSION

Our objective was to know what actions the teacher should take to improve her teaching practice of visual representations. We identified the elaboration of the MN as a tool that facilitates the process of professional development at the self-reflection level. Indeed, the analysis of MN in the first intervention (MN1) allowed better knowledge of the use of visual representations, which conditioned the succession of interventions, giving a new orientation of the study.

Analysis of the multimodal narrative of the succession of interventions (MN2) allowed us to know in detail the characteristics of the representations used by the students and to identify the actions that the teacher should take during the class preparation and in her teaching to promote a situation with multiple representations. That way, the MN proved an efficient and useful document as a basis for self-reflection and to investigate exhaustively, because its elaboration allowed maximization of the knowledge of the impact of visual representations in the teaching and learning of mathematical content. The use of MN could permit and facilitate self-reflection and be available to other teachers.

The teacher carried out her self-reflection by detailed analysis of her practice through the use and analysis of multimodal narratives. The study of research works in the area and the possibility of working together with other researchers may alert teachers to aspects of the practice that they had not yet realized. In this way, the teacher promoted her professional development.

REFERENCES

Ainsworth, S. (2006). DeFT: A conceptual framework for considering learning with multiple representations. *Learning and Instruction, 16*(3), 183–198. doi:10.1016/j.learninstruc.2006.03.001

Arcavi, A. (2003). The role of visual representations in the learning of mathematics. *Educational Studies in Mathematics, 52*(3), 215–241. doi:10.1023/A:1024312321077

Bargagliotti, A. E., & Anderson, C. R. (2017). Using learning trajectories for teacher learning to structure professional development. *Mathematical Thinking and Learning, 19*(4), 237–259. doi:10.1080/10986065.2017.1365222

Bogdan, R., & Biklen, S. (1994). *Investigação qualitativa em educação. Uma introdução à teoria e aos métodos* [Qualitative research in education. An introduction to theory and methods]. Porto: Porto Editora.

Collins, A., Joseph, D., & Bielaczyc, K. (2004). Design research: Theoretical and methodological issues. *Journal of the Learning Sciences, 13*(1), 15–42. doi:10.120715327809jls1301_2

Dreyfus, T. (1991). On the status of visual reasoning in mathematics and mathematics education. In F. Furinghetti (Ed.), *Proceedings of the 15th Conference of the International Group for the Psychology of Mathematics Education* (vol. 1, pp. 33–48). Assisi, Italy: Università di Genova.

Duval, R. (2006). Transformations de représentations sémiotiques et démarches de pensée en mathématiques [Transformations of semiotic representations and thought processes in mathematics]. In *Actes du XXXIIe Colloque COPIRELEM* (pp. 67–89). Strasbourg, France: ULP. IREM.

Fischer, C., Fishman, B., Dede, C., Eisenkraft, A., Frumin, K., Foster, B., ... McCoy, A. (2018). Investigating relationships between school context, teacher professional development, teaching practices, and student achievement in response to a nationwide science reform. *Teaching and Teacher Education, 72*, 107–121. doi:10.1016/j.tate.2018.02.011

Goldin, G. A. (2008). Perspectives on representation in mathematical learning and problem solving. In L. English (Ed.), *Handbook of international research in mathematics education* (pp. 178–203). New York, NY: Routledge.

Goldin, G. A., & Kaput, J. J. (1996). A joint perspective on the idea of representation in learning and doing mathematics. In L. P. Steffe, P. Nesher, P. Cobb, G. A. Goldin, & B. Greer (Eds.), *Theories of mathematical learning* (pp. 397–430). Mahwah, NJ: Erlbaum.

Hargreaves, A. (2007). Sustainable leadership and development in education: Creating the future, conserving the past. *European Journal of Education, 42*(2), 223–233. doi:10.1111/j.1465-3435.2007.00294.x

Heinze, A., Star, J. R., & Verschaffel, L. (2009). Flexible and adaptive use of strategies and representations in mathematics education. *ZDM*, *41*(5), 535–540. doi:10.100711858-009-0214-4

Hevner, A. (2007). A three-cycle view of design science research. *Scandinavian Journal of Information Systems*, *19*(2), 87–92.

Lopes, J. B. (2017). Researching science teaching practices in the classroom over two decades: What matters? In J. B. Lopes, J. P. Cravino, E. S. Cruz, & A. Barbot (Eds.), *Teacher science: Contributions of research for planning, practice and professional development* (pp. 1–26). New York, NY: Nova Science Publishers.

Lopes, J. B., & Cunha, A. E. (2017). Self-directed professional development to improve effective teaching: Key points for a model. *Teaching and Teacher Education*, *68*, 262–274. doi:10.1016/j.tate.2017.09.009

Lopes, J. B., Silva, A. A., Cravino, J. P., Santos, C. A., Cunha, A., Pinto, A., ... Branco, M. J. (2014). Constructing and using multimodal narratives to research in science education: Contributions based on practical classroom. *Research in Science Education*, *44*(3), 415–438. doi:10.100711165-013-9381-y

Lopes, J. B., Viegas, C., & Pinto, A. (2018). Melhorar Práticas de Ensino de Ciências e Tecnologia – Registar e Investigar com Narrações Multimodais [Improving Science and Technology Teaching Practices - Registering and Investigating Multimodal Narratives]. Lisboa: Edições Sílabo.

Montenegro, P., Campos, H., & Aires, A. P. (2017). Tasks involving perimeter and area of plane figures: Analyses of a mathematical textbook. In L. Gómez Chova, A. López Martínez, & I. Candel Torres (Eds), *EDULEARN2017: Proceedings 9th International Conference on Education and New Learning Technologies* (pp. 2469–2475). Barcelona, Spain: IATED Academy. 10.21125/edulearn.2017.1518

Montenegro, P., Costa, C., & Lopes, B. (2017a). Transformações de representações visuais na exploração de múltiplos e divisores de um número [Transformations of visual representations in the exploration of multiples and divisors of a number]. *Comunicações Piracicaba*, *24*(1), 55–68.

Montenegro, P., Costa, C., & Lopes, B. (2017b). The use of visual representations in mathematics classes: A comparative case study. In L. Gómez Chova, A. López Martínez, & I. Candel Torres (Eds.), *INTED2017: Proceedings of 11th International Technology, Education and Development Conference* (pp. 6845–6852). Valencia, Spain: IATED Academy. 10.21125/inted.2017.1590

Montenegro, P., Costa, C., & Lopes, B. (2017c). Multiple representations in problem solving with divisors and multiples. In L. Gómez Chova, A. López Martínez, & I. Candel Torres (Eds.), *INTED2017: Proceedings of 11th International Technology, Education and Development Conference* (pp. 6884–6890). Valencia, Spain: IATED Academy. 10.21125/inted.2017.1599

Montenegro, P., Costa, C., & Lopes, B. (2017d). Visual representation (table) to recreate a routine mathematical exercise. In L. Gómez Chova, A. López Martínez, & I. Candel Torres (Eds.), *EDULEARN2017: Proceedings 9th International Conference on Education and New Learning Technologies* (pp. 2514–2520). Barcelona, Spain: IATED Academy. 10.21125/edulearn.2017.1525

Montenegro, P., Costa, C., & Lopes, J. B. (2018). Transformations in the visual representation of a figural pattern. *Mathematical Thinking and Learning*, *20*(2), 91–107. doi:10.1080/10986065.2018.1441599

Morgan, C., Craig, T., Schuette, M., & Wagner, D. (2014). Language and communication in mathematics education: An overview of research in the field. *ZDM Mathematics Education, 46*(6), 843–853. doi:10.100711858-014-0624-9

National Council of Teachers of Mathematics (NCTM). (2008). *Princípios e normas para a Matemática escolar.* Lisbon, Portugal: APM.

Norton, A. H., & McCloskey, A. (2008). Teaching experiments and professional development. *Journal of Mathematics Teacher Education, 11*(4), 285–305. doi:10.100710857-008-9076-x

Ponte, J. P. (1994). O desenvolvimento profissional do professor de Matemática [The professional development of the Mathematics teacher]. *Educación Matemática, 31*, 9–12, 20.

Ponte, J. P. (2004). Investigar a nossa prática: Uma estratégia de formação e de construção do conhecimento profissional [Investigating our practice: A strategy of training and building professional knowledge]. In *Actas do VIII Simpósio do SEIEM*. Coruña: Universidad de la Coruña.

Sandoval, W. A., & Reiser, B. J. (2004). Explanation-driven inquiry: Integrating conceptual and epistemic scaffolds for scientific inquiry. *Science Education, 88*(3), 345–372. doi:10.1002ce.10130

Saraiva, E. (2017). *Estudo do papel da representação visual no contexto da mediação dos professores de ciências físicas* [Study of the role of visual representation in the context of the mediation of teachers of physical sciences] (Unpublished doctoral dissertation). Universidade de Trás-os-Montes e Alto Douro, Vila Real, Portugal.

Sedig, K., & Sumner, M. (2006). Characterising interaction with visual mathematical representations. *International Journal of Computers for Mathematical Learning, 11*(1), 1–55. doi:10.100710758-006-0001-z

Shagrir, L. (2010). Professional development of novice teacher educators: Professional self, interpersonal relations and teaching skills. *Professional Development in Education, 36*(1–2), 45–60. doi:10.1080/19415250903454809

Soebari, T. S., & Aldridge, J. M. (2015). Using student perceptions of the learning environment to evaluate the effectiveness of a teacher professional development programme. *Learning Environments Research, 18*(2), 163–178. doi:10.100710984-015-9175-4

Sparks, D., & Loucks-Horsley, S. (1989). Five models of staff development. *Journal of Staff Development, 10*(4), 40–57.

Yin, R. K. (2010). *Estudo de caso—Planejamento e métodos* [Case study—Planning and methods] (4th ed.). Porto, Portugal: Porto Editora.

Chapter 11
Assessing Teaching Practices Development With Multimodal Narratives

Ana Edite Cunha
Escola Secundária S. Pedro de Vila Real, Portugal

ABSTRACT

This chapter focuses on teachers' professional development, on the task design and on experimental work, as well as on the role of the teachers' mediation in the quality of student learning. The research problem was how the teacher can promote self-directed professional development, namely, improving the quality of teaching practices to influence the quality of students' learning, in their engagement in experimental tasks and epistemic practices. A longitudinal research methodology was followed during 10 years, based on a qualitative case study, from a curricular approach in secondary education. The analysis of data collection on teaching practices and students' learning over time and the teachers' professional pathways allow to formulate the following conclusions: (1) new traits of teaching practices were identified that promote students' productive engagement; (2) changes to the task design were enough to trigger differences in teachers' mediation, with consequences for students' epistemic practices and their productive engagement.

INTRODUCTION

The essence of this work was devoted to the professional development of a teacher, focused on improving the quality of his teaching practices (planning and teaching practice), in particular considering the conception of experimental tasks and effective mediation in the classroom.

In this way, a curriculum (tasks) of physics and chemistry for secondary education was designed, implemented and evaluated, based on the execution of experimental work in the physics component, taught in the 10th and 11th years of schooling, based on teacher mediation, to promote students' productive engagement in tasks and students' quality epistemic practices.

DOI: 10.4018/978-1-5225-8570-1.ch011

During the execution of this work, reflections were made on the performance of the professor and subsequent changes were made, when necessary. In particular, the study of the role in learning of tasks in experimental teaching – namely, its characteristics, classroom management, and student appropriation, as well as its use as an instrumental tool – was undertaken to develop knowledge, skills and attitudes, as well as to promote epistemic practices in students.

In teaching, by epistemic practice is understood the work that the student does to accomplish a task with a view to the construction of scientific knowledge, having as reference the activity of scientists. To engage in an epistemic practice, the student has to mobilize prior knowledge and has to build scientific knowledge. Epistemic practices should be induced and improved by the teacher, even those that, in some cases, may occur "spontaneously."

The question of research that is considered to help solve the research problem is presented below: What aspects of teaching practices (in task design and effective mediation) have changed over time? The main objectives of this research are:

- To study the influence of increasing the quality of teaching practices on student engagement and on epistemic practices.
- To characterize the influence of the use of multimodal narratives (MNs) on professional development with an impact on teaching practices.

BACKGROUND

Being a teacher requires that one constantly adapt to the reality of the moment and continuously update the knowledge to be taught.

After acquiring competences inherent to the activity of teaching, teachers' professional development and the process of improvement are done continuously, always keeping in mind the goal of achieving excellence. In order to develop professionally, teachers must be careful to update themselves by staying informed about the most appropriate methodologies, as well as all that appear recently.

It was necessary to reflect on methods or strategies that aim to increase teacher-student and student-student interaction (Berland, et al., 2016) and increase the taste for learning science in general and physics in particular. This practical problem was assumed as a research problem and was based on the professional development of a teacher focused on improving the quality of his teaching practices (planning and teaching practice) to improve student learning. It was necessary to realize that curricular modifications had to be carried out, to design pertinent tasks that aimed at the development of learning objectives and to design the way in which they would be implemented. From year to year, more modifications were introduced, taking into account not only the literature but also the reflections on the practice carried out until then. The detailed analysis of the practice developed within the classroom is a method well known and used over time for the professional development of teachers. In order to analyze in detail how this lesson has been developed in various possible ways, teachers should record their class (audio or video). The classroom space is a place where teachers and students (who are human) express emotions in interpersonal relationships of the most varied possible sensations, making teaching and learning unique in each class. A simple audio or video recording, and its simple transcription, is not enough to allow description and study of the relationships established within the classroom between teachers and students as well as the quality of the teacher's mediation (Lopes, Cravino, Cruz, & Barbot, 2017). An attempt was made to find an instrument that would provide as complete a document as possible, detailing the context of the lesson, the teacher's intentions and perceptions, the reactions of the students and the teacher, the work proposed and the work actually done by the students. This instrument is based on the narration of what happens in the classroom, according to the chronological course of time, allowing an easy understanding of events, in a way that preserves the integral understanding of the phenomena and the complexity of the class. Comparison of narrations of several teachers is possible since the narration is done by following a protocol to have the same focus (each task and the teacher's and students' action and discourse) and the same structure. In structure, the narration comprises two essential parts. The first part includes a summary of the lesson and its contextualization, including all the elements related to the lesson, the spatial organization of the classroom and students, the number of students, the total duration of the lesson and each episode, lesson objectives, etc. The second part contains a detailed description of each episode, beginning with the presentation of a task and ending as soon as another task is proposed. (Please see Part 1 for MN examples and Chapter 1 for MN details.) The narrative is also verifiable; that is, it is accompanied by supporting documents (which served as the basis for its construction) so that it can be accepted by other researchers. This instrument was called multimodal narration (Lopes et al., 2014; 2010). Although it uses multimodal data, it must be self-contained (but may have attached files that can prove the descriptions made). Although MN offers the description of a lesson – i.e., it is a fundamentally descriptive narrative – it focuses on the actions and language used by the teacher, and it must be understanding and self-contained, genuine and unique. The variety of data that are included in MN and that complement one another enriches the narration, making it more authentic, and therefore the narrative can be easily verified (through the analysis of accompanying documents on which its construction was based).

In this sense, MNs can be an important training tool for the teacher, since they are relevant from the point of view of mediation, as they narrate concrete action in the classroom.

Teachers each construct narration according to their own points of view, under the influence of their own orientations or educational and epistemological choices. This influences their intentions and decisions during the systematic response to the learning challenges of their students in accordance with the stated objectives. However, in making this description without value judgment, the narrator allows readers to read the narrative without being influenced.

In order for a teacher to produce a change in the level of her or his professional practices, it is necessary for the teacher to be able to take decisions and act in a positive way in future situations. Sharing reflections with colleagues or researchers will provide added value to professional development (Lopes, 2017). MNs are like engines for a learning that transforms the teacher as a professional and fosters his or her growth, either through self-reflection or through shared reflection (Lopes et al., 2017). This initial self-reflection increases the teacher's willingness to improve her or his actions in the classroom, regarding the various dimensions of mediation, trying to adjust them to the learning needs of the students, as well as to respond as best he or she can to the demands of curricular activities. Therefore, this process of self-reflection is crucial for professional development (Cunha, 2015).

The phase of awareness of his or her own actions, taking into account the MN episodes, awakens in the teacher the need to correct his or her actions in the future and the desire to seek continuous improvement, with particular emphasis on the various dimensions of mediation. The MN allows the teacher, through reflection, the ability to accept all possibilities, allowing the confrontation of beliefs and a readiness to change them. It also requires the teacher to make use of reflection to learn and grow professionally.

A teacher who recognizes his or her own class is confronted with this and can start a whole process of reflection allowing him or her to become aware of important aspects in his or her professional life about which he or she had never yet reflected and that need evolution. Any MN that relates to classroom events in the sciences is also useful in the professional development of other teachers, as they can compare the narratives with their own classes, often reviewing them in these classes, to recognize aspects of their own classes that lack learning and evolution (Lopes et al., 2010).

Reading and analyzing an MN allows the teacher who built it or provided the data for its construction to better see the reality of his or her classes (because sometimes one does not have the idea of the realities of the facts). With the collaboration of a research team, the use of MN can actually improve the teacher's teaching practices (Lopes & Cunha, 2017). It can also help the teacher find more effective ways of presenting the intended learning. It can increase teachers' perception and experience and help them develop practical knowledge, leading to professional development (Zeichner, 1993). For other teachers (those who did not provide the data), access to the MN allows them to review their own practices in light of the MN and use the examples in positive or negative ways to improve their own practices.

Teachers need to feel that they are heard, able and supported in building their professional development. Teachers must have the time, resources and ideas to create their own professional learning goals, which in turn will lead to professional development. If done in isolation, professional development experiences most often do not lead to any changes in classroom practices. These experiences need to be part of a teacher-initiated process and need to have preparation goals and follow-up on implementation (Osborne, 2014).

METHODOLOGY

To identify what can serve as a trigger in the evolution of mediation, it is necessary to make a macro analysis of what happens in the teacher's professional career and mediation. From the first classes that evolved, through the formulation of the tasks and the teacher's reflection, as well as during the classes, there was the progressive appearance of characteristics of the teacher's effort to improve students' epistemic practices (EP): the use of visual representations and artefacts, explicitly epistemic, valuing epistemically and synthesizing.

The teacher's reflection and the use of artefacts and visual representations, as well as the formulation of tasks in mediation in order to obtain epistemic practices and the productive engagement of the students, make it possible to improve the quality of teaching practices. This case study consists of an intervention implemented incrementally over 10 successive years from 2005/2006 to 2011/2012, with interventions in the classroom in 2005/2006, 2007/2008 and 2011/2012 in groups of the 10th and 11th years of schooling, alternating according to the year that the teacher was teaching when collecting data. The study was done in a school in the north of Portugal, where the socioeconomic class of students is predominantly middle/upper. The colleagues who collaborated in the reflections with the professor of the study taught in other schools in the region of Trás-os-Montes and in the same years of schooling. In this way, the researchers designed, implemented and evaluated a curriculum (tasks and mediation) of physics and chemistry for secondary education, based on the execution of experimental work in the physics component, taught in the 10th and 11th years of schooling, for 10 consecutive years, based on the teacher's mediation, so that the students were productively engaged in the tasks and it was possible to promote quality epistemic practices. A literature review of the research in physics didactics (especially in secondary education and whenever possible dedicated to the classroom context) and reflection on the results obtained from successive interventions in the classroom were taken into account.

The present investigation consisted of a longitudinal and action-research case study that developed over 10 academic years (Lopes, 2017) between 2004 and 2014. The action-research method is a systematic and cyclic process to improve practice quality by using research procedures. Cohen, Manion and Morrison (2011) propose the following steps: review current practice, identify aspects to be improved, design a way to do it, experiment, monitor and reflect on the repercussions, changing practice design in light of these reflections, and, finally, evaluate the modified action.

The extension of the study for the period of 10 years was due to its nature, since it was an incremental construction (Cunha, 2015; Lopes, 2017), which has been iterative over time, in a search for ideas, with testing and the incorporation of cases that prove to be productive. This is a well-understood course in the light of a research-action study with two iteration cycles (Kemmis, 1989; McNiff & Whitehead, 2006) that had a direct influence on the professional development of this teacher. It was a time-consuming construction with decision making, implementation and evaluation at interpolated moments, with a dynamics very subject to both the characteristics and conceptions of the researcher/trainer and the need to adjust to the new training approach. It was investigated on the teacher's own practice in order to improve it (Cohen et al., 2011; Lumpe, Vaughn, Henrikson, & Bishop, 2014); therefore, it assumed characteristics of the research-action type in the definition and implementation of a training program. In addition, it fulfilled two complete cycles of action research (Kemmis, 1989; McNiff, & Whitehead, 2006), which consisted of:

- Collect data on teaching and learning practices;
- Analyze the data and reflect on the results;
- Adjust a new action plan (reformulation of tasks and traits of mediation);
- Begin a new implementation, restarting the cycle once more.

This analysis took into account the different cycles of research-action, in particular the different phases by which it is constituted (Cohen et al., 2011; Lumpe et al., 2014).

In order to analyze the changes in the design of the tasks during the didactic intervention, the researchers considered as categories of analysis aspects that were particularly relevant in the continuation of this study, in order to better elucidate the modifications that have occurred over the years.

Thus, this analysis was of a macro and meso nature, centered on the curricular modifications and modifications to the tasks and planned mediation, based on reflection on the teaching practice as presented in the MN of the classes themselves. It took into account the analysis of mediation in action, over time, to promote the productive engagement of students' EP. A comparative analysis was carried out between the task as presented in 2006, 2008 and 2012 to verify whether or not there was evolution in the design and application of the tasks to the students. For this, dimensions and categories of analysis, as well as their definition, were created based on Lopes et al., 2014.

In order to analyze the actions of teacher mediation to promote students' productive engagement, three dimensions of MN analysis were determined by the theoretical basis referred to above: (a) teacher mediation to involve students in tasks, (b) engagement of students in the discipline and (c) indicators of student productivity. Through the obtained evidences, the first attempt at codification was made. Each coding was given a succinct designation as well as its definition. The unit of analysis considered was the interaction between teacher and students. After the coding was revised, its definitions were refined, and whether the selected excerpts from the MN were well described by each coding was checked. During this process, each encoding and its definition are not definitive; some encodings could be grouped into one, or they could be implemented. Each excerpt could be described by several codes. After this process was completed, each encoding obtained the dichotomous variable status. That is, each excerpt was characterized by the presence or absence of each of the dichotomous variables found in the open code analysis. This analysis was reviewed by two more independent researchers. The degree of agreement was 84%. In the remaining cases, refinement was made until agreement was reached. After this phase, the investigator made the categorization, which corresponded to reanalyzing the MN by using the dichotomous variables. The dichotomous variables determined by the open code analysis were presented and defined succinctly.

Variables related to the evidences of the students' epistemic practices during the execution of the task in the classroom (experimental work and/or the use of MN) and the teacher's efforts to induce these epistemic practices were also identified from each MN. By using qualitative analysis software (NVivo 8) and through the evidence, the first coding attempt was made. Each coding was given a succinct designation as well as its definition. During this process, each encoding and its definition were not definitive; some encodings could be grouped into one, or they could be implementation. Each excerpt could be categorized by several encodings. After this process was completed, each encoding obtained the dichotomous variable status. That is, each excerpt was characterized by the presence or absence of each of the dichotomous variables found in the open code analysis. This analysis was reviewed by two more independent researchers, one of whom was the professor-researcher. The unit of analysis considered was the interaction. The analysis dimensions derived from the literature review, namely Lopes *et al.* (2014), and the categories of analysis operationalized the dimensions of analysis and also derived from the literature review (Lopes et al., 2017). The degree of agreement was calculated for each case. In the remaining cases, refinement was made until agreement was reached. After this phase, the researcher performed the categorization.

The mediation dynamics in an episode of MN are represented by a temporal sequence of interactions. The interaction between teacher and students corresponds to an action of the teacher and the EP of the students that occur; it ends when the teacher makes another action to promote EP. Teacher actions in student EP promotion and student engagement were categorized. Obviously, the effort (actions) of the teacher can be successful or unsuccessful in the promotion of the students' EP. For each episode (beginning with the presentation of a task and ending with the students' execution of the task), a diagram of the dynamics of teacher and EP student mediation was elaborated. To create the teacher's mediation scheme, the analysis made in Nvivo was used. In the diagram (see Figure 1), the rectangles correspond to the teacher's efforts, the ovals correspond to the students' EP, and these are linked by arrows to indicate the interaction of the sequence of the teacher's mediation dynamics. Figure 1 presents a dynamic of interaction of the teacher's mediation of an episode of a MN from 2012, where the teacher's performances are represented by rectangles. The application of one of these efforts triggers the students' epistemic practices, represented by ellipses.

The formulation of the tasks in the interventions between 2006 and 2012 were also analyzed, taking into account the formulation of the tasks themselves and how to use them.

Figure 1. Diagram of the dynamics of the interaction between teacher and students of an MN from 2012

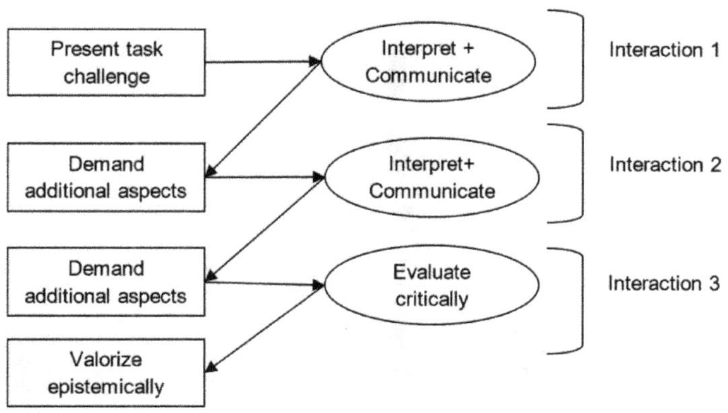

RESULTS AND ANALYSIS

The characterization of mediation actually effected by the teacher was performed, identifying the efforts that the teacher made to produce EP. These efforts composed five categories of analysis, which stood out for helping to explain the efforts of the teacher that had greater impact on students' EP and that had more evolution over the period of the study. These efforts were: (a) synthesizing, (b) asking for additional aspects of student work, (c) epistemically valuing students' ideas, (d) made explicit epistemically the work done by the students or what remained to be done, and (e) introducing the artefact, and introducing and using the visual representation. Overall, these teacher efforts were related to the didactic intent to increase students' EP and develop their epistemic skills. Often, the teacher asked for additional aspects from the students as a way of alerting them to the essential points in the evolution of their work or suggesting actions when they did not know what to do.

By realizing how different teachers teach the same curriculum in the classroom, one can better understand possible differences that may arise in student learning.

The teacher introduces the artefact, and introduces and uses the visual representation (see Figure 3). Through the analysis of MNs from 2006 to 2012, it was found that these efforts are present with the challenge component in greater quantity as time goes on, in which students developed more EP.

The way in which the teacher introduced the artefacts in the task was an important factor to improve the influence of the teacher's effort focused on "presenting the task as a challenge" (see Figures 4 and 5). This aspect promotes the use of visual representations and artefacts by the students and even the introduction of new artefacts and new visual representations.

The following figures show the introduction of artefacts as well as their use from 2006 to 2012. The EP obtained with the use of the artefacts can also be verified.

Figure 4 shows an excerpt from a 2006 MN, with an artefact introduced by the teacher and used by the teacher and some students, allowing only "observing" and "factually manipulating" by the students. Figure 5 presents an excerpt of a MN from 2008, with an artefact introduced by the students, used to "conceptualize" and "communicate autonomously," as well as the teacher's use of the artefact to "make explicit epistemically" the work of the students.

Figure 2. Teacher effort to promote students' epistemic practices in 2006, 2008 and 2012

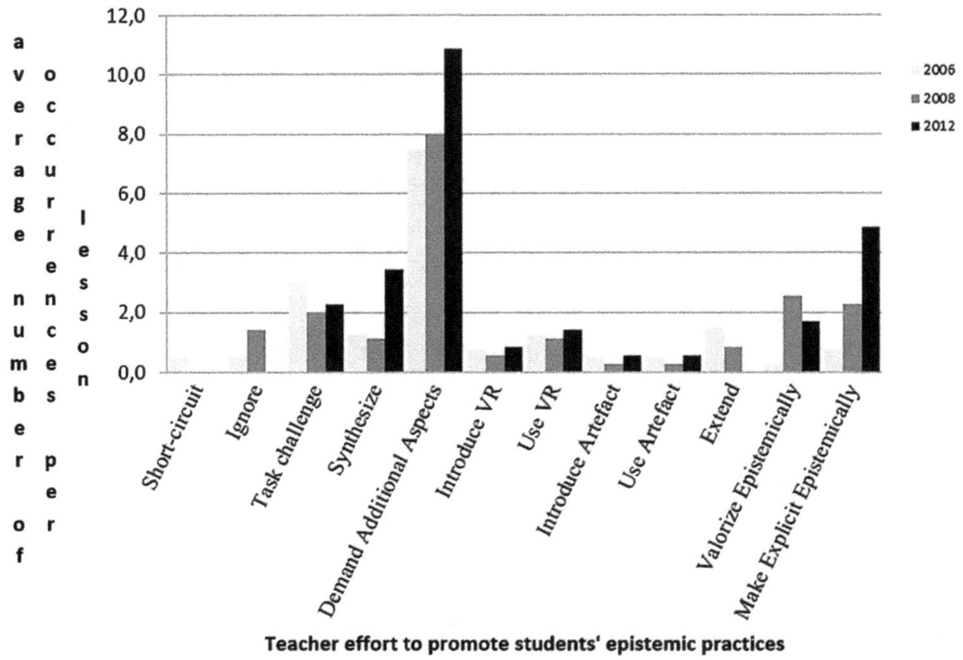

Figure 3. Introduction and use of artefact and visual representation in 2006, 2008 and 2012

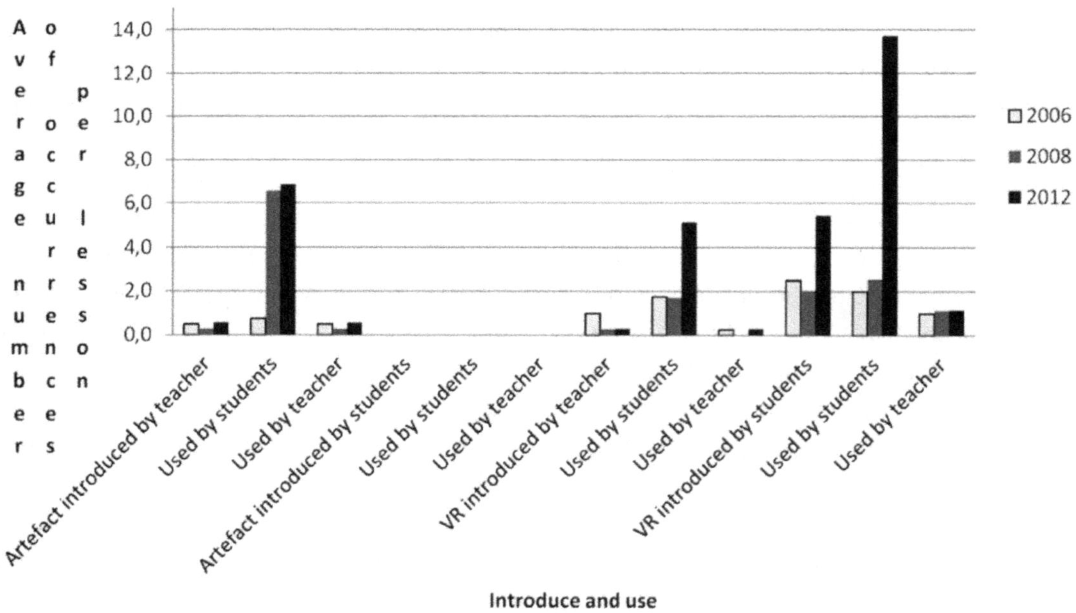

Figure 4. Excerpt from an MN analysis done in NVivo8 (2006) regarding the teacher's task to "introduce artefact" and its use by the students

[...]

I spent to show and introduce the modulator circuit:

Teacher: It got an antenna, a microphone and power supply. We have the radio.

[...]

Maria and Ana put up to go outside.

I went outside with the students to explain how they should do. The students that were on the classroom created a little bit of noise due to their curiosity. I request to students talked to the microphone of the circuit. The students were very amusing, it was noted that they were enjoying the activity.

The students that were in the classroom could hear my voice through the radio, because I talked louder than Maria and Ana.

Teacher: This happens because the laughing and playing. Do you need to concentrate – after my observation the students were silent.

Teacher: Now you can hear. Who wants to come out here now? Catarina, you want to come?

Catarina stands up enthusiastic.

Teacher: You have to speak louder! You speak very low!! Do you want to see how to do?!

Students are very amusing. And I tried several ways to transmit the information.

Teacher: Do you receive the information?

Students: Yes – they answered in choir.

Coding Density

TE Introduce artefact

EP Use artefact

EP Observe

EP Handle equipment factually

Figure 6 presents an excerpt from a MN from 2012, showing that, in this case, the introduction of an artefact allowed the students to "use representation," "use the artefact," "conceptually manipulate," "introduce representation" and "interpret."

That is, in 2008 and in 2012, the teacher's effort is not only to initiate students' EP but especially to improve the students' epistemic work, as illustrated by the excerpts in Figures 4, 5 and 6.

The introduction of artefacts and/or manipulative mediators influenced the mediation of this teacher because each year the introduction of and work with the artefacts and visual representations was done in a more refined way, so that there was more production and learning on the part of the students.

This teacher, in presenting the tasks, gives the students the authority to work autonomously, in groups or individually, in order to be able to build their learning, based on their own experience. The teacher would circulate through the groups during class and give occasional support when he felt the material already made sense to the students (see Figure 7). Through the analysis of MNs from 2006 to 2012, it was found that these efforts are present with the challenge component in greater quantity as time goes on, in which the students developed more EP.

Figure 5. Excerpt from a MN analysis in NVivo8 (2008) regarding the teacher's task to "introduce artefact" and its use by the students

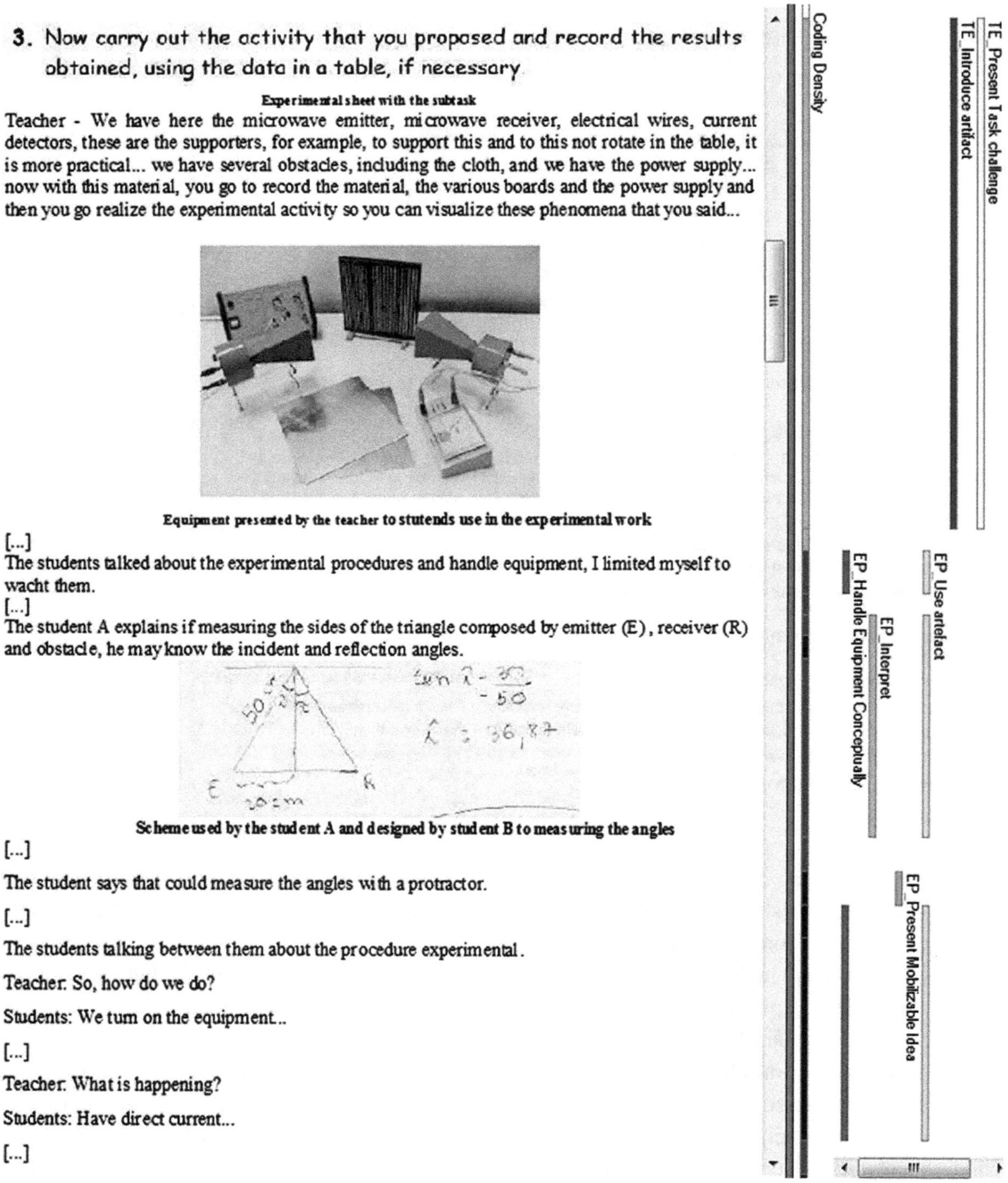

Figure 6. Excerpt from a MN from 2012 and its analysis in NVivo8 regarding the teacher's task to "introduce artefact" and its use by the students

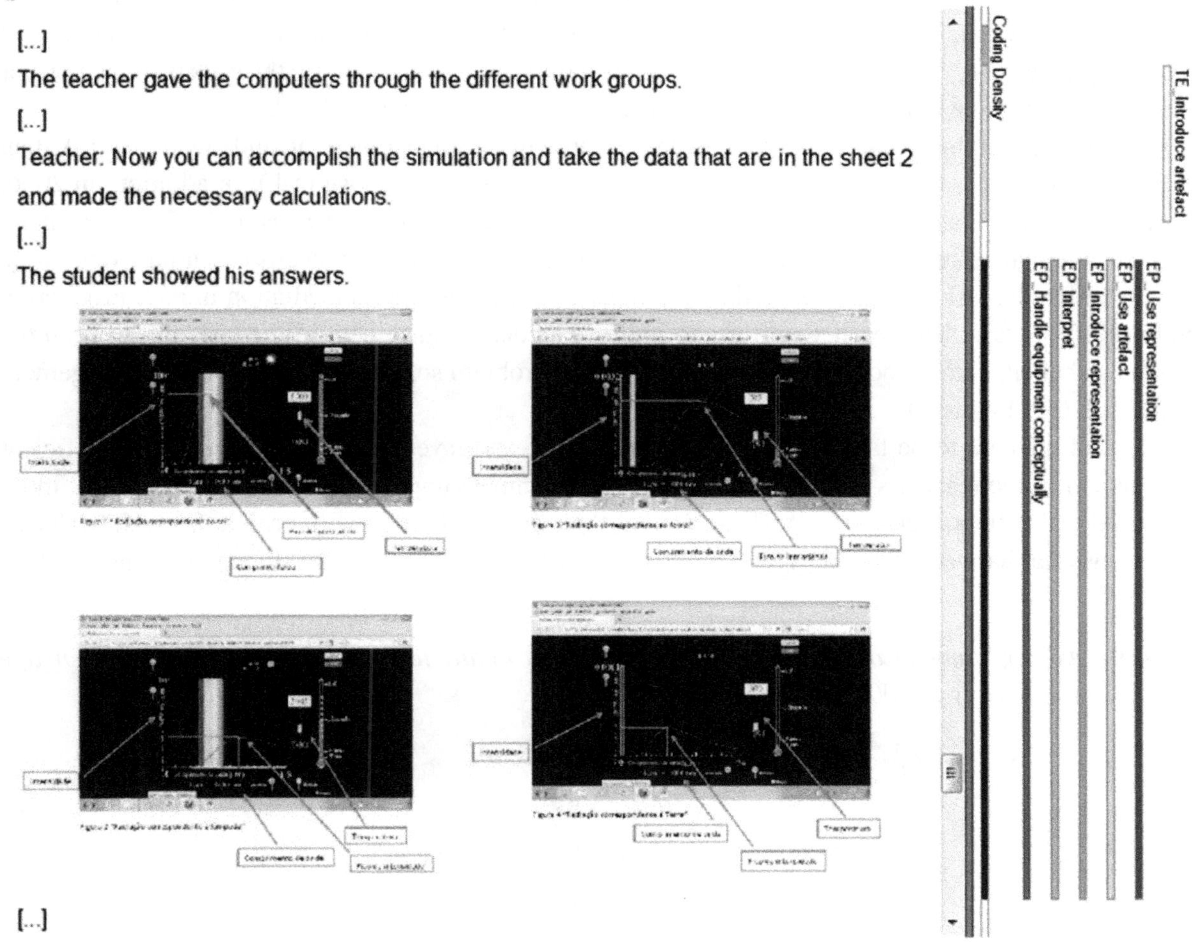

Other efforts of this teacher that became progressively more frequent (from intervention to intervention) and that had great influence on the development of the students' EP were the efforts to "value epistemically," "make explicit epistemically" and allow the students to introduce and use artefacts; these increased considerably throughout the teacher's interventions (see Figure 2). The effort to "value epistemically" was accomplished through taking advantage of students' ideas, schemes or actions, recognizing their epistemic value. This effort on the part of the teacher will trigger another set of epistemic practices on the part of the students. For example, in an excerpt from a 2008 MN where this effort was used, the students' EPs appeared to "handle equipment conceptually," "interpret" and "use artefact." Also in 2012, because the teacher struggled to value epistemically, he allowed the students to "use the artefact," "conceptually manipulate," "observe" and "interpret." The teacher's effort to "make explicit epistemically," which consists of explicitly stating the epistemic status of the work done or in progress, has triggered a new set of student EP. For example, in a MN from 2006 where the epistemically explicit epithets are triggered in the epistemic practices of "using manipulative mediator or visual representation," "introducing manipulative mediator, i.e., introducing visual representation," "conceptualizing" and "communicating not autonomously," the teacher works to "give the student autonomy." This effort allows students to develop epistemic practices such as "conceptually manipulating," "using visual representation," "using artefact," "interpreting," "establishing relationships," "communicating autonomously" and "critically assessing." It is evident that these efforts have been associated with the efforts of the teacher that actually promote quite a few student EPs and have a higher epistemic level.

It turns out that the teacher in 2006 must invest more in the engagement of students in the task than in 2008 to "compensate" for their lower quality (see analysis of the above task). In addition, in 2006, the teacher's effort is to trigger the engagement of the students, and in 2008, as in 2012, the teacher's effort is to enhance student engagement and not to unleash it. This difference is essentially due to the way in which the task is conceived and to the alteration of the teacher's mediation in the introduction and use of this task. The teacher's efforts to improve student engagement in the task are: to keep the task as a challenge, to empower students, to encourage problem solving and to encourage engagement in the task (see Figure 7).

Beyond the care to be taken in the design of the task and investment for its increasingly efficient change and in the teacher's way of introducing and accompanying it so that students perform it more effectively, also important are the reflections that the teacher (alone or with critical friends) performs regarding what happened, and thus how to proceed with a view to improving student learning.

Figure 7. Average number of occurrences by lesson of the characteristics of teacher's effort to engage students productively in 2006, 2008 and 2012

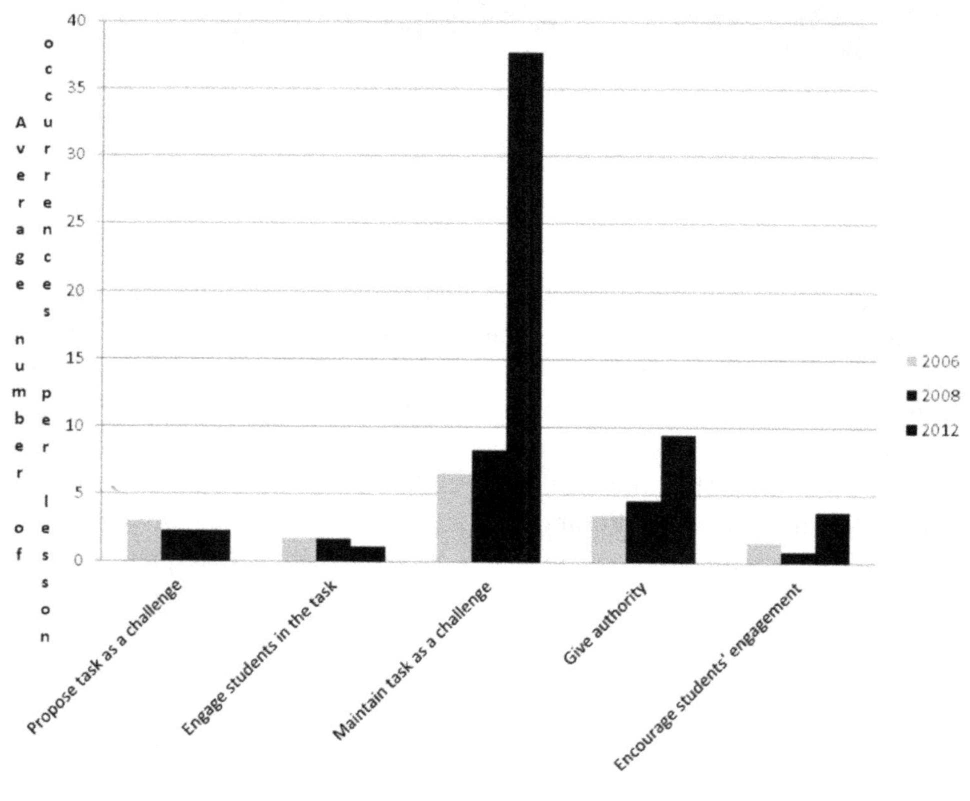

Teacher Reflection

In the second, fourth, fifth and sixth years of the study (among the three implementations), the teacher analyzed the lessons from 2006, reflecting on his teaching practice and the work done by the students. In this reflection were included the efforts in the preparation of the classes, the formulation of the tasks and the attempts to promote EP. There was collaboration in this reflection, since the discussions and advices on the part of its doctoral advisor and colleagues of the professional provided useful suggestions and tips, which this professor took into account and gradually implemented in his teaching practice.

These reflections were based on what happens in the classroom, taking as reflection material the MN of 2006 and 2008 made in these years by the teacher.

The MN being a live report intended to give as correct an idea as possible of what happened in the classroom, it includes everything that was discussed or explored as well as resources, students' reactions to the teacher's expressions and even explanations of actions taken in some cases. Thus, the narrative allows not only the teacher himself to be inside the class but also any other teacher who reads the narrative.

The richness of the narrative is that it allows even a group of teachers to reflect and discuss about moments better achieved and others less. After analyzing and discussing these narratives, it is possible to improve the teacher's own and others' classes, as well as to produce publications that can be transposed to a whole universe of teaching professionals. In 2007, the teacher in question began tentatively modifying his practices, in particular in the planning of these and in the formulation of the tasks, as well as in their implementation. He did the same in the years between the second and third implementation, in 2009, 2010 and 2011. The task, as already mentioned, has become more open, allowing greater autonomy for students. The teacher's mediation began by using different efforts to promote higher sequences of EP produced by the students without the intervention of the teacher.

During the six years of study, the teacher in this case had a notable evolution in professional development in terms of his action and mediation and the learning of students in the classroom, which was accompanied by an academic reflection translated into an increasingly consistent publication activity. This change occurred because the teacher wanted to improve and was willing to take risks inherent in educational innovations and to face the insecurity of new approaches.

Reflections About Mediation

The teacher-researcher, based on hearing the lessons and narratives and with the analysis of the presented data, may not only perform self-reflection but may also compare the mediation done by his or her critical friends.

This analysis, by clarifying some important aspects about student learning, can help teachers to understand what traits of mediation they can practice and how they can improve.

Thus, after construction and analysis of the narratives in this study, it was verified that:

- The importance in caught the attention of the students before starting the debate on a new subject.
- The narrative elucidates in a remarkable way the importance of this approach, making it much easier to show to third parties.
- Mediation can function as systematic moments of great importance in the classroom, leaving the students attentive to their outcome.

Multimodal Narrative Excerpt 1: Teacher A (How Teacher Caught the Students' Attention in 2008)

After 6 minutes:

Teacher: *"How do you get this registered? Isn't it better to have a white sheet underneath to register?" (This is for students to measure the angles of the receiver and the sender with the normal one, in order to know which angle provides better reflection or transmission, etc.)*

Student responds by saying it's a better idea. However, another student said he could register with chalk on top of his wallet.

Miguel explains that by measuring the sides of the triangle formed by the emitter, receiver and obstacle, one can know the angle of incidence and the angle of reflection. This is shown below in a drawing by Rita.

Multimodal Narrative Excerpt 2: Teacher A (How Teacher Caught the Students' Attention in 2006)

P: *"So you do not know in scientific terms. Go back to the hypothesis."*

A student in Group 3 read it again.

P: *"Group 1 does not mind repeating?"*

They gave an explanation of what Group 3 said, giving a more scientific explanation.

P: *"And what, Group 2, did you say? Can you read it again?"*

A student in Group 2 read the hypothesis again.

P: *"Why is it not quantified in amplitude?"*

In the same line as the previous reflection, it will also be important for the teacher to take a look at what the students are saying, because there are arguments that are not immediately pursued (because the teacher considers it not to be an ideal moment), but before closing of the discussion these should be explained and erased from the table, so that no student is unclear about what they have been learning.

Figure 8. Excerpt from a MN from 2008 to ilustrate how students solved the task

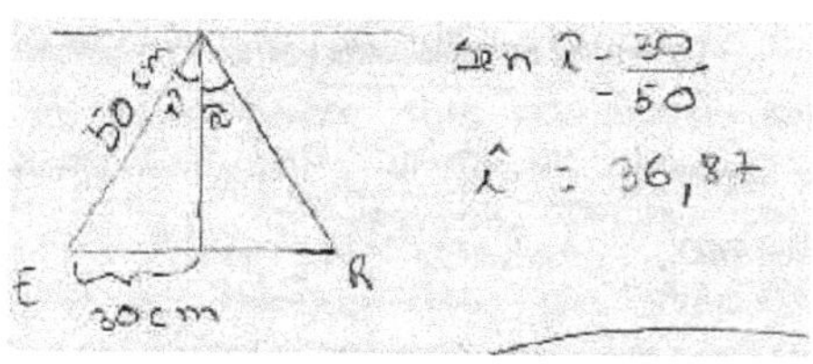

Multimodal Narrative Excerpt 3: Teacher A (how teacher caught the students' attention in 2012)

Teacher: *"Observe the results of all groups in the table and check whether they agree with the results of the simulation or not – whether there are failures or not."*

It is important to allow students time to discuss proposed problems. So that the time factor is not an impediment to doing so, it may be more productive to propose fewer discussion problems, but to allow and encourage the students to solve problems autonomously and in collaboration with colleagues.

Multimodal Narrative Excerpt 4: Teacher A (2008, time factor)

Teacher: *"Can we begin to discuss your hypotheses? Let's go then. They are already going to other fields. You are restless. There are still people to answer. ... Let's start? Is Sara finished yet?"*

Sara: *"I'm really finished ..."*

Teacher: *"It's really over ... Let Sara finish?" (after a few minutes)*

Teacher: *"So let's start at this extreme, for example, or by the way it's indifferent ... José will read your answer ... all attentive."*

José: *"I started at the end, teacher."*

Teacher: *"It does not hurt ..."*

Launch questions (of synthesis) near the end of the class, so that the students synthesize and summarize what they learned during the class and discuss it as a group.

Figure 9. Records of the results of each group in the table.

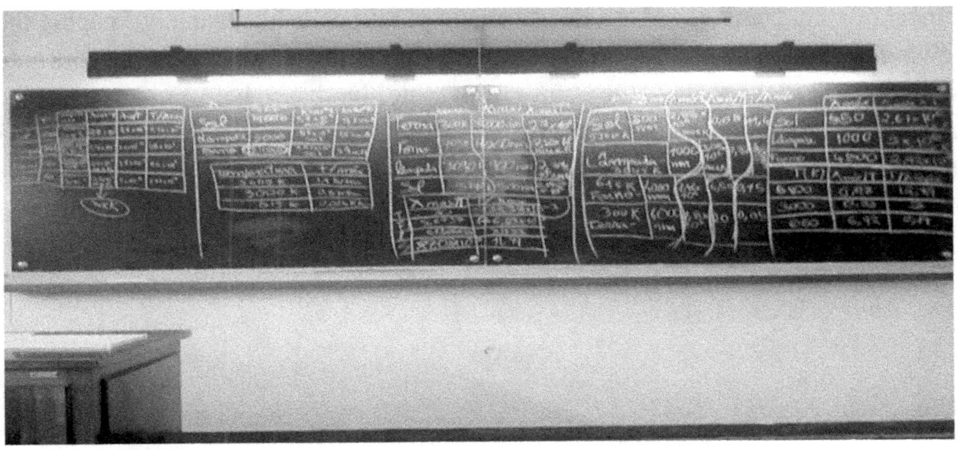

Figure 10. Excerpt from a MN from 2008 about the conclusions that students achieved in the task

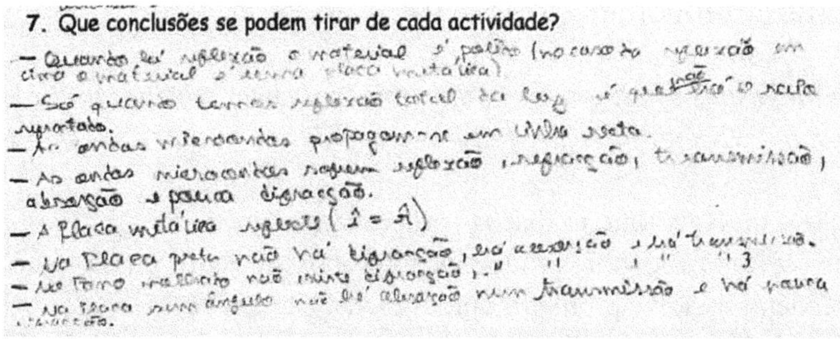
7. Que conclusões se podem tirar de cada actividade?

Multimodal Narrative Excerpt 5: Teacher A (2008, Students' Summary of Conclusions)

Teacher: *"I do not want an introduction, I want the final ones, and it's fast; I'm not going to collect it, and they do not have a conclusion in the correction of the protocol. ... Okay, did everybody already? Rui Branches, have you? So read fast that this will count for note."*

Rui read.

From this analysis, the care that a teacher can take to enable students to enjoy a truly active role within a classroom can be valued not only for the students' engagement, their motivation and interest, but also for their competences and developed epistemic practices.

The teacher presents a closer perspective of an attempt to build knowledge with students, making the classroom environment a social learning environment. The debates that are stimulated between the students serve as motor for learning and for the following discussions. The results obtained point to the students' better perception of teaching and learning as well as the development of skills in a wider group of students.

The reflection produced through the analysis of the teacher's mediation makes it possible to clarify the importance that some details end up having in the classroom, and this reflection may be a wake-up call for the teacher to improve his or her practice, confirming the importance of the traits identified for the development of competencies and the improvement of school results, namely, keeping the tasks as a challenge (giving students the possibility of building knowledge), allowing students to work autonomously (allowing students time to think about the issues), providing clue-based help, guidance and explanations only at crucial times (allowing the teacher's explanations to arise only after the students have thought about the subject), encouraging collaborative work (discussing with colleagues), leading students to mobilize and interpret issues (introducing conceptual issues) and encouraging students to produce real work (asking them to perform small tasks).

Multimodal Narrative Excerpt 6: Teacher A (2008, Providing Clue-Based Help)

Teacher: *"No. It's just a schematic representation; you know what that is? We have not refracted, but now I challenge you to imagine what refraction would look like, a picture on a cloth or so. What would the scheme look like ...? Say it? I do not accept? I accept, I accept ..."*

Teacher: *"Yes, for example ... imagine with a cloth or metal plate."*

A student calls the teacher to the place.

Teacher: *"Not yet complete, is it?"*

DISCUSSION

As previously mentioned, for the teacher's efforts to take place, care is required in the design of the task, along with a critical attitude on the part of the teacher regarding his or her mediation. These results are consistent with Thompson (2015) and Cunha (2015), according to which the teacher should encourage students to think and work without their direct influence, in particular:

- In the study of experimental work in the classroom, exploring relationships with the real and promoting modes of learning influenced by scientific works, favoring epistemic practices (Pinto et al., 2014; Santos, 2017).
- In the evaluation of conceptual evolution, promoting the mobilization of knowledge (Lopes, 2017) in order to develop skills (Lopes, Branco, & Jiménez-Aleixandre, 2011).
- In the use of collaborative learning (Cunha, 2015), seeking to create a classroom climate that facilitates interactions between students and between them and the teacher.

Reflection helped the teacher in this study to see and verify aspects of his action and discourse that were less successful and/or better achieved. The teacher has been modifying his practices, in particular in their planning and in the formulation of the tasks, as well as in their implementation, promoting greater productive engagement and better quality epistemic practices (Lloyd, 2017). This makes it possible to verify that professional development is influenced by the reflections of the teacher on his or her classroom practice. Through the recording of the lessons, it was possible to refine the reflection about these with the construction and analysis of the MN. This instrument allowed the teacher to visualize in detail and at a distance aspects better or worse achieved in the class in question, in the sense of being altered so that they are actually better (Lopes et al., 2014). The iterations between each implementation supported by the daily reflections on the classes and other information acquired by the teacher allowed professional development to occur (Lopes, 2017) that was advantageous not only for the teacher but also and especially for the students, allowing them to develop more and better productive engagement and EP (Cunha, 2015; Cunha et al., 2014).

MNs are compatible with qualitative analysis instruments (such as NVivo, for example), since these instruments make it possible to discern numerous associations between the variables that are under study. They also make it possible to find patterns that can be associated with results related to student learning. With NVivo and some MN, it is possible to cross-analyze factors (e.g., the same teacher or different teachers, the same level of education or different levels, the same content or different content, etc., depending on the research objective) and the role played by particular actions of the teacher.

In this sense, MN can be an important formative tool for the teacher, since MNs are relevant from the point of view of mediation, as they narrate the concrete action in the classroom. The MN allows the teacher, through reflection, the ability to accept all possibilities; it allows the teacher to confront his or her own beliefs and become ready to change them. It also requires the teacher to make use of reflection to learn and grow professionally. Reading and analyzing an MN allows the teacher who built it or provided the data for its construction to better see the reality of her or his classes and to improve the respective practices (Lopes & Cunha, 2017). For other teachers (other than the actors in this process), access to the MN allows them to review their own practices and use the examples in positive or negative ways to improve their own practices.

However, the use of MN still has to be worked on to streamline the entire process of class data collection and storytelling development. First of all, it requires the authorization of all those involved in the action. The exhaustive description of what goes on in the classroom is a very laborious and time-consuming process. It is necessary to develop technologies to obtain all the data about a class almost immediately. It is pertinent to disseminate this instrument so that the means to enjoy this method of reflection and consequent change of teaching practice reaches all.

The reflection of the teacher through the analysis of the mediation carried out clarified the importance of some details in the classroom, which give meaning to the alteration of its practice in order to improve it, namely, to keep the tasks as a challenge; to allow students to work autonomously; to give aid based on guidance and explanations only at crucial times; to allow and encourage problematization; to value students' responses and actions (Cunha, 2015); and to improve the design of the tasks, making them clearer and more open in order to give more authority to students (Thompson, 2015).

With alteration of mediation in order to diversify and intensify efforts to involve students in learning, students have more opportunities to perform real tasks. This transforms the role of the teacher to a different character, promotes discussions among students, gives them more authority and promotes the acquisition of skills (Lopes, 2017). The introduction and use of artefacts, as well as the introduction and use of visual representations, allows for the emergence of higher-level EP of "conceptually manipulating," "autonomously communicating," "manipulating artefact," "organizing information on the results obtained," "identifying empirical conditions present in the activity" and "critically evaluating the results," constituting an evolution in the practice of the teacher and consequently contributing to the teacher's professional development.

CONCLUSION

This work of professional development on the part of the research professor is meant to offer an example of reference for other educational professionals with identical aims who, in this regard, face the same challenges.

The professional development and teaching practices of this teacher proved to be student-appropriate and didactically coherent.

A mediation that incorporates the various dimensions (support and authority given to students, keeping the task challenging, encouraging engagement, allowing and encouraging problem-solving) achieves better results in the students' productive engagement. The incorporation of dimensions related to the epistemic intentionality of the teacher (valorizing epistemically, making explicit epistemically, introducing artefacts) allows the student a higher quality and quantity of EP. However, some mediation traits in particular are critical to achieving better learning outcomes across all dimensions.

The fact that there are several ways of maintaining the challenge task allowed the teacher to develop with more frequency and value the students' productive engagement. It was noticed that this monitoring can be enhanced if the teacher can simultaneously clarify to the students the objectives of the tasks and keeps the task as a challenge. This study showed that there are several specific ways of doing this. The integration of the design of the tasks and the mediation in their actual practice within the classroom, along with successive iterations and reflections on their obtained results, is fundamental to produce a more effective alteration, making it possible to act, modify, reinforce, complete and manage the design of the task in order to more and more effectively achieve the goal of engaging students in learning in an intrinsic way and encourage students' development of desired epistemic practices. The challenge task over time provokes very different reactions, allowing students to work autonomously and in control. In addition, from 2006 to 2012, autonomy increased the epistemic work, since the average number of student EPs between each intervention of the teacher is greater. The reflections made by the researcher about his classroom practice of incremental modifications to tasks and mediation were actually used to modify and adapt them to the found reality. The results indicate that these reflections were useful not only to improve the design of the tasks or the mediation in the application of these tasks in the classroom but also to highlight improvements in effective mediation.

Thus, it is clear that not only can the tasks be redesigned to better fulfill the development of the intended learning, but also the associated mediation can be adjusted, in order to implement these tasks in more appropriate ways in the classroom and to achieve better learning.

REFERENCES

Berland, L. K., Schwarz, C. V., Krist, C., Kenyon, L., Lo, A. S., & Reiser, B. J. (2016). Epistemologies in practice: Making scientific practices meaningful for students. *Journal of Research in Science Teaching*, *53*(7), 1082–1112. doi:10.1002/tea.21257

Cohen, L., Manion, L., & Morrison, K. (2011). *Research methods in education* (7th ed.). London: Routledge.

Cunha, A. E. (2015). *Construção de práticas de referência no ensino da física para o ensino secundário.* Tese de doutoramento. Universidade de Trás-os-Montes e Alto Douro, Vila Real, Portugal.

Cunha, A. E., Saraiva, E., Santos, C. A., Dinis, F., & Lopes, J. B. (2014). Teacher mediation actions and students' productive engagement during the use of computer simulations in physical science classrooms. *Procedia Technology*, *13*, 76–85. doi:10.1016/j.protcy.2014.02.011

Kemmis, S. (1989). Research in action. In T. Husen & T. N. Postlehwaite (Eds.), *International Encyclopedia of Education* (Vol. 6, pp. 3330–3337). Barcelona: Vicens – Vives/MEC.

Lloyd, C. A. (2017). Exploring the real-world decision-making of novice and experienced teachers. *Journal of Further and Higher Education*, 1–17. doi:10.1080/0309877X.2017.1357070

Lopes, J. B. (2017). Self-directed professional development to improve effective science teaching. In J. B. Lopes, J. P. Cravino, E. S. Cruz, & A. Barbot (Eds.), *Teaching science: Contributions of research for planning, practice and professional development* (pp. 247–264). New York: Nova Science Publishers.

Lopes, J. B., Branco, J., & Jiménez-Aleixandre, M. P. (2011). "Learning experience" provided by science teaching practice in a classroom and the development of students' competences. *Research in Science Education, 41*(5), 787–809. doi:10.100711165-010-9190-5

Lopes, J. B., Cravino, J. P., Cruz, E., & Barbot, A. (2017). *Teaching science: Contributions of research for planning, practice and professional development*. New York: Nova Science Publishers.

Lopes, J. B., & Cunha, A. E. (2017). Self-directed professional development to improve effective teaching: Key points for a model. *Teaching and Teacher Education, 68*, 262–274. doi:10.1016/j.tate.2017.09.009

Lopes, J. B., Silva, A. A., Cravino, J. P., Santos, C. A., Cunha, A., Pinto, A., ... Branco, M. J. (2014). Constructing and using multimodal narratives to research in science education: Contributions based on practical classroom. *Research in Science Education, 44*(3), 415–438. doi:10.100711165-013-9381-y

Lopes, J. B., Silva, A. A., Cravino, J. P., Viegas, C., Cunha, A. E., Saraiva, E., … Santos, C. A. (2010). Investigação sobre a mediação de professores de ciências físicas em sala de aula [Research on physical sciences teachers' mediation in the classroom]. Vila Real, Portugal: Universidade de Trás-os-Montes e Alto Douro.

Lumpe, A., Vaughn, A., Henrikson, R., & Bishop, D. (2014). Teacher professional development and self-efficacy beliefs. In R. Evans, J. Luft, C. Czerniak, & C. Pea (Eds.), *The role of science teachers' beliefs in international classrooms* (pp. 49–63). Rotterdam, The Netherlands: SensePublishers.

McNiff, J., & Whitehead, J. (2006). *All you need to know about action research*. London: Sage Publications.

Osborne, J. (2014). Teaching scientific practices: Meeting the challenge of change. *Journal of Science Teacher Education, 25*(2), 177–196. doi:10.100710972-014-9384-1

Pinto, A., Barbot, A., Viegas, C., Silva, A. A., Santos, C. A., & Lopes, J. B. (2014). Teaching science with experimental work and computer simulations in a primary teacher education course: What challenges to promote epistemic practices? *Procedia Technology, 13*, 86–96. doi:10.1016/j.protcy.2014.02.012

Santos, C. A. (2017). The role of teacher decisions in the classroom to develop students' epistemic activity. In J. B. Lopes, J. P. Cravino, E. S. Cruz, & A. Barbot (Eds.), *Teaching science: Contributions of research for planning, practice and professional development* (pp. 119–134). New York: Nova Science Publishers.

Thompson, I. (Ed.). (2015). *Designing tasks in secondary education: Enhancing subject understanding and student engagement*. New York: Routledge.

Zeichener, K. M. (1993). *Reflective teacher training: ideas and practices*. Lisbon: Educa.

Section 4

Researching Science and Technology Practices Using Multimodal Narratives

Section 4 of this book reports research on Science and Technology teaching practices based on academic research criteria using as empirical base the multimodal narratives (MNs) in their final public and shareable form, available in a collection of MNs (see Section 1). The set is representative, but not exhaustive, of the several research studies that have already been carried out with MNs.

As the reader can verify the MN can be used for different purposes, resorting to different theoretical approaches and analytical techniques. Moreover, a same subset of MN can be used for different research objectives.

Chapter 12 gives a historical point of view about the MNs usage for research purposes and offer perspectives and examples of how MNs can be used to carry out studies in science and technology education research.

According to the focus, scope, and research questions of the study, each group of researchers (author[s] of the respective chapter) considered a subset of MN from the collection available based on criteria defined at the outset. In Chapter 13, the teacher-researcher uses the MNs of his own teaching practices as well as MNs of the teaching practices of other teachers at same teaching levels of and similar contexts. In Chapter 14, MNs of teaching practices of different teachers are used to study a particular aspect transversal to the teaching practices according to the focus of the study. Chapter 15 uses MNs from teacher-researcher classes that study a particular aspect of their own teaching practices and how they have evolved over time. Chapter 16 uses several MNs of a teacher to study a specific aspect of teaching practices along a curricular unit.

Thus, the set of chapters in this part of the book aims to illustrate the potentialities of using MNs in a context of scholar research: Chapter 13 addresses a key feature of our research activity—epistemic practices—and is focused on the way MNs help to study these phenomena; Chapter 14 analyses teachers' decisions characteristics using MNs and its influence in students' epistemic practices; Chapter 15 investigates the role of MNs in the research of own teaching practices in order to improve them; Chapter 16 investigates the semiotic registers in Chemistry teaching practice in Brazil.

Chapter 12
Using Multimodal Narratives in Science and Technology Education Research

José Paulo Cravino
Universidade de Trás-os-Montes e Alto Douro, Portugal & Centro de Investigação em Didática e Tecnologia na Formação de Formadores, Portugal

ABSTRACT

This chapter begins by presenting how multimodal narratives were developed and how they have evolved in time as a research tool. Their characteristics are explained, as well as their potential for use in research studies. One important aspect discussed in this chapter is the versatility of this tool, which allows for many possible investigative approaches. In particular, it addresses how multimodal narratives can be used to carry out studies in science and technology education research. Several examples are provided of such studies that were developed within a research community which has been working with multimodal narratives for over a decade.

INTRODUCTION

This chapter starts by introducing the need for a new research tool and how a research group went about developing it in the context of a research project. Over time, this tool became known as multimodal narratives. Their characteristics are described, how they are produced, and how they are validated, as well as some of the difficulties in this process. Since one of the main characteristics of multimodal narratives is their versatility, how they may be used in different research approaches is discussed. To support this claim, and to provide a better insight into how multimodal narratives have been used, several examples are presented and discussed to highlight the contribution of this research tool. The goal of this chapter is to inform the reader what a multimodal narrative is, how it is produced and validated, and how it may be used to study various research problems. It is hoped that this chapter may inspire others to use this tool in their own research studies.

DOI: 10.4018/978-1-5225-8570-1.ch012

BACKGROUND

This section describes the development and evolution of multimodal narratives as research tools.

Multimodal narratives (MNs) were developed within a research group based at the University of Trás-os-Montes and Alto Douro in Vila Real, Portugal, dedicated at the time to a research project on the teaching of physical sciences in the classroom. This project was about the teacher mediation of student learning in physical sciences lessons (project 'Guiding principles and tools for fostering teacher mediation in physical sciences' classes', funded by Fundação para a Ciência e Tecnologia under grant number PTDC/CED/66699/2006). One of the goals of the project was to contribute to improve both teachers' mediation and the quality of students' learning. Relevant empirical data were therefore necessary, in particular empirical evidence from the classroom. This approach focused on the classroom posed a particular problem in collecting, organizing, and processing data on classroom-based teaching. The intention was to collect data about teaching and learning taking place in the classroom that were reliable and useful for research, but with minimum disruption to the actors in the classroom, and preserving the holistic character of the classroom in all its complexity.

The obvious solution—and widely described in numerous studies and research methods handbooks—is to video or audio record lessons. Then the recordings should be transcribed for analysis. It should be noted that at the time some computer programs were already available to assist in qualitative data analysis, and in particular in the content analysis of texts. For example, NVivo 7 was at that time a recent evolution of the NUD*IST software (Numerical Unstructured Data Indexing Searching and Theorizing). There were also other software programs, such as ATLAS.ti, MAXQDA, and QDA MINER. Later, a web-based software was created by Portuguese researchers from the University of Aveiro (webQDA). All these programs allowed the analysis of content based on text files, although it was already possible to work directly on audio or video files with some of them.

However, the main problem that the research group faced was the need to capture the lesson in its integrity and not just the dialogues between teacher and students, which would be easy to capture with audio recordings. The video recording makes it possible to better capture the lesson globally, using multiple camcorders, but this strategy poses some problems: it is quite intrusive and there are many obstacles to recording images in the classroom due to the need for multiple authorizations and privacy issues. Schools and students' parents are very sensitive to these problems. On the other hand, to capture the lesson in a more holistic way, it is usually necessary to use multiple cameras in order to capture both global views and the level of groups or individuals, whether in terms of sound or of image.

Excluding the possibility of collecting data through video recordings, the remaining option was therefore to use only audio recordings. However, these record only the most audible interactions in the classroom—typically the teacher's speech and the interactions between the teacher and his or her students. Of course, it is possible to use several audio recorders strategically placed around the classroom to maximize the capture of what is said, but it is complex to synchronize all interactions and it increases intrusion, potentially compromising the spontaneity of the interactions.

On the other hand, though audio recordings capture (if conditions are appropriate) what is said in the classroom, they do not capture what is written or drawn, the gestures, the expressions, and everything else that occurs in a classroom that is not spoken speech. And a lot of what happens in the classroom is not spoken speech, as the examples just cited show.

Other possibilities were considered: diaries (Hundhausen, 2002; Stein, Isaacs, & Andrews, 2004), noticing (Mason, 2002; Tan & Towndrow, 2009) and other types of account (Alsop, Bencze, & Pedretti, 2005), and narratives (Clemente & Ramírez, 2008; Craig, 2011; Georgakopoulou, 2006; Kramp, 2004; Labov, 2006; Melville, 2008), or daily structured logs to record information about classroom events or interactions (Rowan, Camburn, & Correnti, 2004).

However, in the end, none of these instruments was found to thoroughly capture crucial features, such as descriptions of teacher decisions, their intentions and context, and descriptions of students' efforts, and the work that they actually did. Moreover, these instruments do not have a characteristic that was revealed to be essential to the research project being conducted: a specific focus and structure to allow a more rigorous, precise, and powerful comparability between different situations and cases (Lopes et al., 2014).

After some trials and a lot of reflection, the first attempts at what has come to be called multimodal narratives were born. The main idea was to link the audio recording of classroom interactions with a set of other documents, which together allowed for the construction of a narrative of what happened in a particular lesson. These documents add information to spoken speech, which is captured in the audio recording, and they can be pictures of what was written or drawn on the whiteboard or in the students' notebooks, materials that were distributed, filled, or produced in the classroom, and materials used, among others.

Based on all these collected documents and on the audio recordings of the lesson, the teacher can then write the narrative for that lesson. This narrative has two important characteristics:

1. It is multimodal, because it has attached to it a whole set of data related to the lesson, in different formats and media, which support and complement the narrative, simultaneously allowing an assessment of whether the narrative is reliable and if it reflects the lesson in its complexity and completeness.
2. It benefits from the privileged knowledge of the narrator, who, being the teacher who taught the lesson, can incorporate into the narrative aspects that are not present in other data, such as the teacher's own intentions or decisions, non-verbal reactions of students, gestures or attitudes, silences, etc.

It should also be said that it is not strictly necessary for the MN of a lesson to be elaborated by the teacher who taught it; however, in this case, there is a need for a very close collaboration of the narrator (the person who actually writes the MN) with the teachers who taught the lesson in order to ensure the necessary information for the preparation of the MN and for its validation. One example of this possibility is a study already published (Branco, Magalhães, Ferreira, Lopes, & Cravino, 2009), which was carried out on the basis of MNs written by a researcher (who did not teach the lessons being narrated). In this study, the focus was on the professional development of teachers, stimulated by a researcher who supplied a set of tools to help teachers change their teaching practice and interacted with the teachers based on MNs from a set of their lessons. These MNs were written by the researcher in close collaboration with the teachers, who supplied all the primary data and later validated the MN. Other cases are presented in sections below.

Thus, the MN is not just or simply the transcription of the audio recording of the lesson, but a richer and more documented account of the lesson as a whole.

On the other hand, by organizing the narrative into a single document (the MN itself), although linked to the other data collected (audio recording and all the other documents), the analysis phase is made easier, since the researcher has to analyze, in principle, only this document, typically an illustrated text file (containing photos or drawings).

Since the process of developing the MN as a tool for research was done in the context of teaching and learning physical sciences (in the early stages, mostly physics and chemistry, later including biology, geology, and other sciences and technology subjects), it should be noted that authors such as Kress, Jewitt, Ogborn, and Tsatsarelis (2014) have argued that science learning takes place in a multimodal environment and that reading and writing can only be understood in relation to the larger semiotic system (such as actions, gestures, visuals, and three-dimensional models). Furthermore, Adger and Wright (2015) say that, in science classrooms, there may be times when reading and writing are secondary to other semiotic modalities, which clearly emphasizes the importance of taking the multimodal characteristics of the teaching and learning environment. This, together with the experience of our research group, reinforces the notion that MNs are particularly appropriate to capture data from science classrooms.

This does not mean that the use of MNs is restricted to this type of setting. Indeed, one interesting recent development is that of multimodal interviews (Woolhouse, 2017) in the context of life history narrative, which use a multimodal format to facilitate recollection and active engagement during the process of storytelling. In this case, the multimodal elements are mostly visual representations, but also discursive transitions, including switches in style or topic of conversation, expressions of emotion, pauses, and extended silences.

The subsequent phases of the development of the MNs—their testing and refinement as a tool to support research—led to the need to standardize their structure (described in Chapter 1 of this book and in Lopes et al., 2014), and to develop a protocol for making and validating a MN. This standardization of the structure and of the way of constructing and validating a MN is fundamental to allow its use in research projects. Otherwise, it would be very difficult to compare data between different MNs or to use the same set of MNs for studies with a different focus or research questions.

Precisely to facilitate access to MNs developed in accordance with this protocol, the research group at the Universidade de Trás-os-Montes e Alto Douro has prepared an electronic collection, which is available on the internet, to enable any interested teacher or researcher to access all MNs produced and validated so they can be used in their own research work or for professional development (see Chapter 7 in this book). This repository is available online at the following website: http://multimodal.narratives.utad.pt/

The community that has been developing and using MNs currently encompasses researchers and teachers of various nationalities, with work carried out in Portugal, Brazil, and Angola. The studies developed include lessons and other educational activities in the areas of physics, chemistry, mathematics, biology, geology, informatics, and engineering. Therefore, the studies performed are generally described as being in the field of science and technology education research or, in other European languages, science and technology didactics.

In terms of data analysis, MNs usually undergo content analysis (see, for example, Bardin, 1993; Krippendorff, 2004; Weber, 1990; and, more recently, Cohen, Manion, & Morrison, 2018, Ch. 34).

Again, it should be noted that the biggest advantage of working with an MN is having only one document to analyze for each lesson, because the MN is a self-contained document, although it is linked to all the data that has been collected. It is also relevant to note that the possibilities for including multimodal elements in qualitative research have been acknowledged, and it is recognized that computer aided qualitative data analysis software (CAQDAS) provides opportunities to integrate and interrogate a variety of data types that include text, images, audio, and video data in increasingly complex and novel ways, thereby facilitating a step change in fine-grained studies of social interaction (Qualitative Research, 2016), as is the case for teaching and learning in an educational setting.

In some studies, after the content analysis, cluster analysis was also performed (see, for example, Cohen et al., 2018, Ch.43; Kaufman & Rousseeuw, 1990) to facilitate the detection of patterns in the data already treated by content analysis.

THE ADVANTAGES OF USING MULTIMODAL NARRATIVES IN SCIENCE AND TECHNOLOGY EDUCATION RESEARCH

The history of the development and the characteristics of MNs, described in the previous section, show that the MN is an instrument particularly suitable to collect and organize classroom data, preserving its complexity and its holistic character.

One great advantage of the MN is that it can be used by different researchers to investigate different research questions. That is to say, the same set of MNs can be used by several researchers to study different aspects of their lessons. For example, an investigator can use MNs to study the use of visual representations in the classroom by teachers and students, while another can focus on aspects of teacher mediation, and yet another can study the decisions made by the teacher in real time in the classroom. All these are real examples which will be detailed in the following sections.

There are many possibilities in using MNs in research studies. Some studies may use MNs of the same or different teachers, of the same level of education or of different levels, where lessons are on similar content or similar strategies are used (for example, computational simulations), according to what each researcher intends to study.

Therefore, the MNs allow carrying out studies of very variable dimensions, with foci which can be very distinct, as long as they are about what goes on in the real environment of a classroom (or, for that matter, any other educational setting).

In the next section, some examples are presented of studies conducted using MNs to make it clearer how they are used and to show that they are a tool that is versatile and useful, which can be employed in very different ways depending on the purpose of the actual study.

RESEARCH STUDIES CARRIED OUT USING MULTIMODAL NARRATIVES

As already mentioned, MNs have been used since 2006 by several researchers and teachers in various science and technology subject matters, with studies carried out in educational settings in Portugal, Brazil, and Angola.

Some examples will be given below to illustrate the use of MNs in different contexts (such as levels of education, content covered, teaching strategies) and to study different aspects of the teaching and learning of science and technology.

It should be noted that, although this is not detailed in the following descriptions (to avoid repetitions), in all the studies cited, the elaboration of the MNs was always preceded by the collection of a series of data that, depending on the activities developed in the lesson, included audio recordings of the lesson, documents produced by the teacher and/or students, photos of students' notebooks and of whiteboards, equipment used, and computer screens, and teachers' notes. All the MNs followed the protocol for elaboration and validation published in Lopes et al. (2014).

Study 1: Multimodal Narratives in the Study of Teaching Introductory Physics in Higher Education

This research project, developed on the teaching of an introductory physics course in higher education (in engineering degree programs), aimed at understanding which teachers' practices promote quality learning (in terms of both knowledge and skills development), and what iterative curricular changes, put into practice in the classroom, are productive (or not) in promoting teaching success and the academic success of students.

The results of this project have been published in detail elsewhere (Lopes, Viegas, & Cravino, 2010; Viegas, Lopes & Cravino, 2009, 2010a, 2010b, 2011). In this section, we review briefly how MNs were used in this project in order to show the relevance of their contribution to the success of this project.

In the first part of this study, seven cases were studied, corresponding to seven teachers (one of whom was also the researcher in this study), who taught two consecutive courses in two semesters. Since one of the aspects studied was the teaching practices of the seven teachers and how they helped promote quality learning, it was necessary to collect data about how teachers actually taught their lessons. This was done initially (in the first course/semester) by asking teachers to take notes after teaching their lessons and then meeting with the teachers to obtain information on how they taught specific lessons. Other data were collected, especially documents used by the teacher and the students in preparation for or during the lessons. This approach showed mixed results, since each teacher decided how to take notes from his lesson, and they did so with different levels of detail and scope. In the meetings with teachers, there was also a lot of variation in the information that it was possible to obtain from each teacher. This problem persisted in the next course/semester, but one of the teachers (the teacher-researcher) began making audio recordings of her lessons.

Even though it was not possible to obtain the same kind of information from every teacher, or because the level of detail was different, it was nevertheless possible to obtain clues about what did or did not work in how teachers taught their lessons; however, it was very difficult to actually compare the data from different teachers and even from different lessons. The note-taking was not systematic enough and relying on teachers' ability to recall details from their lessons is not efficient.

This set of cases studies was used mostly to develop a curriculum design for the course and test the impact on student learning. The curriculum design involved both what was taught and how it was taught (tasks presented to students and teacher mediation of student learning).

The results from this phase of the project were important to design the next phase of the study, namely the preparation for teaching in the classroom, by providing valuable information to incorporate changes into the activities planned and to guide the teacher-researcher in her mediation in the classroom.

It was in this third implementation of the introductory physics course that MNs were finally used to try to overcome the difficulties in obtaining reliable and comparable data about how teachers actually taught their lessons. Indeed, this final part of this research was based on an action research design and consisted of the implementation of curricular modifications (changes in the activities planned for the course) and changes to the teacher-researcher's actual teaching practice in the classroom. These changes were based on the data collected in the seven case studies and the teacher-researcher's reflection on the results and conclusions.

This part of the study used MNs from several lessons taught by two teachers. Since one was again the teacher-researcher, she wrote the MNs of her own lessons, using audio recordings from her lessons, along with documents used in the lessons and her own knowledge of what happened in the classroom. The MNs of the lessons taught by the other teacher (who was not part of the initial group of seven and served essentially as a comparison case) were written by the teacher-researcher from the oral narrative that the teacher gave of his own lesson. This oral narrative was aided by notes that the teacher wrote during and after his lesson, and was further supported by the documents used by the teacher and students in each lesson. There was a big difference in the amount and detail of the information obtained in this case, because it was necessary to follow the pre-determined structure and protocol for writing the MN. The enhanced structure of the process and of the document produced was crucial to achieve a thorough picture of what happened in each lesson and of the role played by the teacher. There were still differences in the MNs from the two teachers' lessons. The MNs written by the teacher-researcher on her own lessons were richer in detail. The fact that these lessons were audio recorded also helped. It was not possible to record the lessons from the other teacher. However, the information in the MNs about the lessons and the role of the teacher was much richer than was possible to obtain in the previous parts of the study.

MNs were therefore instrumental in collecting data on how effective the teaching practice of different teachers in the classroom was using the same curriculum (that is, the same syllabus and instruction or teaching methods). The MNs were key to allow identification and comparison of the characteristics of these teachers' mediation in the classroom.

Indeed, the multimodal narrative tells the story of the lesson from the viewpoint of the teacher. It is richer than a transcript because it combines what is said, discussed, or explored, the way in which this is done, the resources used, the students' reactions and expressions, the expressiveness of the teacher (tone of voice, relative position to the students), as well as her sensitivity to what happened in class, with no intention to justify her acts, but simply to describe what led her to certain attitudes at particular moments.

By presenting the story of the lesson, the MN can give a faithful picture of what happens in the classroom. This enables the teacher-researcher to analyze the mediation performed by the teacher in various respects, to reflect upon it, and to perceive more easily some possibilities for improvement. This means that, in the final stage of the investigation, the MNs were also fundamental in providing information for the teacher-researcher's permanent reflection and for the adjustments that were being iteratively introduced in how she taught her lessons.

The MNs were fundamental to realize how the reflection on teacher practice helps ground the curriculum reconstruction in the search for a curriculum that is more adequate to meet students' learning needs and in improving the design of the tasks from the observations collected in the classroom. The MNs also allow the development and improvement of teacher mediation by providing evidence for the identification of mediation traits that were determinant and ways of incorporating them into the curriculum design or ways of enhancing formative evaluation in the tasks proposed to students (Lopes, Viegas, & Cravino, 2010; Viegas et al., 2009, 2010a, 2010b, 2011).

To provide a clearer view of what were identified in this study (Viegas et al., 2011) as effective teacher mediation traces and as possibilities for improvement, a summary is listed.

- One aspect identified was the importance of capturing students' attention before the beginning of the discussion of a new topic. This was very clear in the multimodal narratives, which revealed the importance of this aspect to other teachers more easily.
- Sometimes, during a classroom discussion, the teacher does not hear every student's comments (because they speak at the same time or very softly). If this happens to the same student more than once, it may not be very encouraging. The teacher must develop a more efficient way of collecting students' ideas and contributions—for instance, by forming discussion groups and then asking each group to deliver their conclusion.
- All arguments presented by the students should be considered; even when the teacher decides not to explain them at that particular moment, they should not be forgotten. Each contribution should be taken into account and its relevance to the problem discussed. They can be simply recorded on the whiteboard and erased if and when no longer necessary.
- The importance of allowing enough time when proposing a discussion or task to be developed was highlighted. It is more productive to discuss fewer issues but to let the students do some actual autonomous thinking about them.
- It works when, shortly before the end of the lesson, the teacher asks a conceptual question related to daily life in order to help students leave the classroom thinking about physics and discussing the question.
- Conceptual questions work best when used in two steps: first, by only showing the question, forcing students to mobilize their knowledge, and then perhaps presenting some hypotheses, allowing students to choose and discuss their choices before the teacher's intervention.

In very general terms, this research project led to the conclusion that the integration between the sequence design of teaching based on recent research in didactics of physics, its practice, and effective reflection and successive reflection on its results, is essential to cohesive curricular reform, allowing one to act, modify, reinforce, complete, and manage the curriculum design to achieve more effectively the objective of involving students in their learning and in the desired development of their knowledge and skills.

Study 2: Multimodal Narratives in the Study of the Role of Graphical Representations in Teaching and Learning Physics

This study starts from realizing that physics uses the symbolic codes of three types of language (natural, graphical, and mathematical) and that there is a need for articulation among them. Besides, in most physics lessons, a variety of documents are used containing a series of iconic messages. Their meanings, most of the time, are not understood by students, because they do not master the codes characteristic of this type of language. Thus, it was important to make explicit the knowledge that the understanding of iconic messages mobilizes, since the schemes, diagrams, and other formats of graphic language may contribute to a better understanding of phenomena and facilitate conceptual learning (Saraiva, Lopes, Cravino, & Santos, 2012).

The research questions addressed have focused on how teachers use visual representations in the classroom, what functions can be attributed to these representations, and what is the impact on the epistemic practices carried out by the students.

Epistemic practices are defined in this work as social activities of production, communication, and evaluation of knowledge that give the student the opportunity to develop positive attitudes about science and how it is built and, at the same time, make them more competent in the use of scientific knowledge in a real context (Saraiva, 2017, p. 26).

This research involved five case studies, corresponding to five teachers of physics and chemistry in the third cycle of basic education (7th to 9th grades—that is, students from 13 to 15 years old). These teachers had different levels of professional experience (from 15 to 28 years' teaching experience). Three had research experience; two had not.

Again, MNs were a valuable tool, because it is very easy to integrate the graphical representations used in the classroom into the global story of each lesson, therefore providing important information on how these graphical representations were introduced, what exactly was represented, and how. This, in turn, led to valuable insights about the functions attributed to these representations and, in many instances, about the direct impact on the epistemic practices carried out by the students.

This is one more example of a study that combines MNs that were written by the teachers who taught the lessons and MNs that were written by a researcher based on information supplied by the teachers about each lesson. In this study, three teachers wrote the MNs from their own lessons, while the other two teachers had the MNs of their lessons written by the same researcher. In both cases, it was possible to incorporate all or most of the graphical representations used in the classroom by using photos of the board or of students' work, as well as all those that were present in the materials used in the classroom. Each teacher either wrote in the MN or reported to the researcher (depending on the case) information about how each graphical representation was introduced, by whom, in what context, for what purpose, how it was used, and what developed from there.

Of course, it was always possible to simply collect all the graphical representations, but for the purpose of this study it was essential to have all this other information, provided by the MNs, about what happened in the classroom to be able to understand how graphical representations were actually used by teachers and students and what kind of epistemic practices were developed by using them.

The data on the lessons were organized in MNs, which were subsequently submitted to content analysis (with the help of qualitative data analysis software, in this case NVivo 8) and to cluster analysis to aid in pattern detection (using the Statistica software package).

The results of this analysis led to the conclusion that it is the teachers' efforts that trigger the use of visual representations by students, particularly when they perform specific tasks. Typically, visual representations are used as an illustration of something (which does not induce students to perform epistemic practices) or as epistemic mediators (in this case, increasing the epistemic practices by the students). Epistemic mediators are artifacts manipulated to produce knowledge, either through actions and operations carried out solely at the symbolic level, or by adding other possibilities through other material operations that may be possible.

It was also found that the teacher's experience (defined in terms of the number of years of teaching experience, but also in terms of experience of research in education, through postgraduate training or participation in research projects) makes a lot of difference. Indeed, more experienced teachers use the visual representations more intensively, and with epistemic functions, typically when students perform inquiry-based activities or when tasks are proposed as challenges to students. The teachers with less experience tend to introduce many visual representations, but only as an illustration; they do not require any action from the students, therefore reducing their learning potential.

MNs were fundamental in the identification of the visual representations used and in the identification of the person who introduced them, in what context, and with what consequences in terms of the epistemic practices developed by the students. In most cases, the MNs contained the visual representations that were used in each lesson as some of their multimodal elements. This facilitated the analysis and allowed insights that would otherwise be very difficult to achieve.

Study 3: Multimodal Narratives in the Study of the Role of Computer Simulations in Teaching and Learning Physics and Chemistry

MNs were also used for several studies on the use of computer simulations in the teaching and learning of physics and chemistry. Two examples will be presented (Lopes et al., 2012; Saraiva, Lopes, Cravino, Santos, & Cunha, 2015).

In the first of these studies, the MNs from two teachers' lessons were used to analyze the characteristics of each teacher's mediation that were found to promote the productive engagement (Engle & Conant, 2002; Forman, Engle, Venturini, & Ford, 2014) of the students during the use of computer simulations in 10th grade physics lessons (15–16 year old students).

In this case, the MNs of each lesson were written by the teacher who taught that lesson. The MNs worked very well in this case, because each MN incorporates a lot of information—what the teacher did in the classroom (the teacher mediation), as well as information on how students used the computer simulations (including actual screenshots). This facilitated the analysis by providing ample information in a single document to find indicators of students' productive engagement.

A system of categories resulted from this analysis that allowed characterization of the teacher's effort to involve students during the use of computer simulations in the classroom, as well as the students' own involvement and what they actually produced.

It was also concluded that, for students to become productively engaged in the tasks that they have to perform, it is necessary that they use computer simulation in a systematic way. It is also necessary that the teacher mediates the students' work in specific ways. Most prominently, the teacher must systematically monitor the students' involvement while allowing them to take the initiative. In contrast, it was found that withdrawing authority from students and not being able to keep the task as a challenge reduce the productive engagement of students.

Although this research encompasses the study of only two cases (two teachers), MNs allowed the obtainment of much information with a degree of detail, leading to the conclusions referred to above. It was also possible to see that the greater experience of one of the teachers, in particular by integrating results from education research into his teaching practices, allowed this teacher to conduct a more effective and consistent mediation of the students' work, with a consequent greater productive engagement of students.

The second study, published more recently (Saraiva et al., 2015), focused on the importance of the visual representations present in computer simulations used as symbolic mediators or mediators of action to promote students' epistemic practices.

Indeed, visual representations displayed by computer simulations can work as important tools that may increase, improve, and scaffold the students' epistemic work in the classroom. Visual representations displayed by computer simulations can work as symbolic mediators when used in a communicational context to inform, or illustrate, an idea or a concept. But visual representations can also function as mediators of action when used in interaction with the computer simulation context, to allow the students to manipulate the computer simulation, enabling them to change its configuration, or offering reformulations of or new visual representations of a concept or phenomenon.

The main goal of this study was to highlight the importance of visual representations displayed by computer simulations used in the classroom as symbolic mediators or mediators of action to promote students' epistemic work. The research questions addressed were: a) what is the impact on students' epistemic practices when teachers and students use visual representations as symbolic mediators or as mediators of action? and b) what teacher mediation actions, concerning the use of visual representations, promote the development of students' epistemic practices in the classroom?

This research involved two case studies, corresponding to two teachers of physics and chemistry in secondary school (students aged 15–18 years). Each teacher produced three MNs, corresponding to three of their own lessons.

The content analysis of these MNs led to the following results.

1. The use of visual representations by the students is directly related to the occurrence of a large number of students' epistemic practices. If visual representations are used only as symbolic mediators, they have a reduced impact on the occurrence of students' epistemic practices. Also, the use of visual representations only by the teacher, not complemented by other mediating actions, seems to have no influence on the students' epistemic practices.
2. There are teacher mediation actions related to the use of visual representations which have great impact on the promotion of students' epistemic practices: the proposal of tasks as challenges, giving students autonomy, requesting additional aspects, doing synthesis, and clarifying the epistemic status of work in progress or carried out by students. Taking into account the differences in the use of visual representations displayed by computer simulations (as mediators of action or as symbolic mediators) can contribute to improving the quality of students' epistemic work and of the teacher's mediation.

In this study, each teacher taught three lessons and students, organized into small groups, developed inquiry work. Several types of data were collected from the lessons (e.g., audio recordings, documents produced by the teacher or students, screenshots, photos, indications of silences and gestures, explicit teacher's intentions and decisions, students' notebooks, teacher notebook). These data were used by the teachers to produce multimodal narratives of their own lessons.

Given the multimodal characteristics of the MNs, it was possible, using content analysis, to identify evidences of: (a) the use of visual representations by students and teachers, and their role as symbolic mediators or as mediators of action; (b) the students' epistemic practices that take place; and (c) the teacher's actions concerning the use of visual representations that promote or enable such epistemic practices.

The MN is very useful in this analysis, because it is a single structured document that organizes and systematizes the relevant data, enabling rich and deep analysis of what happens in the classroom. The inclusion of screenshots from the students' computers when they were working with the computer simulations was particularly useful, enabling quick identification of the visual representation used.

Study 4: Multimodal Narratives in the Study of Teachers' Decision-Making in the Classroom

This study (Santos, Lopes, & Cravino, 2011) investigated didactic decisions made in the classroom by teachers of physics and chemistry. It aimed to understand what triggers the didactic decision-making of teachers in the classroom and to detect patterns in these decision-making processes.

This is an important issue, because the classroom is a dynamic and complex environment (Larrivee, 2000) and it is impossible for a teacher, no matter how good she is, to be fully aware of everything that happens, paying attention to all the language and actions of all the students, and also making decisions in real time to improve their learning. Also, there are many things that draw a teacher's attention, as well as several decisions that teachers take in the classroom, and these are different according to the teacher's intentionality and what actually happens in the classroom.

Therefore, this study focused on decisions made by teachers related to what draws their attention in the classroom. The purpose was detecting patterns in teachers' didactic decisions in classrooms, and their didactic functions, with the aim of understanding their potential role in improving students' knowledge.

The research (Santos et al., 2011) involved five teachers of basic, secondary, and higher education, whose lessons were documented using MNs. Each MN was written by the teacher who taught the lesson. The content analysis of the MN was directed toward identification of the didactic decisions taken in the classroom, trying to understand what led to each of them. After this first level of analysis, cluster analysis was used to try to detect patterns associated with the decisions identified.

From the analysis, it was inferred that teachers' decisions in the classroom are directly related to what draws their attention. This has great relevance to professional practice, because what draws a teacher's attention depends on his training and preparation to teach, his ideas about teaching and learning, and his world vision. This is a complex relationship, because the same aspect that draws a teacher's attention may lead the same teacher to make different decisions depending on the details of the context. Similarly, different aspects that draw the teacher's attention may lead her to make the same decision. For any decision, there are many possible options. For instance, when a student asks a question, the teacher can use it to involve the whole class in a discussion, to support the student, or to give information or guide the students to perform a specific task. The decision actually taken depends on the teacher's intentionality or the teacher's desire to incorporate a new idea, students' contributions, or the motivational state in the learning environment.

In this study, it was found that each teacher has a singular profile of decision-making. This profile is a particular combination of three elements: patterns of decision-making, what draws the teacher's attention, and the specific decisions taken.

The results of this study led to the identification of four patterns of decision-making, applicable to all teachers and the various teaching levels. These patterns are related to decisions concerning the adaptation of working conditions in the classroom, or are related to teaching decisions that result from the didactic intentionality of the teacher.

For example, the teaching level may become less important in certain patterns of decision-making, such as the pattern of decisions about management of work/classroom rules. Indeed, in higher education it is not so noticeable, certainly due to the maturity of students in this level of education. Nevertheless—and contrary to Rich and Hannafin (2008), who found that classroom management actions do not directly affect students' learning—the results of this study (Santos et al., 2011) suggest that some of the teacher's classroom management actions may provide benefits for collective work, therefore contributing directly to the students' learning. This was found from differences in the work of each student or group, as identified in the MNs.

From this brief description, it can be seen that MNs were fundamental to identifying and understanding decision-making patterns and their causes. The fact that each MN was written by the teacher who taught the lesson means not only that it describes the decisions that are being taken, but often there are clues as to the intentions of the teacher. This information, coupled with information from what the students do and how they react to the decisions of the teacher, provides an overall picture of the teaching and learning taking place in the classroom that is richer than the usual recording of the lesson or the self-reported perceptions of either teachers or students. This overall picture is much more insightful and may be very useful in research studies and in the design of professional development activities for teachers.

DISCUSSION

Even though this may not be evident from the example studies presented, the elaboration of an MN is always a complex and time-consuming process, since it involves collecting and organizing all the information on which it is based and that gives it its multimodal character. Then it is necessary to write the MN itself, incorporating the multimodal elements considered essential and linking to the other documents. The upside of this effort is to gather in one single document the essential information, allowing concentration of the data analysis efforts (although it is always possible to access any document or information that is the basis of the MN).

As shown in the studies presented in this chapter, the MN, especially if written by the teacher who taught the lesson, ties together a lot of information that is easier to analyze and makes it more insightful, especially when studying aspects that are very often not documented during the lesson, such as the actions and speech of teachers and students. As it incorporates, as multimodal elements, photos and computer screenshots, MNs have proved very useful in studies about graphical representations and computer simulations (study 2 and study 3). Another type of research where MNs proved particularly insightful was when studying teachers' decisions in the classroom (study 4). The narrative, written from the point of view of the teacher who taught the lesson, naturally incorporated a lot of information about his decisions.

If the MN is written by someone who is not the teacher who taught the lesson, there is an increased workload on that person, as she must not only gather all the data necessary to write the MN, but also maintain a close contact with the teacher to make sure that the final MN is as reliable and detailed as possible. At the end of this process, the teacher must recognize his or her lesson in full.

When the MN is written by a researcher or the teacher who taught the class (in particular if he does so in the double role of teacher-researcher), there is a potential risk, as was noticed at the beginning of the process of developing what became MNs. In these cases, the narrator may, even involuntarily, tend to focus his attention (and therefore the MN) on aspects that may seem to him to be of greater importance because of the research that he may be conducting. We tried to instruct the researchers to avoid this pitfall, and we did not find it in later MNs (that is, in the studies that we report here as examples of the use of MNs). Being aware of this possibility and incorporating instructions to avoid it in the MN protocol ensures that should not happen. The validation process—also part of the MN protocol—further ensures that the final MN is free from this possible side effect.

The published protocol for the elaboration and validation of MNs (Lopes et al., 2014) is therefore fundamental for its use in research, because following it ensures that the structure of MNs is always identical and that the steps followed in the elaboration and validation are also similar (see Chapter 1). This further ensures that the information contained in the MN is reliable, comparable, and as complete as possible. The use of this protocol means that MNs can be used by any researcher as sources of data to study various research issues, ensuring that the information is reliable and comparable.

CONCLUSION

The diversity of studies discussed in this chapter shows how MNs were used as a tool to support studies in science and technology education research, in particular those that must rely on data from real classroom environments. In fact, as shown in the examples discussed above, MNs allowed the study of particular aspects of lessons from different disciplines and different levels of education. The studies can focus on the MNs of a single teacher or pull together MNs of several teachers. On the other hand, MNs can be written by the teacher teaching the lessons or by someone else, though in close collaboration with the teacher.

The protocol for the elaboration and validation of MNs (Lopes et al., 2014) is fundamental for their use in research, because following it ensures a comparable structure and that the steps followed in the elaboration and validation are also similar (see Chapter 1). The use of this protocol means that already existing MNs can be used by any researcher as sources of data to study many possible research questions, ensuring that the information used is both reliable and comparable.

Readers who are interested in finding out more can read about MNs and their development, such as how they can be elaborated and validated, as well as their possible uses, in the following publications: a first book in Portuguese (Lopes, Silva, Cravino, Viegas, Cunha, Saraiva, et al., 2010), an article published in an international research journal (Lopes et al., 2014) and, more recently, an international edited book (Lopes, Cravino, Cruz, & Barbot, 2017).

Since this is an ongoing work, more publications will certainly appear in the literature and new developments will be posted on the website (http://multimodal.narratives.utad.pt). This website also hosts the repository of validated MNs that teachers and researchers kindly make available to the research and educational community.

REFERENCES

Adger, C. T., & Wright, L. J. (2015). Discourse in educational settings. In D. Tannen, H. Hamilton, & D. Schiffrin (Eds.), *The handbook of discourse analysis* (pp. 858–879). Chichester, UK: Wiley Blackwell.

Alsop, S., Bencze, L., & Pedretti, E. (Eds.). (2005). *Analysing exemplary science teaching*. Maidenhead, UK: Open University Press.

Bardin, L. (1993). *L'analyse de contenu*. Paris, France: Presses universitaires de France.

Branco, M. J., Magalhães, R., Ferreira, A., Lopes, J. B., & Cravino, J. P. (2009). Narrações de aulas: Um estímulo ao desenvolvimento profissional? [Classroom narratives: A stimulus to professional development?]. In F. Paixão & F. R. Jorge (Eds.), Educação e Formação: Ciência, Cultura e Cidadania. Actas XIII Encontro Nacional de Educação em Ciências (pp. 740–749). Castelo Branco: Escola Superior de Educação, Instituto Politécnico de Castelo Branco.

Clemente, M., & Ramírez, E. (2008). How teachers express their knowledge through narrative. *Teaching and Teacher Education*, *24*(5), 1244–1258. doi:10.1016/j.tate.2007.10.002

Cohen, L., Manion, L., & Morrison, K. (2018). *Research methods in education*. London: Routledge.

Craig, C. J. (2011). Narrative inquiry in teaching and teacher education. In K. Julian, P. Darlene Ciuffetelli, & P. Debbie (Eds.), *Advances in research on teaching* (Vol. 13, pp. 19–42). Bingley, UK: Emerald Group.

Engle, R. A., & Conant, F. R. (2002). Guiding principles for fostering productive disciplinary engagement: Explaining an emergent argument in a community of learners classroom. *Cognition and Instruction*, *20*(4), 399–483. doi:10.1207/S1532690XCI2004_1

Forman, E. A., Engle, R. A., Venturini, P., & Ford, M. J. (2014). Introduction to special issue: International examinations and extensions of the productive disciplinary engagement framework. *International Journal of Educational Research*, *64*, 149–155. doi:10.1016/j.ijer.2013.07.007

Georgakopoulou, A. (2006). Thinking big with small stories in narrative and identity analysis. *Narrative Inquiry*, *16*(1), 122–130. doi:10.1075/ni.16.1.16geo

Hundhausen, C. D. (2002). Integrating algorithm visualization technology into an undergraduate algorithms course: Ethnographic studies of a social constructivist approach. *Computers & Education*, *39*(3), 237–260. doi:10.1016/S0360-1315(02)00044-1

Kaufman, L., & Rousseeuw, P. J. (1990). *Finding groups in data: An introduction to cluster analysis*. Chichester, UK: John Wiley & Sons, Inc. doi:10.1002/9780470316801

Kramp, M. K. (2004). Exploring life and experience through narrative inquiry. In M. Kathleen De & S. D. Lapan (Eds.), *Foundations for research: Methods of inquiry in education and the social sciences* (pp. 103–122). Mahwah, NJ: Erlbaum.

Kress, G., Jewitt, C., Ogborn, J., & Tsatsarelis, C. (2014). *Multimodal teaching and learning: The rhetorics of the science classroom*. London: Bloomsbury.

Krippendorff, K. (2004). *Content analysis: An introduction to its methodology*. Thousand Oaks, CA: SAGE Publications.

Labov, W. (2006). Narrative pre-construction. *Narrative Inquiry*, *16*(1), 37–45. doi:10.1075/ni.16.1.07lab

Larrivee, B. (2000). Transforming teaching practice: Becoming the critically reflective teacher. *Reflective Practice: International and Multidisciplinary Perspectives*, *1*(3), 293–307. doi:10.1080/713693162

Lopes, J. B., Cravino, J. P., Cruz, E., & Barbot, A. (Eds.). (2017). *Teaching science: Contributions of research for planning, practice and professional development*. New York, NY: Nova Science Publishers.

Lopes, J. B., Cunha, A. E., Santos, C. A., Saraiva, E., Cravino, J. P., & Dinis, F. (2012). Envolver produtivamente os alunos em aulas de Física e Química durante o uso de simulações computacionais: Dois professores com mediações distintas e uso distinto [Productively engaging students in physics and chemistry lessons during the use of computer simulations: Two teachers with different mediations and different use]. *SENSOS*, *2*(2), 121–137.

Lopes, J. B., Silva, A. A., Cravino, J. P., Santos, C. A., Cunha, A., Pinto, A., ... Branco, M. J. (2014). Constructing and using multimodal narratives to research in science education: Contributions based on practical classroom. *Research in Science Education*, *44*(3), 415–438. doi:10.100711165-013-9381-y

Lopes, J. B., Silva, A. A., Cravino, J. P., Viegas, C., Cunha, A. E., Saraiva, E., ... Santos, C. A. (2010). Investigação sobre a Mediação de professores de Ciências Físicas em sala de aula [Research about teacher mediation of physical sciences in the classroom]. Vila Real, Portugal: Universidade de Trás-os-Montes e Alto Douro.

Lopes, J. B., Viegas, C., & Cravino, J. P. (2010). Improving the learning of physics and development of competences in engineering students. *International Journal of Engineering Education*, *26*(3), 612–627.

Mason, J. (2002). *Researching your own practice: The discipline of noticing*. New York, NY: Routledge.

Melville, W. (2008). Mandated curriculum change and a science department: A superficial language convergence? *Teaching and Teacher Education*, *24*(5), 1185–1199. doi:10.1016/j.tate.2007.03.004

Qualitative Research. (2016). Editorial. *Qualitative Research*, *16*(1), 3–8.

Rich, P. J., & Hannafin, M. J. (2008). Decisions and reasons: Examining preservice teacher decision-making through video self-analysis. *Journal of Computing in Higher Education*, *20*(1), 62–94. doi:10.1007/BF03033432

Rowan, B., Camburn, E., & Correnti, R. (2004). Using teacher logs to measure the enacted curriculum: A study of literacy teaching in third–grade classrooms. *The Elementary School Journal*, *105*(1), 75–101. doi:10.1086/428803

Santos, C. A., Lopes, J. B., & Cravino, J. P. (2011). Patterns of didactic decisions made by teachers in physical sciences classrooms. *Problems of Education in the 21st Century*, *36*, 76–89.

Saraiva, E. (2017). *Estudo da Linguagem gráfica no contexto da mediação da aprendizagem em Ciências Físicas* [Study of graphical language in the context of learning mediation in physical sciences] (Unpublished doctoral dissertation). Universidade de Trás-os-Montes e Alto Douro, Vila Real, Portugal.

Saraiva, E., Lopes, J. B., Cravino, J. P., Santos, C., & Cunha, A. E. (2015). Teacher mediation of students learning using visual representations displayed by computer simulations. In *INTED2015 Proceedings* (pp. 4996–5005). Madrid, Spain: IATED.

Saraiva, E., Lopes, J. B., Cravino, J. P., & Santos, C. A. (2012). How do teachers of physical sciences with different professional experiences use visual representations with epistemic functions in the classroom? *Problems of Education in the 21st Century*, *42*, 97–114.

Stein, S. J., Isaacs, G., & Andrews, T. (2004). Incorporating authentic learning experiences within a university course. *Studies in Higher Education*, *29*(2), 239–258. doi:10.1080/0307507042000190813

Tan, A. L., & Towndrow, P. A. (2009). Catalyzing student-teacher interactions and teacher learning in science practical formative assessment using digital video technology. *Teaching and Teacher Education*, *25*(1), 61–67. doi:10.1016/j.tate.2008.07.007

Viegas, C., Lopes, J., & Cravino, J. (2010a). Introductory physics teaching in engineering education: How can the reflection on teaching practice be used to improve teaching and learning? In *ICEE 2010 Proceedings: International conference on engineering education*. Gliwice, Poland: iNEER. Retrieved from http://www.ineer.org/Events/ICEE2010/papers/R9A/Paper_961_1175.pdf

Viegas, C., Lopes, J. B., & Cravino, J. P. (2009). Incremental innovations in a physics curriculum for engineering undergraduates. In W. Aung, K.-S. Kim, J. Mecsi, J. Moscinski, & I. Rouse (Eds.), Innovations 2009: World innovations in engineering education and research (pp. 175–186). Arlington, VA: iNEER.

Viegas, C., Lopes, J. B., & Cravino, J. P. (2010b). Learning through assessment and feedback in and out of class. In W. Aung, K. Kim, J. Mecsi, J. Moscinski, & I. Rouse (Eds.), Innovations 2010: World innovations in engineering education and research (pp. 33–46). Arlington, VA: iNEER.

Viegas, C., Lopes, J. B., & Cravino, J. P. (2011). Reflection on teachers' practice to improve teaching and learning of introductory physics in engineering education. In W. Aung, V. Ilic, J. Moscinski, & J. Uhomoibhi (Eds.), Innovations 2011: World innovations in engineering education and research (pp. 211–225). Potomac, MD: iNEER.

Weber, R. P. (1990). *Basic content analysis* (2nd ed.). Newbury Park, CA: Sage Publications. doi:10.4135/9781412983488

Woolhouse, C. (2017). Multimodal life history narrative. *Narrative Inquiry*, *27*(1), 109–131. doi:10.1075/ni.27.1.06woo

Chapter 13
Researching Epistemic Practices Development With Multimodal Narratives

Elisa Saraiva
Escola D. Maria II de Vila Nova de Famalicão, Portugal & Politécnico do Porto, Portugal

ABSTRACT

This chapter describes an empirical study using multimodal narratives for research into students' development of epistemic practices in the classroom. Multimodal narratives can give access to classroom events, preserving their complex and holistic nature. Through content analysis, they allow a good comprehension of the multimodal nature of teaching and learning practices. The results of this work highlight the importance of multimodal narratives as a research instrument. Their importance is based on the richness of elements they contain that allow the identification, categorization, and characterization of teacher mediation actions that promote, scaffold, and enlarge students' epistemic practice development. This chapter seeks to describe both their multiple potentialities as an instrument and their limitations when researching the development of students' epistemic practices in the physical sciences classroom.

INTRODUCTION

This chapter describes an empirical study using multimodal narratives (MNs) as an instrument for research about teachers' mediation actions that influence the development of students' epistemic practices in the classroom. The study intends to identify and characterize teacher mediation actions that promote, sustain, and enlarge students' epistemic work. The use of MN as a research instrument allows the achievement of a good level of understanding about teacher mediation actions that can promote and scaffold students' work in the classroom when engaged in social practices that lead to knowledge construction in the classroom. This study intends to highlight the role played by MN as a research instrument, elaborating its multiple potentialities and limitations when used for the identification and characterization of teacher mediation actions that can promote and enlarge the development of students' epistemic practices in the classroom.

DOI: 10.4018/978-1-5225-8570-1.ch013

Researching students' epistemic practices as they occur in the classroom remains an important issue: it is important to know more about the cognitive, discursive, and social activities in which students are engaged to develop their understanding of scientific knowledge and to produce, communicate, and evaluate knowledge (Kelly, 2008; Sandoval, 2005). For research in this field of knowledge, and for teachers' professional development, it is very important to know more about teachers' mediation actions that influence the quality and variety of students' epistemic practices in the classroom. The teacher's mediating role is crucial in promoting and sustaining a learning environment that enables the acquisition of rich and diverse conceptual and epistemological knowledge: teachers need to develop professional knowledge about their actions and about their impact on students' epistemic work in the classroom. That is why the main focus of this research is the identification of teachers' mediation actions that influence the quality and variety of students' epistemic practices. This research work intends to investigate teachers' mediation actions and the occurrence of students' epistemic practices, but without interfering with the natural environment of the classroom. The research question focuses on the identification of teacher mediation actions and their influence, and the occurrence and development of students' epistemic practices in the classroom. For that, MN was adopted as the research instrument. The richness of the different elements that MN contains allows the identification of "different modes, modalities, means, languages, resources and spatial and temporal organization adopted by teachers in the classroom" (Lopes et al., 2010, p. 16) and students' engagement in investigative activities in the classroom (Sandoval & Millwood, 2005).

RELEVANCE OF EPISTEMIC PRACTICES FOR STUDENTS' LEARNING

Students' engagement in epistemic practices, as claimed by some authors (Berland et al., 2016; Duschl & Grandy, 2013; Jiménez-Aleixandre & Crujeiras, 2017; Krist, 2016; Osborne, 2014), is crucial for the development of a deeper understanding about the nature of scientific knowledge and the process of its construction. It is essential that learning about scientific concepts should not be separated from learning about the practices used in the production of scientific knowledge (Duschl & Grandy, 2013). According to the authors listed, there must be a change from an education focused on just what you know (e.g., facts and skills) to an education focused on how we know what we know.

Teaching and learning science requires significant and productive student engagement (Engle & Conant, 2002; Mortimer et al., 2014) in social practices for the construction of knowledge, including the elaboration of theories and models, the construction of arguments, and the use of different forms of communication and representations of concepts or phenomena (Duschl & Grandy, 2013). Such epistemic practices emerge from the investigative activities carried out by students in the classroom (Sandoval & Millwood, 2005) and have as a reference the work done by scientists in the context of producing scientific knowledge. The epistemic practices must emerge from students' attempts to solve a problem or task, "mobilizing prior knowledge and using procedures like description, representation, prediction and so on" (Lopes, Branco, & Jiménez-Aleixandre, 2011, p.2).

Thus, epistemic practices may be defined as the social activities of the production, communication, and evaluation of scientific knowledge (Kelly, 2008; Kelly & Duschl, 2002; Kelly, McDonald, & Wickman, 2012; Sandoval, 2005) that give students the opportunity to develop positive attitudes about science and ways of doing science. Students' engagement in scientific practices in the classroom gives them more authority and makes them more competent in the application of scientific knowledge in the real context (Lopes et al., 2011).

RESEARCHING EPISTEMIC PRACTICES WITH MULTIMODAL NARRATIVES

The importance of epistemic practices for science and for learning science underlines the need for teachers' development of professional knowledge that enables them to adopt mediation actions that promote, scaffold, and enlarge these practices in the classroom. Thus, the investigation should produce more knowledge in this field, based on data collected in the natural environment of the classroom, which can be shared and placed at the service of teachers. That is why our research team carried out several studies around this topic (Cunha et al., 2012; Lopes, Cravino, Silva, & Viegas, 2012; Lopes et al., 2013; Lopes, Silva, et al., 2012; Saraiva, 2017b; Saraiva, Cunha, Santos, Lopes, & Cravino, 2012; Saraiva, Lopes, Cravino, & Santos, 2012) and found that MN can play a determinant role as a research instrument that fulfills the double need to preserve the natural environment of the classroom and give access to multiple and multimodal elements of classroom events.

MNs are an effective way to gather, organize, and transform data, avoiding confusing and time-consuming manipulation of data while minimizing the natural subjectivity of the narrator (Lopes et al., 2014). Through the content analysis of an MN, it is possible to identify the occurrence of students' epistemic practices and teachers' mediation actions. Since MNs are a vivid description of all events that take place in the classroom, focused on the actions and language adopted by the teacher and students during the accomplishment of a task, they provide the researcher with relevant information about the tasks presented to students and the work really demanded from them to accomplish the tasks successfully; the way the teacher organizes the work in the classroom; what resources the teacher uses and what resources he provides to his students; the work that is actually produced by the students; the reactions of all actors in the classroom; the level of autonomy conceded; how students perceive the proposed tasks; what investigative actions students develop; and in what ways students are effectively engaged in social practices of knowledge construction. Another argument in favor of adopting the MN as a research instrument is the fact that they all maintain the same structure, even when elaborated by different teachers or researchers, which is very helpful for the development of the multiple case study. They allow the use of the same methodology of analysis and of empirical verification and comparability between situations and cases in the study (Lopes et al., 2014). Additionally, they are reliable and verifiable, since they act as a repository of all data collected through several instruments and methods, including teacher diaries, student notebooks, audio recordings, interviews with teachers, resources supplied for students' work, and others.

METHODS

Taking into account that the main porpoise of this research study is to highlight the important role that the MN can play as a research instrument—in particular, to identify and characterize teacher mediation actions that promote, sustain, and enlarge students' epistemic work—a qualitative and interpretative approach (Cohen, Manion, & Morrison, 2000, p. 22) was adopted, using a multiple case study (Bogdan & Biklen, 1994, p. 97; Stake, 2013).

Participants

This study involved five Portuguese teachers working in public schools. We analyzed a total of 15 multimodal narratives, drawing upon data collected from classes they taught during two school years. The teachers had different professional experiences, taught in different contexts; during their classes, students have developed different types of activity. This diversity of cases for study allowed us to illustrate a greater variety of environments and subjects (Bogdan & Biklen, 1994, p. 97) and to compare or contrast the presence or absence of certain characteristics in each (Stake, 2013, p. 6).The main characteristics of the five cases are summarized in Table 1 and were the same as those adopted in a wider research (Saraiva, 2017a) focused on studying the role of visual representations in the promotion of students' epistemic work. The research was developed during two school years. During the first data collection (school year 2007/2008), we worked with all five teachers (A, B, C, D, and E); during the second data collection (school year 2011/2012), we worked with only two of the teachers (B and E).

The teachers were selected based on their teaching experience, defined as a combination of teaching years and experience in research on science education. In terms of professional development, we

Table 1. Main characteristics of the five case studies

		Teacher A	Teacher B	Teacher C	Teacher D	Teacher E
Gender		Female	Female	Male	Female	Female
Academic background		Ph.D. student in physical sciences education	Ph.D. student in physical sciences education	Degree in physics and chemistry teaching	Degree in physics and chemistry teaching	Ph.D. student in physical sciences education
Teaching experience		28 years	15 years	15 years	15 years	23 years
Research experience		Yes	Yes	No	No	Yes
Grade and topic of the lessons	1st year	Physics 9th grade Electrical Circuits	Chemistry 8th grade Chemical Reactions	Physics 7th grade Energy	Physics 8th grade Optics	Chemistry 8th grade Chemical Bonding
	2nd year		Chemistry 12th grade Electrochemical Cells			Physics 10th grade Blackbody Radiation
Activity developed in the classroom	1st year	Experimental work	Experimental work	Lecture with some demonstrations performed by the teacher	Lecture with some demonstrations performed by the teacher	Work with molecular models
	2nd year		Computer simulation as pre-experimental work			Computer simulation replacing experimental work
No. of students	1st year	19	22	23	29	26
	2nd year		20			23
Number of lessons and total time	1st year	2 (90 + 45 min)	2 (90 + 90 min)	2 (90 + 90 min)	2 (90 + 45 min)	1 (45 min)
	2nd year		3 (90 + 90 + 135 min)			3 (135 + 90 + 90 min)

can consider teachers A and E to be the most experienced and teachers C and D the least experienced. Teacher B is in the middle of the two groups, because she has almost the same number of years' teaching experience as teachers C and D, but has higher academic studies and some experience in research in the field of physical science education.

The content of the lessons, the topics taught, and the activities developed in the classroom were also different for the five cases. During the first school year, none of the teachers used computer simulations in the classroom. However, during the 2011/2012 school year, teachers B and E taught in secondary education and both used computer simulations in their lessons.

Data Collection

Since the teacher plays a central role, we selected him as the central actor, capturing his intention and preserving the complex and holistic nature of teaching and learning. Over two school years, we collected several types of data from the lessons (e.g., audio recordings, documents from the teacher and/or students, print screens from computers, photos, indications of silences and gestures, student reactions, explicit teacher intentions and decisions, students' notebooks, teacher's notebook, schemas of spatial organization of the classroom, schemas on blackboard by the teacher and/or by students). Then we used MNs to organize and systematize data. They describe all relevant episodes of teacher mediation based on audio recordings of the lessons, several documents, and multimodal elements obtained by the teacher. All MNs had the same structure and the same focus (the development of the tasks in the classroom) in order to allow comparability, and all were constructed by the teachers themselves. The accuracy and readability of each multimodal narrative was validated by other researchers from our team using the data collected. The MN corresponding to the lessons taught by case teachers A, B, and E were drawn up by the teachers themselves, but those relating to cases C and D were prepared by a researcher who had worked with the teachers during a research about professional development (Lopes, Silva, et al., 2012).

In both situations, the classroom's events were presented with objectivity but enriched by the multimodal elements included and a set of information to which only the teacher had access. In the case of MNs made by a researcher, in addition to the data collected to compose the narrative (e.g., audio recordings, students' books, teacher diaries, photographs, worksheets provided, etc.), it was necessary to talk to and meet with teachers to have access to the same information, mainly related to his decisions, perceptions, and adoption of given actions.

Data Analysis

The MNs were subjected to a process of content analysis (Bogdan & Biklen, 1994, pp. 220–239; Krippendorff, 2004) with the help of NVivo 8 software®, in steps of codification and categorization. Through analysis of the MN content, it was possible to identify the occurrence of epistemic practices of students and the actions of teacher mediation related to them, since these are organized in two category systems that subsequently formed variables in a cluster analysis. Thus, it was possible to identify relationship patterns between the actions taken by teachers and epistemic practices in which students were involved.

We started the process by identifying and encoding parts of the MN that presented evidences about the engagement of students in epistemic practices and about the actions and language adopted by the teacher during mediation of students' learning. This process enabled the categories to emerge, which were subsequently organized into "lists of categories of encoding" (Bogdan & Biklen, 1994, p. 221) or "systems of categories" (Kelly et al., 2012, p. 146). The categories were not previously defined; they emerged as a result of the analogic and progressive process of classification of elements, based on evidence regarding the "topics present in the data" (Bogdan & Biklen, 1994, p. 221). However, the criteria used to identify and codify excerpts of MNs that contained evidence about the occurrence of students' epistemic practices were the verification that, in a given moment, students were developing investigative activities in the classroom in order to solve a problem or question, taking as a reference the work done in the context of scientific production (Sandoval, 2005; Sandoval & Millwood, 2005). The investigative activity developed by students understands their involvement in the planning and implementation of strategies to respond to a question or challenge launched by the teacher, in the development of activities of research, experimentation, data analysis, reporting, and validation of knowledge produced among their peers (Kelly, 2008). Also, we intended to identify the moments where students used mediator artifacts, made their own models, used different ways to represent situations and concepts, argued with their peers or with the teacher, and other important practices that lead to knowledge construction in the classroom. Having these important theoretical aspects in mind, we encoded the excerpts of MNs that contained evidences about the logical processes adopted by students in the development and verification of knowledge, as is the case in the identification and mobilization of guiding concepts, the elaboration of conjectures, and the adoption of strategies for research or clarification of thought patterns (Jiménez-Aleixandre & Bustamante, 1997). The categorization process was reiterated by another researcher on our team, more experienced and with a higher level of expertise. Whenever there was a disagreement in relation to the category or definition, there was a discussion in order to solve it, since each category was assigned a name and a brief description (see Table 2).

Through the content analysis of the MNs, it was possible to identify the moments when the students were involved productively in tasks that enabled them to build knowledge in the classroom and develop epistemic skills, whether they were induced by the actions and language adopted by the teacher, or a result of students' initiatives. Concerning teacher mediation actions, during the process of content analysis of the MNs, we also looked to identify and code those excerpts that contained evidences about tasks proposed, ways of presenting tasks to the students, contexts selected, resources provided, mediators introduced and used, language adopted, epistemic support given to students, evaluation of students' work, and autonomy granted, among others. We sought to identify the actions and language adopted by the teacher to promote, develop, extend, or even inhibit students' epistemic practices (Saraiva, Cunha, et al., 2012). Such actions are important epistemic movements since they guide the student, outlining the knowledge to reach and the appropriate forms of its construction. Such actions help students understand the nature and value of the knowledge produced (Reiser, 2004). As examples of these actions, we can have moments when the teacher provides some 'tips' or 'clues' to the student at key moments, or when he requests some additional aspects to help students progress during their work.

The content analysis of the MNs also allowed the possibility of quantifying the occurrence of epistemic practices for each case under study, as well as teacher mediation actions that are more frequent and the characteristics of each case. In order to accomplish that, the number of epistemic practices identified during the MN content analysis was counted. For each case study, since the number of lessons taught was different, we took into account the number of epistemic practices identified and considered 45-minute units for each teacher. A similar procedure was adopted to count the number and diversity of teacher mediation actions.

RESULTS

Based on the categorization process previously described, for each of the dimensions in the study (i.e., epistemic practices of students that actually take place in the classroom and mediation actions of the teacher who promotes and sustain them), we analyze each of the episodes present in the MN, organizing the evidences collected. To better understand the role played by MNs as a research instrument and the richness of elements that they contain, we present a small excerpt from one episode from the lesson taught by teacher B during the second year of our research. During this lesson, students made use of a computer simulation to prepare a project to build an electrochemical cell with the materials available. The request was that the cell needed to display a voltage of 1.5 V. The computer simulation allowed students to identify some of the empirical conditions relating to the construction and operation of an electrochemical cell. The excerpt in question is related to the moment when the students were working with the computer simulation after the teacher had presented the task and explained what students were expected to produce. Since the episode is extensive, we present here only a small extract where the teacher's interactions with a group of students become clear.

Multimodal Narrative Excerpt 1: Teacher B (Electrochemical Cells, Chemistry 12th Grade)

The students start by choosing the zinc electrode for the anode and silver electrode for the cathode, which raised a doubt on the part of Sonia, Helena, and Tatiana.

Student: Teacher, we chose to build the cell with zinc and silver. It gives 1.5 V, but with a negative value. Is it possible?

Teacher: The value seems good to me!

Student: The cell is working in the opposite way?

Another student: But it works... see the electrons... (student is pointing to electrons moving on screen)

Teacher: Have you already tried to look for the standard electrode potentials? Between these two metals, which one will be reduced? Think about that...

The students continue with their work...

During this episode, we can see that students are using the visual representation displayed by the computer simulation as a mediator to reason and communicate. When students start operating with the simulation, they verify that the difference in electric potential registered on the voltmeter screen has a negative value. Since they observe the movement of the electrons on screen, they conjecture about the current direction. This allows students to change the focus of their attention to access other elements relating to the physical situation. The teacher is demanding an additional aspect, and such intervention allows them to think about other possibilities. The teacher does not present an immediate answer to the students, but allows the students to work autonomously and gives them the space to think about the answer by themselves. These important actions (i.e., requesting additional aspects and giving autonomy to students) have a clear and important relationship with the students' epistemic practices that took place. It becomes clear from the episode that students are handling equipment (i.e., computer simulation) conceptually; they are establishing relationships; they raise questions informed by their previous knowledge; they are mobilizing ideas and using different formats of representation.

In the same way that MNs allow us to identify teacher mediation actions and the impact on students' epistemic practices, they can also give evidence about teacher actions that have less potential in promoting students' epistemic work. In the same way that some teacher mediation actions can promote, sustain, and even extend the epistemic practices of students, there are others which do not have the same effect on the promotion of students' epistemic work. There are teacher mediation actions that can even inhibit such work, as, for example, the dispersion on the approach to the physical situation whenever the verbal interactions do not focus on a single situation under study. In order to illustrate the extent to which MNs also contain evidence about teachers' actions with less potential to promote and scaffold students' epistemic practices, here is an excerpt from an episode in a lesson taught by teacher C in the first year of our research. In this excerpt, the dispersion of the approach to the physical situation, and the fact that the teacher does not focus students' attention on a manageable number of physical situations, is clear.

Multimodal Narrative Excerpt 2: Teacher C (Energy, Physics 7th Grade)

Teacher: If I open the windows and the door, there could be a movement of matter? Or not? Some air could come out and some air could come in. Someone, badly behaved, could exit through the window (says, in a playful tone), or someone with good manners could come through the door… (continues in the same joking tone). Then there could be exchanges of matter…

Student 1: Exchanges of energy…

The teacher seems to ignore student intervention and continues with his reasoning.

Teacher: There could be an exchange with the outside world. Could or could not?

Student 2: Yes, it can.

Teacher: You can also get energy out there?

At this moment, the teacher is looking for the student that said energy before; however, the teacher continues talking.

Teacher: For example, the energy that comes from the sun. Can it come into the room if you have open windows?

Student 3: No!

Student 4: Doesn't warm the room.

Teacher: Doesn't warm inside the room?!

Student 3: Yes. The thermometer will rise soon.

Student 4: But only if the window is closed.

Student 5: Especially if you are near the thermometer!

Student 6: If there is a thunderstorm.

Teacher: A thunderstorm?

Teacher: What kind of energy is that? What will be? …

Student 4: Electrical.

Teacher: Will certainly contain electrical energy, isn't it?

No one says anything, and teacher continues.

Teacher: Well, returning to what we were discussing. There will be exchanges of energy if we have the doors open? Yes or no?

In this case, the dispersion in the approach to the physical situation in question is clear, since the verbal interactions that take place do not focus on a single form of energy, or on a single process for transferring energy, or even on a single example of the system under study. The concept of the system that the teacher tries to introduce at the beginning of this discussion with the students does not become clear. Students present some mobilizing ideas that are ignored by the teacher.

The content analysis of MNs allows the identification of teacher mediation actions that promote, scaffold, and enlarge the occurrence of students' epistemic practices, but also those that diminish their occurrence, or even disable the possibility of their development. All epistemic practices developed by students and teacher mediation actions that were identified during the content analysis of the MNs are organized into systems of categories (Tables 2 and 3). Table 2 contains several categories about students' epistemic practices and a brief definition, which emerged from the analogic process of content analysis of MNs already described in the previous section.

The content analysis of MNs also makes it possible to quantify the occurrence of each category of epistemic practice that takes place for our cases (Figure 1).

The information in Figure 1 shows that several epistemic practices take place during the lessons taught by our five case study teachers. There is a set of epistemic practices common to the five cases: present mobilizing ideas, identify empirical conditions, and question factually. These epistemic practices are closer to the world of observable facts, objects, and events. Such epistemic practices can be considered low level epistemic practices. There are also some practices, closer to observable facts and events, that were only identified during the lessons from teachers A, B, and E, and that are related to observation (i.e., observe and describe, observe and interpret). These teachers, as presented in Figure

Table 2. Categories identified in MNs about students' epistemic practices

Epistemic Practice	Description
Observe and describe	The student uses scientific language to report observations.
Present mobilizing idea	The student suggests a possible way of solving a problem or questions, mobilizing prior knowledge.
Identify empirical conditions	Based upon available evidence, the student identifies the conditions under which a given phenomenon takes place.
Handle factually	The student handles artifacts without recourse to any guiding knowledge—follows the instructions provided by the teacher, but without understanding the purpose.
Question factually	The student formulates questions to clarify terms or comments related to the empirical conditions of a given phenomenon.
Establish relations	The student is able to make arguments or inferences that allow him to relate data, variables, and/or concepts to different situations.
Observe and interpret	The student is able to decode the content of elements represented, even when partially represented.
Handle conceptually	The student manipulates artifacts and/or equipment, mobilizing prior knowledge, knowing the purpose or purposes of their actions. The student formulates issues by mobilizing prior knowledge for a better understanding.
Question conceptually	The student formulates questions to clarify terms or comments, mobilizing prior knowledge.
Communicate non-autonomously	The student presents ideas or results, but in accordance with the suggestions or requests from the teacher.
Make predictions	On the basis of logical knowledge, is able to predict experimental results.
Change representation format	The student is able to move between more concrete or abstract forms of representation, regardless of the language used; includes the meaning of the elements represented.
Conceptualize	The student produces a symbolic representation of a phenomenon, given the context of use.
Argue	The student produces statements based on their own ideas or understanding about the phenomena. The student presents claims or assertions based on available data (empirical or theoretical)
Evaluate critically	The student analyzes and delivers value judgments about results, ideas, or language used, evaluating the plausibility of the same.
Validate	The student evaluates the validity of the knowledge produced by classmates and teacher.

2, are the ones that, more often, demand additional aspects in relation to students' work; they are also the ones that present most of the tasks as challenge to students. These actions, combined with the fact that they concede more autonomy to students, are related to a higher number of epistemic practices and greater diversity of epistemic practices.

From the information in Figure 1, it is also possible to identify a set of epistemic practices which we can consider high level practices, that only took place during the lessons taught by teachers A, B, and E. Such epistemic practices are handling equipment conceptually, questioning conceptually, changing representation format, and conceptualizing. During these lessons, students developed experimental work and/or used computer simulations. There are other epistemic practices, closer to theories and models, that only were developed during the lessons taught by teachers B and E (e.g., arguing and evaluating critically). Also, the practices of validating and making predictions were only identified during teacher E's lessons. It is important to note that this teacher was the one who more often adopted the actions of valorizing and making explicit epistemically the work developed by students, as shown in Figure 2. This teacher made more use of students' ideas, schemes, or actions, recognizing their epistemic value; he was also the one that more often valorized and made explicit the epistemic nature of the work done

Figure 1. Number of epistemic practices that take place in each case study lesson

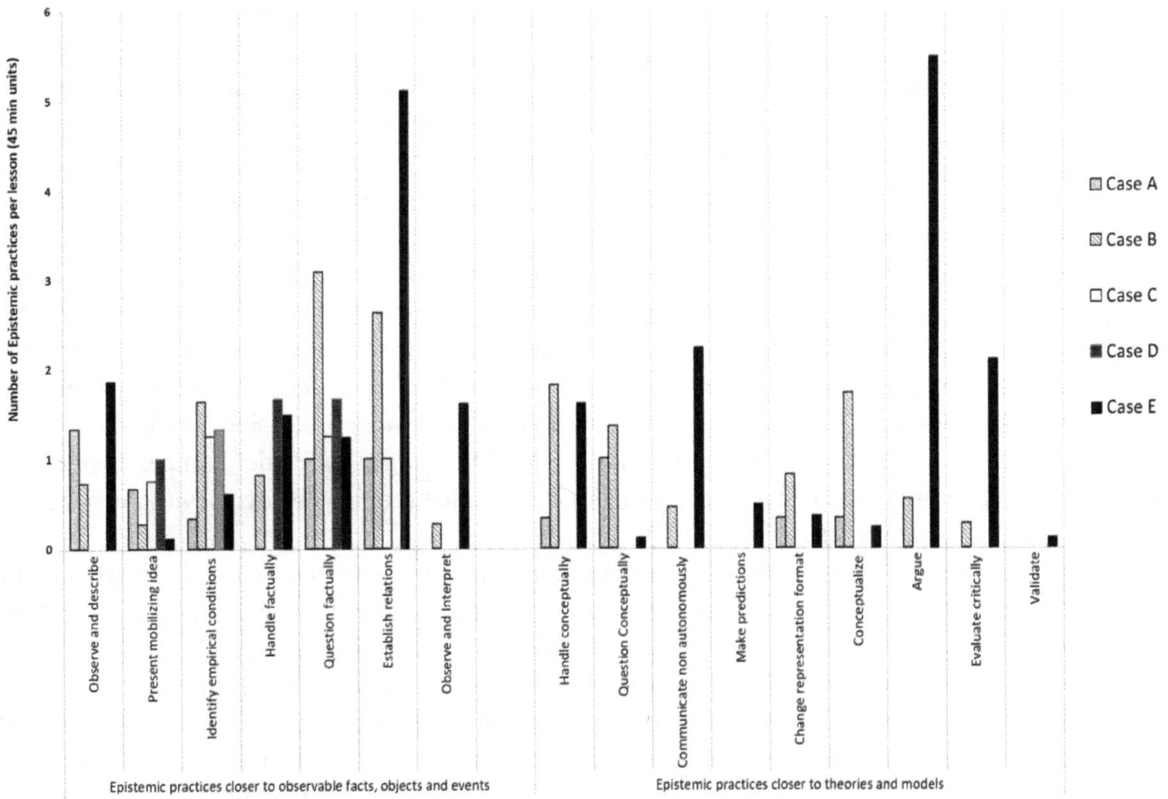

Figure 2. Teacher mediation actions adopted during each lesson for each case study teacher

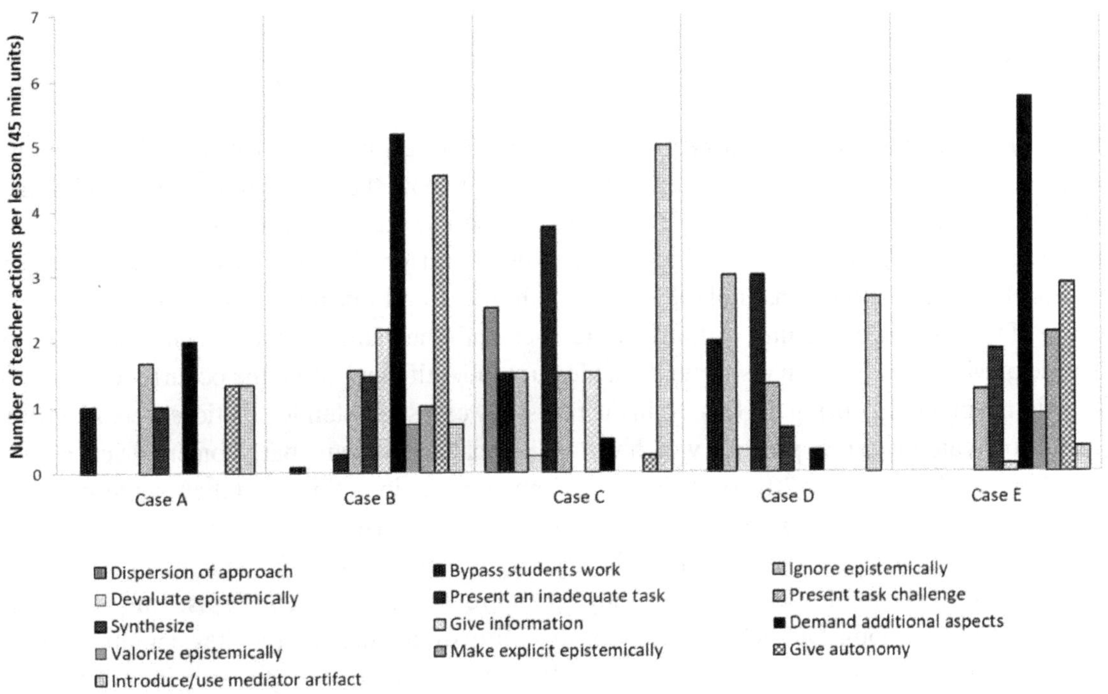

or ongoing. Also, this teacher, like teacher B, conceded more autonomy and opportunities for students to develop their work independently, providing information only when absolutely necessary. This is important to scaffold the students' work and ensure the necessary epistemic support.

Through analysis of Figure 2, it is possible to learn more about teachers' mediation actions that were more frequent during the lessons where students only developed epistemic practices that could be consider low level. Such practices were more frequent during the lessons of teachers C and D (e.g., dispersion of the approach to physical situations, bypassing students' necessary work, ignoring epistemically the student initiatives or even devaluing them from an epistemic point of view) and are related to the occurrence of few epistemic practices and/or practices more distant from the theories and models. During these teachers' lessons, only practices closer to the observable facts and events were identified.

Considering the differences in terms of professional experience between these five teachers (see Table 1) and the data presented in Figure 1 about the occurrence of students' epistemic practices, it becomes clear that during the lessons taught by our teachers with more professional experience (cases A, B, and E), the students have more opportunities to develop a larger and wider number of practices. These teachers more often adopted a set of mediation actions that provided students with the necessary epistemic scaffold to their work.

From our data analysis, it becomes clear that the most determinant actions, with more impact on the development of students' epistemic practices, are related to the presentation of the task as a challenge and the level of autonomy conceded to students. If students have, from the beginning of the task, an acceptable control over their actions and the teacher adopts a set of actions that nurtures their curiosity and helps them to achieve deeper levels of comprehension and conceptualization, the number and quality of practices will increase. Conversely, when the teacher adopts some dispersion of the approach to the physical situations and does not avoid the temptation to be too 'nosey' about the students' work, the students do not follow the necessary path in order to achieve higher levels of conceptualization and the number and variety of epistemic practices developed decreases.

DISCUSSION

Regarding the epistemic practices, it becomes clear from the results presented that MNs contain evidences and elements that allow the development of research about their occurrence and development in the physical sciences classroom. They provide a way to gather, organize, and transform data, avoiding confusing and time-consuming manipulation, since they describe all the events that take place in the classroom while minimizing the natural subjectivity of the narrator (Lopes et al., 2014). Also, they preserve the nature of teaching and learning without interference with the natural environment of the classroom.

The use of MNs as a research instrument enabled us to identify and count the occurrence of a diverse set of epistemic practices during the lessons in the five study cases. Epistemic practices were identified on the basis of our categorization process, which was mainly data-driven and based on empirical evidences contained in the MN (Saraiva, 2017b). When describing teacher intentions and their perceptions of the events in the classroom, it is possible to have access to more information than by recourse to another instrument, namely the transcription of an audio recording. This aspect is fundamental for understanding the role that MNs can play as a tool for studying and researching students' epistemic practices that take place in the classroom. The richness and multimodality of the information they contain, focused on students' development of each task, allowed us to identify what students were doing in order to accom-

Table 3. Categories identified in MNs about teacher mediation actions

Teacher's actions	Definition
Dispersion of the approach to physical situations	The teacher does not focus on a number of physical situations that can be easily managed or operationalized in the classroom.
Bypass students work	Teacher proposes a task with the potential to develop students' epistemic practices. However, teacher does not allow students to develop their own reasoning or solve the tasks alone. The students are just invited to perform routine procedures.
Ignore epistemically	Teacher ignores students' ideas, questions, or initiatives. The teacher does not take into account the mobilizing ideas or students' proposals to solve a problem.
Devalue epistemically	The teacher expresses his epistemic representation, devaluing a potential student epistemic work.
Present an inadequate task	There is an implicit task (e.g., requesting students' attention to approach a new subject) or there is an explicit request for some kind of activity that does not require epistemic work.
Present task challenge	The teacher proposes a task that students face as a challenge to develop investigative activities.
Synthesize	Teacher incorporates students' results to reinforce their understanding.
Demand additional aspects	The teacher demands additional aspects from students, helping them to move on in their work or helping them to clarify some aspects.
Give autonomy	The teacher gives students opportunities to develop their work independently, but providing the necessary epistemic support.
Give information	The teacher provides the students with some information on the operational aspects of the task.
Valorize epistemically	The teacher uses ideas, schemes, or actions of students, recognizing their epistemic value (e.g., hypotheses, predictions, etc.).
Make explicit epistemically	Teacher makes explicit the epistemic nature of the work done or ongoing.
Introduce and/or use a mediator artifact	Teacher introduces and/or uses a computational simulation or another artifact with materiality to be handled and worked to improve the conceptual understanding.

plish the tasks and if their actions were similar to what scientists do in the context of science production (Sandoval, 2005; Sandoval & Millwood, 2005). The MN content analysis allowed the identification, based on empirical evidences, of several epistemic practices that took place in the classroom. Some of these epistemic practices are similar to others already identified in literature about epistemic practices (Berland et al., 2016; Kelly, 2008), but some are more detailed and based on an understanding of what students are actually doing and/or discussing with their peers. Good examples of such practices are those that distinguish between the handling of equipment factually or conceptually. In the first case, we found elements that allowed us to understand that when students adopt some technical procedures with equipment, mobilizing prior knowledge is quite different from doing it only guided by teacher instructions, for example. Also, the distinction between questioning in order to clarify terms or issues related to empirical conditions of a given phenomenon and questioning to clarify terms or comments, mobilizing prior knowledge, is very important to clarify that are epistemic practices that are closer to facts and events and others to theories or models.

For the five cases in this study, the occurrence of several epistemic practices was identified, but most of them are closer to the world of observable facts, objects, and events; as some authors claim, they are from the world of objects and events (Tiberghien, 1996). Only during the lessons taught by teachers that adopted some teacher mediation actions—like presenting the task as a challenge, demanding additional aspects, or valorizing the work that students were developing—did the number of epistemic practices, and the diversity of such practices, increase. As the results presented demonstrate, the occurrence of a more significant number of epistemic practices that allow students a higher abstraction or withdrawal in relation to the observable facts is connected with teacher mediation actions that allow an epistemic scaffold to students' work. From our analysis, during the lessons where the teacher adopted mediation actions from that type, the number and diversity of epistemic practices was greater, and a good proportion of them were closer to the world of theories and models (Tiberghien, 1996). A good example of such epistemic practices is the moments when students, either independently or supervised by a teacher, analyze, argue, and make a critical evaluation of hypotheses or resources.

Taking into account the differences identified among the five study cases in terms of the development of students' epistemic practices, it becomes clear that teachers with more professional experience more often adopt actions that can promote, sustain, and enlarge students' epistemic practices. During their lessons, students have more opportunities to develop a larger number of epistemic practices, since they provide them with a classroom environment more conducive to the development of a larger number and diversity of practices of knowledge construction. These teachers also adopt more actions associated with the occurrence of a larger number of epistemic practices in the field of production, communication, and evaluation of knowledge. The presentation of the task and the autonomy they grant to their students are the most crucial actions they can adopt toward the promotion of students' epistemic work, as already published in previous studies (Lopes, Cravino, Branco, Saraiva, & Silva, 2008; Lopes, Cravino, et al., 2012; Saraiva, 2017b; Saraiva, Lopes, et al., 2012). The presentation of a task as a challenge to students allows students to have an acceptable control over what you need to do in order to accomplish it (Lopes et al., 2008). From the content analysis of MNs, we identified that when teachers do not adopt mediation actions like presenting the task as challenge to students, demanding additional aspects that will help students to go further in their work, or providing the necessary autonomy that will allow an acceptable control over what they need to do, the number of epistemic practices decreases, as does their diversity. Also, when that happens and teachers do not valorize the students' work from an epistemic point of view, they only develop lower level epistemic practices and do not develop the necessary pathway that will lead them to the development of epistemic practices closer to the world of theories and models. This only allows students to produce knowledge in reference to the objects and events that they are working with (Pinto et al., 2012). Teacher mediation actions have a large impact on students' epistemic practices and on the development of a set of epistemic practices that approach the knowledge constructed according to the world of theories and models (Tiberghien & Buty, 2007), which requires a higher level of abstraction. These results are consistent with those reported by several authors (McNeill & Krajcik, 2009; McNeill, Lizotte, Krajcik, & Marx, 2006; Pinto et al., 2012) suggesting that teachers' mediation actions are determinant in scaffolding students' development of an epistemic pathway that includes practices that are closer to the facts and events that student already know, and practices that are gradually more distant from that and are closer to the world of theories and models (Tiberghien & Buty, 2007). Different teacher mediation actions will lead to the occurrence of a different number, diversity, and level of epistemic practices (Pinto et al., 2012; Saraiva, 2017a).

CONCLUSION

As widely described during this chapter, MNs can be useful as a research tool and not only as an organizing element for all the data collected from the classroom. Through the use of MNs as a research instrument, we can achieve a good level of understanding about what students actually do in the classroom and what they are doing in order to progress their conceptual understanding. Also, MNs allow researchers access to teachers' requests and demands for students' actions, how they propose the tasks, and if they give them the necessary autonomy and support that will allow them the construction of knowledge in the classroom. Considering these important facts highlighted in the description of this empirical study, the use of MNs as a research instrument allows us to comprehend how teacher actions can promote, scaffold, and enlarge students' epistemic work. The content of MNs also provides researchers with sustainable evidences about what students are effectively doing during each moment of their task development. This is very important to achieve a better understanding of the factors that influence students' development of epistemic practices in the classroom.

FUTURE RESEARCH ABOUT STUDENTS' EPISTEMIC PRACTICES

The importance of students' epistemic practices for the teaching and learning of science, as extensively discussed throughout this chapter, justifies the need for teachers' development of professional knowledge that enables them to develop their mediation efforts toward the promotion of these practices in their classroom. Thus, research should produce more knowledge in this field based on data collected from the natural environment of the classroom and from the national context. This knowledge needs to be shared and placed at teachers' service.

MNs relate the events which take place in the classroom, easily allowing readers' identification with the facts that are described and narrated, some of them in the first person, when the MN is elaborated by the teachers themselves. The MN, as a research instrument, allows the identification of all actions adopted by the teacher in the context of their mediation, including those with less potential for promoting students' epistemic work. In this way, an MN can amplify knowledge about teachers' mediation actions and their influence on students' epistemic practices and can be an important contribution to research for teachers' professional learning and development.

Research already developed in this field (Cunha et al., 2012; Lopes, Cravino, et al., 2012; Lopes et al., 2013; Lopes, Silva, et al., 2012; Saraiva, Cunha, et al., 2012; Saraiva, Lopes, et al., 2012) has used the MN as a research instrument and concerns the teaching of physics and chemistry at all school education levels, referring to lessons taught by teachers with different levels of professional development in terms of education, teaching experience, and research experience in the field of science education. The lessons that we analyze through the correspondent MNs have distinct objectives and didactic intentionality. We have lessons in which students develop inquiry, with and without recourse to computer simulations; we also have content which is more expositive, where teachers only introduce demonstrations with instrumental apparatus. However, we do not yet have MNs that cover a complete didactic unit, which could allow us to assess students' learning by recourse to other instruments, in order to complete the information collected by the narrations. It would be interesting for us to assess student learning outcomes in a situation of pre-and post-education, taking into account the teaching of a complete didactic unit. Also, it could be interesting to assess whether the learning is perpetuated and strengthened over time. In this way, it would be possible to combine the information collected through MNs about the epistemic competences developed by students.

Another potentiality of MNs as a research instrument about students' epistemic practices would be their use in the context of supervision of teaching practices or training programs for teachers' professional development. The MN is a powerful instrument for reflective professional development, as it combines writing as a reflective activity while acting as a catalyst for the reflective process and consequent professional learning (Lopes et al., 2010, p. 22). Thus, another line of research with potential interest for the use of MNs is research into professional development, concerning the change in teacher practice focused only on the transmission of facts and skills to teacher practice that seeks to lead students to better understand how scientific knowledge is built (Duschl & Grandy, 2013; Krist, 2016; Osborne, 2014) and the conditions that make that knowledge valid.

REFERENCES

Berland, L. K., Schwarz, C. V., Krist, C., Kenyon, L., Lo, A. S., & Reiser, B. J. (2016). Epistemologies in practice: Making scientific practices meaningful for students. *Journal of Research in Science Teaching, 53*(7), 1082–1112. doi:10.1002/tea.21257

Bogdan, R., & Biklen, S. (1994). *Investigação qualitativa em educação*. Porto, Portugal: Porto Editora.

Cohen, L., Manion, L., & Morrison, K. (2000). *Research methods in education* (5th ed.). London: Routledge.

Cunha, A. E., Saraiva, E., Dinis, F., Anacleto, A., Pires, C., Agostinho, A., . . . Lopes, J. B. (2012). The influence of the teacher's research experience in his mediation of secondary school students learning using computer simulations. In R. Pintó, V. López, & C. Simarro (Eds.), Learning science in the society of computers (pp. 11–19). Barcelona, Spain: Centre for Research in Science and Mathematics Education (CRECIM).

Duschl, R. A., & Grandy, R. (2013). Two views about explicitly teaching nature of science. *Science & Education, 22*(9), 2109–2139. doi:10.100711191-012-9539-4

Engle, R. A., & Conant, F. R. (2002). Guidding principles for fostering produtive disciplinary engagement: Explaning an emergent argument in a community of learners classroom. *Cognition and Instruction, 20*(4), 399–483. doi:10.1207/S1532690XCI2004_1

Jiménez-Aleixandre, M. P., & Bustamante, J. D. (1997, March). *Analysing classroom discourse: Practical work in the biology laboratory*. Paper presented at the American Educational Research Association (AERA) annual meeting, Chicago, IL.

Jiménez-Aleixandre, M. P., & Crujeiras, B. (2017). Epistemic practices and scientific practices in science education. In K. S. Taber & B. Akpan (Eds.), *Science education. New directions in mathematics and science education* (pp. 69–80). Rotterdam, The Netherlands: Sense Publishers. doi:10.1007/978-94-6300-749-8_5

Kelly, G. J. (2008). Inquiry, activity, and epistemic practice. In R. Duschl & R. Grandy (Eds.), *Teaching Scientific Inquiry: Recommendations for Research and Implementation* (pp. 99–117). Rotterdam: Sense Publishers.

Kelly, G. J., & Duschl, R. A. (2002, April). *Toward a research agenda for epistemological studies in science education*. Paper presented at the Annual meeting of the National Association for Research in Science Teaching, New Orleans, LA.

Kelly, G. J., McDonald, S., & Wickman, P.-O. (2012). Science learning and epistemology. In B. Fraser & K. Tobin (Eds.), *Second international handbook of science education* (pp. 281–291). Dordrecht, The Netherlands: Springer. doi:10.1007/978-1-4020-9041-7_20

Krippendorff, K. (2004). *Content analysis: An introduction to its methodology*. Newbury Park, CA: SAGE Publications.

Krist, C. (2016). How a 6th grade classroom develops epistemologies for building scientific knowledge. In C.-K. Looi, J. Polman, U. Cress, & P. Reimann (Eds.), *Proceedings of the international conference of the learning sciences (ICLS 2016)* (Vol. 2, pp. 306–313). Singapore: International Society of the Learning Sciences.

Lopes, J. B., Branco, M. J., & Jiménez-Aleixandre, M. P. (2011). 'Learning experience' provided by science teaching practice in a classroom and the development of students' competences. *Research in Science Education, 41*(5), 787–809. doi:10.100711165-010-9190-5

Lopes, J. B., Cravino, J. P., Branco, M., Saraiva, E., & Silva, A. A. (2008). Mediation of student learning: Dimensions and evidences in science teaching. *PEC 2008: Problems of Education in the 21st Century, 9*(9), 42–52.

Lopes, J. B., Cravino, J. P., Silva, A. A., & Viegas, C. (2012, June). *The role of teacher mediation using computer simulations to improve students' epistemic competences in physical sciences: A theoretical framework.* Paper presented at the 10th International Conference on Computer Based Learning in Science (CBLIS), Barcelona, Spain.

Lopes, J. B., Cunha, A. E., Santos, C. A., Saraiva, E., Cravino, J. P., & Dinis, F. (2013). Envolver os alunos produtivamente em aulas de física e química durante o uso de simulações computacionais: Dois professores com mediações distintas e uso distinto das simulações. *Sensos, 4*(2), 121–137.

Lopes, J. B., Silva, A. A., Cravino, J. P., Santos, C. A., Cunha, A. E., Pinto, A., ... Branco, M. J. (2014). Constructing and using multimodal narratives to research in science education: Contributions based on practical classroom. *Research in Science Education, 44*(3), 415–438. doi:10.100711165-013-9381-y

Lopes, J. B., Silva, A. A., Cravino, J. P., Viegas, C., Cunha, A. E., Saraiva, E., & Santos, C. A. (2010). *Investigação sobre a mediação de professores de Cências Físicas em sala de aula.* Vila Real, Portugal: Universidade de Trás-os-Montes e Alto Douro.

Lopes, J. B., Silva, A. A., Cravino, J. P., Viegas, C., Cunha, A. E., Saraiva, E., ... Santos, C. A. (2012). Instrumentos de ajuda à mediação do professor para promover a aprendizagem dos alunos e o desenvolvimento profissional dos professores. *Sensos, 2*(1), 125–171.

McNeill, K. L., & Krajcik, J. (2009). Synergy between teacher practices and curricular scaffolds to support students in using domain-specific and domain-general knowledge in writing arguments to explain phenomena. *Journal of the Learning Sciences, 18*(3), 416–460. doi:10.1080/10508400903013488

McNeill, K. L., Lizotte, D. J., Krajcik, J., & Marx, R. W. (2006). Supporting students' construction of scientific explanations by fading scaffolds in instructional materials. *Journal of the Learning Sciences, 15*(2), 153–191. doi:10.120715327809jls1502_1

Mortimer, E. F., & Araújo, A. O. (2014). Using productive disciplinary engagement and epistemic practices to evaluate a traditional Brazilian high school chemistry classroom. *International Journal of Educational Research, 64*, 156–169. doi:10.1016/j.ijer.2013.07.004

Osborne, J. (2014). Teaching scientific practices: Meeting the challenge of change. *Journal of Science Teacher Education, 25*(2), 177–196. doi:10.100710972-014-9384-1

Pinto, A., Barbot, A., Viegas, C., Silva, A. A., Santos, C. A., & Lopes, J. B. (2012). Teacher education using computer simulations pre and in-service primary school teacher training to teach science. In R. Pintó, V. López, & C. Simarro (Eds.), Learning science in thes of computers (pp. 28–36). Barcelona, Spain: Centre for Research in Science and Mathematics Education (CRECIM).

Reiser, B. J. (2004). Scaffolding complex learning: The mechanisms of structuring and problematizing student work. *Journal of the Learning Sciences, 13*(3), 273–304. doi:10.120715327809jls1303_2

Sandoval, W. A. (2005). Understanding students' practical epistemologies and their influence on learning through inquiry. *Science Education, 89*(4), 634–656. doi:10.1002ce.20065

Sandoval, W. A., & Millwood, K. A. (2005). The quality of students' use of evidence in written scientific explanations. *Cognition and Instruction, 23*(1), 23–55. doi:10.12071532690xci2301_2

Saraiva, E. (2017a). *Estudo do papel da representação visual no contexto da mediação dos professores de ciências físicas* [Study of the role played by visual representation in the context of teacher mediation of physical sciences] (Doctoral thesis). Universidade de Trás-os-Montes e Alto Douro, Vila Real, Portugal.

Saraiva, E. (2017b). Visual representations in the classroom when teaching physical sciences. In J. B. Lopes, J. P. Cravino, E. S. Cruz, & A. Barbot (Eds.), *Teaching science: Contributions of research for planning, practice and professional development* (pp. 227–246). New York, NY: Nova Publishers.

Saraiva, E., Cunha, A. E., Santos, C. A., Lopes, J. B., & Cravino, J. P. (2012). Papel da mediação do professor na promoção de trabalho epistémico dos alunos durante o uso de simulações computacionais. In C. Portela & T. Peña (Eds.), *Física 2012: Atas do XXII Encontro Ibérico para o Ensino da Física* (pp. 51–58). Lisboa, Portugal: Sociedade Portuguesa de Física.

Saraiva, E., Lopes, J. B., Cravino, J. P., & Santos, C. A. (2012). How do teachers of physical sciences with different professional experience use visual representations with epistemic functions in the classroom? *PEC 2012: Problems of Education in the 21st Century, 42*(42), 97–114.

Stake, R. E. (2013). *Multiple case study analysis*. New York, NY: Guilford Press.

Tiberghien, A. (1996). Construction of prototypical situations in teaching the concept of energy. In G. Welford, J. Osborne, & P. Scott (Eds.), *Research in science education in Europe* (pp. 100–114). London: The Falmer Press.

Tiberghien, A., & Buty, C. (2007). Studying science teaching practices in relation to learning: Time scales of teaching phenomena. In R. Pintó & D. Couso (Eds.), *Contributions from science education research* (Vol. 2, pp. 59–75). Dordrecht, The Netherlands: Springer. doi:10.1007/978-1-4020-5032-9_5

Chapter 14
Researching Science and Technology Teachers' Decisions Through Multimodal Narratives

Carla Aguiar dos Santos
Universidade de Trás-os-Montes e Alto Douro, Portugal & Centro de Investigação em Didática e Tecnologia na Formação de Formadores, Portugal

ABSTRACT

This chapter aims to identify certain interaction dynamics between pedagogical decisions and students' epistemic practices (EPs) that occur during science and technology lessons conducted by teachers at two different teaching levels. A content analysis was undertaken of multimodal narratives (MNs) of lessons based on two case studies of secondary and higher education teachers. MN excerpts are used to illustrate the interaction dynamics between pedagogical decisions and students' EPs for each teacher. Results show that the secondary education teacher makes more pedagogical decisions than the higher education teacher and that the secondary school students engage in fewer EPs than the higher education students. The results also show that it is possible to use MNs as an instrument to develop research on teachers' pedagogical decisions. Teachers' pedagogical decisions are an important asset for teacher professional development as they have an impact on students' epistemic work in the physical science classroom.

INTRODUCTION

Decisions are present in almost everything that humans do, either in their personal or professional lives (Edwards & Fasolo, 2001). Teachers are no exception and as noted by Shavelson (1973), any act of teaching is a decision, so it is easily inferred that the teacher makes many decisions throughout the entire teaching process (Shavelson & Stern, 1981). Teachers' decisions include three distinct phases, which take place at different moments in the teaching process (e.g., Shavelson & Stern, 1981). They can occur in the lesson planning phase, during the interaction in the classroom, or after it has happened (i.e., after the lesson) (e.g., Shavelson & Stern, 1981). In this study, we focus only on the teachers' decisions that take place inside the classroom, which for simplification we call "pedagogical decisions."

DOI: 10.4018/978-1-5225-8570-1.ch014

Students' epistemic practices (EPs) are important in developing competences (Lopes, Branco, & Jiménez-Aleixandre, 2011), and for promoting the construction of scientific knowledge (e.g., Berland et al., 2016; Driver, Asoko, Leach, Scott, & Mortimer, 1994) in the science and technology classroom; this can effectively be used in a huge variety of contexts or situations (Kirschner, van Vilsteren, Hummel, & Wigman, 1997).

The study of pedagogical decisions and students' EPs requires data analysis related to the practice of teaching, which generates a large amount of data, making its analysis very time consuming. Grouping all the data collected in the classroom and turning them into a single analytical instrument is possible through the use of multimodal narratives ([MNs], Lopes et al., 2014), facilitating research. Similarly, one may exploit MNs as a tool preserving the holistic nature of teaching and the natural environment of the classroom (Lopes et al., 2014). The great added value of MNs, which we wish to highlight in the context of this work, is that they have the same structure, they are focused on the development of the proposed tasks in the classroom, they are narrated from the perspective of the teacher, and they enable comparison of multiple case studies (Lopes et al., 2014). Even if they are elaborated by different teachers or researchers, as they have the same structure, they are comparable and verifiable (Lopes et al., 2014), facilitating the adoption of methodological approaches such as content analysis. In this particular case, we intend to investigate the ways in which the use of MNs allows access to information regarding all the actions and decisions of each of the teachers and the students' EPs at the different teaching levels in the context of the classroom.

Teachers' pedagogical decisions are an important asset for teacher professional development since they have an impact on students' epistemic work in the physical science classroom. We studied lessons that took place in identical teaching contexts with privileged inquiry methodologies, and in which experimental work took place. The interpretative analysis (Cohen, Manion, & Morrison, 2011) of this research was undertaken through the MNs related to the lessons taught by two teachers in the area of physical sciences. The study of the interactions between teacher's pedagogical decisions and the students' EPs at different teaching levels can help to understand the mutual influence of these interactions.

Thus, the objective of this study was to identify the interaction between pedagogical decisions adopted by teachers and the students' EPs in two different teaching contexts. We present excerpts from the MNs that highlight these pedagogical decisions and the students' EPs.

BACKGROUND

According to Kennedy (2006), during the planning phase, the teacher imagines how the class will be conducted and how all associated actions will occur. However, no matter the extent to which a teacher plans the lesson, there are always decisions that have to be taken in the classroom to address challenging situations that arise (Aho, Haverinen, Juuso, Laukka, & Sutinen, 2010), regardless of the teaching level. However, given the specificities of each teaching level, there may be a set of decisions more characteristic of one level than others (Santos, Lopes, & Cravino 2011).

Pedagogical decisions are not always interpreted by teachers as a decision, but as their way of acting in the classroom. Sometimes the teacher takes decisions without being aware of the act, but even when taken spontaneously, such action is characterized as a result of a decision (Shavelson, 1973). This can be explained by the fact that teachers during their work act "without thinking about it" (Roth, Masciotra, & Boyd, 1999, p. 771). It may also happen because the act of deciding is inherent to every human and it is so ingrained that one does not realize a decision is being taken. Thus, we can say that the decisions that teachers take, conscious or unconscious, can affect their own behavior (Duschl, & Wright, 1989), and will be representative of their beliefs (e.g., Bryan, 2012). Several studies have shown that in general science teachers have beliefs about teaching and learning that influence their classroom practice (Bryan, 2012). Through the use of MNs, teachers can be aware of the pedagogical decisions taking place in their lesson, demystifying their beliefs, and opening up new opportunities to improve teaching practices.

Pedagogical decisions are made in real time to provide answers to certain situations that occur in the classroom (e.g., Cooper, 2014; Zodik & Zaslavsky, 2008). The information that the teacher collects during the lesson, through observations, is related to the students' learning, behavior, or motivation, as well as to the operation of the class itself. This information that the teacher collects needs to be processed very quickly because the teacher's pedagogical decisions are made quickly and under uncertain conditions (Cooper, 2014). However, even if they are quickly taken, these decisions can be useful (Klein, 2008).

Pedagogical decisions are taken to give an answer to observations (Clark & Peterson, 1986) of unexpected situations (e.g., Cooper, 2014; Zodik & Zaslavsky, 2008), i.e., situations that the teacher did not anticipate would occur during the lesson; through the analysis of the MNs, it was possible to verify this aspect. Thus, MNs are an asset not only for research, but also for professional development (Schwarz, 2001), especially if they are written by the teacher him/herself, since they are a narrative of what happened in the classroom (Lopes et al., 2010), allowing reflection (Cunha, 2015; Lopes et al., 2010). The use of an MN allows a process of self-reflection, which increases the teacher's willingness to improve his/her actions in the classroom.

Teachers of science and technology need to develop professional knowledge that allows them to be attentive to the various aspects that occur in the classroom, collecting at all times information that allows them to make real-time and assertive pedagogical decisions in order to provide better help to their students (Kennedy, 1998). Being more attentive and gaining a better understanding of what motivates and is involved in making pedagogical decisions can help teachers to improve their practice and be more effective in the support they provide to their students (Schoenfeld, 2015).

To increase students' scientific knowledge, it is necessary for their engagement to be meaningful and productive (Berland et al., 2016). This aspect is essential for students to develop a deep understanding of the nature of science (Krist, 2016; Osborne, 2014). That is, to improve students' learning, it is necessary to engage in epistemic activities. For this to happen, the role of teacher mediation is critical in creating and sustaining a learning environment for students (Osborne, 2014). Understanding their pedagogical decisions can help teachers to be more effective (Schoenfeld, 2015), and consequently improve students' learning (Prachagool, Nuangchalerm, Subramaniam, & Dostál, 2016). There is much more to learn about the process of decision making and how better to use this knowledge (Jonassen, 2012). Also, one can evaluate the effectiveness of teachers' pedagogical decisions by studying, in the context of the physical science classroom, which of these decisions promote certain (or intended) EPs among students (Santos, 2017).

As students interact with the teacher, they have an impact on the decisions made (Badreddine & Buty, 2011). Thus, the interactions between teacher's pedagogical decisions and students' EPs that occur in the classroom can be mutually influenced.

Using MNs to narrate every event and/or incident in the classroom has great potential to develop teachers' awareness of their own actions and pedagogical decisions. Since MNs can give us an account of teaching in the teacher's own voice, they can be a powerful tool for teachers' professional development. MNs are also a powerful tool for research and their content can have a great "significance for professional learning and instructional decision making" (Little, 2007, p. 217).

Everything that a teacher does in the classroom is very important since it influences the students' learning process (Wright, Horn, & Sanders, 1997). Teacher's pedagogical decisions can promote more and better EPs among students (Santos, 2017). Therefore, being aware of the pedagogical decisions that are taken may open up new opportunities to improve teaching practices and be of great relevance for teachers' professional development, consequently improving the students' learning through the impact on their EPs. When one understands how something is done, one can show others how to do it (Schoenfeld, 2015).

METHODOLOGY

As this study intended to identify the interaction between the pedagogical decisions adopted by teachers and the students' EPs at two different teaching levels, we selected MNs from lessons taught by teachers with comparable academic and professional experience. Also, we selected MNs from lessons in which the teachers adopted similar teaching methodologies: Students were involved in experimental work and worked collaboratively in small groups to reach the solution to a problem. We also selected teachers with similar research experience, since they had all participated in research projects. The MNs analyzed were all constructed by the teachers, using data collected using several methods and techniques. Adopting this procedure, we intended to control for important variables, since we were focused on identifying the decisions of science and technology teachers at different teaching levels (i.e., secondary and higher education).

The teaching methodology used was research-based and inquiry-based teaching, during which students performed experimental activities. The study comprised an interpretative and qualitative analysis of physical science lessons through two case studies (Cohen, Manion, & Morrison, 2011). After collecting data on the lessons of the two teachers using several techniques and instruments, each teacher wrote MNs (Lopes et al., 2014). The MNs were then subjected to a content analysis process in NVivo 8® software. Since the purpose of our research was to identify the pedagogical decisions that occur during science and technology teachers' lessons at different teaching levels and the respective students' EPs, these were made explicit using some excerpts from the MNs analyzed.

Participants

In this study, the lessons of a secondary teacher and a higher education teacher, both in physical sciences, were analyzed. In each of the teachers' classes, students undertook research work organized in small groups. The main characteristics of the lessons of these two teachers are presented in Table 1.

Data Collection

Over several lessons, each teacher collected multimodal elements, such as audio records, photocopies of students' notebooks, teaching documents, explicit teachers' intentions and decisions, students' reactions (silence, gestures, etc.), amongst others. After this, each teacher developed his own MN (Lopes et al., 2014). The MN is used as a central component of the hermeneutic unit that encompasses all types of

Table 1. Main characteristics of the classes in the study

	Teacher A (students in year 11)	Teacher B (students in higher education)
Class topic	Long-distance communication	Didactics of the nature of science
Teaching context	Physics (inquiry and laboratory)	Teachers' training in general sciences (inquiry and laboratory)
Average number of students per lesson	18	14

data collected inside and outside the classroom (Lopes et al., 2014). Also, the MN is a narrative thread that can be verified by consulting the audio record of the lesson and other multimodal elements (Lopes et al., 2014). The MNs were also submitted to a validation conducted by external experts, who used the multimodal elements and the MN construction protocol to assess the accuracy and readability of the MN, and verify whether the MNs were self-contained or not. If necessary, the MNs were improved until an acceptable final version was obtained.

Data Analysis

The teachers' pedagogical decisions and students' EPs were used as dimensions of analysis to identify the interaction dynamics. Based on content analysis (Krippendorff, 2004), each MN was analyzed by a researcher using qualitative analysis software (NVivo 8®). From the MNs, variables were identified using open code analysis (Böhm, 2004; Cohen, Manion, & Morrison, 2011), taking into account other categories identified in previous studies on pedagogical decisions (e.g., Santos, 2017; Santos, Lopes, & Cravino, 2011) and students' EPs (e.g., Pinto et al., 2014; Santos, 2017). Although there are existing studies that provide such categorizations, new categories can emerge when analyzing an MN. Each excerpt can be described by various encodings and provides information about the teacher's pedagogical decisions taken at a given moment of students' work. After accomplishing this process, each encoding is given the status of a category; that is, each excerpt is characterized by the presence or absence of each of the categories found by the open code analysis.

After the content analysis of the MNs, the occurrences of each category of pedagogical decision were counted, i.e., for all interactions, the number of pedagogical decisions adopted was identified. We also counted the number of students' EPs taking place in the classroom.

Finally, the interpretative analysis of the interaction dynamics between the pedagogical decisions and students' EPs were illustrated through excerpts. In this study, we present some excerpts from the MNs containing the categories of different pedagogical decisions and the interactions with students' EPs as an example. However, since any act of teaching is the result of a decision (Shavelson, 1973), many more excerpts could have been presented, but the large number of excerpts generated by the analysis make this impractical.

Results

First, we present the categories determined by the open code analysis for teachers' pedagogical decisions and students' EPs in Tables 2 and 3 respectively. We then present two graphics showing the number of occurrences for pedagogical decisions (Figure 1) and students' EPs (Figure 2). This is followed by

Table 2. Categories of teachers' pedagogical decisions identified through the MNs

Categories	Brief definition
Allow autonomous pathway	The teacher allows the students to follow an autonomous pathway when accomplishing the task.
Allow student's influence	The teacher allows the work of a student or group of students to influence the class work.
Ask for clarification	The teacher asks the students to clarify what was communicated, or what they did, or to clarify the reasoning used.
Ask presentation of work	The teacher asks a student or group of students to present work or ideas.
Encourage class	The teacher encourages the class to complete the task or helps the students find a way to execute the task and persist in its execution.
Encourage student	The teacher encourages a student or a group of students to complete the task, or helps the student or group find a way to execute the task and persist in its execution.
Expose to class	The teacher transforms a doubt/question from a student to a question for the whole class, or addresses the class concerning aspects found by some students.
Give clues	The teacher gives hints, clues, or suggestions that he considers necessary for the students to progress independently in accomplishing the task, or uses questioning to lead the students to understand their reasoning or what they are doing.
Give information	The teacher provides information to students, but without removing the autonomy in task accomplishment.
Make synthesis	The teacher provides a synthesis of students' answers.
Manage conditions of task accomplishment	The teacher manages the conditions to accomplish the task, for example, improving or changing the resources for the task, or providing more resources, or changing the course of the task based on the students' behavior to ensure task accomplishment.
Manage task	The teacher supplies information to the students in order to expedite their actions or allow the collective work to potentially benefit from differences found in the work of each student or group.
Manage time	The teacher manages the accomplishment of the task by giving more time or skipping to another task due to lack of time, or reminding the students to manage their time to accomplish the task.
Manage work	The teacher suggests guidelines to the students to enable more efficient ways of performing the work and expedite task accomplishment, or requires the students to participate actively.
Monitor	The teacher follows the students' actions to verify their engagement and/or to lead their learning.
Rebuke	The teacher rebukes a student for anomalous behavior or draws attention to the distraction of the student.
Return the issue	The teacher remands to student question, problems, or requests for help.
Set or reformulate task	The teacher asks something specific in the context of the task or reformulates the task.

exemplar excerpts to illustrate the interaction dynamics between pedagogical decisions and students' EPs for each teacher through the use of MNs.

Figure 1 presents the number of occurrences of pedagogical decisions for each of the teachers during a didactic sequence.

From Figure 1 we can observe that the teachers in this study, and for these lessons, put into practice 18 types of pedagogical decisions during their lessons. Of these, 16 are common to both teachers, although the number of occurrences differs. The total number of occurrences of pedagogical decisions made by teacher A is much higher than for teacher B. Also, for these two teachers "ask for clarification," "monitor," and "set or reformulate task" are the pedagogical decisions that show a greater discrepancy in the number of occurrences, likely due to the different teaching levels. The secondary students need more monitoring than the students of higher education, and therefore teacher A needs to intervene more frequently. Since the secondary education students have a more limited background than those in higher

Table 3. Categories of students' EPs identified through the MNs

Categories	Brief definition
Communicate autonomously	The students present ideas, results, or conclusions about their epistemic work.
Communicate non-autonomously	The students present their ideas strongly related to the teacher's suggestions or requirements.
Establish relations	The students make connections between data, variables, and/or concepts in different situations.
Evaluate critically	The students analyze, using relevant argumentation, the hypothesis, resources, models, results, or language, etc., in use.
Find information	The students search for, detect, or address relevant values of variables or parameters, and/or characteristics of the reality under study to characterize it.
Formulate hypothesis	The students make a statement based on previous knowledge to guide the epistemic work.
Handle equipment conceptually	The students manipulate the equipment in their practice guided by knowledge.
Handle equipment factually	The students manipulate the equipment following instructions given by the teacher or tentatively without any guidance based on knowledge.
Identify empirical conditions	The students recognize, refer to, or use variables or parameters that are important.
Make predictions	The students make statements about possible experimental or theoretical results, taking into account the explanation of the parameters and conditions on which the reasoning is based.
Organize information	The students register, classify, and/or structure data collection.
Present mobilizable idea	The students mobilize prior knowledge in a global idea that gives a possible way to solve the problem.
Question conceptually	The students formulate questions and problems based on knowledge, for focus on new perspectives to gain an insight of the reality under study.
Question factually	The students formulate questions to clarify the terms, observations, or empirical conditions of a phenomenon.
Use representation	The students operate with an available symbolic representation to characterize the reality under study.
Validate	The students test, experimentally or conceptually, the knowledge constructs using discussion between peers and teacher, or an experimental design.

education, teacher A felt a greater need to "ask for clarification" to force the students to think about the teaching subject. Teacher A also makes more use of "set or reformulate task" in order to provide students with more achievable tasks and enable them easily to achieve the intended goals.

Figure 1 shows that the most frequent pedagogical decisions of teacher A are "monitor," "ask for clarification," and "give clues," while for teacher B they are "give clues," "make synthesis," and "allow autonomous pathway." There are two types of pedagogical decisions taken only by teacher A: (a) "encourage student" and (b) "rebuke."

Figure 2 presents the number of occurrence of students' EPs for each of the teachers during the lessons.

From Figure 2 we can observe that the students in this study, and for these lessons, put into practice 15 types of EPs, of which 13 are common to both groups of students, despite the number of occurrences differing. The total number of occurrences of EPs for teacher A is lower than for teacher B. Also, for these two teachers "communicate autonomously" and "identify empirical conditions" show the greatest discrepancy in the number of occurrences.

Figure 2 verifies that the most frequent EPs for teacher A's students are "communicate non-autonomously," "handle conceptually," and "present mobilizable idea." Those for teacher B's students are "communicate autonomously," "communicate non-autonomously," "evaluate critically," and "identify

Figure 1. Number of pedagogical decisions per teacher

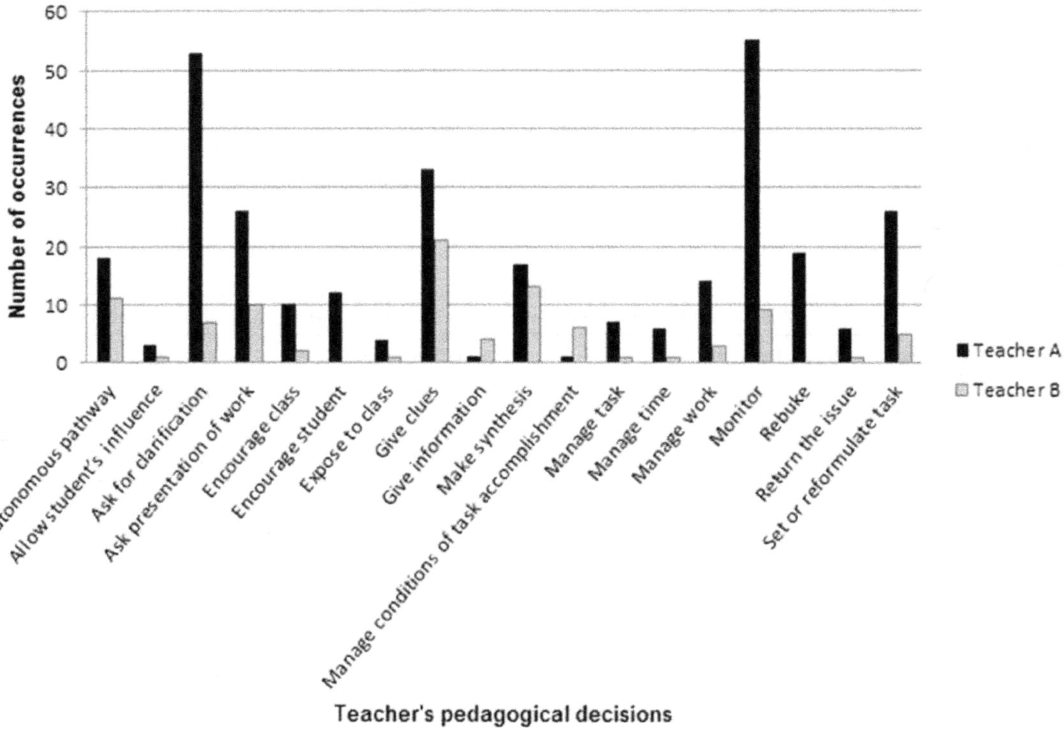

Figure 2. Number of students' epistemic practices

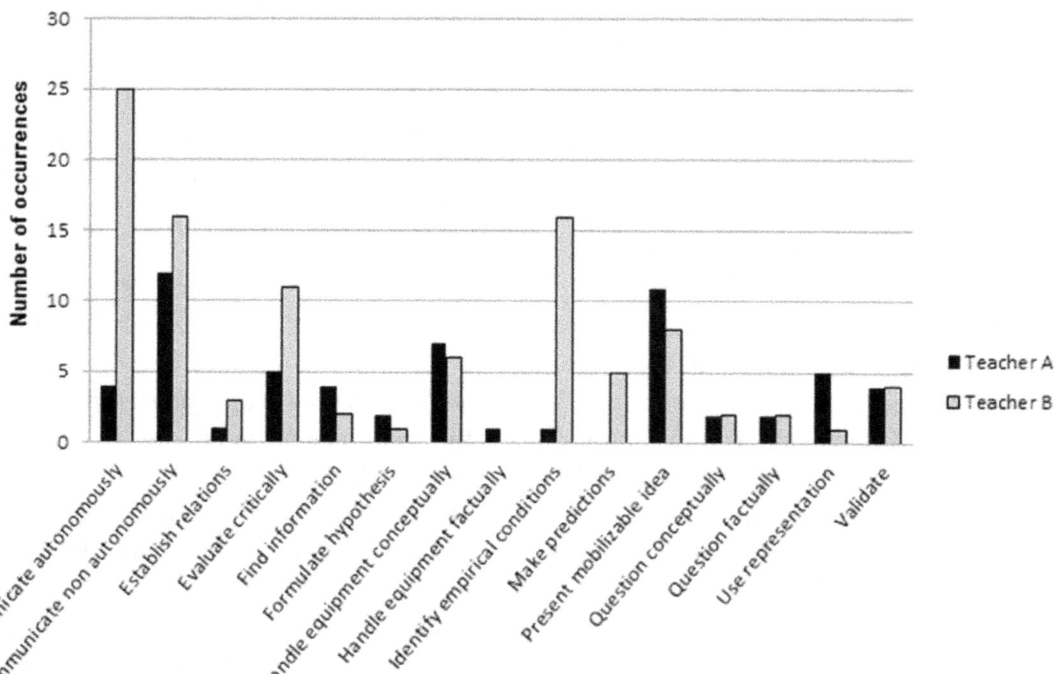

empirical conditions." The EP "handle equipment factually" occurs only once in teacher A's lesson, and the EP "make predictions" occurs only in the lessons of teacher B, likely due to the fact that higher education students have more background than secondary education students.

Dynamic Interactions: MN Excerpts

This section presents the MN excerpts that illustrate the interaction dynamics between pedagogical decisions and students' EPs for each teacher through the use of MNs. Each excerpt represents part of the MN. Before the presentation of each excerpt, the respective categorizations of the interaction dynamic are given. In the following paragraphs, the MN excerpts for teacher A are presented first, followed by those for teacher B.

In terms of the interaction dynamic in MN excerpt 1A, initially the teacher makes one pedagogical decision and there is one student EP; then the teacher asks the students what they are doing (1A) and the students do nothing in response (no EP). The respective categorizations for this interaction dynamic are follows: the teacher A initially "manages work," while students "handle equipment factually," and then teacher A "asks for clarification."

MN Excerpt 1A: Teacher A (Long-Distance Communication)

[...] I insisted students accomplish the experiment and record the results in the table for them to be discussed in class. The students were making several attempts to place the material.

Teacher: And now? I do not understand what you are doing, clarify there! I prompted the students to say what they were doing, but the students didn't say anything. [...]

MN excerpt 2.1A and MN excerpt 2.1B show that although both teachers began their pedagogical decisions with "allow autonomous pathway," and then "monitor," the students' EPs differed. In response to this epistemic work, the following teachers' pedagogical decisions differed for each teacher. In these two excerpts, we also note that teacher A needs to make two decisions simultaneously, but that there is only one EP, while teacher B only takes one decision and the students present two EPs.

The second excerpt from teacher A shows that the teacher makes six decisions and there are two students' EPs. The respective categorizations for this interaction dynamic from excerpt 2.1A are as follows: teacher A "allow autonomous pathway" and "monitor" what the students are doing ("handle equipment conceptually"). In excerpt 2.2A, teacher A "monitors" and "ask presentation of work," and the students "communicate autonomously" and "use representation." Finally, in excerpt 2.3A, the students "present mobilizable idea," and the teacher A "encourage class" and then "make synthesis."

MN Excerpt 2.1A: Teacher A (Long-Distance Communication)

[...] The students talked about the experimental procedures and handled the equipment; I limited myself to watching them. [...]

Figure 3. Scheme used by student A and designed by student B to measure the angles

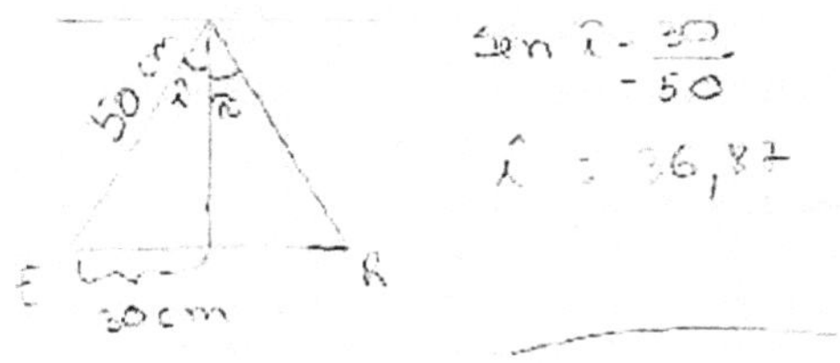

MN Excerpt 2.2A

[...] Teacher: All of you are following what your colleagues are doing? Please, go and explain to your colleagues because they may not agree...

Student A explains that by measuring the sides of the triangle composed of emitter (E), receiver (R), and the obstacle, he can identify the incident and reflection angles.

Figure 3 shows the scheme used by student A and designed by student B to measure the angles [...]

MN Excerpt 2.3A

[...] The student says that it is possible to measure the angles with a protractor.

I also accepted the second suggestion and I began to clarify each of the schemes. [...]

MN excerpts 3A and 4A show "rebuke," only employed by the secondary teacher, illustrating two types: to address anomalous behavior and to draw attention to the distraction of the student.

MN Excerpt 3A: Teacher A (Long-Distance Communication)

[...] Teacher: My dear, did you do the task? So, now be quiet and do not disrupt the lesson. [...]

MN Excerpt 4A: Teacher A (Long-Distance Communication)

[...] Teacher: Can't you anything? Are you expecting your colleagues to finish it, so you can copy it later? You know how to think by yourself, do you not? You always wait for your colleagues and just copy from them; that is very ugly, think by yourself... what does that mean? [...]

We now turn to three excerpts from teacher B. In MN excerpt 1B, the teacher from higher education "allows student influence," and there are three student EPs. The respective categorizations for this interaction dynamic from excerpt 1B are as follows: teacher B "allows student influence," giving the student the opportunity to "communicate autonomously," "use representation," and "formulate a hypothesis"; then the teacher "manages the task" to potentiate the collective work.

Figure 4. Scheme done by student on the blackboard

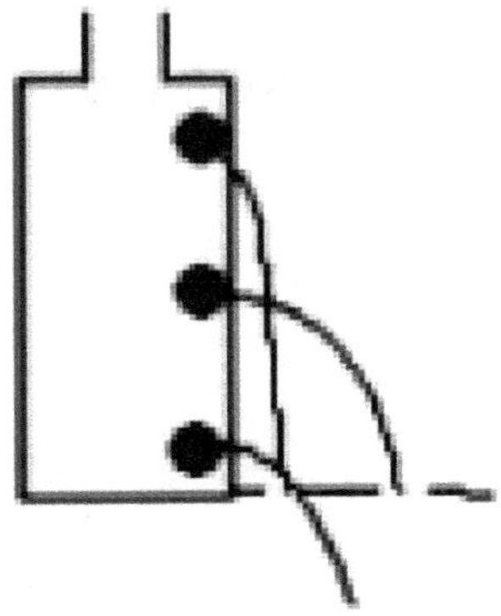

MN Excerpt 1B: Teacher B (The Longest Fountain)

[…] Student D – May I go to the blackboard?

Teacher – You can, go, draw there.

Student D did the drawing shown in Figure 4.

I took the opportunity and asked student D to write in the corner of the blackboard the concepts already mentioned. […]

In excerpts 2, teacher B makes four pedagogical decisions and there are six student EPs. The respective categorizations for this interaction dynamic are as follows: in excerpt 2.1B, teacher B initially "allows an autonomous pathway," then "monitors" the students and verifies that they are engaged in the task, "presenting a mobilizable idea" and "communicate non-autonomously"; after that, in excerpt 2.2B, teacher B "gives clues," and the student "presents a mobilizable idea" and "communicates non-autonomously"; in excerpt 2.3B, the teacher "asks for clarification," and the students "communicate non-autonomously" and "present a mobilizable idea."

MN Excerpt 2.1B: Teacher B (Shower Efficiency)

[…] The students went to the benches to work. I was following the discussions in each group.

In a group:

Student A – We are thinking of these criteria: price, quality of shower, comfort, and efficiency in saving water. We think that the last one is the most important because if it is efficient, it saves water and also saves money.

MN Excerpt 2.2B

Teacher – You have already thought about the concepts of effectiveness and efficiency?

About efficiency, it means to conjugate…

Student A – Comfort with quality of shower.

MN Excerpt 2.3B

Teacher – What do you mean by shower quality?

Student A – If it is weak material…

Student C – So it is another criterion […]

In MN excerpt 3, initially the students present three EPs and then they "make predictions." These EPs do not occur in teacher A lessons at secondary level. The respective categorizations for this interaction dynamic are as follows: in excerpt 3.1B, teacher asks students from different groups to presented their way of doing the experimental work ("ask for presentation of work"), and the students "communicate non-autonomously," "identify empirical conditions," and "evaluate critically"; in excerpt 3.2B, after listening the students, teacher B "sets or reformulates the task." Then, in excerpts 3.3B, 3.5B and 3.7B, the students "make predictions." In excerpts 3.4B and 3.6B, the teacher "gives clues"; finally, in excerpt 3.8B, the teacher "manages the conditions of task accomplishment."

MN Excerpt 3.1B: Teacher B (Shower Efficiency)

[...] Student A – We noticed that we have to mark up where we opened the water tap to keep the water pressure constant in all tests. […]

[…] Student B – We start with the bottle, but gravity does not give any pressure, nobody will take a shower like that. […]

[…] Student C – We did the same as the other group. But we opened the water tap all the way.

MN Excerpt 3.2B: Teacher B (Shower Efficiency)

Teacher – Fully opening the water tap can also take more or less time and that also introduces differences. Another issue that we have been talking about here is about regulating the pressure on the water tap knob. Let's imagine that I put a balloon on this water tap and open it in such a way that it only drips, and on the other water tap I put a similar balloon, but I open it up more. What do you think is going to happen over time?

MN Excerpt 3.3B

Student A – In the open water tap the balloon will burst. In the other it will not.

Student B – I think it's going to burst the same.

MN Excerpt 3.4B

Teacher – Why?

MN Excerpt 3.5B

Student B – Because if you wait three or four days it will be full and it won't be able to hold it any longer.

MN Excerpt 3.6B

Teacher – But you said that if you just open the water tap by a little, the water will have no pressure.

MN Excerpt 3.7B

Student B – Yeah, it won't burst.

MN Excerpt 3.8B

Teacher – Please continue the experiment so you can choose the showerhead.

I'm going to set up the balloons here on the water taps. [...]

Although it is not within the scope of this work, through these excerpts we can verify that is possible to make analyses using MNs and that they can be used for research.

DISCUSSION

The fact that all the MNs have the same structure, being focused on the tasks that students accomplish, provides information on all the actions, decisions, and intentions of the teachers; this allows comparisons between cases, even though the teachers were teaching different teaching levels, since MNs are comparable and verifiable (Lopes et al., 2014). Moreover, according Lopes et al. (2014), MNs can be analyzed with regard to several analytical teaching dimensions, for example, the students' productive engagement and epistemic work, among others.

The classroom is dynamic and complex (e.g., Larrivee, 2000; Seidel, Stürmer, Blomberg, Kobarg, & Schwindt, 2011), and there are some aspects to which the teacher needs to particular attention (Kennedy, 2006). Thus, it is expected that the MN will not contain everything that happens. In the elaboration of the MN, each narrator has his/her own writing style, emphasizing information to different degrees about the pedagogical decisions made (Santos et al., 2011). The fact that the teacher from secondary school needs to "monitor" (due to the students' age) and verify that the students cannot developing their knowledge (EPs) "obliges" the teacher to make more decisions to try to improve the students' epistemic pathway. These aspects, related to the teaching level, explain, for each teacher, the differences in the total numbers of occurrences of pedagogical decisions and students' EPs for each of the categories. This is in line with the proposal of other authors (e.g. Hattie, 2003), who note differences associated with the age level of the students; in the case of teacher A, who teaches at secondary level, this resulted in an "obligation" to pay more attention to the students individually, and to solve problems and situations raised by each student individually, and at every moment. This requires a greater number of adjustments and corrections, resulting in a greater number of pedagogical decisions being taken during the lesson, and consequently the students not being engaged in epistemic activity.

In the case of higher education, the teacher has access to more information about the class as a whole more homogeneously. Thus, it is expected that students in higher education will engage in a greater number of EPs. The higher education students are more autonomous and can develop more competences because they have more background knowledge than secondary level students. This not only determines the occurrence of fewer situations liable to lead to pedagogical decisions (hence they are less frequent), but also gives rise to the difference in the number of students' EPs for each of the variables between the different teaching levels. In this case the teacher of higher education makes less pedagogical decisions than teacher of secondary. Indeed, sometimes the teacher needs make multiple pedagogical decisions (Jonassen, 2012; Santos, 2017), until his/her influence on students is concretized (Santos, Lopes, Saraiva, & Cunha, 2015).

Another aspect that may influence the number of pedagogical decisions is the number of students. In this study, this is not significant, since the average difference in the number students per class is small (four). In addition, as shown in an earlier work (Santos et al., 2011), there are certain decisions that are more related to the teaching level than others, as is the case for "rebuke" (the teacher draws attention to anomalous behavior, or to the distraction of the student). This decision was only identified at the secondary level, most likely associated with the lack of maturity of these students versus those in higher education. In contrast, the students in higher education were found to take better advantage of the autonomy granted to them by the teacher. In the case of secondary students, even if they are granted autonomy, they need much more follow-up from the teacher to maintain engagement in work (Cunha, Saraiva, Santos, Dinis, & Lopes, 2014).

Despite this, it was possible to identify, based on the empirical evidence, a series of pedagogical decisions that could be of great value for professional development through consciousness raising. This can help teachers reflect, and according to Lloyd (2017), help them learn from their own practices. MN can work as powerful resource for teacher learning regarding pedagogical decisions and for research on these decision-making processes.

CONCLUSION

In this study, we have shown that it is possible through multimodal narratives (MNs) to identify types of teachers' pedagogical decisions and students' epistemic practices (EPs). This is accomplished through content analysis, once the events have been narrated and presented in the voice of the teacher, explaining intentions and decisions. This does not happen with other types of instruments, such as audio or video recordings.

This study aims to contribute to the enrichment of knowledge concerning science teaching practices in relation to the teacher's pedagogical decisions and the students' EPs, based on empirical evidence collected in the classroom and studied through MNs. The findings highlight that it is possible to identify pedagogical decisions through the use of MNs, allowing the teaching community to be more informed of such decisions, and enabling teachers to reflect on moments when they have been faced with similar situations in their classes. Despite the different teaching levels, the results show that these teachers adopt many types of pedagogical decisions that are common during their lessons. The results also show that the students adopt many types of EPs that are common, regardless of the different teaching levels. Although the types of decisions are identical, the teacher at secondary level needs to make more decisions than the teacher in higher education. Moreover, secondary school students present fewer EPs than higher education students.

This study illustrates how MNs can be used for research. We observe that it is possible to undertake qualitative and interpretive research using MNs as tools to serve not only the organization of the different data collected in the classroom, but also categorical content analysis because of the level of detail of the information that is narrated.

The results of this study are only related to content analysis of the MNs. Clearly, further study is needed to broaden the analysis, for example, exploring and correlating teachers' pedagogical decisions that can promote or enable students' EPs in the classroom. This can be used to gain better understanding of the students' learning and the effectiveness of pedagogical decisions. For this, it will be necessary to employ an additional type of analysis (e.g., cluster analysis) that crosses the analyses between teacher's pedagogical decisions and the students' EPs.

ACKNOWLEDGMENT

We acknowledge the doctoral fellowship SFRH/BD/101005/2014 funded by the FCT (Fundação de Ciência e Tecnologia), within the scope of POCH (Human Capital Operating Program), co-financed by the European Social Fund and by national funds of the MCTES.

REFERENCES

Aho, E., Haverinen, H.-L., Juuso, H., Laukka, S. J., & Sutinen, A. (2010). Teachers' principles of decision-making and classroom management; a case study and a new observation method. *Procedia: Social and Behavioral Sciences*, *9*, 395–402. doi:10.1016/j.sbspro.2010.12.171

Badreddine, Z., & Buty, C. (2011). Discursive reconstruction of the scientific story in a teaching sequence. *International Journal of Science Education*, *33*(6), 773–795. doi:10.1080/09500693.2010.496475

Berland, L. K., Schwarz, C. V., Krist, C., Kenyon, L., Lo, A. S., & Reiser, B. J. (2016). Epistemologies in practice: Making scientific practices meaningful for students. *Journal of Research in Science Teaching*, *53*(7), 1082–1112. doi:10.1002/tea.21257

Böhm, A. (2004). Theoretical coding: Text analysis in grounded theory. In U. Flick, E. von Kardorff, & I. Steinke (Eds.), *A companion to qualitative research* (pp. 270–274). London: Sage.

Bryan, L. A. (2012). Research on science teacher beliefs. In B. J. Fraser, K. G. Tobin, & C. J. McRobbie (Eds.), *Second International Handbook of Science Education* (pp. 477–495). New York: Springer International Handbooks of Education. doi:10.1007/978-1-4020-9041-7_33

Clark, C., & Peterson, P. (1986). Teachers' thought processes. In M. Wittrock (Ed.), *Handbook of research on teaching* (3rd ed.; pp. 255–296). New York: Macmillan.

Cohen, L., Manion, L., & Morrison, K. (2011). *Research methods in education* (7th ed.). London: Routledge.

Cooper, J. (Ed.). (2014). Classroom teaching skills (10th ed.). Cengage Learning.

Cunha, A. E., Saraiva, E., Santos, C. A., Dinis, F., & Lopes, J. B. (2014). Teacher mediation actions and students' productive engagement during the use of computer simulations in physical science classrooms. *Procedia Technology, 13*, 76–85. doi:10.1016/j.protcy.2014.02.011

Cunha, A. E. R. M. (2015). *Construção de práticas de referência no ensino da Física para o Ensino Secundário (Doutoramento)*. Universidade de Trás-os-Montes e Alto Douro, Vila Real.

Driver, R., Asoko, H., Leach, J., Scott, P., & Mortimer, E. (1994). Constructing Scientific Knowledge in the Classroom. *Educational Researcher, 23*(7), 5–12. doi:10.3102/0013189X023007005

Duschl, R. A., & Wright, E. (1989). A case study of high school teachers' decision making models for planning and teaching science. *Journal of Research in Science Teaching, 26*(6), 467–501. doi:10.1002/tea.3660260602

Edwards, W., & Fasolo, B. (2001). Decision technology. *Annual Review of Psychology, 52*(1), 581–606. doi:10.1146/annurev.psych.52.1.581 PMID:11148318

Hattie, J. (2003). Teachers make a difference: What is the research evidence? In M. Meiers (Ed.), *Building teacher quality: What does the research tell us?* (pp. 1–17). Melbourne: Australian Council for Educational Research.

Jonassen, D. H. (2012). Designing for decision making. *Educational Technology Research and Development, 60*(2), 341–359. doi:10.100711423-011-9230-5

Kennedy, M. M. (1998). Education reform and subject matter knowledge. *Journal of Research in Science Teaching, 35*(3), 249–263. doi:10.1002/(SICI)1098-2736(199803)35:3<249::AID-TEA2>3.0.CO;2-R

Kennedy, M. M. (2006). Knowledge and vision in teaching. *Journal of Teacher Education, 57*(3), 205–211. doi:10.1177/0022487105285639

Kirschner, P., van Vilsteren, P., Hummel, H., & Wigman, M. (1997). The design of a study environment for acquiring academic and professional competence. *Studies in Higher Education, 22*(2), 151–171. doi:10.1080/03075079712331381014

Klein, G. (2008). Naturalistic decision making. *Human Factors, 50*(3), 456–460. doi:10.1518/001872008X288385 PMID:18689053

Krippendorff, K. (2004). *Content analysis: An introduction to its methodology*. SAGE Publications.

Krist, C. (2016). How a 6th-grade classroom develops epistemologies for building scientific knowledge. In C.-K. Looi, J. Polman, U. Cress, & P. Reimann (Eds.), *Proceedings of the International Conference of the Learning Sciences (ICLS 2016)* (Vol. 2, pp. 306–313). Singapore: International Society of the Learning Sciences.

Larrivee, B. (2000). Transforming teaching practice: Becoming the critically reflective teacher. *Reflective Practice: International and Multidisciplinary Perspectives*, *1*(3), 293–307. doi:10.1080/713693162

Little, J. W. (2007). Teachers' accounts of classroom experience as a resource for professional learning and instructional decision making. In P. Moss (Ed.), *Evidence and decision making* (pp. 217–240). Malden, MA: John Wiley & Sons. doi:10.1111/j.1744-7984.2007.00103.x

Lloyd, C. A. (2017). Exploring the real-world decision-making of novice and experienced teachers. *Journal of Further and Higher Education*, 1–17. doi:10.1080/0309877X.2017.1357070

Lopes, J. B., Branco, J., & Jiménez-Aleixandre, M. P. (2011). 'Learning experience' provided by science teaching practice in a classroom and the development of students' competences. *Research in Science Education*, *41*(5), 787–809. doi:10.100711165-010-9190-5

Lopes, J. B., Silva, A. A., Cravino, J. P., Santos, C. A., Cunha, A., Pinto, A., ... Branco, M. J. (2014). Constructing and using multimodal narratives to research in science education: Contributions based on practical classroom. *Research in Science Education*, *44*(3), 415–438. doi:10.100711165-013-9381-y

Lopes, J. B., Silva, A. A., Cravino, J. P., Viegas, C., Cunha, A. E., Saraiva, E., … Santos, C. A. (2010). Investigação sobre a mediação de professores de ciências físicas em sala de aula [Research on physical sciences teachers' mediation in the classroom]. Vila Real, Portugal: Universidade de Trás-os-Montes e Alto Douro.

Osborne, J. (2014). Teaching scientific practices: Meeting the challenge of change. *Journal of Science Teacher Education*, *25*(2), 177–196. doi:10.100710972-014-9384-1

Pinto, A., Barbot, A., Viegas, C., Silva, A. A., Santos, C. A., & Lopes, J. B. (2014). Teaching science with experimental work and computer simulations in a primary teacher education course: What challenges to promote epistemic practices? *Procedia Technology*, *13*, 86–96. doi:10.1016/j.protcy.2014.02.012

Prachagool, V., Nuangchalerm, P., Subramaniam, G., & Dostál, J. (2016). Pedagogical decision making through the lens of teacher preparation program. *Journal for the Education of Gifted Young Scientists*, *4*(2), 41–52.

Roth, W.-M., Masciotra, D., & Boyd, N. (1999). Becoming-in-the-classroom: A case study of teacher development through coteaching. *Teaching and Teacher Education*, *15*(7), 771–784. doi:10.1016/S0742-051X(99)00027-X

Santos, C. A. (2017). The role of teacher decisions in the classroom to develop students' epistemic activity. In J. B. Lopes, J. P. Cravino, E. S. Cruz, & A. Barbot (Eds.), *Teaching science: Contributions of research for planning, practice and professional development* (pp. 119–134). Nova Science Publishers.

Santos, C. A., Lopes, J. B., & Cravino, J. P. (2011). Patterns of didactic decisions made by teachers in physical sciences classrooms. *Problems of Education in the 21th Century, 36*, 76–89.

Santos, C. A., Lopes, J. B., Saraiva, E., & Cunha, A. E. (2015). Teacher decisions and didactical intentionality using computer simulations in physical science classrooms. In L. G. Chova, A. L. Martínez, & I. C. Torres (Eds.), *9th International Technology, Education and Development Conference* (pp. 5347–5356). Madrid: IATED Academy.

Schoenfeld, A. H. (2015). How we think: A theory of human decision-making, with a focus on teaching. In S. J. Cho (Ed.), *The Proceedings of the 12th International Congress on Mathematical Education* (pp. 229–243). Seoul, South Korea: Springer International Publishing.

Schwarz, G. (2001). Using teacher narrative research in teacher development. *Teacher Educator, 37*(1), 37–48. doi:10.1080/08878730109555279

Seidel, T., Stürmer, K., Blomberg, G., Kobarg, M., & Schwindt, K. (2011). Teacher learning from analysis of videotaped classroom situations: Does it make a difference whether teachers observe their own teaching or that of others? *Teaching and Teacher Education, 27*(2), 259–267. doi:10.1016/j.tate.2010.08.009

Shavelson, R. J. (1973). *The basic teaching skill: Decision making*. Stanford, CA: Stanford University School of Education, Center for Research and Development in Teaching.

Shavelson, R. J., & Stern, P. (1981). Research on teachers' pedagogical thoughts, judgments, decisions, and behavior. *Review of Educational Research, 51*(4), 455–498. doi:10.3102/00346543051004455

Wright, S. P., Horn, S. P., & Sanders, W. L. (1997). Teacher and classroom context effects on student achievement: Implications for teacher evaluation. *Journal of Personnel Evaluation in Education, 11*(1), 57–67. doi:10.1023/A:1007999204543

Zodik, I., & Zaslavsky, O. (2008). Characteristics of teachers' choice of examples in and for the mathematics classroom. *Educational Studies in Mathematics, 69*(2), 165–182. doi:10.100710649-008-9140-6

Chapter 15
Researching One's Own Practice:
Reflections on the Teaching of Linear Algebra Using Multimodal Narratives

Ricardo Gonçalves
Instituto Politécnico do Cávado e do Ave, Portugal

Cecília Costa
Universidade de Trás-os-Montes e Alto Douro, Portugal & Centro de Investigação em Didática e Tecnologia na Formação de Formadores, Portugal

ABSTRACT

This chapter aims to link the known research results on the teaching and learning of linear algebra to research on teaching practices. The same person, as a teacher and as a researcher, plans and implements a teaching sequence, constructs multimodal narratives (MNs) for some classes, and reflects on his own practice, based on the emerging categories from content analysis of the MNs. Regarding the methodology and focusing on the role of MNs in this research, it is worth highlighting the value attributed to MNs because they function as an instrument that embodies the practices of the teacher and aid him in reflecting on his own practice.

INTRODUCTION

This chapter's theme is in the field of higher mathematics education, in particular linear algebra teaching, and the reported research is based mainly (1) on comprehensive knowledge of the literature regarding teaching and learning of linear algebra, but also (2) on the study of teaching practices of undergraduate mathematics teachers.

In the 20th century, linear algebra became a branch of mathematics and began to be taught mainly in university and colleges. At the end of the century, the learning difficulties observed in the students, the frustration of the teachers, the technological advances, and the needs of client disciplines led to interest

DOI: 10.4018/978-1-5225-8570-1.ch015

in improving the undergraduate linear algebra curriculum in several countries (Carlson, Johnson, Lay, & Porter, 1993; Dorier, 1998; Dubinsky, 1997; Tucker, 1993). The work that became better known was that conducted in the United States by the *Linear Algebra Curriculum Study Group* (LACSG) (Carlson et al., 1993). Over the last three decades, the earlier lack of studies in the field (Dubinsky, 1997) has been remedied with an extensive body of research on the teaching and learning of linear algebra, first concerning student learning and later, in the search for answers to these difficulties, with the development of new teaching strategies.

Another matter is the recurrent reference to a small number of investigations for describing and analyzing the teaching practices of undergraduate mathematics teachers (Hannah, Stewart, & Thomas, 2013; Schoenfeld, Thomas, & Barton, 2016; Speer, Smith, & Horvath, 2010). Teachers constantly investigate their own practice, reflecting on their activity. However, when they reflect uncritically on their teaching practices (or without the rigor of scientific or academic investigations), it only results in "tacit" knowledge (Cochran-Smith & Lytle, 1993) or "craft" knowledge (Artigue, 2001). This is why there is a lack of research aimed at knowing, understanding, and improving "what teachers do and think daily, in class and out, as they perform their teaching work" (Speer et al., 2010, p. 99). In this regard, the research problem we considered was how to link teaching practices to the orientations of the international movement on the teaching and learning of linear algebra. In the search for contributions to an answer, we developed an investigation focusing on the teaching practices of the teacher himself (first author).

In our opinion, the Multimodal Narrative (MN) approach is the best way of making known the work done by the teacher (and by the students). The MN is a relatively recent instrument, but has already been validated in the scientific community (Lopes et al., 2010, 2014). MNs have been used mostly to investigate physical science teaching practices, some in higher education (e.g., Lopes, Cravino, Branco, Saraiva, & Silva, 2008; Pinto et al., 2014; Santos, Lopes, & Cravino, 2011). Our research is the first to use MNs in mathematics in higher education.

Chapter 1 of this book presents MNs in detail, and in the "MN: Value, Construction, and Validation" subsection we will show how we used them in our investigation, now in the context of linear algebra. With specific reference to possible ways of teaching the concept of linear combination, and considering the suggestions in the literature ("Theoretical Background"), we will seek in the "Results" and "Conclusion" sections to uncover the emerging opportunities and difficulties faced by a linear algebra teacher. This is based on the teacher's own reflection on his practice and the content analysis of the MNs constructed for some classes.

THEORETICAL BACKGROUND

This research builds on extensive knowledge of the research results on the teaching and learning of linear algebra published over the last three decades.

Concerning learning difficulties related to linear algebra in general, Dorier and Sierpinska (2001) organized them, in a review of literature, according to conceptual difficulties (related to the nature of linear algebra) and cognitive difficulties (related to the kind of thinking required for the understanding of linear algebra). On specific learning difficulties, and pointing to the particular concept of linear combination students look at linear algebra as a catalogue of abstract concepts, commonly defined in generic terms (Harel, 1997), and "submerged under an avalanche of new words, new symbols, new definitions, and new theorems" (Dorier, Robert, Robinet, & Rogalski, 2000, p. 95). For these researchers, the problem of formalism appears mainly to be associated with the concepts of the theory of vector spaces, in which the concept of linear combination occupies a central position. In addition to the difficulty of dealing with the abstract nature of the concept, students reveal difficulties in interpreting it in more intuitive contexts, such as Cartesian geometry (Dorier et al., 2000). Insufficient knowledge of analytic geometry compromises the development of geometric intuition that can be associated with the concept of linear combination of vectors of R^n (Dorier, 1998).

It is possible to explore the concept of linear combination according to geometric, algebraic, and formal languages. However, students are not always able to explain the definition of this concept in each of the languages (Stewart & Thomas, 2009), or notice how their geometrical representation relates to algebraic representation (Sierpinska, 2000).

Students usually reveal difficulties in establishing relations between the concept of linear combination and other concepts of the theory of vector spaces. In contrast, they easily relate it to the solution of systems of linear equations, because it is part of the associated process. As a disadvantage, students force themselves on the algorithm procedure and do not recognize the advantage of understanding the concept according to the formal language (Hillel, 2000).

Regarding the teaching of linear algebra, we highlight two challenges that teachers should consider. One of these is how to link linear algebra to the digital age and the current needs of the target audience (Carlson et al., 1993). Thus, it is justified to teach the topics of linear algebra through the matrix approach, instead of the traditional axiomatic approach. Another challenge for teachers lies in teaching subjects seen as completely new. Unlike calculus, students will learn subjects that do not (apparently) continue naturally from what they have learned before (Harel, 1997). In order to help students overcome difficulties, several researchers have proposed a set of didactic recommendations for the teaching of linear algebra. According to our interpretation, they can be grouped into seven categories:

- Use of technology. Technology allows new teaching models, for example, the flipped classroom (Love, Hodge, Grandgenett, & Swift, 2014; Talbert, 2014). In addition, the use of computer algebra systems and dynamic geometry software programs allow new ways of approaching concepts, making the solution of realistic applied problems and developing geometric intuition feasible (e.g., Carlson et al., 1993; Klasa, 2010).
- The matrix-oriented approach. Synonymous with "proceed from concrete, and in many cases practical, examples to the development of general concepts, principles, and the concomitant theory" (Carlson et al., 1993, p. 42), in opposition to the axiomatic approach.
- Problem solving and excursions into applications. The goal is to introduce implicitly new linear algebra concepts through the use of modeling situations or as a tool for solving realistic applied problems, giving an indication of the pervasive use of linear algebra in many client disciplines (e.g., Cárcamo, Fortuny, & Gómez, 2017; Carlson et al., 1993; Domínguez-García, García-Planas, & Taberna, 2016; Trigueros & Possani, 2013).
- Knowledge of the prerequisites. This entails considering students' knowledge and skills previously developed as a basis for learning new concepts (e.g., Rogalski, 2000).
- Connections between the linear algebra concepts. In the presentation of definitions, it is necessary to look for the different relations between the concepts, for example, based on the construction of genetic decompositions (e.g., Trigueros & Possani, 2013) or conceptual maps (e.g., Lapp, Nyman, & Berry, 2010; Stewart & Thomas, 2009).

- Assigning a specific role to formalism. Formal concepts are introduced as the last stage of a process that begins with intuitive approaches and with a focus on prerequisites (e.g., Carlson et al., 1993; Dorier, 1998).
- Attention to language and multiple representations of concepts. This involves using, with criteria, the geometric language, articulate it with the algebraic and formal languages, and exploring the different representations of the concepts according to the language (e.g., Carlson et al., 1993; Hannah et al., 2013; Hillel, 2000).

Trigueros and Possani (2013), in the particular case of teaching the concept of linear combination, show how it can be explored based on the solution of a mathematical modeling problem. As for the different ways of relating it to other concepts, the work of Dorier (1998) constitutes one of the main references. For example, from a historical and epistemological analysis, he presents a perspective on how to look at possible linear combinations between vectors, and consequently how the concept of linear dependence and independence can become a unifying and generalizing concept.

One of the recommendations of the LACSG points to the strong emphasis on the geometric representation of the concepts of vector space theory in the context of R^n (Carlson et al., 1993), and hence the geometric exploration of the concept of linear combination in bi- and three-dimensional spaces, for example using dynamic geometry software. However, the literature argues that the exploration of the concept of linear combination is not exhausted (nor does it assume a central position) in the geometric approach (or in a single language). To demonstrate this, Aranda and Callejo (2010) give an example of how the simultaneous use of languages can be reflected in the discussion of properties, linking the idea of the collinearity and coplanarity of vectors to the possibility that one vector could be written as a linear combination of others.

Knowing the difficulties already identified led us to anticipate some of the students' behaviors in learning linear algebra and to try to mitigate some of these difficulties. At the same time, knowing the didactic recommendations allowed us to have a basis for interpretation and orientation in the development of a teaching sequence (i.e., the outline of a strategy to teach the main topics of linear algebra according to known didactic recommendations), and to form a theoretical framework for the analysis of this research, presented in the "Methodology" section.

METHODOLOGY

This research grew out of the first author's recent PhD thesis (Gonçalves, 2018), which aimed to make more consistent the results already known from the literature on research concerning the teaching and learning of linear algebra through the teacher's own reflection on his practices in teaching this subject. Employing the action research method, in the context of a qualitative approach and the critical theory paradigm, we developed one cycle in accordance with the planning–action–analysis–reflection sequence. MN was the instrument used to show the practices of by the teacher. In the research that we present in this chapter, and for the particular case of the concept of linear combination, the analysis of some MNs was developed in the search to identify means of taking advantage of the recommendations for the teaching of linear algebra.

Planning Phase

The construction of an original text of linear algebra (Gonçalves, 2015) was the way we found to show our interpretation of the didactic recommendations and to present a possible concretization. The role of this text was be to reify the teaching sequence to be implemented by the teacher in the classroom. It covered four main topics on linear algebra—matrices, systems of linear equations, determinants, and vector spaces—and the concepts that are usually included in each based on the matrix approach. In the particular case of the concept of linear combination, we highlight some aspects that we aimed to include in the text:

- Emphasis on inter-clump links, that is to say, in the early introduction to the concept, relating it to other topics (matrix multiplication and the resolution of systems of linear equations).
- Exploration of the geometrical meaning of the concept, with the support of GeoGebra and Mathematica (in R^2 and R^3, respectively).
- An initial focus on the algorithmic processes as a way of looking at the concept as a *procept* (Gray & Tall, 1994), that is to say, due to the fact that it includes an expression that simultaneously reveals its meaning and the associated procedure.
- An emphasis on intra-clump links, that is to say, within the topic of vector spaces.

Action Phase

Contextual Information and Participants

The aforementioned linear algebra text guided the teacher's practice in the classroom. The experience was undertaken with a group in the first year of a course on electrical engineering in a public polytechnic institution. Eighteen students regularly attended the sixteen classes and each class lasted one hour and forty-five minutes. The teacher gave the linear algebra text to the students and sought to have them use it as a textbook. To this end, the teacher used the examples and exercises proposed, and explored the different concepts in the same way in which they were presented.

The use of technology was a characteristic of the teaching sequence outlined. Consequently, the use of the computer and different software packages was widely included, both in the linear algebra text and in the classes. The students only used Scilab on their laptops because it is a free computer algebra system (no rooms with computers or licenses for students to use any software were available at the institution). From his laptop, the teacher projected his use of Scilab, and as dynamic geometry software GeoGebra and the Mathematica.

With regard to data collection, all classes were audio recorded using a single device capturing the teacher's speech, and interaction with students and groups as he moved around the classroom (in some cases it was also possible to hear some interactions among the students). As a complement, some photographs and daily notebooks provided a demonstration of the teacher's records on the board. The productions of the students in the class were also collected: group work, individual exercises, etc. Before each lesson, the teacher elaborated a lesson plan, including the summary and objectives, the moments in which he intended to use the different software programs, and the line-up of the subjects and tasks. At the end of each class, the teacher recorded his impressions of the work done in class, and the facts and occurrences that he considered relevant. Henceforth designated as independent data ("MN: Value, Construction, and Validation"), these elements were collected with a view to constructing MNs.

MN: Value, Construction, and Validation

We considered the construction of an MN as "making known" the teaching practices achieved by the teacher. Based on its characteristics, presented in more detail in Chapter 1, the MN was configured as an instrument fulfilling that purpose for the following reasons: (1) it collects data on teaching practices in the classroom; (2) it is a storyline describing the events that took place throughout the class; (3) it allows the inclusion of other kinds of information (dependent data) missed in audio or video recordings, such as the teacher's intentions and decisions, the teacher's and students' attitudes and reactions, gestures, silences, etc. (Lopes et al., 2014).

The value and status that we attribute to the MN resulted from three characteristics that we consider relevant. The first, of a technical nature, is that it allows the gathering of all data in a single instrument, with the advantage of circumventing the difficulty of dealing with multiple data sources (Lopes et al., 2014). It also dispenses future references to the data considered in its construction. The second, due to the importance in our study, is that it allows the inclusion of data depending on the teacher (Chapter 1). For us, it was crucial that we knew the teacher's initial expectations and the impressions he was getting throughout the class, or even after it ended, depending on whether or not his expectations were reached. The third concerned the methodological options, to be unalterable following validation (Chapter 1). In our investigation the same person assumed the dual role of teacher and researcher. We looked at the MN as a way of mitigating the promiscuous performance in the two roles and isolating them. First, he constructed the MNs in the role of teacher, and then, *a posteriori*, also analyzed them in the role of researcher. Above all, the research was restricted to what was "written."

We proceeded to construct the MNs of five classes, with the criterion of choice referring to the topics discussed. The representativeness and the predominance of the didactic recommendations established in the "Theoretical Background" of this chapter defined the choice of classes. Following the protocol for the construction of an MN (Chapter 1), the teacher, with the function of narrator, began by gathering the aforementioned independent data (audio recordings, photographs, daily notebooks, students' productions, lesson plans, and teacher's notes). Then, each MN involved the summary report of the entire lesson and the detailed narration of each episode. In addition to enriching the narrative, he added multimodal elements that came not only from independent data, but also from dependent data, including teacher and student attitudes, teacher intentions, and student reactions and behaviors (Chapter 1).

One of the advantages of the MNs in implying "external" validation (peer review) was its contribution to the improvement of the narrative. Regarding the initial versions, the narrator agreed with the recommendations of the two validators (among them, the second author), with special emphasis on the requirement to add more multimodal elements to evidence the facts narrated and enrich the "story," for example, pictures to check the teacher's records on the board.

Analysis and Reflection Phases

The analytical technique we chose was content analysis (Bardin, 1997), as proposed by Lopes et al. (2014). This is a technique that can start with large amounts of text, with the objective of reduction (Cohen, Manion, & Morrison, 2011).

We initiated the categorization process by defining the seven items related to the didactic recommendations as the dimensions of analysis. With the support of the qualitative analysis software NVivo, and influenced by the work of Lopes et al. (2014), we proceeded with the analysis of the MNs through three distinct stages:

- Selection of all excerpts consistent with each of the dimensions of analysis in a first reading of the MNs.
- *Open code* categorization (Cohen et al., 2011) guided by the dimensions of analysis.
- Qualitative analysis of categories.

We subsequently examined the MNs at two levels of depth. First, more superficially, we analyzed the categories that emerged in each of the seven dimensions, which enabled us to unveil the set of practices employed by the teacher in the classroom. Second, we undertook a deeper interpretation of the meaning of the collected data, considering them as new data resulting from the content analysis of the MNs. The next strategy was to put the resulting category system and analysis into a "dialogue" with the linear algebra text (as data anticipating the expectation of what could be done in the classroom), and with the literature.

RESULTS

The MNs made it possible to demonstrate that the work of the teacher in the classroom was coherent with the outlined teaching sequence. In Table 1 we synthesize the results obtained. What we termed "pedagogical options expressed by the linear algebra text" reflects the didactic recommendations that we aimed to materialize in the constructed linear algebra text, both in general and in particular ("Theoretical Background" and "Planning Phase"). However, the MNs also showed that sometimes the teacher did not follow the plan. For this reason, we added the item "Other teacher's pedagogical options in class."

From content analysis of the MNs, a set of categories emerged that are presented and subtitled in Table 1. The analysis and reflection on these categories presented the possibility of dividing them into opportunities for teaching the concept of linear combination, and into difficulties to be overcome, according to the different pedagogical options considered by the teacher. We observed the predominance of a number of categories that constituted opportunities for teaching the concept of linear combination. As we discuss below, the same category may be related to both opportunities and difficulties.

This section is organized based on the discussion of categories associated with the different didactic options identified in the MNs: revisiting the relationship with other concepts (matrix operations and systems of linear equations), and the relation with other concepts of the theory of vector spaces ("Inter- and Intra-Clump Links with Other Concepts"); formal definition of and importance attributed to the formal definition ("Concept Formalization Process"); geometric interpretation and algebraic exploration of the concept by practical means ("New Paths to Geometric and Algebraic Interpretation"); application in problem solving ("Application in Problem Solving").

Table 1. Relationship between the didactic options and the analysis of MNs concerning the teaching of the concept of linear combination

Pedagogical Options Expressed by the Linear Algebra Text	MN Analysis	
	Opportunities	Difficulties
Revisiting the relations with other concepts: matrix operations	C-cp, L-m-ga, Ct-res, Cs-res	
Revisiting the relations with other concepts: resolution of systems of linear equations	Sf-d-vn, C-cp, Cs-p, Fm-s, Ct-res, Cs-res, L-m-ga	
Geometric interpretation and algebraic exploration of the concept by practical means	Sf-d-vn, Fm-a, Fm-tp, L-m-ga	Fm-a
Formal definition	C-s/s, F-s-mi	
Application in problem solving	Sf-ca-ts, A-i, Fm-ps, Fm-s, C/T-s, C/T-cta, C/T-cd	E-pc
Relations with other concepts of the theory of vector spaces: subspace spanned	Ct-rel, L-m-gaf, C-cp	
Other teacher's pedagogical options in class		
Importance attributed to the formal definition	F-s-mi, F-s-fc	
Legend:		
Sf-d-vn: software function-to discover new concepts-visual nature Sf-ca-ts: (…)-to calculate-time saving A-i: attitudes-initiative C-s/s: configuration-syllabus/sequence of subjects C-cp: (…)-concept presentation Fm-ps: factors motivated by the matrix approach-problem solving Fm-s: (…)-consideration of the software Fm-a: (…)-application of algorithms Fm-tp: (…)-form of teacher's performance C/T-s: context and aspects of the underlying theory-real or scientific situation C/T-cta: (…)-concepts and theoretical aspects involved	C/T-cd: (…)-conceptual development E-pc: externalization of the exploration of tasks and applications-possible constraints Ct-res: connections established by the teacher-resource Cs-res: connections made by students-resource Cs-p: (…)-procedural Ct-rel: (…)-relational F-s-mi: strategies that work around formalism-motivation of an idea through case discussion or practical approach F-s-fc: (…)-formalization of concepts L-m-ga: concepts framed in multiple languages-geometric/algebraic L-m-gaf: (…)-geometric/algebraic/formal	

Inter- and Intra-Clump Links With Other Concepts

As in Gonçalves (2015), the syllabus included as core topics, in this order, matrices, systems of linear equations, determinants, and vector spaces. In the study of matrix operations and resolution of systems of linear equations, the concept of linear combination was introduced at an early stage. This strategy was motivated by: (1) the attempt to highlight its importance because we wanted to place it as a central concept for learning the other concepts of the theory of vector spaces; (2) the attempt to break the tendency of students to see only the intra-clump relations of this concept (Lapp et al., 2010), i.e., only between the concepts of the theory of vector spaces; (3) the attempt to mitigate students' difficulties in interpreting formal concepts when related to intuitive contexts (Dorier et al., 2000).

According to Table 1, the MN analysis permitted identification of a vast set of categories that provide opportunities for teaching the concept of linear combination according to this strategy. The categories Ct-res and Cs-res emerged from the relations between the concepts revealed in the MNs, from the point of view of the teacher and the students, in terms of concepts that constitute a resource for approaching other concepts. The early approach to the concept of linear combination allowed us to show an alternative way of multiplying matrices, exemplified in the following excerpt:

Multimodal Narrative Excerpt 1: Teacher R (Connections With the Concept of Linear Combination)

Reviewing the cases, I began by saying:

"You even did this exercise in class, which was 1.13 [see Figure 1], when I gave you here ... look here is the concept of linear combination. I then asked you ... to try to adapt this information in the matrix multiplication." [...]

Referring back to the exercise, I went on to say:

"We can say that this matrix ... or the product AX is a linear combination of the columns of A. Notice that the format is nothing more nothing less than what? [...] then it stayed and it was what you wrote ... 2 times the first column plus -3 times the second column plus 1 times the third column." [...]

$$\begin{bmatrix}5\\1\end{bmatrix} = 2\begin{bmatrix}2\\3\end{bmatrix} - 3\begin{bmatrix}1\\1\end{bmatrix} + 1\begin{bmatrix}4\\-2\end{bmatrix}$$

"That is, ... what is the format of a linear combination?" I asked and tried to finish the approach. I added that it was possible to identify the vectors and the scalars. "A certain vector may result from operations with others," I said finally.

Similarly, the early reference to the concept of linear combination in the study of systems of linear equations allowed us to add a new representation of the latter concept. This was achieved through the geometric and algebraic resolution of the vector equation $xu+yv=t$, with u, v, and t representing the vectors associated with the coefficients of the unknowns and the independent terms. These vectors were written in the form of a matrix column, reinforcing the matrix approach (C-cp).

The revealing of the algebraic expression of the concept, provided by these two instances, had positive implications *a posteriori*. In the study of vector spaces, students demonstrated recognition of the connection between matrix addition and scalar multiplication, as well as systems of linear equations with the concept of linear combination, as suggested by the Multimodal Narrative Excerpt 2:

Multimodal Narrative Excerpt 2: Teacher R (Connections With the Concept of Linear Combination)

"From what you already know, and did well, tell me how can I get u, using v_1 and v_2?"

"Inverting and adding them," Fernando began.

"As a linear combination is always the same thing. By how much I multiply one of them and how much I do multiply the other so that together they give u?" I reminded him.

"It's for -1 and 1," Fernando answered again.

In the particular case of the relation with the systems of linear equations, and according to the category Cs-p, the students anticipated that the procedure for determining whether a vector was a linear combination of the others would include the resolution of a system of linear equations to determine the existence of scalars. In relation to other concepts, namely subspace spanned, the teacher reinforced the central position of the concept of linear combination (Multimodal Narrative Excerpt 3) and related it to that of subspace spanned based on the definition (Ct-rel), that is, as the set of all linear combinations (Multimodal Narrative Excerpt 4):

Multimodal Narrative Excerpt 3: Teacher R (Connections With the Concept of Linear Combination)

Throughout the supporting text, there was some evidence of the particular attention I had given to this concept, motivated by the fact that it is a nuclear concept in the study of vector spaces.

Multimodal Narrative Excerpt 4: Teacher R (Connections With the Concept of Linear Combination)

Looking at the next image, I asked students to indicate the linear combination of vectors u and v using vectors v_1 *and* v_2*. The main suggestion was based on the application of the triangle rule or parallelogram rule for the addition of vectors geometrically represented.*

I considered that the time had come to pose the question that I considered central to unveiling the concept of subspace spanned.

"Using v_1 *and* v_2*, can I reach any of the points marked? Is there a linear combination for this?"*

Figure 1.

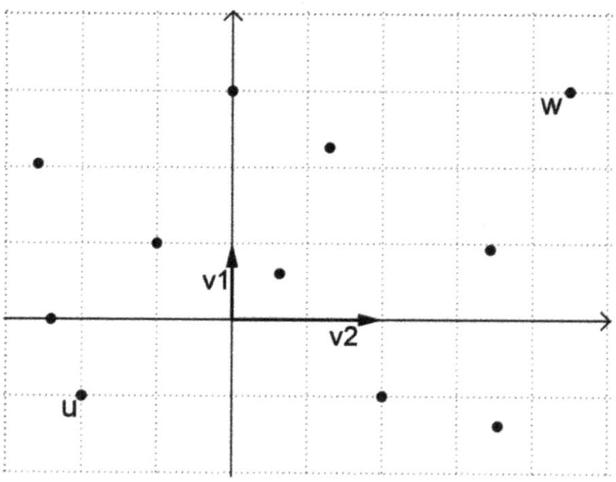

In the early approach to the concept of linear combination, the teacher began to reveal the possibility of exploration according to the geometric and algebraic languages (L-m-ga). Related to matrix multiplication, the algebraic expression for "AX as a linear combination of the columns of A" (Figure 2) served the purpose of starting to function as a *procept* (Gray & Tall, 1994), giving the first indications of the definition and the process. Related to systems of linear equations, the scalar requirement for the addition of two vectors of R^2 to result in another induced the first opportunity to show the possibility of the concept of linear combination being explored geometrically (Figure 3).

We recapitulate that the concept of linear combination was explored in the geometric and algebraic languages, and that the concept of subspace spanned was introduced from the generalization of the set of all linear combinations of the given vectors. The L-m-gaf category reinforces that this last concept was approached according to the three languages of linear algebra (geometric, algebraic, and formal). Our conception of formal language is close to the analytic-structural thinking mode proposed by Sierpinska (2000), according to which an object is defined by a set of properties.

Figure 2. Exercise 1.13 from the linear algebra text

Exercise 1.13: *Let $A \in M_{m \times n}(\mathbb{K})$ be a matrix with columns $v_1, \ldots, v_n$ and $X \in M_{m \times 1}(\mathbb{K})$ a column matrix such that*

$$X = \begin{bmatrix} x_1 \\ x_2 \\ \vdots \\ x_m \end{bmatrix}.$$

We have $AX = x_1 \times v_1 + x_2 \times v_2 + \cdots + x_m \times v_n$ and AX is a linear combination of the columns of A.

Let $A = \begin{bmatrix} 2 & 1 & 4 \\ 3 & 1 & -2 \end{bmatrix}$ and $B = \begin{bmatrix} 2 \\ -3 \\ 1 \end{bmatrix}$. Write the product AB as a linear combination of the columns of A and determine it.

Figure 3. Geometric resolution of a system of linear equations based on the concept of linear combination

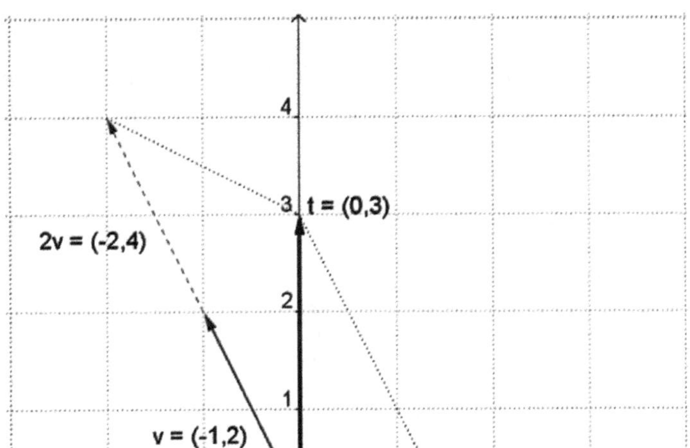

Concept Formalization Process

Related to the C-s/s category, the following excerpt locates the chosen moment for the formalization of the concept of linear combination:

Multimodal Narrative Excerpt 5: Teacher R (Formalization of the Concept of Linear Combination)

As we have explored the concepts of vector space and the subspace of a vector space in the previous class, this lesson [...] continues with the exploration of the concept of linear combination [...] and the approximation to the concept of subspace spanned.

The formalization of the concept of linear combination was prepared so that the formal definition appeared in the last moment. In this context, and following the linear algebra text, the teacher began by recalling the two moments in which the concept of linear combination had appeared earlier ("Inter- and Intra-Clump Links with Other Concepts"), and the algebraic and geometric languages, exploring three questions to study several possibilities for writing vectors of R^3 as a linear combination of others (Figure 4).

This work of anticipation constituted the strategy outlined to circumvent the problem of formalism (F-e-mi). If the students already knew the algebraic expression associated with the concept of linear combination, the answer to these questions would allow the teacher to introduce a procedure associated with the concept. However, due to the students' performance, the teacher decided at this point not to follow the lesson plan. According to the following excerpts (Multimodal Narrative Excerpts 6, 7, and 8), the teacher assumed that the students had already conceptualized the concept and decided to omit their formal presentation (F-s-fc):

Multimodal Narrative Excerpt 6: Teacher R (Formalization of the Concept of Linear Combination)

I told the students that I would continue to approach the linear combination concept without formally defining it for now, but exploring the algebraic procedure and geometric interpretation.

Figure 4. Excerpt from the linear algebra text

Consider the following vectors of $\mathbb{R}^3$:

$$v_1 = \begin{bmatrix} 1 \\ 2 \\ -1 \end{bmatrix}, v_2 = \begin{bmatrix} -2 \\ 0 \\ 1 \end{bmatrix} \text{ and } v_3 = \begin{bmatrix} -4 \\ 4 \\ 1 \end{bmatrix}.$$

Q1: Show that $v_3 = 2v_1 + 3v_2$.

Q2: Verify if the vector $u = \begin{bmatrix} -1 \\ -10 \\ 3 \end{bmatrix}$ is a linear combinations of vectors v_1, v_2 and v_3.

Q3: Verify if the vector $w = \begin{bmatrix} 3 \\ -2 \\ 3 \end{bmatrix}$ is a linear combination of vectors v_1, v_2 and v_3.

Multimodal Narrative Excerpt 7: Teacher R (Formalization of the Concept of Linear Combination)

This was the time I had planned to show how to check algebraically the possibility of a vector being written as a linear combination of others.

Multimodal Narrative Excerpt 8: Teacher R (Formalization of the Concept of Linear Combination)

I understood that at this moment the students already had the notion of the concept and were able to carry out the associated algebraic procedure.

New Paths to Geometric and Algebraic Interpretation

The use of technology has become relevant in the exploration of the concept of linear combination. GeoGebra and Mathematica allowed the students to develop the geometric intuition of the concept by allowing the manipulation of images. Scilab began to assist in the development of algebraic procedures, permitting the quick row reduced echelon form of the augmented matrix to be found during the resolution of a system of linear equations for determining the existence of scalars. In the process of formalizing the concept of linear combination, dynamic geometry software performed the function of discovering new concepts (Sf-d-nv), because they allowed more in-depth understanding of the concept before corresponding formalization. By their kinesthetic character, they made it possible to make more concrete some relations with objects of a more abstract nature (Multimodal Narrative Excerpt 9), as anticipated by Sierpinska, Trgalová, Hillel, and Dreyfus (1999).

Multimodal Narrative Excerpt 9: Teacher R (Geometric and Algebraic Procedures)

To interpret the situation geometrically, I began by asking what one could conclude about the relative position of vectors in space. Then I tried to illustrate the situation, using pens to represent vectors intuitively and using Mathematica.

"Geometrically, those vectors represented in space, their position ... want to risk some position for those three vectors, it always being possible to write one vector as a linear combination of the others?" I asked.

The algebraic procedure associated with the concept of linear combination involves, as an auxiliary calculation, the resolution of systems of linear equations. In most cases, it is only important to verify if it is a consistent or inconsistent system. At this point, attention is no longer focused on whether students can solve systems of linear equations by hand, applying the Gauss or Gauss–Jordan elimination methods. To speed up the resolution of these, leveraging the use of technology, Scilab acquires the calculation function to save time (Sf-ca-ts), as suggested by the Multimodal Narrative Excerpt 10:

Multimodal Narrative Excerpt 10: Teacher R (Geometric and Algebraic Procedures)

Then Hélder called me:

"Do I have to write this down in equations and solve it?" asked the student, pointing to the augmented matrix in the row reduced echelon form obtained from Scilab [Figure 5].

"Just keep it. It's consistent!" I said.

"Just to say if it's consistent?"

"Is it not? The system is consistent, with infinitely many solutions," I replied. I finished, suggesting that the student round the values [...].

MN revealed the tendency for the teacher to act in terms of the valorization of algorithmic procedures (categories Fm-a and Fm-tp), concretizing the exploration of a linear combination between vectors in a sequence of steps to be taken: write the algebraic expression; solve a system of linear equations; decide, depending on the classification of the system, on whether or not the vector can be written as a linear combination of others (Multimodal Narrative Excerpts 11, 12, and 13):

Multimodal Narrative Excerpt 11: Teacher R (Geometric and Algebraic Procedures)

I put the letter u, followed by the equal sign, on the board and asked the students to think of the "format" seen above for a linear combination in order to complete the expression.

Multimodal Narrative Excerpt 12: Teacher R (Geometric and Algebraic Procedures)

"And now what do we do? What are a, b, and c?" I asked. "Or rather, are there a, b, and c?" I said next, as I added this information on the board, with the purpose of schematizing the procedure.

[...] I added that the next step in the procedure would be the resolution of a system of linear equations by the Gaussian elimination method and I said that it was an auxiliary calculation. To that end, I reminded them that they could use Scilab to expedite resolution.

Figure 5.

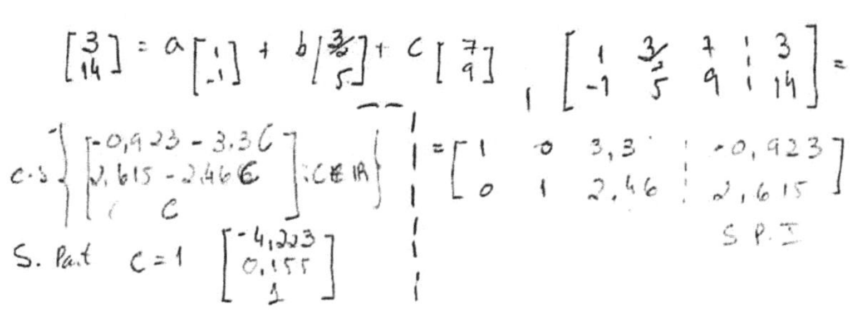

Multimodal Narrative Excerpt 13: Teacher R (Geometric and Algebraic Procedures)

"[…] what kind of system of linear equations would we have to find in order to conclude that one vector can be written as a linear combination of others?"

Confining the approach to vector space R^n, we believe is a strategy that helps to solve the related exercises and contributes to the conceptualization of the concept. We would not say the same if we had also considered more abstract vector spaces. This action of the teacher can also be a disadvantage: resulting in the mechanical reproduction of what the teacher has taught (Vinner, 1997) and restricting students to practical thinking (Dorier & Sierpinska, 2001).

Application in Problem Solving

Scilab maintains its status in resolving tasks and exploiting applications. The task "On the way to district capitals," which we created for the application of the concept of linear combination and the introduction of later ones, revealed the possibility of relating it to concrete situations and provided relevant opportunities for teaching. In addition to leveraging the use of technology, we highlight its advantage in fostering conceptual development (C/T-cd), as it provided computational situations that do not appear in the usual examples and exercises. According to the next excerpt (Multimodal Narrative Excerpt 14), the fact that an atypical situation arose led to the need for students to evaluate the effectiveness of the application of the usual method.

Multimodal Narrative Excerpt 14: Teacher R (Application in Problem Solving)

"Professor, can we solve it this way?"

"Such as?"

"Old fashioned [by the substitution method]," Juliana said.

"Old fashioned?! Two equations and two unknowns? Yes, you can," I said, understanding that the time expenditure would be similar if the option were to find the echelon form of the augmented matrix.

"That's a solution then," said the student.

This task also constituted an example that embodied some of the difficulties of performing practical work in the classroom (E-pc). In none of the group tasks assigned did the students not solve them in the expected time. This led to changes in the teacher's performance. For example, it was only from the MN (Multimodal Narrative Excerpt 15) that the teacher realized how the delay in their resolution led him to reduce the activity of the students given the importance attributed to the fulfilment of the lesson plan.

Multimodal Narrative Excerpt 15: Teacher R (Application in Problem Solving)

Noting that there was not much time remaining before the end of the class and that the students were addressing the subtask more slowly than expected, he said that it was enough to structure the answer, asking them to finish the work at home, determining the linear combinations of the vectors.

CONCLUSION

In this chapter, through the use of MNs and highlighting their role, we have aimed to emphasize the importance of researching the teaching practices of linear algebra teachers and enhancing understanding of their potential work in the classroom. This is based on the attempt to materialize the main didactic recommendations known for the teaching of linear algebra. According to Latorre (2003), there is no quality teaching practice that does not rely on research results or research that does not find the way and the natural space to investigate, analyze and apply its results.

The process of constructing and analyzing the MNs had implications for the personal practice of the teacher (first author) in the teaching of linear algebra. It was observed that the immediate perceptions about the way in which each class takes place fall far short of those that emerge after the construction of an MN, and even more so after the moments of analysis and reflection. For us, it became very evident how the construction of an MN led us from the tacit reflection (when the teacher took some notes at the end of the class) to the act of awareness of the practices implemented (when the teacher listened to the audio recording, compared it with the other independent data, and wrote the document). After that, the analysis and the reflection on the categories emerging from the content analysis of the MNs led us to see the "big picture."

Cumulatively, after the different phases inherent in MN, the teacher verified in a fundamental way how his practices influenced the way in which students internalized (and related) certain concepts and procedures, how they induced situations of conceptual development, or how they led to different functions attributed to the use of technology. The teacher also became aware of the typical ways in which he behaved in the classroom (e.g., emphasis on algorithmic procedures), and how of the actions and options resulted in aspects that needed to be corrected (e.g., reducing student activity in solving some tasks).

Three contributions are included in the systematization of the research results related to teaching and learning of linear algebra, and incorporated in a new proposal for its teaching:

- The instruments supporting the empirical part comprised pedagogical materials that could be considered by the professional community. In addition to a new linear algebra textbook (Gonçalves, 2015), which materialized a possible teaching sequence for a first course in linear algebra, we highlight the MN as a resource that allows professionals to acquire concrete ideas on ways of approaching the concepts of linear algebra (among them, the concept of linear combination).
- The MN has considerable methodological value. For us, in the context of the action research method, it contributed to clarifying the roles played by the same person, sometimes with the role of teacher, sometimes with the role of researcher. The MN has a construction protocol to be followed: data collection, construction, and validation. Peer review affords improvement and as the final version is attained, the MN is locked and becomes inalterable. For this reason, the same

person can look at what has been written in the analysis and reflection phases in a way that is completely neutral.

- Due to the relationship between the opportunities and difficulties highlighted (Table 1), and demonstrated by the categories emerging from the content analysis of the MNs, the way in which the concept of linear combination was introduced and exploited based on research results has proven to be beneficial to teaching. We emphasize the effect of the use of technology, in particular the use of computer algebra systems and dynamic geometry software; the connection of the concept to applications through the resolution of tasks; the means of avoiding the formalism associated with the concept; the interconnection between algebraic and geometric languages.

REFERENCES

Aranda, C., & Callejo, M. L. (2010). Construcción del concepto de dependencia lineal en un contexto de geometría dinámica: Un estudio de casos [Construction of the concept of linear dependence in a context of dynamic geometry: A case study]. *Relime, 13*, 129–158.

Artigue, M. (2001). What can we learn from educational research at the university level? In D. Holton (Ed.), *The teaching and learning of mathematics at university level: An ICMI study* (pp. 207–220). Dordrecht: Kluwer Academic.

Bardin, L. (1997). *L'analyse de contenu* [Content analysis]. Paris: Presses Universitaires de France.

Cárcamo, A., Fortuny, J. M., & Gómez, J. (2017). Mathematical modelling and the learning trajectory: Tools to support the teaching of linear algebra. *International Journal of Mathematical Education in Science and Technology, 48*(3), 338–352. doi:10.1080/0020739X.2016.1241436

Carlson, D., Johnson, C. R., Lay, D. C., & Porter, A. D. (1993). The Linear Algebra Curriculum Study Group recommendations for the first course in linear algebra. *The College Mathematics Journal, 24*(1), 41–46. doi:10.1080/07468342.1993.11973504

Cochran-Smith, M., & Lytle, S. L. (1993). *Inside-outside: Teacher researcher and knowledge.* New York, NY: Teacher's College Press.

Cohen, L., Manion, L., & Morrison, K. (2011). *Research methods in education* (7th ed.). Abingdon, UK: Routledge.

Domínguez-García, S., García-Planas, M. I., & Taberna, J. (2016). Mathematical modelling in engineering: An alternative way to teach linear algebra. *International Journal of Mathematical Education in Science and Technology, 47*(7), 1076–1086. doi:10.1080/0020739X.2016.1153736

Dorier, J.-L. (1998). The role of formalism in the teaching of the theory of vector spaces. *Linear Algebra and Its Applications, 275/276*, 141–160. doi:10.1016/S0024-3795(97)10061-1

Dorier, J.-L., Robert, A., Robinet, J., & Rogalski, M. (2000). The obstacle of formalism in linear algebra. In J.-L. Dorier (Ed.), *On the teaching of linear algebra* (pp. 85–124). Dordrecht: Kluwer Academic.

Dorier, J.-L., & Sierpinska, A. (2001). Research into the teaching and learning of linear algebra. In D. Holton (Ed.), *The teaching and learning of mathematics at university level: An ICMI study* (pp. 255–273). Dordrecht: Kluwer Academic.

Dubinsky, E. (1997). Some thoughts on a first course in linear algebra at the college level. In D. Carlson, C. R. Johnson, D. Lay, A. D. Porter, A. E. Watkins, & W. Watkins (Eds.), *Resources for teaching linear algebra* (Vol. 42, pp. 85–105). Washington, DC: Mathematical Association of America.

Gonçalves, R. (2015). *Álgebra linear: Teoria e prática* [Linear algebra: Theory and practice]. Lisbon: Sílabo.

Gonçalves, R. (2018). *A investigação em didática da álgebra linear: Uma perspetiva do seu contributo para o ensino por meio da reflexão sobre a prática* [Research on the teaching and learning of linear algebra: A perspective of its contribution to teaching through reflection on practice] (Unpublished doctoral dissertation). UTAD, Portugal.

Gray, E. M., & Tall, D. O. (1994). Duality, ambiguity and flexibility: A proceptual view of simple arithmetic. *Journal for Research in Mathematics Education, 26*, 115–141.

Hannah, J., Stewart, S., & Thomas, M. (2013). Emphasizing language and visualization in teaching linear algebra. *International Journal of Mathematical Education in Science and Technology, 44*(4), 475–489. doi:10.1080/0020739X.2012.756545

Harel, G. (1997). The Linear Algebra Curriculum Study Group recommendations: Moving beyond concept definition. In D. Carlson, C. R. Johnson, D. Lay, A. D. Porter, A. E. Watkins, & W. Watkins (Eds.), *Resources for teaching linear algebra* (Vol. 42, pp. 107–126). Washington, DC: Mathematical Association of America.

Hillel, J. (2000). Modes of description and the problem of representation in linear algebra. In J.-L. Dorier (Ed.), *On the teaching of linear algebra* (pp. 191–207). Dordrecht: Kluwer Academic.

Klasa, J. (2010). A few pedagogical designs in linear algebra with Cabri and Maple. *Linear Algebra and Its Applications, 432*(8), 2100–2111. doi:10.1016/j.laa.2009.08.039

Lapp, D. A., Nyman, M. A., & Berry, J. S. (2010). Student connections of linear algebra concepts: An analysis of concept maps. *International Journal of Mathematical Education in Science and Technology, 41*(1), 1–18. doi:10.1080/00207390903236665

Latorre, A. (2003). *La investigación-acción* [Action research]. Barcelona: Graó.

Lopes, J. B., Cravino, J. P., Branco, M., Saraiva, E., & Silva, A. A. (2008). Mediation of student learning: Dimensions and evidences in science teaching. *Problems of Education in the 21st Century, 9*, 42–52.

Lopes, J. B., Silva, A. A., Cravino, J. P., Santos, C. A., Cunha, A., Pinto, A., ... Branco, M. J. (2014). Constructing and using multimodal narratives to research in science education: Contributions based on practical classroom. *Research in Science Education, 44*(3), 415–438. doi:10.100711165-013-9381-y

Lopes, J. B., Silva, A. A., Cravino, J. P., Viegas, C., Cunha, E., Saraiva, E., ... Santos, C. A. (2010). *Investigação sobre a mediação de professores de ciências físicas em sala de aula* [Research on physical science teachers' mediation in the classroom]. Vila Real: UTAD.

Love, B., Hodge, A., Grandgenett, N., & Swift, A. W. (2014). Student learning and perceptions in a flipped linear algebra classroom. *International Journal of Mathematical Education in Science and Technology, 45*(3), 317–324. doi:10.1080/0020739X.2013.822582

Pinto, A., Barbot, A., Viegas, C., Silva, A. A., Santos, C. A., & Lopes, J. B. (2014). Teaching science with experimental work and computer simulations in a primary teacher education course: What challenges to promote epistemic practices? *Procedia Technology, 13*, 86–96. doi:10.1016/j.protcy.2014.02.012

Rogalski, M. (2000). The teaching experiment in Lille. In J.-L. Dorier (Ed.), *On the teaching of linear algebra* (pp. 133–149). Dordrecht: Kluwer Academic.

Santos, C. A., Lopes, J. B., & Cravino, J. P. (2011). Patterns of didactic decisions made by teachers in physical sciences classroom. *Problems of Education in the 21st Century, 36*, 67–89.

Schoenfeld, A. H., Thomas, M., & Barton, B. (2016). On understanding and improving the teaching of university mathematics. *International Journal of STEM Education, 3*(4), 1–17.

Sierpinska, A. (2000). On some aspects of students' thinking in linear algebra. In J.-L. Dorier (Ed.), *On the teaching of linear algebra* (pp. 209–246). Dordrecht: Kluwer Academic.

Sierpinska, A., Trgalová, J., Hillel, J., & Dreyfus, T. (1999). Teaching and learning linear algebra with Cabri. In O. Zaslavsky (Ed.), *Proceedings of the 23rd Conference of the International Group for the Psychology of Mathematics Education* (Vol. 1, pp. 119–134). Haifa, Israel: Program Committee.

Speer, N., Smith, J. III, & Horvath, A. (2010). Collegiate mathematics teaching: An unexamined practice. *The Journal of Mathematical Behavior, 29*(2), 99–114. doi:10.1016/j.jmathb.2010.02.001

Stewart, S., & Thomas, M. (2009). A framework for mathematical thinking: The case of linear algebra. *International Journal of Mathematical Education in Science and Technology, 40*(7), 951–961. doi:10.1080/00207390903200984

Talbert, R. (2014). Inverting the linear algebra classroom. *PRIMUS Problems, Resources, and Issues in Mathematics Undergraduate Studies, 24*, 361–374.

Trigueros, M., & Possani, E. (2013). Using an economics model for teaching linear algebra. *Linear Algebra and Its Applications, 438*(4), 1779–1792. doi:10.1016/j.laa.2011.04.009

Tucker, A. (1993). The growing importance of linear algebra in undergraduate mathematics. *The College Mathematics Journal, 24*(1), 3–9. doi:10.1080/07468342.1993.11973500

Vinner, S. (1997). Scenes from linear algebra classes. In D. Carlson, C. R. Johnson, D. Lay, A. D. Porter, A. E. Watkins, & W. Watkins (Eds.), *Resources for teaching linear algebra* (Vol. 42, pp. 155–171). Washington, DC: Mathematical Association of America.

Chapter 16
Researching Language and Semiotic Registers in Chemistry Lessons Using Multimodal Narratives

Carolina José Maria
Universidade Metodista de Piracicaba, Brazil

ABSTRACT

The main objective of this chapter is to extend the discussion of language and the use of semiotic registers in chemistry lessons. Audio and video recordings, students' notebooks, photographs of content, and activities recorded on the blackboard in chemistry lessons for high school students taught in a public school in the interior of the State of São Paulo enabled the construction of multimodal narratives (MNs). The qualitative analysis of the MNs allowed the identification and understanding of the semiotic registers present in eight lessons conducted by a chemistry teacher. The study revealed several semiotic registers present in the lessons, but there is little exploration of the processes of conversion between semiotic registers by the teacher and therefore by the students. The use of different semiotic registers without the necessary understanding of them can result in difficulties in the teaching and learning processes of chemical concepts.

INTRODUCTION

Since the earliest times, scholars have sought explanations for natural phenomena that are part of everyday life. However, explaining something that is not directly perceptible through the senses is an extremely complex task. In the case of atoms and molecules, which have a real structure that is not perceptible by the senses, this a great challenge for chemistry, and consequently the teaching of this science. One of the major issues is how to enable students to perceive the relationship between microscopic and macroscopic levels, that is, how to ensure they understand microscopically the phenomena of their daily lives.

DOI: 10.4018/978-1-5225-8570-1.ch016

One of the main objectives of the teaching of chemistry is that the student appropriates the diversity of representations that belong to this area, that is, it is necessary that the student appropriates the specific language of this field of knowledge characterized by a diversity of semiotic registers. The relevance of the study of the semiotic registers is due to the fact that the development of the sciences is linked to semiotic systems that are increasingly specific and independent of the natural language. To overcome this challenge, the discussion of the appropriation of the specific language of chemistry is fundamental. According to Roque and Silva (2008), the chemist understands the world through models, graphs, figures, and equations, namely idealized and simplified representations of the real world (semiotic registers). It is thus understood that for students to understand natural phenomena, they must learn the specific language of chemistry. Understanding this language can facilitate the study of this science, since for the teaching and learning of chemical concepts it not necessary to have a concrete macroscopic model; indeed, through symbols, students can imagine the substances and their behavior.

Comprehension of chemistry is subject to the understanding and meaning of symbolic representations. Thus, it becomes necessary to study semiotics - the study of any and all language. According to Peirce (2005), the processes of signification include written symbols, gestures, and written or spoken words, and these can occur whenever something has meaning for an individual. Lemke (1998, p. 6) states that "the natural language of science is a synergistic integration of words, diagrams, pictures, graphs, maps, equations, tables, charts and other forms of visual and mathematical expressions." Therefore, it is understandable that chemistry lessons must be permeated by diverse semiotic representation registers that will help the student to develop the ability to "read" the world.

Thus, in a general way, this study aims to understand how the use of semiotic registers in chemistry lessons are employed considering one case study, without aiming at generalizations, in order to broaden the discussion of the language and the reflection on how the use of these registers can contribute to understanding of the chemical concepts.

BACKGROUND

Based on semiotic science, recent research on science education, mainly with a cognitive orientation (Zompero & Laburú, 2010), has provided important advances in understanding aspects related to representations, including mental, internal, and external (semiotics), involving the learning of concepts (Duval, 2003). For this author, the mental representations comprise a set of images and conceptions that an individual can have concerning an object; the internal aspects are those characterized by the automatic execution of information, and the external, or semiotics, are productions constituted by the use of signs belonging to a system of representation, used as a means of externalizing the mental representations for communication purposes.

According to Duval (1993) semiotic registers can be understood as any form of enabling the symbolic representation of the construction of cognitive thinking. In the case of sciences, we can highlight mathematical formulae, Cartesian graphs, and algebraic writing, among other semiotic representations present in teaching. Understanding of scientific concepts requires the engagement of several systems of representations called registers, such as verbal language, and graphical and numerical representations, etc. Each register can contribute in a specific way to a wider understanding of the context and therefore to the organization of the student's mental representations. This means that, when learning chemistry, the student must also learn the specific language.

Concerning the external representations (semiotics), according to Duval (2003), there are three cognitive activities: the formation of representations, the treatment (under the same register), and the conversion (between registers). The formation of representations consists of the elaboration of a representation of a given register. This activity implies the selection of a set of characters and determinations that constitute what one wants to represent. In chemistry, an example of such activity would be the naming of substances. The treatment of representations is characterized by the transformation of representation in the same register, that is, it is an internal transformation of a register, allowing an amplification of the information, or the improvement and consolidation of the intelligibility of what is being treated. Finally, the conversion is understood as the transformation of the representation into a representation of another register, of a different semiotic system, namely an external transformation to a register. The conversion process requires a mechanism for coordination between registers by the individual performing the process. Among the cognitive activities, this conversion is presented as the most complex. The representation of a written statement through chemical equations is an example of a conversion activity in chemistry (Duval, 2003, 2017). The students, according to the author, had difficulty in the conversion between semiotic registers since this involves less spontaneous cognitive activity. Santos (2014) presents a very clear example of conversion: the expression "equilateral triangle" (verbal natural language → oral or written) can be converted into the image of an equilateral triangle. This example shows the conversion of verbal natural language to a figurative representation.

A scientific concept can be presented to students using a few or several semiotic registers. Studies such as that of Zompero and Laburú (2010) point out that the performance of the students is qualitatively different, when they have the possibility of working with several semiotic registers, compared to the work done without this variety of semiotic registers.

Chemistry lessons are full of symbols and the meaning of these symbols needs to be appropriated by students. If this appropriation does not happen, the students experience difficulties in understanding the concepts that are related to the symbols, formulae, and equations. This idea is reinforced when Santos (2014) affirms that the understanding of what represents a symbol is fundamental to communicative relations, and to the development of oral and written competences. In chemistry teaching, when using molecular models, the teacher makes use of an external representation that serves as a support for the construction of a mental representation by the student. Roque and Silva (2008, p. 922) emphasize: "The molecular model provides a way of studying and understanding better the molecular structure invisible to our eyes [...]. Learning about this way of representation, the students no longer need to have the concrete model in their hands, they can imagine it from the structural formula drawn."

For Duval (2009), the various characteristics of an object cannot be represented by a single semiotic register, which justifies the importance of using different registers in the lesson. Therefore, it is understood that all representations are partial; i.e., different registers represent different aspects of content. The teaching of chemical concepts requires conversion processes between semiotic registers. The lack of activities that value these processes can lead to difficulty in learning. For Gilbert (2010), a chemical equation is a good example of a system that requires many symbolic representations. When working with chemical equations, students need to use several auxiliary representations (McGee & Martinez-Planell, 2014) so that they can deal with the problems posed.

However, beyond internal and external representations, from the perspective of social semiotics it is also fundamental to understand the role that semiotic modes - gestures, facial expressions, and voice intonation - can play in the discursive process of knowledge construction in the classroom. The classroom is a social and cultural space, characterized by interpersonal relations mediated by language, and by several semiotic registers and semiotic modes in the sense-building movement. Several authors, such as Lemke (1998), have emphasized the various ways of presenting information that one wishes to communicate. According to Kress, Jewitt, Ogborn, and Tsatsarelis (2001), these modes are developed through social use, constituting networks of resources for the construction of signs that are of specific interest in a given context for communicative effect in a particular situation.

Given the importance of the issue addressed by this study, it is necessary to emphasize that the presence of semiotic registers in lessons is not enough to ensure the meaningful learning of students since this involves other variables; that is, it does not depend only on how the teacher works in the classroom or on the quality of the teacher's mediation. Rather, the mastery of the teacher in the different semiotic registers and the knowledge of the importance of the conversion between them can improve learning.

METHODOLOGY

This study was carried out in a public school located in the suburbs of a city in the state of São Paulo. The school has around 800 students, distributed between the elementary school (students aged between 11 and 15 years) and high school (students aged between 16 and 18 years). The research was approved by the Research Ethics Committee of the UNIMEP (n. 09/13). Those responsible for the students, the teacher and the School Board, allowed the video recordings. They were informed that the results of the filming and recordings would be transcribed, analyzed by researchers, and used only for research purposes. Two lessons on the subject "Chemical Bonds," taught to a second-year high school class, provided the data for the analysis and discussion of the question proposed in this study.

Participants

The class participating in the study comprised 37 students (21 female and 16 male), attending the second grade of high school; no students declined participation in the research. Most were from the neighborhoods near the school and the average age was 16 years. In terms of the training of the chemistry teacher responsible for the class, she has undertaken training in the teaching of science and had a qualification in chemistry from a private university. She also had 14 years of experience in basic education (elementary and high school). During the development of the study, the teacher was conducting physics lessons for middle school and science lessons for elementary school students in addition to teaching chemistry.

Data Collection

The data that allowed reflection on the issue in this study were obtained through the monitoring of the two aforementioned chemical lessons taught to the students. The data construction was enabled through the audio and video recording of these lessons, obtained by a camera positioned at the end of the classroom and a voice recorder placed on the teacher's desk (located at the front of the classroom), as well as the contribution of other materials: the researcher's field notebook and the students' personal notebooks, and the teacher's notebook and the students' notebooks distributed by the state government. These data formed the basis for the elaboration of the MNs (Lopes et al., 2014). Each of the two lessons allowed the elaboration of an MN, aiming to encompass all didactic action.

According to Lopes et al. (2014), an MN comprises the description of a sequence of events (including the actions and language used by teachers and students in the classroom), that starts with the task proposal and finishes with its accomplishment. This instrument is intended to inform the dialogue between teachers and students, and among students, the actions and reactions, the dialogues, the postures, the silences produced by teacher and students, as well as the work developed by students (Lopes et al., 2014), as illustrated in detail in Chapter 1. It is important to point out that for the MNs to be complete and clear, they were submitted to a validation process by a researcher who had not followed the lessons, and therefore had no knowledge of the case studied. In this process, this second researcher used all the documents associated with the MNs (audio and video recordings, photos, student and teacher notebooks, among others) to validate them, that is, to verify if all the possible details portraying the lessons were incorporated in the document. This validation procedure gives greater validity and reliability to the study.

Data Analysis

After being elaborated and validated, the two MNs were imported for qualitative analysis in NVivo 8®. This tool offers a wide range of resources facilitating the analysis of data, for example, the management and the codification process of the MN. One of the major benefits of the use of this software is the organization of a large amount of data, as in the case of this study, since the MNs are extensive documents providing a wealth of information about the lessons. Other advantages are also highlighted by several authors (Ozkan, 2004; Rettie, Robinson, Radke, & Ye, 2008) who have chosen to use NVivo, such as: it allows the researcher time to make inferences and important proximity for analysis between the data and the researcher. Also, in this study content analysis (Krippendorff, 2004) was conducted in two stages. The first stage, called first-order analysis, entailed a process of open categorization (Böhm, 2004; Cohen, Manion, & Morrison, 2011), such that categories related to the topic of interest (language and semiotic registers) emerged from the data (MN). In the second stage, the categorization process was complemented by literature on the subject. Each category was given a name and a brief description. These categories also underwent a validation process by a second researcher. Cases of disagreement were discussed until there was full agreement among researchers.

RESULTS AND DISCUSSION

The analysis of MNs allowed the observation of a certain diversity in semiotic registers. These registers, along with the sequence of lessons, were used by the teacher and also by the students. The verbal language (oral and written) is the predominant semiotic register during the lessons, but this register was highlighted to indicate every time language was associated with other semiotic registers. The teacher and students, in addition to using oral and/or written language, also used semiotic registers important for the teaching of chemical concepts, such as formulae. The semiotic registers identified as used by the teachers (Table 1) and the students (Table 2) are presented with a brief description.

The first point to highlight when looking at the tables is that the semiotic registers used by students (Table 2) are the same as those used by the teacher (Table 1). Students, especially during the accomplishment of exercises, take the registers used by the teacher as a model. The second point concerns the semiotic registers termed the “macroscopic level” and the “microscopic level.” From the MNs and with

Table 1. Semiotic registers used by the teacher

Semiotic Register	Description
Formula	In explanations or solving the exercises, the teacher uses a chemical formula (symbols, equations, etc.).
Verbal language	The verbal language–oral and written – is used as a semiotic register during the explanations of concepts, and in the development of various activities in the classroom.
Macroscopic level	The explanation of the concept is based on what is visible and/or tangible to the student, for example, the lighting of a lamp to explain electric current conduction.
Microscopic level	The explanation of the concept is grounded in elements that refer to objects not directly visible and/or tangible to the student, for example, the idea of atoms, electrons, protons, etc.
Visual representation	Some visual representations–rules and facilitator schemata, for example–are used in the explanations of chemical concepts.

the help of NVivo 8® software, it was possible to identify the frequency with which these registers, used by both the teacher and the students, were present in lessons - for the microscopic and macroscopic levels these frequencies are, respectively, 209 and 30. Terms such as atoms, ions, and electrons, clearly recur in lessons. However, these terms are not particularly related to the aspects of macroscopic phenomena and other semiotic registers related to such phenomenon are also rarely used. This aspect could contribute to a better understanding of chemical concepts by students.

The episode in which the teacher works on the theme of chemical bonds, ionic bonds, provides a good illustration of the emphasis on only the microscopic aspects of the theme. Initially, the teacher asks students to copy a text the title of which is "Stability of the atom" (student's notebook excerpt):

Multimodal Narrative Excerpt 1

Translation of Figure 1:

27/05 Chemical bonds

Stability of the atom

Table 2. Semiotic registers used by students

Semiotic Register	Description
Verbal language	The verbal language - oral and written–is used as semiotic register during the development of various activities in the classroom.
Formula	Use of chemical formulae (symbols, equations, etc.) as semiotic registers.
Macroscopic level	The written answer of the student is supported at the macroscopic level, in other words, the answer shows visible and/or tangible elements.
Microscopic level	The written answer of the student is supported at the macroscopic level, in other words, the answer shows visible and/or tangible elements.

Figure 1. Stability of the atom – student's notebook

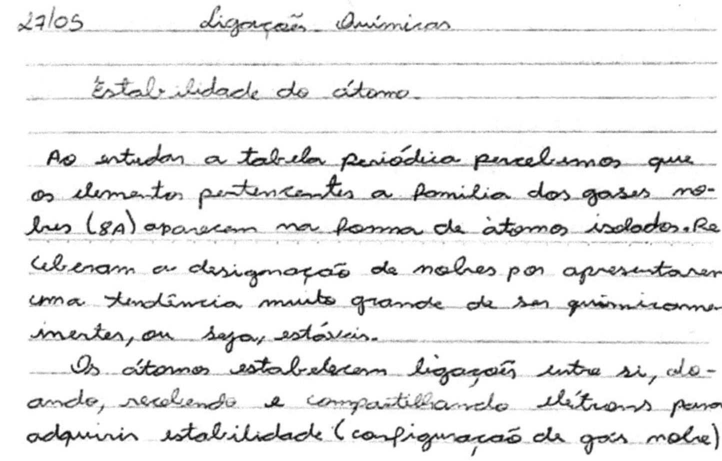

27/05 Ligações Químicas

Estabilidade do átomo.

Ao estudar a tabela periódica percebemos que os elementos pertencentes a família dos gases nobres (8A) aparecem na forma de átomos isolados. Receberam a designação de nobres por apresentarem uma tendência muito grande de ser quimicamente inertes, ou seja, estáveis.

Os átomos estabelecem ligações entre si, doando, recebendo e compartilhando elétrons para adquirir estabilidade (configuração de gás nobre)

By studying the periodic table, we notice that the elements belonging to the family of noble gases (8A) appear in the form of isolated atoms. These are given the designation of noble as they present a very strong trend for being chemically inert, that is, stable.

The atoms establish bonds among each other, giving, receiving and sharing electrons to acquire stability (noble gas configuration))

To address the theme "ionic bonding," the teacher uses as an example the compound sodium chloride (NaCl) (from MN excerpt 1 (continuation)):

MN Excerpt 1 (Continuation): (Stability of the Atom)

[…] Teacher: We started talking about the cooking salt that was NaCl, where there was a connection called ionic bonding between Na and Cl, and I put the example up for you. What was that ionic bonding? What does ionic bonding mean? Do you remember? What's going on with the elements so we have ionic bonding?

Student 1: It is donating the electrons to those who want to receive.

[…]

Teacher: An ionic bonding is related to this donation and receipt of electrons. So we got there, the example that I used most was NaCl. I wrote on the blackboard a representation of $_{11}Na$ elements and $_{17}Cl$ and I showed the electronic distribution in layers for the sodium and chlorine.

[...]

Figure 2. Electronic distribution of sodium (Na) and chlorine (Cl) elements

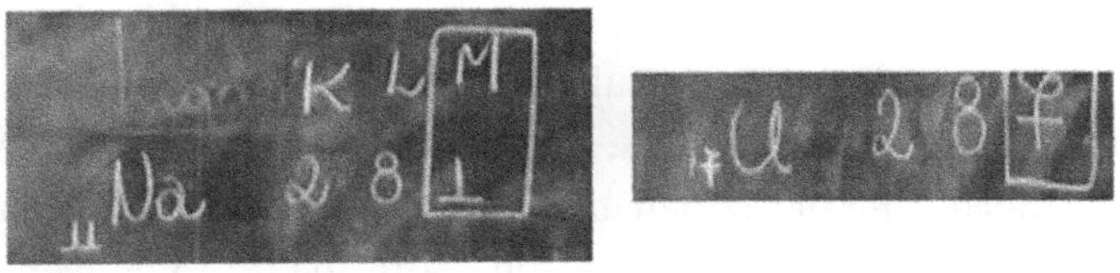

Teacher: ... The Na and the Cl, when they join, one donates and the other receives an electron, one is positive and the other is negative, what happens to the number of electrons of these? What is it going to be? What's going to happen? It will stay neutral, electrically neutral.

[...]

Teacher: If it has one electron, the trend is giving, remember that little rule I put on the blackboard? One, two and three give, is positive, that is ... The one that has 5 receives 3, the one that has 6 receives 2, and the one that has 7 receives 1, to get 8, right?

Teacher: So you have there, Na and Cl... The tendency is to donate—I referred to the tendency of the element sodium. The ionic bonding is represented by this shape, an arrow.

Teacher: The little arrow is indicating that the electron from Na is going towards the Cl, okay? So it is represented there this bonding, and the ionic compound formed is NaCl.

Figure 3. "Rule" presented by the teacher

Figure 4. Electronic formula of NaCl

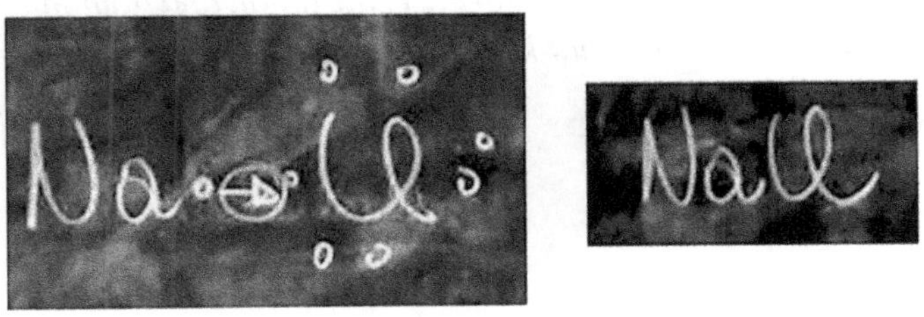

The analysis of the excerpts presented allows some inferences. It can be said that the entire speech of this lesson is based on the microscopic level, and the relationships that could lead the students to understand the phenomena related to the concept of chemical bonds were not established.

Regarding the diversity of semiotic registers, in this first excerpt from the MN, we can notice the presence of different registers during the explanation of the concept of ionic bonding—verbal language (oral and written), formulae and visual representation. With regard to conversion processes between semiotic registers, it is possible to affirm that this happens in a few moments during the two classes on chemical bonds. For example, when the teacher is talking about sodium chloride (oral and written language), and then presents the symbols of the chemical elements that make up this compound, and also represents it through the electronic formula of the compound formation. The same process of conversion between semiotic registers is required from the students when they do the exercises proposed by the teacher (Figure 5).

Multimodal Narrative Excerpt 2 (Exercise)

Teacher: Does the bonding between fluorine and aluminum happen? How do I know if chemical bonding is going to happen? What are the parameters to know about the bonding?

No student showed up, so I continued:

Teacher: For example, ionic bonding. What does ionic bonding need?

Once again the students are silent. So I insisted on the question.

Teacher: To make the ionic bond what is needed?

Still, no student showed up and I reminded them that we had studied this theme in yesterday's class.

Teacher: Yesterday we talked about it.

Student 4: You have to pass the electron to the other...

Teacher: The ionic bond what does it do, does it just pass the electron? It donates its electron. It will donate the electron, the cation, the cation is always the donor, the cation it donates, it will donate the electrons to the anion, because the anion is negative, it is what you need to receive.

Teacher: I'm going to make the bonding between fluorine and aluminum, which is the cation and which is the anion between fluorine and aluminum? Fluorine is the anion and aluminum is the cation. So when I have a cation and an anion the bond is what kind?

Student 8: Ionic.

[...]

Teacher: Aluminum has three electrons, it will donate three and I want to bond it to fluorine. How many does fluoride have? In total it has seven, it needs to gain one. One, two, three, four, five, six, seven.

Student 11: What do you mean?

Teacher: It has seven electrons in the last layer, it needs to gain one, so I represent here the number of electrons it has, except that I know that it needs to gain one. So what's going to happen? It makes that bonding here, the aluminum donates.

Teacher: The arrow is the donating, donation to fluoride, only it did not get eight (I referred to aluminum), to keep eight it has to do what?

Student 8: You have to put two more fluorine there.

Teacher: Aluminum has to lose all three of it so have to put what? More fluoride.

Teacher: One more, I made the donation. I placed one more arrow indicating that aluminum is donating one more electron to the second fluorine atom.

Teacher: There is still one left, nothing can be left, one more fluoride and I make the donation. Finally, I added another fluorine and the arrow indicating that the aluminum was donating its last electron to the third fluorine atom.

Teacher: So I formed the substance Al...

Student 4: $F_{3.}$

Teacher: F_{3}.

Teacher: I made the bonding there. Therefore, here is the representation of Lewis. Do you understand?

Examining excerpt 2, it can be affirmed that in general, the students are led to the memorization and application of rules for the resolution of exercises. Excerpt 2 shows an application of what was worked on in the previous lesson (excerpt 1). The exercise proposed to the students, although requiring the use of different semiotic registers (formula, verbal language), only asks them to reproduce what was said during the explanations. The main issue is that these explanations do not expand with the use of new semiotic registers, enabling the understanding of the concept of chemical bonds.

Figure 5. Exercise proposed by the teacher

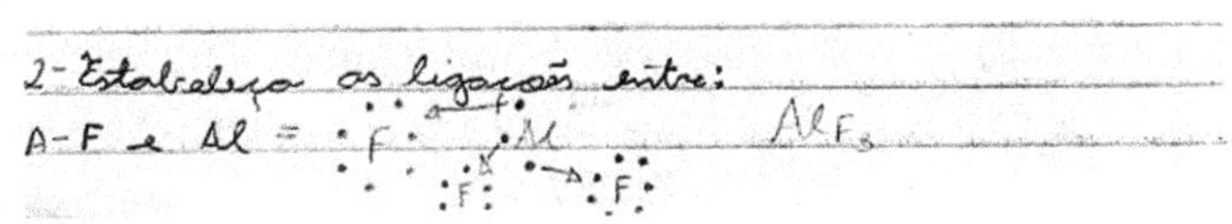

The analysis of the results of this study points out that in these chemistry lessons, different semiotic registers are used by the teacher. Moreover, especially during the resolution of exercises, these registers are also used by the students. Although a certain diversity in registers has been identified, this fact alone does not guarantee meaningful learning of the students.

The teacher's mediation in the classroom is essential when it comes to the use of semiotic registers. It is up to the teacher to select the register, and to encourage the students to work with different semiotic registers and conversions, allowing the opportunities to relate and understand different levels of representation of the phenomena studied. We agree with Gilbert (2005), who considers that teachers need to develop professional knowledge that will allow students to be engaged in activities requiring the use of semiotic representation registers so that they can benefit from meaningful learning. The use of different semiotic registers in the classroom has the important function of pointing out that the same phenomenon can be represented in different ways (Duval, 1993). Each representation offers specific information on the phenomenon without describing it completely, that is, the representations are cognitively partial in relation to what they represent. Hence the importance of thinking about the work that will be carried out in the classroom in terms of different semiotic registers.

Activities that require conversion between semiotic registers are little emphasized in the lessons. When these occur, they are limited to two or three registers, and thus do not give the students the opportunity to appropriate the different ways of representing the same phenomenon. This is also highlighted by the teacher's acknowledgment of the importance of the subject for chemistry teaching. For Duval (2009), the conversion between semiotic registers is an essential ability for understanding and gaining control of scientific language. Not offering the students the opportunity to develop such skills will certainly lead to greater difficulties when it is necessary to use the knowledge in situations of greater complexity.

Specifically in relation to the teaching of the concept of chemical bonds, according to the guidance in the didactic materials distributed by the government to teachers (teacher's notebook), the octet rule can be difficult for students to understand. It is common during the lessons to use phrases such as "an atom acquires stability when it has eight electrons in the outermost shell or two electrons when presenting only the K shell." The octet rule is certainly not sufficient to explain the formation of ionic compounds (Mortimer, Mol, & Duarte, 1994). However, the authors of textbooks "favor the emphasis on aspects such as the tendency to 'donate' or 'win' and the 'octet rule'" to explain certain chemical transformations and stability of substances, according to Oliveira, Mendonça, and Justi (2012, p. 10). An alternative to highlighting the octet rule might be to highlight the attraction between ions of opposite charges that causes the ions to connect.

According to Pazinato (2016), studies carried out on the topic of chemical bonds show that, in general, students do not understand the nature of the different types of chemical bonding. For students, ionic bonding occurs due to the transfer of electrons, the covalent bond by the sharing of electrons, and the metallic by the transfer or sharing of electrons. Moreover, students cannot relate the macroscopic aspects and properties of the compounds to the microscopic representation of chemical bonds.

The notebooks indicate that the concepts to be taught must allow an understanding of the physical world, social, political, and economic, organized from measurable facts so that students can understand the issues at stake, and only then should explanations requiring abstractions be introduced. In the case of chemical bonds, an energetic approach would be more advisable than using rules. However, it is common in chemistry courses to use practical rules that take the place of the chemical principles giving rise to them (Mortimer et al., 1994). Edward and Mercer (1987, as cited by Mortimer et al., 1994) differentiate this knowledge, which they term ritual (related to know-how), from knowledge of the principles (how the procedures and processes work). The simple actions of learning rules, and memorizing symbols and representations arbitrarily by the students, can lead to misunderstandings about the concept in question.

Although the literature points out that the use of the octet rule is insufficient to understand the concept of chemical bonds, it should be noted that the teaching of this topic using the octet rule is a classic in the chemistry lessons of Brazilian schools. Even in teacher training courses, this type of teaching is often emphasized. This "tradition" in the teaching of chemistry could explain, in part, the events that occurred in lessons on chemical bonds.

When the teaching of chemical concepts is based on the use of a single semiotic register, there is a risk of hindering students' learning, such that they will not be able to relate different phenomena that are supported by the same concept (Maturano, Aguilar, & Nuñez, 2009; Roque & Silva, 2008). In the highlighted excerpts, the semiotic registers identified may allow partial learning of the concept of ionic bonds, but they do not guarantee that the students will appropriate the concept broadly, that is, students will be able to use such a concept to explain several phenomena present in their day-to-day lives. It is fundamental that the activities developed in the classroom focus on conversion between several semiotic registers, as this would give students opportunities to develop the skills and abilities related to the representation and communication of chemical phenomena. This idea is reinforced by Duval (2003), who argues that a concept is only assimilated when the individual is able to make the conversion between different semiotic registers. It is possible to say that a student has a meaningful learning experience concerning a certain concept when he/she performs the activity of conversion and expresses equivalence of meanings between different representations, while being able to integrate these meanings in a multimodal articulation of representation, so that the student does not remain dependent on a specific sign or exclusive mode of communication (Duval, 2009).

The results obtained through the analysis of the lessons on chemical bonds do not give us indications that significant learning actually occurred. The actions performed by the students are restricted to the resolution of exercises such as those in Figure 2. This is a process of conversion (verbal written language → formula), but it is insufficient to understand the concept. There is no complete meaning in itself; meaning depends on different sources of information in a context of experiences and meanings (Lemke, 1998). Chemical science has a certain specificity in teaching and learning in terms of working with models, symbols, representations, that is, working with abstract ideas, which requires the teacher to use various strategies and resources in an attempt to reach their diverse students, since each student is unique. Depending on how the lesson is planned and conducted, their learning will be more or less effective.

The semiotic registers used to teach chemical bonds in the lessons may contribute to the introduction of the topic. However, other representations can be used to help the students in the construction of knowledge, such as the use of images and undertaking experiments that may generate discussions about the properties of different substances due to the nature of their bonding, among others. The concept of chemical bonding is fundamental in the teaching of chemistry, since it is necessary for the understanding of other concepts, and consequently for the understanding of the various phenomena present in everyday life.

CONCLUSION

This study has sought to contribute to the discussions on the importance of language and the use of semiotic registers, both by the teacher and by students, to promote meaningful learning in chemistry lessons for high school students. From the application of multimodal narratives (MNs) it was possible to identify a certain diversity in semiotic registers during the sequence of lessons in this study, for example figures, symbols, formulae, and other ways of representation used by the teacher and students. Although this diversity of registers is perceived during the lessons, the treatment of semiotic registers as well as conversion between them do not reflect adequate exploration. It was possible to identify these aspects because MNs capture all the information in a single instrument, including the students' notebooks, the registers present on the blackboard by the teacher, and so on.

Few opportunities were given to the students to work on the process of conversion between registers, a fundamental activity for the learning of scientific concepts. In the lessons considered, the predominant conversion was between two registers—verbal written language → formula—which might result in difficulty for students in understanding and establishing relationships between what was studied in the classroom and daily phenomena.

The cause of this might lie in the training of chemistry teachers. In general, in the process of training, issues of language and the use of semiotic registers for teaching and learning are poorly or superficially treated. Therefore, teachers often do not work on these issues because they do not know their relevance for practice. Thus, it is fundamental to discuss the issue of teacher's mediation in initial and continuing training courses, since the teacher is responsible for helping the student in the learning process. Teachers need to understand the importance of promoting classroom work that prioritizes the use of different semiotic registers and emphasizes the conversion between these registers, since learning centered on the conversion of different types of semiotic registers and the different representations they provide can develop effective understanding.

It is expected that the results obtained by this study and the emerging discussions can contribute to reflection on teaching and learning processes, as well as raising important questions about initial and continuing training for chemistry teachers.

REFERENCES

Böhm, A. (2004). Theoretical coding: Text analysis in grounded theory. In U. Flick, E. von Kardorff, & I. Steinke (Eds.), *A companion to qualitative research* (pp. 270–274). London: Sage.

Cohen, L., Manion, L., & Morrison, K. (2011). *Research methods in education* (7th ed.). London: Routledge.

Duval, R. (1993). Registres de représentation sémiotique et fonctionnement cognitif de la pensée. *Annales de Didactiques et de Sciences Cognitives*, *5*, 37–65.

Duval, R. (2003). Registros de representação semiótica e funcionamento cognitivo da compreensão matemática [Registers of semiotic representation and cognitive functioning of mathematical understanding]. In *S. D. A. Machado (Ed.), Aprendizagem em matemática: registros de representação semiótica* [Learning in mathematics: Semiotic representation register] (pp. 11–33). Campinas, São Paulo: Papirus.

Duval, R. (2009). *Semiósis e pensamento humano: registros semióticos e aprendizagens intelectuais* (Vol. 1; L. F. Levy & M. R. A. da Silveira, Trans.). São Paulo: Livraria da Física.

Duval, R. (2017). Registers of semiotic representations and analysis of the cognitive functioning of mathematical thinking. In T. M. M. Campos (Ed.), *Understanding the mathematical way of thinking – The registers of semiotic representations* (pp. 45–71). Cham: Springer. doi:10.1007/978-3-319-56910-9_3

Gilbert, J. K. (2005). Endpiece: Research and development on visualization in science education. In J. K. Gilbert (Ed.), *Visualization in science education* (pp. 333–335). Holland: Springer. doi:10.1007/1-4020-3613-2_16

Gilbert, J. K. (2010). The role of visual representations in the learning and teaching of science: An introduction. *Asia-Pacific Forum on Science Learning and Teaching, 11*(1), 1–19.

Kress, G., Jewitt, C., Ogborn, J., & Tsatsarelis, C. (2001). *Multimodal teaching and learning: The rhetorics of the science classroom.* London: Continuum.

Krippendorff, K. (2004). *Content analysis: An introduction to its methodology.* Thousand Oaks, CA: SAGE Publications.

Lemke, J. L. (1998). Teaching all the languages of science: Words, symbols, images, and actions. In *Conference on Science Education in Barcelona.* Retrieved from http://www-personal.umich.edu/~jaylemke/papers/barcelon.htm

Lopes, J. B., Silva, A. A., Cravino, J. P., Santos, C. A., Cunha, A., Pinto, A., ... Branco, M. J. (2014). Constructing and using multimodal narratives to research in science education: Contributions based on practical classroom. *Research in Science Education, 44*(3), 415–438. doi:10.100711165-013-9381-y

Maturano, C., Aguilar, S., & Nuñez, G. (2009). Conversion de imagenes al linguaje escrito. Un desafio para el estudiante de ciencias naturales. *Revista Eureka sobre Enseñanza y Divulgación de las Ciencias, 6*(1), 63–78.

McGee, D. L., & Martinez-Planell, R. (2014). A study of semiotic registers in the development of the definite integral of functions of two and three variables. *International Journal of Science and Mathematics Education, 12*(4), 883–916. doi:10.100710763-013-9437-5

Mortimer, E. F., Mol, G., & Duarte, L. P. (1994). Regra do octeto e teoria da ligação química no ensino médio: Dogma ou ciência? [Octet rule and chemical bond theory in high school: Dogma or science?]. *Quimica Nova, 17*(2), 243–252.

Oliveira, D. K. B., Mendonça, P. C. C., & Justi, R. (2012). Argumentar para aprender ligações químicas: contribuições de uma atividade investigativa [Argumentation to learn chemical bonds: Contributions of a research activity]. XVI Encontro Nacional de Ensino de Química (XVI ENEQ), Salvador – Bahia.

Ozkan, B. C. (2004). Using NVivo to analyze qualitative classroom data on constructivist learning environments. *Qualitative Report, 9*(4), 589–603.

Pazinato, M. S. (2016). Ligações químicas: investigação da construção do conhecimento no ensino médio [Chemical bonds: Investigation of the construction of knowledge in high school] (Tese de Doutorado). Universidade Federal de Santa Maria, Santa Maria.

Peirce, C. S. (2005). *Semiótica* [Semiotics] (3rd ed.). São Paulo: Perspectiva.

Rettie, R., Robinson, H., Radke, A., & Ye, X. (2008). CAQDAS: A supplementary tool for qualitative market research. *Qualitative Market Research*, *11*(1), 76–88. doi:10.1108/13522750810845568

Roque, N. F., & Silva, J. L. P. B. (2008). A linguagem química e o ensino da química orgânica [The chemical language and the teaching of organic chemistry]. *Quimica Nova*, *31*(4), 921–923. doi:10.1590/S0100-40422008000400034

Santos, G. L. S. (2014). Os registros de representação semiótica mobilizados por acadêmicos de um curso de ciências contábeis em resolução de problemas [Semiotic representation registers mobilized by academics from a course in accounting sciences in problem solving] (Tese de Doutorado). Universidade Estadual de Londrina, Londrina.

Zompero, A. F., & Laburú, C. E. (2010). As relações entre aprendizagem significativa e representações multimodais [The relationships between meaningful learning and multimodal representations]. *Revista Ensaio*, *12*(3), 31–40.

Section 5

Using Multimodal Narratives in Other Contexts

In this section of the book we try to blur boundaries and to point out contexts that are less exploited in the initial moments of multimodal narrative (MN) tool development. Seeking to go beyond familiar circumstances, we are moving forward with the conviction that the characteristics of MN are sufficient grounds to support our confidence that they will be usable and useful in several professional contexts. The two approaches we bring in this section of the book are, first, illustrations of the diversity we foresee for the use of MN. In the following two chapters, we present work done using MN in two very distinct contexts. The first context presents the work done on the Science Museum Educators' Discourse and the second on Nursing Education. It is intended that these examples, in addition to the intrinsic and relevant issues they present, constitute a challenge to the reader to reflect on the MN's potentialities in multiple contexts.

Chapter 17
Accessing Science Museum Educators' Discourse Through Multimodal Narratives

Susana Afonso
University of Exeter, UK

Ana Sofia Afonso
Universidade do Minho, Portugal

ABSTRACT

Museum educators play an important role in science communication, as they connect elements of an exhibit with visitors through emotion-driven experiences that are meaningful to them. Language is their main modus operandi in face-to-face interactions, but little is known of how they use it to communicate science, in part because little attention has been given to their practices and professional development. Nowadays, museums are changing, and science communication has become more demanding as these institutions exhibit hot themes of science. In this context, it is important that museum educators become aware of how they communicate science with an intended audience and reflect on how their practice can be improved. In this chapter, the authors focus on the way multimodal narratives can be used as a tool to access museum educators' discourse as well as how to promote museum educators' reflection about their practice and their professional development.

INTRODUCTION

Science museums, initially created to preserve and study collections of objects of scientific interest, have expanded their mission to include an educational purpose that can contribute to supporting the process of adults' lifelong learning. This has led to the reinvention of the museum space so that exhibitions can be framed around hot scientific themes that are pressing society (e.g. global warming, sustainable development), or around processes of science (e.g. modeling or contemporary scientific research), that attend to the visitors' needs, interests and motivations (e.g. Dillon et al., 2016), in contrast with tradi-

DOI: 10.4018/978-1-5225-8570-1.ch017

tional exhibitions that provide friendly interpretations of the message and do not encourage visitors to go beyond what they already know. Exhibitions on hot themes cover complex scientific ideas, requiring visitors to think critically about science (Rennie, 2014). As a result, science museums face the challenge of how to represent science in a version that encourages engagement and that is meaningful to an intended audience (Tlili, Cribb, & Gewirtz, 2006). One way to move forward is to present key ideas in a clear and simple story (Rennie, 2013), told mainly through objects and interactive exhibits explicitly linked together. However, because resources do not talk for themselves, front-line museum educators (museum educators, henceforth) may aid visitors to go beyond what they already know, and to create emotional links with what is being presented (Rennie, 2014). Hence, they can impact on visitors' learning experiences in multiple ways, including inspiration and creativity, attitudes and values, aesthetic appreciation, knowledge and understanding in science, among others, as has been accounted in a diversity of frameworks (see for example the six strands of science learning in Bell, Lewenstein, Shouse, and Feder (2009), or the Generic Learning Outcomes in Hooper-Greenhill (2007)).

Communicating science in museums is a challenging and complex task for museum educators, who are asked to support visitors' learning in diverse types of activities, ranging from guided visits to unstructured interactions in the exhibition area. They act between unseen or hidden science of the curatorial department and the visitors (Anderson, Cosson, & McIntoshosh, 2015). What they say is, to some extent, shaped by the institutions' missions and rules (Clark, 1996), but their discourse is not prescriptive. Indeed, they need to adjust the content of the message and the form of the language used to engage in successful science dialogues with an intended audience, i.e. dialogues where any participant can ask questions, give, and evaluate an answer (Gilbert, 2013). The emergent dialogues are unplanned and unfold as a museum educator and an audience coordinate their individual actions to negotiate meaning (i.e. a joint activity), which is influenced by visitors' feelings, beliefs, desires, and the role they assume in the conversation (Clark, 1996). There are some good practices that museum educators need to have in mind to enhance the coordination between what they mean and what their addressees understand them to mean (e.g. Laszlo, 2006; Stocklmayer & Rennie, 2017). In particular, museum educators need to take time to know the audience (Stocklmayer & Rennie, 2017); to carefully select the content (e.g. key ideas to convey and their level of precision, complexity, and abstraction (Yeo & Gilbert, 2014)); and to adjust the form of language, both at a macrolevel (e.g. the structure of the discourse itself) and at a microlevel (e.g. lexicon, use of analogies), to an intended audience. As a result, dialogues on hot science themes may have different levels of complexity and formality depending on the audience: from those that focus on an awareness of scientific issues, to those in which participants reflect on the impact of science on their lives. However, little is known on how museum educators use language to communicate science in museums, in part because little attention has been given to their professional development (Patrick, 2017). The majority of known studies focus on museum educator-student interactions during guided school visits. While some studies suggest that museum educators tend to transfer their knowledge to students, to use scientific terminology without clarification, and to neglect students' needs (Cox-Peterson, Marsh, Kisiel, & Melber, 2003; Tal & Morag, 2007), other studies show that museum educators are receptive to students' needs, employ and adjust the depth and content of the message by using a diversity of strategies (Tran, 2007). Additionally, Pattison and Dierking (2012, 2013), in studies with families in a science museum, pointed out that museum educators employed several strategies (e.g. checking in, guiding, modeling, encouraging) to facilitate family learning. These studies provide few insights on how museum educators use language in science communication, as in most of them data were collected by observing and/or interviewing museum educators about their pedagogical approaches. Additionally, language for communicating science in museums is often associated with verbal language, rather than with a broader concept to include the strategies by which a participant in a conversation means something for the other, independently of the mode of communication used (Clark, 1996). Gestures, facial expressions, or body actions, are important in conversation and are intertwined with talk to produce meaning (vom Lehn, 2006). These nonverbal elements can enhance science communication by calling attention to an exhibition element; drawing a physical analogy; or indicating patterns (Gilbert, 2013).

Since language is the major tool, and often the only one, employed by museum educators in face-to-face interaction with an audience (King & Tran, 2017), it is of paramount importance to know how it is used for science communication. In particular, which language strategies are employed to know the audience (e.g. to perceive their level of expertise, their beliefs, assumptions, and other information); how museum educators build on that information to transform science into a version that is accessible to visitors, and how they react to visitors' responses. Additionally, because language is based on the same conceptual system that underlines all human activities (Lakoff & Johnson, 1980), accessing museum educators' language allows an inquiry into how science ideas are conceptualized in the museum context. Thus, studying how museum educators use language for science communication in the museum provides insights into their practices, which can lead to a more effective use of language in science communication, promote museum educators' professional development, and ultimately improve museums' educational provision.

Access to museum educators' discourse requires a very complex descriptive tool that cannot only record verbal language but also other elements that are intimately related to it, such as gestures. Multimodal narratives (MNs) is an instrument that emerged in the context of formal science education (Lopes et al., 2014). It is a public document that includes multimodal data to describe teachers' practices in the classroom whose relevance for professional development has been recognized by several studies (e.g. Lopes, 2017).

In this chapter, we discuss to what extent MNs may be a useful tool to assess museum educators' discourse and how it can facilitate their professional growth.

LANGUAGE IN SCIENCE COMMUNICATION

The nature of language varies according to the different schools of thought. The perspective on language in this chapter is one in which language is not an autonomous cognitive ability. Rather, it is ruled by the same principles that underlie all other cognitive abilities. Hence, through language it is possible to access the conceptual system that is embodied, i.e. it arises from the bodily experience of the external world (Lakoff & Johnson, 1980). However, language is not only a reflection of our experience with the outside world, it also construes it depending on the perspective speakers wish to adopt when describing a particular event. For the same event – for instance a glass breaking – various linguistic realizations are possible, e.g. *I broke the glass, the glass broke, the glass was broken.* In the first case, the speaker is the actor of the event, whereas in the two other examples, the speaker chose to highlight the breaking and the end point or result of the breaking event, backgrounding the agent, regardless the direct involvement of the speaker in the event. Another case is, for example, how referents are conceptualized depending on the perspective taken, e.g. terrorist groups vs. freedom fighters; colonial wars vs. independence wars.

The direct embodied experience of the world also gives rise to abstract concepts through metaphor, a unidirectional mapping between a source domain characterized by the interaction between the human sensorimotor system with the surrounding environment (structuring basic concepts such as containment, substance, travel, etc.), and a target domain characterized by abstract concepts, which are structured metaphorically by the basic concepts (Lakoff & Johnson, 1980). For instance, the basic concept of containment arises from our bodily interaction with entities that are bounded and that can hold other entities inside. The concept of container metaphorically structures abstract concepts such as emotions and states (e.g. *to be in love*). Metaphors, therefore, more than aesthetic devices, allow us to make sense of abstract entities through familiar entities and experiences. The conceptual structure is also, therefore, metaphorical in nature.

In the context of science, scientific discourse has dissociated itself from the embodied and metaphorical nature of scientific concepts. Lakoff and Nuñez (2000), in relation to mathematics, have argued that its disembodied and abstract nature is a myth, providing evidence of embodied and metaphorical mathematical concepts. An example of a metaphorical and embodied mathematics concept is arithmetic, which is conceptualized as construction of objects (ARITHMETIC IS OBJECT CONSTRUCTION). The source domain is composed by the embodied notion of object that is mapped onto the target domain in which abstract concepts such as number, subtraction etc., are found. According to Lakoff and Nuñez (2000), mathematical ideas cannot be disembodied as they are conceptualized by humans who structure mathematical concepts in the same way as any other abstract concepts. Metaphors in science perform a crucial role in creating new terminology (e.g. wormwholes in general relativity and electron clouds) as well as in theory-building (e.g. Bohr's appearance of the atom based on the solar system) (Boyd, 1979). However, and despite the use of metaphorical embodied concepts in science, the use of language in scientific discourse has created a "disjunction between everyday common sense knowledge and the systematized knowledge of the disciplines" (Halliday & Martin, 1993, p. 53).

The examination of language use in contexts in which science is disseminated, e.g. in informal settings such as communicating science in a museum, can shed light on the conceptualization of scientific phenomena. Afonso, Afonso, and Rodrigues (2017) analyzed the way in which museum educators explain light phenomena such as reflection and refraction to 8th-grade students during an unstructured school visit. Museum educators use hybrid lexicon, i.e. a mixture of everyday language (e.g. *light goes through*) and scientific terminology (e.g. *propagation*) and metaphors. Some of the central metaphors used by museum educators are LIGHT IS A SUBSTANCE, encoded in expressions such as *more or less light*, *pieces of light*, *quantity of light*, which conceptualize light as a countable substance that can be itemized, and LIGHT PROPAGATION IS LIGHT TRAVELING encoded in expressions such as *light comes from*, *reaches*, *goes through*, conveying a source, destination and path of motion. Through these embodied metaphors, the conceptualization of light held by museum educators can be accesses, some of which are consistent with the scientific ideas (LIGHT PROPAGATION IS LIGHT TRAVELING), but in other cases (e.g. LIGHT IS A SUBSTANCE), the conceptualization of the nature of light as a countable substance cannot be mapped onto the scientific characteristics of light in the target domain.

Even though it is through the analysis of verbal language at the micro level, i.e. at the level of structures, from lexical items to more complex syntactic structures, that it is possible to access the conceptual system, other equally important elements must be brought into play in order to understand how science is effectively communicated in science museums. First of all, the act of communication is a joint action between a speaker and an interlocutor/ interlocutors who coordinate their individual actions in order to attain not only their individual goals but also a common shared goal (Clark, 1996). Often, participants switch roles – the speaker becomes the hearer and vice versa – in order to make sense of what is being communicated. Communication is multilayered. Interaction between participants may not occur only at a given time, face to face where communication actions are performed, but also in a construed world in which experiences are simulated at a distance. Layering can be found in storytelling (e.g. jokes, anecdotes, etc.), in staged communicative acts (e.g. irony, sarcasm, over- and understatements, rhetorical questions) and in ostensible communication acts (e.g. pro forma requests involving polite ritualized gestures), and they trigger a range of emotions (Clark, 1996). From a social viewpoint, layering is an effective tool to keep social relations equitable.

In the context of science communication, layering is commonly found. Often museum educators tell stories with which the audience can relate to, for example, personal journeys or childhood memories (Luebke & Matiasek, 2013; Pedretti, 2012), creating a synergy between entertainment and education and triggering an emotional response. The efficacy of story storytelling in understanding the exhibit is known (Mortensen, 2011). This has had consequences in relation to the design of the exhibits, in particular the emergence of immersion exhibits, in which science is presented through stories, e.g. make-believe games (Mortensen, 2011). These stories provide the context in which conversations in multiple layers may emerge, and give participants the opportunity to reconstruct the story in a version that is meaningful for them. This means that the whole communication setting must be analyzed, not only at the microlevel but also at the macrolevel, i.e. the level of the structure of discourse – how speech is structured, how and when the different layers occur in discourse, etc.

Also crucial for communication are gestures, which reflect "a complex interplay of representational, interactional and contextual dimensions of communication (Kok, Bergmann, Cienki, & Kopp, 2016). Although verbal language is often analyzed in isolation, "language itself is inseparable from gesture. While gestures enhance the material carriers of meaning, *the core is gesture and speech together*" (emphasis in original) (McNeill, 2015, p. 3). Certain types of gestures could be considered to be part of the linguistic system given their capacity to convey symbolic structures, which, just like a grammatical structure, may become entrenched and conventionalized (Kok & Cienki, 2016). What this means is that particular gestures and their symbolic meaning are perceived and recognized in multiple communicative experiences. For example, in many European languages, the "cyclic gesture" (Ladewig, 2011), in which the hand or finger(s) continuously rotate outwards in a circular way, or rather the family of cyclic gestures, is used in somewhat limited discourse situations and is recognized by a particular speech community as instances of the concept of cycle, therefore showing their symbolic nature (Cienki, 2012). The meaning of gesture is elaborated against the background set by speech (Bressem, 2012), further specifying the content conveyed through language. Considering that some types of gestures can be considered symbolic structures, through gesture analysis, it is therefore possible to access the conceptual structure, as is the case with verbal language. Gestures reveal inner mental processes and points of view that are not encoded in verbal language. McNeill (1992) shows that when describing a cartoon in which one character tries to reach another by climbing up the inside of a drainpipe, speakers produce slightly different gestures showing which aspects of an event were more salient for them (e.g. the upward movement or the climbing, etc.) and the vantage point taken, either the characters' viewpoint, being therefore inside the story, or an observer's viewpoint, outside the story.

In formal science education, gestures have been considered important for learning (Radford, Edwards, & Arzarello, 2009), performing several functions, namely pointing to a real or imaginary object under discussion, drawing a physical analogy of something familiar to explain more abstract concepts, indicating trends, e.g. repetition of events, and adding emphasis (Gilbert, 2013). Sherr (2008) argues that students' thinking about physics concepts, and scientific concepts in general, can be better understood by gesture analysis. In relation to algebra and how problems are described verbally, Alibali, Bassok, Solomon, Syc, and Goldin-Meadow (1999) showed that when gestures reinforced the verbal description, the problem was more likely to be solved, concluding that mental representations are more easily accessible when considering gesture and verbal language together.

THE "MIGRATION" OF MULTIMODAL NARRATIVES FROM THE CLASSROOM TO THE MUSEUM

Face-to-face conversations are instantaneous, and fade quickly without leaving a record or artifact (Clark, 1996). Hence, accessing museum educators' use of language in museums requires a representation of the discourse records, which is often presented in the form of a written transcript. The transcript needs to encapsulate the features of the discourse records it represents.

A discourse record is the public course of events in a conversation, which is mutually known to the participants, and that results in the advance of the conversation (Clark, 1996). It is composed of two parts: a textual representation, and a situational representation (Johnson-Laird, 1983). A textual representation includes verbal language (e.g. the words, intonation, pause), and other signals (e.g. gesture, body language) used in communication. The situational representation is related to the situation being talked about, and includes external representations that situate what is said in the museum context (e.g. physical setting of the exhibition, exhibits and their labels, exhibit guides, spatial distribution of the participants, with whom they engage, their role in a group, and the products that may have been developed by visitors) (Johnson-Laird, 1983). Producing a written transcript of a discourse record is expensive, time-consuming and challenging for many reasons. First, it demands the collection of multiple data (e.g. from field observation, audio, and video recording), which, as any data, must be valid, reliable, and collected with ethical principles.

Second, while the available techniques for data collection (e.g. multimedia-based and computer-based) enable the generation of complex and rich data, they also generate a high volume of data that may drag analysts into getting lost in the detail (Barron, 2007). Hence, the collected data need to be reduced to those that can be useful to represent the situation under study, a procedure that is guided by the questions asked (Barron, 2007). For example, a study that aims at analyzing how body movement is embedded within conversations that take place in the museum (e.g. as in the studies by vom Lehn (2006) and vom Lehn, Heath, and Hindmarsh (2001)) demands a representation of discourse records with a larger amount of detailed data, than a study that aims to analyze the content of the discourses generated in front of exhibits (e.g. as in the study by Ash, 2007).

Third, it is difficult to structure and organize a transcript of a discourse record that expresses both textual and situational representations of the discourse record, which is multimodal by nature. While there are conventional protocols to transcribe parts of a discourse record in which the verbal transcript can be marked, such as gestures, intonation, pauses, action descriptions, elongated pronunciation, there is no consensual agreement on how to articulate it with other multimodal data (e.g. a graph in a label, video vignettes showing the movement of a body), without which the transcript might become incomprehensible (Stevens & Hall, 1998; vom Lehn et al., 2001).

Fourth, when the focus of the inquiry is on museum educators' use of language, there are advantages in enriching the representation of a discourse record with their intentions in order to gain an inside perspective (Barron, 2007). In fact, both joint public goals – e.g. to promote an intended outcome of science communication (see Stocklmayer, 2013) – and individual or personal goals – e.g. to promote a certain museum, be identified as a good professional – are brought to the context of communication (Clark, 1996). These goals will have different consequences on the language that is used. Fifth, the representations of discourse records need to be produced in a friendly format, easily shared among practitioners and researchers.

An MN is the type of instrument that might be worthwhile to explore to access museum educators' discourses in science communication. Created in the context of formal science education, an MN is a written and public document that gathers multimodal data – both independent data (e.g. audio transcripts, video vignettes, organization of the space) and teachers' data (e.g. intentions, attitudes) – to describe teachers' practices in the classroom. Its creation follows a protocol that includes guidelines for data collection, for the organization of the document in a chronological storyline that incorporates actions, teacher–student interactions, the practices through the teachers' eyes, and for the validation of the MN in terms of its internal structure and features (see Lopes et al. (2014) for details on MNs).

The structure of an MN looks promising for reporting museum educators' practices. Its first part provides a summary of the situation being talked about (e.g. the pre-script in a structured visit, the exhibits, a map of the exhibition area, the labels) as well as a summary of the museum educators' practice. Hence, it facilitates the immersion in the context of the museum educator–visitor interaction. The second part of an MN is, in fact, a multimodal representation of the discourse record of an "event," enriched with dependent data (e.g. museum educators' private and public goals, perception of the public reactions and responses). Some adaptations, however, are important to consider for elaborating an MN in the museum context. These refinements do not constitute limitations of the instrument per se, but its movement from the area of formal science education to the research area of science communication. This is because when "concept migration" takes place, an analogy between areas, rather than by literal transfer, must be drawn (Mulkay, 1972). "Concept migration" expresses the displacement of concepts, skills, methodologies, techniques, or solutions to problems from one research area to another area of research, in a way that goes beyond their initial scope and common use.

A central concept in MN is the episode. In the classroom, an episode narrates what happens from the presentation of a task until it is completed or a new one is introduced (Lopes et al., 2014). In the museum, the boundaries of an episode are less well defined. While at first sight an exhibit could be seen as a "task," and therefore clearly delimiting an episode, the fact is that it does not for mainly two reasons. First, through "juicy questions" an exhibit can be explored in different ways that go beyond the designer intention, even when it is close-ended (Gutwill & Allen, 2010). Visitors engage with exhibits in multiple ways regardless the (explicit or implicit) instructions set by the designer. Hence, an exhibit can engage visitors in different "tasks." Second, discussing an exhibit does not necessarily end when the visitor leaves the exhibit. Indeed, some visitors keep discussing the exhibit while they are approaching a new one (Allen, 2002), and may return to the exhibit at a later stage and engage with it again in yet another way. Since an episode is the basic unit of an MN, which in turn focuses on the museum educators' and visitors' discourses, a tentative delimitation of the episode in the museum can be provided in terms of the actual interaction as a joint activity between the museum educator and the audience that occurs continuously at a given time, which is not necessarily bound to any particular exhibit, show, tour, etc. Each joint activity has boundaries: an entry (i.e. participants go from not being in a joint activity to being in it) and exit (i.e. participants go from being in a joint activity to not being there) (Clark, 1996). There are markers that can help in the identification of the boundaries of a joint activity. An entry is often marked by a greeting that triggers a joint commitment to talk. Closing does not happen just by stopping a conversation, but it is prepared by participants. Hence, it is marked by utterances that summarize the content of the conversation, justify the end of the conversation, express pleasure about each other, wish each other well (*good evening*), among others (Clark, 1996). The first stage in creating an MN is to collect the data (Lopes et al., 2014). As in the case of MNs in formal education, several types of data need to be collected (e.g. the language used in face-to-face interaction; the exhibition elements; the track followed by museum educators and visitors; the products developed by visitors with museum educators' support, among others), but the collection of this diversity of data in a museum raises several technical and ethical issues (Heath & vom Lehn, 2004). For example, video-recording has been frequently employed to register face-to-face conversations, but its use is not trivial. The quality of the sound recorded can be compromised by the noisy environment of the museum and the recordings produced are not fully objective, as only a chunk of the interaction taking place due to the visitors' motion is recorded, only a particular angle may be captured, only particular aspects may be zoomed in, etc. (Barron, 2007; Derry et al., 2010). In order to minimize the limitations of the data collected through video-recording and to triangulate the data, it has been suggested that these data can be complemented (Derry et al., 2010), e.g. by audio-recording museum educators and visitors in conversations, by unobstructed observations of participants in a conversation, and by tracking visitors and museum educators' paths. This tracking, traditionally collected through personal observation, has been replaced by highly automated technical systems, like Bluetooth. However, not only are some of these procedures, e.g. asking museum educators to use a microphone, very intrusive, but also there are ethical issues that limit the extent on how unobstructed data can be (Allen, 2002). Consequently, these data may not capture the spontaneous and complex nature of the experiences that occur in the museum. These types of data are generally not recorded in formal contexts for elaborating MNs (see e.g. studies on MNs in Lopes, Viegas, and Pinto (2018)), and even when an observer is in the classroom to take notes (say a teacher's critical friend), after some time students tend to become familiar with his/her presence and therefore the observation is potentially less intrusive than in the museum.

The second stage of the protocol is the elaboration of an MN. When concerned with the teachers' practices, it is desirably elaborated by them as they select the relevant data to include. Thus, the complete transcription of the conversations may not be provided (Lopes et al., 2014). In the case of an MN reporting the practices of museum educators, it is important to transcribe the full conversation by someone experienced, and to include them in the MN, for two main reasons. First, there are code schemes to transcribe the conversation, so that they can include not only the verbal language, but also intonation, gestures or volume, depending on the purpose of use of the transcript (Barron, 2007). Second, there are a number of features that are worthy of consideration in the analysis of communication (e.g. use of metaphors; gestures). However, because they are often used unconsciously by practitioners (Clark, 1996; Lakoff & Johnson, 1980), they are only perceived by a team of experts of different areas (Barron, 2007). Once the transcripts are available, it seems that there are gains when the MN is elaborated by the museum educator in collaboration with a researcher/practitioner who performs unobstructed observations of his/her practice. The latter can encourage museum educators to ask critical questions and reflect on the dependent data to include, and may enhance a better articulation of the different data.

The third stage in the elaboration of an MN is validation. Echoing Lopes et al. (2014), this procedure ensures that relevant and reliable information is included and articulated in the MN. Since it is performed by an independent researcher, it might be needed to reconcile different perspectives on the data included. This procedure seems to be better achieved when there is a community of learning with a strong emphasis in the collaboration between museum learning practitioners and academics from different disciplines (Dillon et al., 2016; Lopes, 2017).

MULTIMODAL NARRATIVE: A TOOL FOR RESEARCH AND PROFESSIONAL GROWTH

MN as an enriched representation of discourse records provides an opportunity to study museum educators' use of language to communicate science. This inquiry can take several dimensions, which are acknowledged in the literature (e.g. Afonso et al., 2017; Dillon et al., 2016; Pattison & Dierking, 2012, 2013; Zhai & Dillon, 2014;). For example, on museum educators' discourse at a microlevel (e.g. in terms of the lexicon, analogies and metaphor used), and at a macrolevel (e.g. in terms of layering in storytelling); on how museum educators conceptualize big ideas in science; on the building of a common ground to make science communication effective with different publics and in different activities; on mediation strategies. Furthermore, the independent data in an MN provides real corpora for studying language. The existence of these corpora to study expressions used to encode abstract concepts (e.g. light) by real speakers in naturally occurring language is of paramount importance to understand through language – a very broad system including gestures as well as verbal language – what the conceptualization of certain scientific concepts is.

Additionally, in the science communication domain, the use of MN allows a description of museum educators' practices, and an analysis of what can be improved to enhance the communication of science. The outcomes of this research are particularly important nowadays to inform practice, as there is an increased interest in the professionalization of museum educators (Tran & King, 2007). It is important to note that current research agendas for museums are framed on an intercommunity of learning in which practitioners, researchers, and museum administrators collaborate (Dillon et al., 2016). This collaboration presupposes more than working together. Rather, it requires participants to engage in identifying problems of practice, and to look at them through the lenses of different academic disciplines so that common plans can be set up (Dillon et al., 2016). In these intercommunities of learning, MN, which itself was elaborated through the collaboration of practitioners and researchers, can play an important role in enhancing the museum educators' professional development in a reflexive model (Patrick, 2017). This reflexive model of professional development, in which teachers' professional development is framed, has also been adopted in the professional development of museum practitioners (e.g. Allen & Crowley, 2014: Ash & Lombana, 2012). Within this model, self-reflection plays an important role in professional development because it leads museum educators to better understand "what they know" about science communication and "what they do" in science communication (King & Tran, 2017). Rather than being a solitary procedure, reflection needs to be shared within an -intercommunity of learning that supports it (Patrick, 2017). This partnership can bring to light elements of the practice that are used unconsciously and are hidden, e.g. visions and norms about how people learn, the design and goals of educational offers (King & Tran, 2017); questioning the science communication that was generated; and support the discussion and implementation of new alternatives and perspectives. Furthermore, because reflection is situated in one's working context, it gives a genuine reason to be engaged, and enhances the coordination between research and practice (Tran, Werner-Avidon, & Newton, 2013). In some successful programmes for enhancing museum educators' professional development, practitioners engaged in reflective practices through video analysis (e.g. Ash & Lombana, 2012; Tran et al., 2013), or through online activities, namely forums and blogs (Tran et al., 2013). The latter give museum educators time to organize their thoughts before posting a contribution. However, spontaneous discourse is difficult to revisit and to reflect upon if the only recorded register is the practitioner's reflective forum or blog. The way they have communicated science is unlikely to be recalled as spontaneous language (including gestures) is unconscious.

Furthermore, even with visualization of a video, speakers may not be aware that certain structures are indeed metaphors as they are conventionalized (and not novel), so their use is also unconscious. They are also certainly unaware how certain concepts are framed through the use of language. Hence, those resources cannot be used to reflect on this aspect of science communication in order to enhance it. On the other hand, MNs do seem a promising tool to engage in that kind of reflection.

The fact that an MN provides an externalized representation of the communication potentially allows a shared reflection with the intercommunity of learning on "What did I communicate?, " "How did I communicate?," "How did I include the mission of my institution in the communication?" and "Did I build a joint project with visits?"

The MN also facilitates the interpretation of the science communication that emerges (e.g. "what does my practice say in terms of the beliefs, assumptions, and values that I brought to my communication?"). Additionally, with the support of an MN it is possible to discuss the potentialities and limitations of science communication that were established, leading to the reconstruction of the practices in science communication. All these elements are part of the reflexive processes proposed by Smyth (1989) and can contribute to practitioners' professional growth. However, even with the support of an MN, reflection in one's communication practices is a complex issue but worthwhile developing as it contributes to the transformation and emancipation of those involved (Patrick, 2017).

RECOMMENDATIONS

MN as a document offers the opportunity for museum educators to report on their practices, intentions, and decisions and to share them with elements of an intercommunity of learning. However, because not all reflections are necessarily public, it seems important that a museum educator keeps a reflexive diary, so that she/he can write for her/himself. Because writing is not evanescent, this diary can be reread several times, challenge the writer, and be re-edited (Clark, 1986) by calling upon the MN as a stimulated recall. Hence, reflexive diaries are potential spaces of transformation (Vieira, 2011). In order to build a relevant diary, some elements need to be included, so as to provide links between theory and practice. The diary needs to present relevant ideas, supported by evidence (e.g. provided by an MN), hold personal perspectives, include a critical analysis and the ability to question, reflect paths of self-learning regulation (Vieira, 2011), and reflect on the alignment of practices with the mission of the museum and educational department (Patrick, 2017).

In some science museums, it is common to find museum educators with diverse professional experience, holding different perspectives on how to broaden visitors' experiences (Dillon et al., 2016). In other museums, the number of museum educators is low and their "isolation" often results in sharing similar practices (e.g. they may use common examples, analogies, and stories). While in the former case, sharing reflections of the MN can provide opportunities to find novel approaches in science communication, in the latter case that may not occur. One possible solution is to create a platform for museum educators, belonging to different museums (Valente de Oliveira, 2018), with a repository of MN (Lopes et al., 2018), which can be used to support discussions on common problems and potential solutions.

FUTURE RESEARCH DIRECTIONS

MNs provide an enriched representation of the discourse, which can be analyzed through different perspectives. Since little is known on how museum educators' use language in science communication, it seems important to enquire into it using MN as a support. A holistic understanding of this issue presupposes that not only verbal language is analyzed but also its combination with body language, particularly gestures, in the production of meaning. This is an aspect in which research is limited, in part because it is difficult to build a transcript that articulates the two models of communication. Again, multimodal narratives, as a multimodal tool, may enhance this inquiry.

Language also mirrors how one conceptualizes ideas. This can also be a line of inquiry in which MN can be used as the corpus of analysis. Finally, since MN can be an instrument to engage museum educators in reflexive practices, it is important to analyze how MNs promote them in comparison with other tools that have been used, namely video analysis. Additionally, it is important to evaluate how discussion forums and blogs can indeed be well articulated with MNs.

CONCLUSION

This chapter analyzed the role that MN can play in studying science museum educators' practices and on how they contribute to their professional development, within a reflexive model. Among the endless enquiries that MN can support: from museum educators' practices in a diversity of activities to how they can promote and sustain visitors' engagement, we emphasized the contribution MN can have on accessing museum educators' discourse. Unlike other instruments, namely diaries and journal writing, MN constitutes an enriched representation of discourse records. It is precisely here that the importance of MN for understanding museum educators' practices lies, namely the use of language to communicate science with an intended audience as it is often the only tool used by these practitioners. Language use was approached from a cognitive perspective, which considers that language, as a joint activity, is a window to the conceptual structure of participants involved in interaction. Analyzing the way that science educators communicate science and how the audience responds to it, it is possible to access how particular phenomena are conceptualized. The "migration" of MN as a tool from the formal education context to the museum context suggests that some adaptations are needed in the protocol for its elaboration, namely in the stages of data collection and elaboration of MN. By building on the concept of intercommunity of learning, several considerations were made in terms of how MN can be articulated with other reflexive instruments, namely diaries, to enhance the professional development of museum educators.

ACKNOWLEDGMENT

The authors would like to thank the reviewers and editors for their valuable contribution.

The contribution of Ana Sofia Afonso to this chapter is funded by CIEd – Research Centre on Education, projects UID/CED/1661/2013 and UID/CED/1661/2016, Institute of Education, University of Minho, through national funds of FCT/MCTES-PT.

REFERENCES

Afonso, A. S., Afonso, S., & Rodrigues, F. (2017, August). *The role of language in building common ground in a science museum*. Poster presented at the 12th Conference of the European Science Research Association (ESERA), Dublin, Ireland.

Alibali, M. W., Bassok, M., Solomon, K. O., Syc, S. E., & Goldin-Meadow, S. (1999). Illuminating mental representations through speech and gesture. *Psychological Science, 10*(4), 327–333. doi:10.1111/1467-9280.00163

Allen, L. B., & Crowley, K. (2014). Challenging beliefs, practices, and content: How museum educators change. *Science Education, 98*(1), 84–105. doi:10.1002ce.21093

Allen, S. (2002). Looking for learning in visitor talk: A methodological exploration. In G. Leinhardt, K. Crowley, & K. Knutson (Eds.), *Learning conversations in museums* (pp. 259–303). Mahwah, NJ: Lawrence Erlbaum Associates.

Anderson, D., Cosson, A., & McIntoshosh, L. (2015). Foreword: Research informing the practice of museum educators: Diverse audiences, challenging topics, and reflective praxis. In D. Anderson, A. Cosson, & L. McIntoshosh (Eds.), *Research informing the practice of museum educators: Diverse audiences, challenging topics, and reflective praxis* (pp. vii–xii). Rotterdam: Sense Publishers.

Ash, D. B. (2007). Using video data to capture discontinuous science meaning making in non-school settings. In R. Goldman, R. Pea, B. Barron, & S. J. Derry (Eds.), *Video research in the learning sciences* (pp. 207–226). Mahwah, NJ: Erlbaum.

Ash, D. B., & Lombana, J. (2012). Methodologies for reflective practice and museum educator research. In D. Ash, J. Rahm, & L. M. Melber (Eds.), *Putting theory into practice: Tools for research in informal settings* (pp. 29–52). Rotterdam, The Netherlands: Sense Publishing. doi:10.1007/978-94-6091-964-0_4

Barron, D. (2007). Video as a tool to advance understanding of learning and development in peer, family, and other informal learning contexts. In R. Goldman, R. Pea, B. Barron, & S. J. Derry (Eds.), *Video research in the learning sciences* (pp. 159–187). Mahwah, NJ: Erlbaum.

Bell, B., Lewenstein, B., Shouse, A., & Feder, M. (2009). *Learning science in informal environments: People, places, and pursuits*. Washington, DC: The National Academies Press.

Boyd, R. (1979). Metaphor and theory change: What is 'metaphor' a metaphor for? In A. Ortony (Ed.), *Metaphor and thought* (pp. 356–408). Cambridge, UK: Cambridge University Press.

Bressem, J. (2012). *Repetitions in gesture: Structures, functions, and cognitive aspects* (Unpublished doctoral dissertation). European University Viadrina, Frankfurt, Germany.

Cienki, A. (2012). Usage events of spoken language and the symbolic units we (may) abstract from them. In K. Kosecki & J. Badio (Eds.), *Cognitive processes in language* (pp. 149–158). Frankfurt am Main: Peter Lang.

Clark, H. H. (1996). *Using language*. Cambridge, UK: Cambridge University Press. doi:10.1017/CBO9780511620539

Cox-Peterson, A., Marsh, D., Kiesel, J., & Melber, L. (2003). Investigation of guided school tours, student learning, and science reform recommendations at a museum of natural history. *Journal of Research in Science Teaching*, *40*(2), 200–218. doi:10.1002/tea.10072

Derry, S. J., Pea, R. D., Barron, B., Engle, R. A., Erickson, F., Goldman, R., ... Sherin, B. L. (2010). Conducting video research in the learning sciences: Guidance on selection, analysis, technology, and ethics. *Journal of the Learning Sciences*, *19*(1), 3–53. doi:10.1080/10508400903452884

Dillon, J., DeWitt, J., Pegram, E., Irwin, B., Crowley, K., Haydon, R., . . . Xanthoudaki, M. (2016). A learning research agenda for natural history institutions. London: Natural History Museum.

Gilbert, J. K. (2013). Helping learning in science communication. In J. K. Gilbert & S. Stocklmayer (Eds.), *Communication and engagement with science and technology. Issues and dilemmas* (pp. 165–179). New York: Routledge.

Gutwill, J., & Allen, S. (2010). *Group inquiry at science museum exhibits: Getting visitors to ask juicy questions*. Walnut Creek: Left Coast Press.

Halliday, M. A. K., & Martin, J. R. (1993). *Writing science: Literacy and discursive power*. London: The Falmer Press.

Heath, C., & vom Lehn, D. (2004). Configuring reception: Looking at exhibits in museums and galleries. *Theory, Culture & Society*, *21*(6), 43–65. doi:10.1177/0263276404047415

Hooper-Greenhill, E. (2007). *Museums and education: Purpose, pedagogy, performance*. New York: Routledge. doi:10.4324/9780203937525

Johnson-Laird, P. N. (1983). *Mental models: Towards a cognitive science of language, inference, and consciousness*. Cambridge, MA: Harvard University Press.

King, H., & Tran, L. (2017). Facilitating deep conceptual learning: The role of reflection and learning communities. In P. G. Patrick (Ed.), *Preparing informal science educators: Lessons for science communication and education* (pp. 67–85). New York: Springer. doi:10.1007/978-3-319-50398-1_4

Kok, K., Bergmann, K., Cienki, A., & Kopp, S. (2016). Mapping out the multifunctionality of speakers' gestures. *Gesture*, *15*(1), 37–59. doi:10.1075/gest.15.1.02kok

Kok, K., & Cienki, A. (2016). Cognitive grammar and gesture: Points of convergence, advances and challenges. *Cognitive Linguistics*, *27*(1), 67–100. doi:10.1515/cog-2015-0087

Ladewig, S. (2011). Putting the cyclic gesture on a cognitive basis. *CogniTextes*, 6. Retrieved from http://journals.openedition.org/cognitextes/406

Lakoff, G., & Johnson, M. (1980). *Metaphors we live by*. Chicago: The University of Chicago Press.

Lakoff, G., & Nuñez, R. (2000). *Where mathematics comes from. How the embodied mind brings mathematics into being*. New York: Basic Book.

Laszlo, P. (2006). *Communicating science: A practical guide*. Berlin: Springer.

Lopes, J. B. (2017). Researching science teaching practices in the classroom over two decades: What matters? In J. B. Lopes, J. P. Cravino, E. Cruz, & A. Barbot (Eds.), *Teaching science: Contributions of research for planning, practice and professional development* (pp. 1–26). New York: Nova Science Publishers.

Lopes, J. B., Silva, A. A., Cravino, J. P., Santos, C. A., Cunha, A., Pinto, A., & Branco, M. J. (2014). Constructing and using multimodal narratives to research in science education: Contributions based on practical classroom. *Research in Science Education, 44*(3), 415–438. doi:10.100711165-013-9381-y

Lopes, J. B., Viegas, C., & Pinto, A. (Eds.). (2018). Melhorar práticas de ensino de ciências e tecnologia: Registar e investigar com narrações multimodais [Improving teaching practices in sciences and technology: Register and investigate with multimodal narratives]. Lisboa: Edições Sílabo.

Luebke, J. F., & Matiasek, J. (2013). An exploratory study of zoo visitors' exhibit experiences and reactions. *Zoo Biology, 32*(4), 407–416. doi:10.1002/zoo.21071 PMID:23740472

McNeill, D. (1992). *Hand and mind: What gestures reveal about thought*. Chicago: University of Chicago Press.

McNeill, D. (2015). *Why we gesture: The surprising role of hand movements in communication*. Cambridge, UK: Cambridge University Press.

Mortensen, M. F. (2011). Analysis of the educational potential of a science museum Learning environment: Visitors' experience with and understanding of an immersion exhibit. *International Journal of Science Education, 33*(4), 517–545. doi:10.1080/09500691003754589

Mulkay, M. (1972). *The social process of innovation*. London: Macmillan. doi:10.1007/978-1-349-01450-7

Patrick, P. G. (2017). Introduction. In P. G. Patrick (Ed.), *Preparing informal science educators: Lessons for science communication and education* (pp. 1–9). New York: Springer. doi:10.1007/978-3-319-50398-1_1

Pattison, S. A., & Dierking, L. D. (2012). Exploring staff facilitation that supports family learning. *Journal of Museum Education, 37*(3), 69–80. doi:10.1080/10598650.2012.11510743

Pattison, S. A., & Dierking, L. D. (2013). Staff-mediated learning in museums: A social interaction perspective. *Visitor Studies, 16*(2), 117–143. doi:10.1080/10645578.2013.767731

Pedretti, E. (2012). The medium is the message: Unraveling visitors' experiences of Body Worlds and the Story of the Heart. In E. Davidsson & A. Davidsson (Eds.), *Understanding interactions at science centers and museums – A sociocultural perspective* (pp. 45–62). Sense Publishers. doi:10.1007/978-94-6091-725-7_4

Radford, L., Edwards, L., & Arzarello, F. (2009). Introduction. *International Journal of Mathematics Education, 70*, 91–95.

Rennie, L. J. (2013). The practice of science and technology communication in science museums. In J. K. Gilbert & S. Stocklmayer (Eds.), *Communication and engagement with science and technology. Issues and dilemmas* (pp. 197–211). New York: Routledge.

Rennie, L. J. (2014). Learning science outside of school. In N. G. Lederman, & S. K. Abell (Eds.), Handbook of research on science education (Vol. 2, pp. 120–144). Routledge.

Sherr, R. (2008). Gesture analysis for physics education researchers. *Physics Education Research, 4*(1), 1–9.

Smyth, J. (1989). Developing and sustaining critical reflection in teacher education. *Journal of Teacher Education, 40*(2), 2–9. doi:10.1177/002248718904000202

Stevens, R., & Hall, R. (1998). Disciplined perception: Learning to see in technoscience. In M. Lampert & M. L. Blunk (Eds.), *Talking mathematics in school: Studies of teaching and learning* (pp. 107–150). New York: Cambridge University Press. doi:10.1017/CBO9780511571251.007

Stocklmayer, S. (2013). Engagement with science: Models of science communication. In J. K. Gilbert & S. Stocklmayer (Eds.), *Communication and engagement with science and technology. Issues and dilemmas* (pp. 19–38). New York: Routledge.

Stocklmayer, S., & Rennie, L. J. (2017). The attributes of informal science education: A science communication perspective. In P. G. Patrick (Ed.), *Preparing informal science educators: Lessons for science communication and education* (pp. 527–544). New York: Springer. doi:10.1007/978-3-319-50398-1_26

Tal, T., & Morag, O. (2007). School visits to natural history museums: Teaching or enriching? *Journal of Research in Science Teaching, 44*(5), 747–769. doi:10.1002/tea.20184

Tlili, A., Cribb, A., & Gewirtz, S. (2006). What becomes of science in a science centre? Reconfiguring science for public consumption. *Review of Education, Pedagogy & Cultural Studies, 28*(2), 203–228. doi:10.1080/10714410600739921

Tran, L. U. (2007). Teaching science in museums: The pedagogy and goals of museum educators. *Science Education, 91*(2), 278–297. doi:10.1002ce.20193

Tran, L. U., & King, H. (2007). The professionalization of museum educators: The case in science museums. *Museum Management and Curatorship, 22*(2), 129–147.

Tran, L. U., Werner-Avidon, M., & Newton, L. R. (2013). What is successful professional learning for informal educators and how do we get there? *Journal of Museum Education, 38*(3), 333–348. doi:10.1080/10598650.2013.11510785

Valente de Oliveira, L. (Ed.). (2018). O papel dos museus [The role of museums]. Porto: Amigos do Museu Nacional Soares dos Reis.

Vieira, F. (2011). Entre o privado e o público: O diário como lugar de transformação. In M.A. Moreira (Eds.), Narrativas dialogadas na investigação, formação e supervisão de professores (pp. 115–136). Mangualde: Pedágio.

vom Lehn, D. (2006). Embodying experience: A video-based examination of visitors' conduct and interaction in museums. *European Journal of Marketing, 40*(11/12), 1340–1359. doi:10.1108/03090560610702849

vom Lehn, D., Heath, C., & Hindmarsh, J. (2001). Exhibiting interaction: Conduct and collaboration in museums and galleries. *Symbolic Interaction, 24*(2), 189–216. doi:10.1525i.2001.24.2.189

Yeo, J., & Gilbert, J. K. (2014). Constructing a scientific explanation – A narrative account. *International Journal of Science Education*, *36*(11), 1902–1935. doi:10.1080/09500693.2014.880527

Zhai, J., & Dillon, J. (2014). Communicating science to students: Investigating professional botanic garden educators' talk during guided school visits. *Journal of Research in Science Teaching*, *51*(4), 407–429. doi:10.1002/tea.21143

Chapter 18
Multimodal Narratives in Nursing Education:
Exploring Their Potentialities

Marília Rua
Universidade de Aveiro, Portugal & Centro de Investigação em Didática e Tecnologia na Formação de Formadores, Portugal

Rita M. F. Leal
Universidade de Aveiro, Portugal & Unidade de Cuidados na Comunidade Aveiro, Portugal

Nilza Costa
Universidade de Aveiro, Portugal & Centro de Investigação em Didática e Tecnologia na Formação de Formadores, Portugal

ABSTRACT

Nursing education is driven by emerging challenges of scientific, technological, and professional advances that require the use of strategies that promote students' development of critical thinking for decision making in different contexts. It also requires that teachers constantly reflect on their pedagogical practices and (re)think them using strategies that allow their enhancement. The use of multimodal narratives (MNs) can be an important tool for teachers' professional development, namely to improve their classroom practices. Given the novelty of the use of MNs in nursing education, this chapter presents an analysis concerning the experience of making a MN and how it has been reflected in the authors' pedagogical practices. With this experience, potentialities of continuing to use MNs in nurse education are explored.

INTRODUCTION

The use of multimodal narratives (MNs) as an instrument to improve teachers' practices has been intensively researched, since 2008, by a Portuguese team under the coordination of J. B. Lopes (Lopes et al., 2012, 2014). Since 2017, there has been a web page available (http://multimodal.narratives.utad.pt) with information for researchers and teachers about all the work that has been developed so far by the

DOI: 10.4018/978-1-5225-8570-1.ch018

team. From this work, there is strong evidence that MNs play an important role in teachers' professional development. However, this research has been mainly conducted in the field of science education in elementary and secondary schools, although its scope has been expanded, more recently, to other areas, like mathematics, information & communication technologies and to higher education (HE). The idea of studying the potentialities of the use of MNs in nursing education in HE was the main reason for conducting the analysis to be described in this chapter.

The chapter starts with an overview on the complexity of nursing education in HE, namely at a postgraduate level, due to the challenges faced by the nursing profession nowadays. This section also explores the professional requirements that nursing educators should have, particularly which professional competences are concerned, and the instruments that have been used to develop them. Then, and given the characteristics of a MN, a comparative analysis is made between the potentialities of the use of narratives and MNs which leads to the relevance of the use of MNs in nursing education. Furthermore, a description and analysis of the use of a MN in nursing education is conducted, specifically in a postgraduate class, and the implications for the professional development of the teacher are explored. The last section includes the main conclusions of this analysis concerning its potentialities, the limitations of this study, and recommendations for future work.

BACKGROUND

Nursing education integrated into higher education (HE), as it is in Portugal, has undergone deep changes, not only to follow the scientific–technological evolution of the area, but also the paradigm's changes in nursing care and the pedagogical-didactics approaches recommended for HE. Today it is expected that a nurse, in the context of clinical practice, is a professional who supports his/her practices with scientific evidence for a contextualized decision-making in order to care for the person, family, groups and community throughout the life cycle, and to follow them up in their transitions and in the management of their health projects (Costa, 2011; International Family Nursing Association/IFNA, 2017; Rua, 2011).

Regarding the health-care contexts, where nursing practices most often occur, and where decision-making based on a set of specific competencies can be observed, these are highly demanding and unpredictable, not only from the point of view of the health/ illness conditions of the users, but also from their involvement in their health projects. This requires nurses' competencies in different dimensions, which are developed from initial education, but which must be extended throughout life, particularly when leading to a specialist level in a specific nursing area (IFNA, 2015; Popil, 2011). Given these assumptions, the training of these professionals, at the specialization level (always in the context of HE), requires a constant movement to improve the pedagogical practices of their teachers based on strategies that promote the development of new and deeper competences, such as critical thinking, considering that they are postgraduate students and, most of them, have been working for some years.

Taking on board the literature about teaching practices in the context of nursing education at the second cycle (postgraduation), where this study took place, it is evident that there is a constant need to improve them through more active strategies and critical-thinking activities, among other competencies.

For years in HE, students have been educated through what is identified today as traditional lectures (expository teaching). Active-learning strategies have been progressively widespread and researched, and they "...contrasted to the traditional lecture where students passively receive information from the instructor" (Prince, 2004, p. 223). A definition of active learning may be "...any instructional method that engages students in the learning process. In short, active learning requires students to do meaningful learning activities and think about what they are doing" (Prince, 2004, p. 223). Active-learning interventions have shown benefits in education, specifically with the improvement in the students' exam scores and better passing rates when compared to traditional lecture classes (Freeman et al., 2014).

Active-learning interventions include a variety of strategies and approaches that encourages students' interactions, and can include inquiry-based learning, case studies, collaborative learning, among others. Even when the teacher's lecture is centered on the presentation of a power-point, the students should not only be looking at the slides, but also debating the teacher's questions, which will also include asking the students to critically analyze the information presented (Center for Research on Learning and Teaching of the University of Michigan, 2016).

As one example of an active-learning strategy is the problem-based approach (PBA). This guides the students to solve practical multifaceted realistic problems and is believed to develop enduring skills and to give the students the role of collaborators in their learning process (Hmelo-Silver, 2004). PBA learning also triggers intrinsic motivation, as it "...occurs when learners work on a task motivated by their own interests, challenges, or sense of satisfaction" (Hmelo-Silver, 2004, p. 241).

Case studies or case methods in teaching are another example of an active and reflective learning strategy that has been proven effective in enhancing clinical problem-solving, promoting critical thinking, making knowledge meaningful, self-directed, enduring and applicable to practical contexts and have become popular in nursing education (Popil, 2011; Tomey, 2003).

Another important dimension to consider in pedagogical contexts is nonverbal language. Loio and Ançã (2006) identified the importance of silence and body gestures (for example, posture, movement of the body in its whole or in parts, facial expression, gestures) in the classroom. This type of language, besides being an information vehicle for students and teachers, assumes great relevance in the classroom, contributing to a greater understanding of the process of interaction between both. The teacher's gestural attitude and students' gestural signs, which are always emitted, are elements that determine the process of teaching and learning (Loio & Ançã, 2006).

In the search for updating and improving their pedagogical practices, the nursing teacher has at his/her disposal a set of approaches and tools that can enhance the development of competences that sustain this improvement (Berger & Karabenick, 2011; Guney & Al, 2012; Ideaslab & Paper, 2011; Lopes et al., 2012).

The reflective approach, initially referenced by Schon (1983), but widely disseminated in the educational context by several authors (Alarcão, 1996; Sá-Chaves, 2002), is one of the tools available to teachers. Its relevance lies on the possibility that teachers have to access a deliberate and active process of self-critical analysis of their practices, so they can change by making them more adequate to the contexts and students' formative process.

Teachers' reflective practices in different moments, namely "*on* and *about* the action," have in themselves a potential for development that will have diverse effects if accompanied by different support tools/instruments (Lopes et al., 2012). Several instruments can/and have been used, namely narratives, in the format of diaries (Dias, Pitolli, Prudencio, & Oliveira, 2013; Zabalza, 2004), reports of significant episodes (Kiely, Davis, & Wheeler, 2010), class audio and/or video recordings, with subsequent analysis (Bateman, 2016), and simulation using debriefing (Jong, Lane, & Sharp, 2012). Notice, however, that each of these instruments use only one way of collecting data about teachers' practices (teacher's written record, audio/video records), which may not preserve the holistic and complex nature of the teaching, learning and assessment process.

In this context, multimodal narratives (MNs) emerge as a tool that can support this reflection process, acting as a promoter of the development of teachers' competences, if configured and analyzed under different perspectives and by different sources. They can also act as mediators of the professional development process, namely if the teacher has the opportunity to reflect together with their peers (Lopes et al., 2012).

Continuing the work of the research group led by Lopes, a MN is a document that makes a chronological and detailed description (without any author opinions), about a certain context (classroom) and about the participation of each person (teacher and students). Using the authors' words: "(...) a MN is of multimodal nature. This is strongly connected to the multimodal nature of teaching and learning process. A MN explicitly depicts context information, actions, decisions, intentions, dialogues, gestures, silences, documents, schemes, objects, images and other multimedia resources and interactions" (Lopes et al., 2014, p.421). The authors also add that a MNs can allow the verifiability of the narrated events and the comparability between MNs written by different teachers.

It is important to identify the differences between narratives and MNs. Table 1, adapted from Lopes et al. (2014)), synthesizes them according to their focus, structures, formal characteristics, data used and format.

According to the MN specific characteristics, it can be stated that their contribution to the development of the teacher's pedagogical–didactic competences depends not only on their reflective ability, but also on the tools that support reflection from the conceptual and procedural point of view, in order to find new ways for their professional practice and increase their knowledge while acting as mediators in the students' learning experiences (Lopes et al., 2012). Concerning the conceptual point of view, and considering the role of the teacher as mediator in the learning contexts, the above-mentioned authors present in a systematized form, theoretical evidence of this process, affirming that the teaching practices can be thought of as having two dynamics that influence and complement each other: (i) an interaction

Table 1. Differences between narratives and multimodal narratives (adapted from Lopes et al., 2014, p. 422)

Differences according to the ...	Narratives	Multimodal Narratives
... focus ...	... is variable according to the narrator.	... is the same among narrators.
... structure takes into account	... the connectedness and coherence; the direction through time.	... the tasks requested of the students and their sequence in a chronological order and based on episodes, each one beginning with the presentation of a task. ... two parts with: the explicit contextual elements and a summary of all episodes (part I); the detailed narration of each episode (part II).
... formal characteristics ...	... has a plot (generally without any protocol). ... only the point of view of the narrator (the teacher) is given.	... has a plot with a protocol, which includes, for example the sequence of episodes, temporal sequences of events. ... teacher's point of view encompasses his/her epistemological, ethical, and educational perspectives and choices.
... data used ...	... one source only.	... several types of independent data (e.g. audio and/or video recording, documents produced by the teacher or students).
... format is...	... a document only.	... a single document but with a storyline composed of sequential events, with multimodal elements.

dynamics with the other (teachers, students, ...inside and outside the room); and (ii) a dynamics of interaction with the epistemic object (Lopes et al., 2012). For these dynamics, the authors define ten dimensions of mediation, five linked to each of the two types of interaction.

The dimensions of mediation linked to the interaction dynamics as the epistemic object (A) correspond to:

A1: Work requested from the students.
A2: Scientific and technological context.
A3: Epistemic and / or axiological practices.
A4: Information.
A5: Teacher awareness and real-time classroom decision-making.

In relation to the dimensions of mediation linked to the interaction dynamics with the other (B), they integrated:

B1: Conversation in the classroom.
B2: Support and authority granted to students.
B3: Productive involvement in the subject.
B4: Evaluation and feedback.
B5: Induced learning.

Considering each of the teachers' ten dimensions of mediation, the identification of pedagogical practice aspects, most evident as facilitators or inhibitors of students' learning process, emerges.

In summary, there are several aspects that underlie the pedagogical competences of the teacher, which can be evidenced in a MN. In this assumption, each MN must contain the essential data for its analysis (Lopes, 2010, 2012, 2014, 2017).

THE USE OF THE MULTIMODAL NARRATIVE IN A NURSING EDUCATION CONTEXT

As mentioned above, to be a nurse educator requires having and maintaining a relevant and up-to-date curriculum in clinical practice, but also in pedagogical competences. These are essential to promote the use of dynamic strategies for teaching, learning and assessment.

Nursing education, as in other teaching areas, has changed substantially due to several challenges (emerging from cultural, economic and scientific features), as well as to student and learning contexts. A shift in nursing education has occurred from a teacher-centered curriculum to a student-centered learning, where critical thinking and problem-solving are the desirable outcome for both teacher and student (Stanley & Dougherty, 2010). Based on the acknowledgment that: "Nurse educators can no longer teach the way they were taught" (Stanley & Dougherty, 2010, p. 380), the authors sought, with this work, a chance to analyze and critique their conduct as nursing educators and develop pedagogical competences. Given the features of MN, stated in the previous section, the authors of this chapter chose it as the instrument to better understand the work carried out in the classroom context of a postgraduate nursing course, with physically present students and others attending online. Thus, they set out this analysis to understand their competencies and weaknesses as educators, with the aim of improving the strategies used in the classroom and adjusting them to the students' needs.

The authors followed the MN protocol published by Lopes et al. (2014), which divides it into three phases, namely: Phase 1 – Data collection; Phase 2 – Construction; and Phase 3 – Validation.

In Phase 1 the teacher sought each student's consent. They referred to feeling more comfortable with audio recording, rather than video recording. Therefore, data collection was audio-recorded, and photographs of the classroom and board were taken. The names used in the MN were changed to guarantee anonymity. The classroom included six physically present students and four assisting remotely. The students' consent, for audio recording and picture taking, for the construction of the MN was unanimously and previously obtained. The session then proceeded as normally as possible, so as not to interfere with the dynamics of the previous sessions. Despite this, the teacher noticed that the class was initially under some pressure, although shortly after, the participants (students and teacher) ended up forgetting that the class was being recorded.

Later, the teacher gathered independent data, basically using a smartphone for the audio recording and the photo taking. Some of these included the classroom and student characteristics, and pictures of what was written on the board (see Figure 1).

The teacher's power-point presentation also served as a multimodal element, which contributed to a better understanding and enrichment of the MN, as evidenced by the example shown in Figure 2.

The authors also gathered pictures of the students' annotations in their notebooks, as shown in Figures 3 and 4.

Later, all the data was downloaded onto a computer, and furthermore, the teacher's perception and intentions during the class (dependent data) were annotated.

In Phase 2 began the construction of the MN by listening to the audio recording, choosing segments to transcribe onto a Word® document, exactly as it was heard thus guaranteeing confidence. To be able to cross-reference, other modal elements were added (examples: Figure 1–5 and dependent data) and posteriorly analyzed. In this process it was necessary to reread the transcription of the audio recording

Figure 1. Picture of what the teacher wrote on the board during class
[Translation: IGg- immunity (g-good); IGm-recent infection (m-bad)]

Figure 2. A picture of a slide presentation

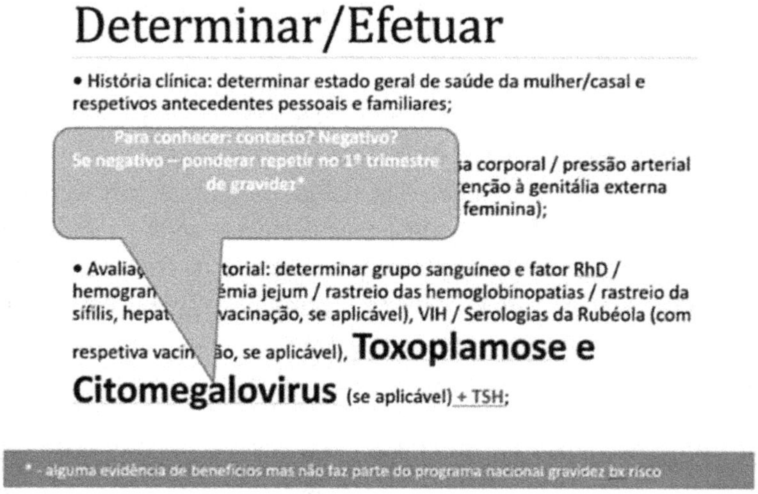

Figure 3. A picture of a student's notebook
[Translation: CMV (illness that does not confer permanent immunity)
There may be infection but the ability to produce damage is ↓ decreasing risks to the fetus
CMV with IGg positive =the existence of contact
TSH – influence of thyroid activity on maternal hormones and the fetus development.
There is a recommendation to supplement with iodine in pregnancy.
We must guarantee a good thyroid function.]

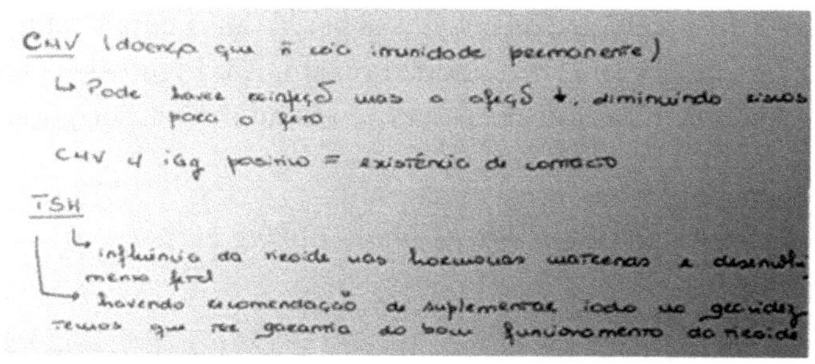

Figure 4. A picture of another student's notebook
[Translation: IGg- immunity (g-good); IGm- recent infection (m-bad)
If both are positive it signifies a recent infection.]

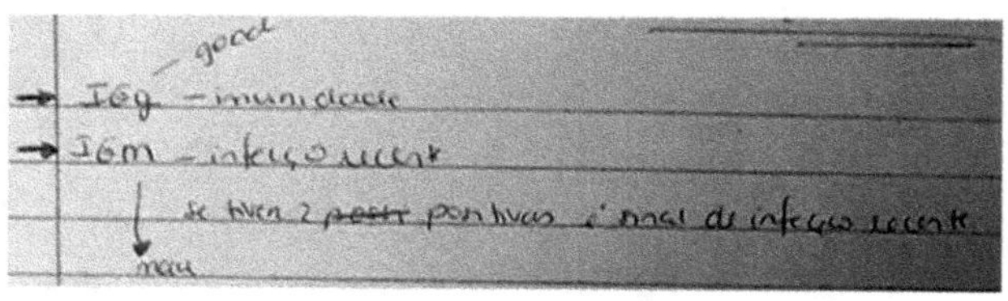

several times to clearly and identify and frame all events, including interaction between the students and the teacher, but also the attitudes and silences experienced during classes (Lopes et al., 2014). This was undoubtedly the first difficulty. The second difficulty was the transposition of the narrative discourse into each episode, contextualizing it in time and space, thus capturing each moment, so that readers can understand it. Although, these difficulties are recognizably transient, as this was the first time constructing a MN.

The authors then sketched a map of the classroom (Figure 5) and listed its resources. The MN was improved and enriched progressively with all the multimodal elements.

In Phase 3 the authors proceeded to the MN validation, through an external expert with a vast experience. This phase guarantees the precision and consistency of the MN. As a result of this external review, the MN was completed with a more in-depth description of the different episodes and the time span between the episodes.

Subsequently the MN served as a valuable empirical material, to pursue a reflexive process.

The MN also enabled the understanding of what it is to be "on the other side," when attending our own class. This reflective process permitted a genuine self-evaluation, allowing new perspectives of professional identity and a broader professional self-awareness. In addition, the reflexive practice was more effective and achieved a higher level of understanding, because it was shared with colleagues, acting as "friendly critics" who permitted a collaborative evaluation (Yeigh, 2008). This process requires maturity and a determination to progress as an educator, due to the MN exposure to colleagues. In this process the colleagues ensured a constructive, honest and a nonjudgmental collaborative evaluation, making the MN analysis a valuable outcome.

Figure 5. Map of the classroom and the students' location

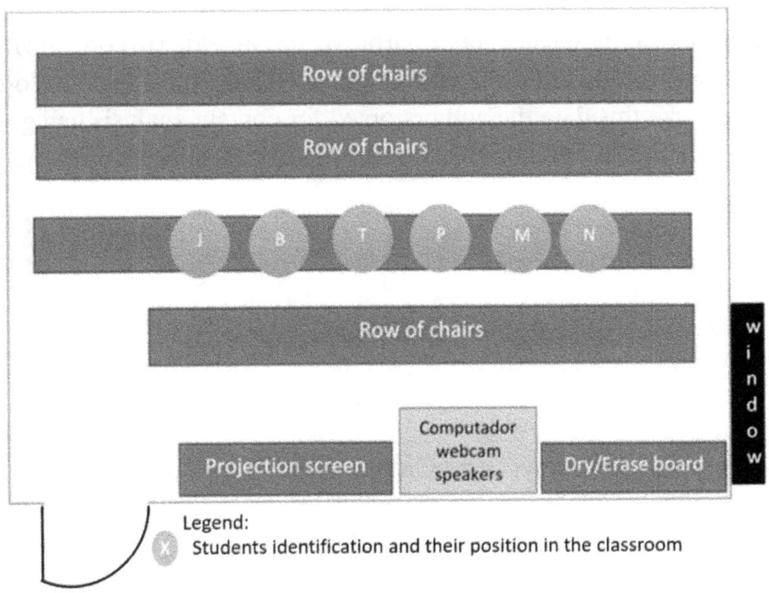

THE MULTIMODAL NARRATIVE ANALYSIS

This analysis, which underlies the construction of a MN, was carried out with the purpose of complementing the author´s own pedagogical competencies within the scope of nursing education. In the expectation of understanding how and to what extent this MN is an important element in professional development, through reflection on practices, it will be analyzed under different "lens" considering also its bioecological dimension (person, process, context and time). As a driving force for professional development (not only in terms of the development of pedagogical practices), the construction of this MN, demanded knowledge of MNs (What they are. How are they constructed.), as well as the influence they may have on the development of pedagogical competencies. This analysis, as referenced before, was supported on the research and experience of the work led by Lopes. Other authors also revealed themselves to be a valuable contribution for the construction and application of MN (Bateman, 2016; Faria, Faria, & Ramos, 2013; Jewitt, 2008). The need to (re)visit concepts and perspectives related to reflective practice was also identified (Alarcão, 1996; Bulman & Schutz, 2013; Caldwell & Grobbel, 2013; Duffy, 2007; Marilia Rua, 2012; Sá-Chaves, 2002; Schon, 1983), with the development of competencies (Anema & Mcoy, 2010; IFNA, 2017; Perrenoud, 2000; Rua, 2011) and educational strategies (Carroll, 2015; Patterson et al., 2017; Russell, Comello, & Lee Wright, 2007; Tavares, 2003; Waterkemper & Prado, 2011). The possibility of constructing a MN, alone, allowed the authors to begin a reflective process on their own pedagogical practices, forcing them to confront what they had experienced, its meaning, not only for their professional development but also for the construction of their students' knowledge. This MN allowed the authors to deepen reflection (reflection-on-action and reflection-on reflection-in action) (Schon, 1983), and made it possible to systematize the most important events.

This MN refers to a midwifery course (at postgraduate level) for registered nurses. The content was prenatal care, and the subject specifically addressed was the screening for infectious diseases, thyroid hormone levels and body mass index in the preconception stage, and the interpretation of these results for pregnancy monitoring.

Given that this is an analysis based on a single narrative, it is intended to compare its elements with the theoretical perspectives and highlight those that may act as enhancers for the development of pedagogical competences. To analyze this data the authors opted for content analysis using webQDA software, which enabled to structure the "reflection on action." Four dimensions emerged, as shown in Figure 6.

Figure 6. Illustration of the dimensions under analysis in the multimodal narrative

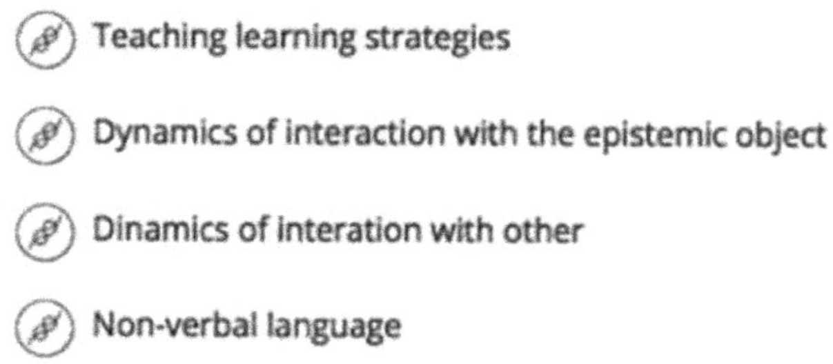

Regarding the dimension "Teaching learning strategies" the following categories emerged:

- Problem-based approach;
- Clinical case study approach;
- Questioning;
- Expository/lecturing approach with slide presentation;
- Explanatory/ illustrative using a dry erase board.

It was evident through the MN that the teacher used different strategies throughout the lesson.

Multimodal Narrative Excerpt 1 (Problem-Based and Clinical Case Approach)

I have a woman whom in her preconception was negative for cytomegalovirus. That is, a IGg and IGm negative. Then, we have a positive result for IGg and IGm in the first trimester of pregnancy. I must alert the risks that exist for the fetus.

Problem-based approach and clinical case study interventions were identified (Excerpt 1), although it was possible to identify that the teacher did not give the students enough time to reflect, discuss and find the solutions. Instead, the teacher gave them the answers before they could do so, as is illustrated below.

MN Excerpt 2 (Questioning)

Why is it important to evaluate the women's weight in preconception? What is weight gain during pregnancy related to? What is advisable? (However, the teacher gave the answers).

Questioning in the classroom as a teaching approach has been used over time. In this MN the authors verified that questioning was the most used teaching learning strategy, despite the need to give more time for students to think, discuss and find the answers.

MN Excerpt 3 (Expository Approach)

So, then I point to the slide (Figure 2) and explain …

The expository teaching approach, with reference to the slide presentation prepared to support the subject, was the second most recorded approach throughout the MN.

Analyzing the MN, and specifically the part when the teacher sketched an illustration on the board (Figure 1), it was possible to identify the use of a mnemonic and the teacher's perception and intention in using it. It was noted, through MN analysis, that the teacher was attempting to clarify the difficulties she perceived the students had, due to their nonverbal communication. In fact, in the MN this mnemonic was found to have been copied into some of the students' notebooks (see Figure 4). However, it was noticed that the teacher only used the board once and proceeded with a single illustration.

Regarding the dimension "Dynamics of interaction with the epistemic object," specifically the teacher as mediator of learning, the following categories are defined:

- Work requested from the students;
- Scientific and technological contexts;
- Epistemic and / or axiological practices;
- Information;
- Teacher awareness and real-time classroom decision-making.

In the MN analysis, the authors found elements that illustrate some aspects of practice that combines the above categories thus allowing to sustain the reflection on pedagogical competences.

MN Excerpt 4 (Work Requested From the Students, Scientific and Technological Contexts)

So, girls, how much can you gain? Each one of you should analyze your BMI and see. At home: Madalena, Andrea and Marli, (catching the attention of the students who are watching by videoconference), How much can you gain during pregnancy, in total?

Regarding the work requested from the students, this only occurred once during the lesson. As for the "scientific and technological contexts," these were present several times in class. This process of mediation allowed the teacher to present "real situations" for students to demonstrate knowledge acquisition.

MN Excerpt 5 (Epistemic and/or Axiological Practices)

In an advanced stage of pregnancy, cytomegalovirus infection is less teratogenic, as it will not interfere with the early embryonic development. On the other hand, at this stage there is no longer any possibility of a legal medically assisted pregnancy interruption.

The epistemic and axiological practices were identified in the MN. This was used by the teacher and described as shown in this excerpt. The description of phenomena and the recognition of phenomena in their context, are according to Lopes et al. (2012), important elements for the contextualization of knowledge, moving from empirical language to conceptual language. This dimension allowed the students to recognize the phenomena in its context.

MN Excerpt 6 (Information)

This slide illustrates what the Directorate-General for Health proposes in the Low Risk Pregnancy Surveillance Program, and specifically in relation to the preconception consultation, in relation to diagnostic tests for toxoplasmosis and cytomegalovirus.

This dimension was present during the lesson, as observed through the MN.

The information was presented to the students, framed in the theoretical references and addressed in class, based on scientific evidence and normative protocols, the national low-risk pregnancy surveillance program, this being referenced orally and simultaneously illustrated with slide presentation.

MN Excerpt 7 (Teacher's Awareness and Real-Time Classroom Decision-Making)

So, I put the acronyms of the immunoglobulins that are investigated in the blood serum, namely IGg and IGm on the board. I found through nonverbal communication that the students appeared to confuse the two acronyms, so I decided to share a mnemonic that might be useful and could help distinguish and clarify them.

Teacher's awareness and real-time classroom decision-making was evidenced in the MN several times, which allowed the teacher to reinforce the information, or to diversify the way it was presented by attempting to enable the students to access and integrate it.

In relation to the dimension "Dynamics of interaction with the other," the categories defined were:

- Conversation in the classroom;
- Support and authority granted to students;
- Productive involvement in the subject;
- Assessment with feedback;
- Induced learning.

MN Excerpt 8 (Conversation in the Classroom and Support and Authority Granted to the Students)

I continued the lesson with Natalia's question: "Wasn't there a study in Aveiro with the use of women's placenta?" To which I replied – "I do not know. I do not know, but it would be interesting to find out."

Natalia insisted: "The study was from the University of Aveiro. It was on television, on the news." Since I could not get her to resume her attention to the subject I had intended to cover in class, which was, weight gain during pregnancy, I decided to give her some attention and try to understand what the study was about.

The conversation in the classroom was a constant, between teacher and students, in an interactive way, thus allowing the observation that there was an interconnection between the teacher's discourse and the way students complemented the given information. As for the support and authority granted to the students, the teacher conducted the class supporting the students in clinical reasoning and integrating their interventions.

Regarding productive involvement and assessment/feedback, these were scarcely used. Although, this may be because the focus is on a single MN and not on a set of classes in which a moment for assessment could also be foreseen. Likewise, the teacher did not refer or suggest additional study topics or reading that could be extended beyond class.

Nonverbal language also emerges from MNs as an important element in the mutual understanding between students and teacher. Within this dimension the following categories emerged:

- Gestures;

- Facial expressions;
- Silences;
- Positioning in the room.

MN Excerpt 9 (Gestures and Facial Expressions)

I shook my head assertively, smiled, and explained: I'm laughing because Natalia is thinking at this time – Here comes another Bible (…) and I added in a joking tone: That's why she chose to be a nurse (everyone laughs ...). It is to be able to escape from sick people.

The gestures followed/complemented oral language and emerged as reinforcement of information, which, in the classroom, can support teachers and students. The most used gestures were without doubt made with the hands, or with the head, often allowing, in and of itself, to regulate the learning environment. In a MN it is possible to identify gestures of the teacher, especially with a sense of agreement with the student, thus encouraging their interlocution. Facial expressions also constituted an important factor in the mediation of learning, reinforcing interventions or results, or trying to draw attention by demonstrating non-compliance with student behaviors. It can also be used to ease the distances between teacher and student, namely through smiles. This facial expression is often associated with tone of voice.

MN Excerpt 10 (Silences)

There was silence for a few seconds. But their attitude showed that they were following the topic, thinking about the answer and that they had some doubts.

Silences are moments that can be indicators of reflection, indecision, transmission of emotions or feelings. In the classroom these silences can be good indicators for the teacher, especially if they happen after the teacher's interpellation, with global questioning to the class or individually. They may be indicators of lack of attention, or that they simply cannot keep up with what the teacher is presenting or need time to process information and build response.

MN Excerpt 11 (Positioning in the Classroom)

Barbara was distracted and talking to the side. I tried to catch her attention by moving myself towards her in the classroom …

Concerning positioning in the classroom, this was often used by the teacher to encourage student participation, or to get their attention by shifting and diverting their focus.

It is evident that in the MN, there were relevant elements found that sustain the authors´ reflection on pedagogical practices.

MULTIMODAL NARRATIVES' IMPLICATIONS ON THE DEVELOPMENT OF PEDAGOGICAL COMPETENCES

It was noted in the MN that the teacher used problem-based interventions to stimulate critical thinking, a characteristic that is fundamental in nursing education (Stanley & Dougherty, 2010). Its application in this type of class was an interesting strategy given that they were postgraduate students and all professional registered nurses, and so had some knowledge of the content although not as profound as necessary for a midwife. In addition, the application of the problems associated with students' experiences is recognized in literature to be a means to support intrinsic motivation (Hmelo-Silver, 2004).

The case study approach was another strategy applied during the class. This type of active learning strategy is documented in the literature to enhance critical thinking and problem-solving analysis (Popil, 2011; Tomey, 2003). In this MN it was relevant that the teacher applied the case method strategy, which is similar to case study but smaller and directed to an explicit objective (Tomey, 2003). Although a meaningful strategy for nursing education (Popil, 2011; Tomey, 2003), it should be stated that through MN analysis the authors consider that the teacher did not give the students much time to reflect, discuss and find solutions, but gave them the answer before they could do so. Therefore, the use of this strategy needs improvement.

The MN analysis revealed that questioning was the most used teaching strategy during the class, even more than the expository/lecturing approach, which differs from research that directs that questioning was found to be used second to lecturing (Cotton, 2001). Indeed, the use of "...questions and answers to challenge assumptions, expose contradictions, and lead to new knowledge and wisdom is an undeniably powerful teaching approach" (Cotton, 2001, p.1).

The expository or lecture teaching was the second most recorded approach in this MN. Although it is recognized in the literature as a traditional method with poor student engagement (Prince, 2004), it is also noted as an efficient learning strategy and serviceable for time-saving (Maheshwari, 2013). In contrast to the findings referenced by Cotton (2001), it was not the most popular strategy found in this MN.

Active learning's interventions, which implies the pursuit of "meaningful learning activities," are known in the literature to enhance students' results when compared to traditional expository/lecture classes (Freeman et al., 2014). It is also known that "...education should provide active learning that promotes critical thinking and analysis and problem-solving skills" (Stanley & Dougherty, 2010, p. 379). On the other hand, according to Prince (2004, p. 7), "Teaching cannot be reduced to formulaic methods and active learning is not the cure for all educational problems." The authors believe that the strategies (active and expository) were used with balance in this class.

With regard to the illustrative strategy with recourse to the mnemonic technique used by the teacher in an attempt to clarify the perceived difficulties of the students, the authors believe it was able to guide the students in their learning process, because it does remain consistent in the literature that "carefully constructed illustrations continue to receive high marks..." (Carney & Levin, 2002, p. 23), and the use of the mnemonic technique has been demonstrated as "...improving both student's memory for and application of central textual information" (Rummel, Levin, & Woodward, 2003, p. 327). In fact, in this MN this mnemonic was found to have been copied into some of the students' notebooks (Osmond & Barker, 2000) as shown in Figure 4, although, through the content analysis it was possible to understand that the teacher only used the board once to illustrate and clarify ideas. The authors of this chapter consider that the teacher could have taken advantage of this resource and used it more often, and not do much the slide presentation which revealed to be used as the main support.

In relation to the "Dynamics of interaction with the epistemic object," the analysis of this MN allowed the verification that although there were interactions in all the categories defined by Lopes et al. (2012) there is still some work to be carried out towards the development of competences adjusted to the students in formation.

The use of *work requested from the students* in this case could have been utilized more often, since they already had a background in nursing that would allow them to integrate this knowledge and perform more complex tasks. Therefore, the *learning by doing* strategy, referred to by John Dewey, cited in Bot, Gossiaux, Rauch, and Tabiou (2005), could be effective.

According to Hill and Smith (2005), cited in Lopes et al. (2012), in the MN references emerged the "scientific and technological contexts" dimension used in the classroom, which is congruent with pedagogical practices to be used in higher education. This process of mediation can still be improved in the authors' competencies, while promoting an effective knowledge acquisition.

The "epistemic and axiological practices" used by the teacher and described in the MN, such as the description of phenomena and its recognition in the context, is seen, according to Lopes et al. (2012), as an important element for knowledge contextualization, moving from empirical to conceptual language. This dimension was present during the lesson, as observed through the MN.

The *information* provided to the students by the teacher during class, as along with *the real-time decision-making in the classroom,* were evident and used in an appropriate way and marked by theoretical and normative references on the one hand and adjusted to the feedback from the students on the other, not only through the interaction, but also by the nonverbal behaviors described in the MN. This is congruent with the current perspectives in HE (Eison & Ph, 2010; Pereira, Caetano, Moreira, & Ataíde, 2015; Rantavuori, Engeström, & Lipponen, 2016).

The dimension *teacher awareness* was present in the MN, converging with the theoretical references that have been presented regarding teaching–learning strategies, and evidencing the pedagogical competence of the teacher at the student level (Berger & Karabenick, 2011; Draganoc & Sanna, 2015)

Concerning the *dynamics of interaction with others*, pointed out by Lopes et al. (2012) as an important element in the mediation of teachers, it was noticeable that throughout the class there was much interaction *(classroom conversation)* with and between the students physically present and the teacher. This interaction revealed a positive learning environment where students felt at ease in discussing, interrupting and asking questions. However, this was not observed with the students who were participating online. This demonstrates the lack of preparation of the teacher to manage mixed classes (with online and face-to-face students). This type of class is widely used nowadays, especially in postgraduate education, which demands from the teacher a (re)constructing of his/her pedagogical practices in order to preserve the same dynamics and interactivity. Now teachers "...have the good fortune and opportunity to break new ground, discover new possibilities, as they reconceptualize the nature of teaching and learning in a technology-rich learning world" (Ideaslab & Paper, 2011, p. 3).

As for the *support and authority granted to the students*, it was evident in the MN that the teacher as mediator of the learning process used it in different situations, being thus congruent with the perspective of Lopes et al. (2012), embodied by other authors (Felder, Woods, Stice & Rugarcia, 2000; Reiser, 2004; Redish, 2003).

The *productive involvement, assessment and feedback*, as well as *inductive learning*, are important for the teacher's approach as a mediator in learning (Lopes et al, 2012; Thurlings, Vermeulen, Bastiaens, & Stijnen, 2013), although they did not have great expression in the pedagogical practices of the teacher.

Although this emerged in the analysis of a single class, this result could indicate a need to develop them in the future.

Finally, regarding nonverbal language and specifically gestures, facial expression, silences and the positioning in the room, the MN analysis showed they were elements that the teacher used and took into account, adjusting the interaction with the students in response to their interpretation. This is in accordance with the perspectives of Loio & Ançã (2006) and Sousa, Leal, & Sena (2010).

This study reveals that MN is indeed a tool that allows the teacher to reflect on own teaching practices, not only because there is a self-confrontation when constructing the narrative while (re) visiting the class, but also because it simultaneously allows a critical evaluation that should be supported with scientific evidence. This reflection, shared with peers acting as "critical friends," also promotes the possibility of understanding teacher performance and support changes necessary to (re)construct an enhanced path.

CONCLUSION

The main conclusion of this analysis is the culmination of a project that, now, allows a new vision on strategies on the development of pedagogical competences, namely the use of MN in HE. The research and the reflection inherent in the construction of the theoretical framework that underlies this work were, as shown, elements that enhance teachers' professional development.

After completing the MN analysis, the following main conclusions are drawn about the potentialities of this tool:

- To increase the recognition of (a) the determinant role of *reflection in the action* and *on action*, in a specific context, concerning not only the pedagogical approaches used but also how often they occur during the class, and (b) the relevant use of active and innovative pedagogical strategies for learning (such as clinical case studies and problem-based analysis), enabling the teacher to reach a high level of students' interaction and motivation, and that in nursing education their use should be elected to stimulate critical thinking for decision-making;
- To allow a look at the teacher's role as mediator of the students' learning;
- To promote the development of the teacher's pedagogical competences, by deepening and (re) structuring pedagogical strategies adjusted to the situation of the class, having considered not only the level of education, but also the students' ability to interact and their motivation. In this situation, the MN analysis revealed that there is a need to improve and acquire competences in the management (a) of the time given to students to think, discuss and find solutions to the teacher's request and (b) of mixed classes, maintaining the level of interaction desirable and balanced (on-line and physically present students). It is also apparent through this MN that nonverbal language during class is powerful but should be carefully explored, although this is difficult to manage with online students.

Thus, MN is reaffirmed as relevant for the development of teacher competences in a teaching domain not explored before – nursing education in HE.

Study Limitations

Although the results go beyond the initial expectations of the authors, it is possible to identify limitations of this study: their inexperience in using MN; the presence/absence and balance of strategies and resources used during the class. Although this last limitation may result from the need of the teacher to develop his/her professional competences, it may also be because this study involved only one teacher and one class.

Proposal for Future Work

Although positive results were found with the use of MN in nursing education in HE, being this a single analysis, it is necessary to increase its use in more classes, by more teachers and in more diverse formative contexts. One suggestion is to apply MN not only in formative classes in HE, but also in clinical practices. To deepen the knowledge about MN in nursing education and practices, the experiences of its use should also be the object of research.

REFERENCES

Alarcão, I. (1996). Reflexão crítica sobre o pensamento de D. Schön e os programas de formação de professores. In *Formação reflexiva de professores. Estratégias de supervisão*. Porto: Porto Editora.

Anema, M., & McCoy, J. (2010). *Competency-based nursing education*. New York: Springer Publishing Company, LLC.

Bateman, J. A. (2016). From narrative to visual narrative to audiovisual narrative: The multimodal discourse theory connection (Invited Talk). *OASIcs-OpenAccess Series in Informatics*, *53*(1), 1–11. doi:10.4230/OASIcs.CMN.2016.1

Berger, J. L., & Karabenick, S. (2011). Motivation and students' use of learning strategies: Evidence of unidirectional effects in mathematics classrooms. *Learning and Instruction*, *21*(3), 416–428. doi:10.1016/j.learninstruc.2010.06.002

Bot, L., Gossiaux, P.-B., Rauch, C.-P., & Tabiou, S. (2005). "Learning by doing": A teaching method for active learning in scientific graduate education. *European Journal of Engineering Education*, *30*(1), 105–119. doi:10.1080/03043790512331313868

Bulman, C., & Schutz, S. (2013). Reflective practice in nursing. In C. Bulman & S. Schutz (Eds.), *Nurse education today* (5th ed.; Vol. 1). Oxford, UK: Wiley-Blackwell.

Caldwell, L., & Grobbel, C. (2013). The importance of reflective practice in nursing. *International Journal of Caring Sciences, 6*(3), 319–326.

Carney, R. N., & Levin, J. R. (2002). Pictorial illustrations still improve students' learning from text. *Educational Psychology Review*, *14*(1), 5–26. doi:10.1023/A:1013176309260

Carroll, J. (2015). *Tools for teaching in an educationally mobile world* (1st ed.). London: Routledge Edition.

Center for Research on Learning and Teaching of the University of Michigan. (2016). *Introduction to active learning*. Retrieved from http://www.crlt.umich.edu/active_learning_introduction

Costa, M. (2011). *Mais saber, Melhor Enfermagem : A repercussão da formação na qualidade de cuidados*. Universidade Lusófona de Humanidades e Tecnologias.

Cotton, K. (2001). Classroom questioning. *School Improvement Research Series, 3*, 14. doi:10.1080/00221347308981289

Dias, V., Pitolli, A., Prudencio, C., & Oliveira, M. (2013). O Diário de Bordo como ferramenta de reflexão durante o Estágio Curricular Supervisionado do curso de Ciências Biológicas da Universidade Estadual de Santa Cruz – Bahia. S. Paulo: IX Encontro Nacional de Pesquisa em Educação em Ciências.

Draganoc, P., & Sanna, M. C. (2015). Avaliação da s competências dos professores de enfermagem para facilitar a aprendizagem de adultos. *Cogitare Enfermagem, 20*(3), 556–564. doi:10.5380/ce.v20i3.41104

Duffy, A. (2007). A concept analysis of reflective practice: Determining its value to nurses. *British Journal of Nursing (Mark Allen Publishing), 16*(22), 1400–1407. doi:10.12968/bjon.2007.16.22.27771 PMID:18361389

Eison, J., & Ph, D. (2010). *Using active learning instructional strategies to create excitement and enhance learning. East Fowler*. University of South Florida.

Faria, Á., Faria, P., & Ramos, M. (2013). Interactive multimodal narrative as an approach to developing emergent literacy in early childhood education. *Sino-US English Teaching, 10*(4), 328–333.

Felder, R. M., Woods, D. R., Stice, J. E., & Rugarcia, A. (2000). The Future of Engineering Education II. Teaching Methods That Work. *Chem. Engr. Education, 34*(1), 26–39.

Freeman, S., Eddy, S. L., McDonough, M., Smith, M. K., Okoroafor, N., Jordt, H., & Wenderoth, M. P. (2014). Active learning increases student performance in science, engineering, and mathematics. *Proceedings of the National Academy of Sciences of the United States of America, 111*(23), 8410–8415. doi:10.1073/pnas.1319030111 PMID:24821756

Guney, A., & Al, S. (2012). Effective learning environments in relation to different learning theories. *Procedia: Social and Behavioral Sciences, 46*, 2334–2338. doi:10.1016/j.sbspro.2012.05.480

Hmelo-Silver, C. E. (2004). Problem-based learning: What and how do students learn? *Educational Psychology Review, 16*(3), 235–266. doi:10.1023/B:EDPR.0000034022.16470.f3

Ideaslab, A. N., & Paper, W. (2011). *Undersanding virtual pedagogies for contemporary teaching & learning*. Pascoe Vale Road: Richard Olsen & ideasLAB. Retrieved from http://www.ideaslab.edu.au/wp-content/uploads/2011/07/Understanding-Virtual-Pedagogies_CKC_ideasLAB.pdf

International Family Nursing Association (IFNA). (2015). *Position statement on generalist competencies for family nursing practice*. Mankato: IFNA.

International Family Nursing Association (IFNA). (2017). *IFNA Position statement on advanced practice competencies for family nursing*. Retrieved from https://internationalfamilynursing.org/2017/05/19/advanced-practice-competencies/

Jewitt, C. (2008). Multimodality and literacy in school classrooms. *Review of Research in Education, 32*(1), 241–267. doi:10.3102/0091732X07310586

Jong, T., Lane, J., & Sharp, S. (2012). The efficacy of simulation as a pedagogy in facilitating pre-service teachers' learning about emotional self-regulation and its relevance to the teaching profession. *Australian Journal of Teacher Education, 37*(3). doi:10.14221/ajte.2012v37n3.6

Kiely, R., Matt, D., & Wheeler, E. (2010). *Investigating Critical Learning Episodes: a practical guide for continuing professional development.* Academic Press.

Loio, C., & Ançã, M. (2006). A dimensão pedagógica do gesto: Contributo para a formação de professores de Português. Aveiro: Universidade De Aveiro.

Lopes, B., & Cravino, J. (2017). *Práticas de Ensino e Tecnologia Acervo de Narrações Multimodais* (B. Lopes & J. Cravino, Eds.). VIla Real: Universidade de Trás-os-Montes e Alto Douro. Retrieved from http://multimodal.narratives.utad.pt

Lopes, J. B., Silva, A. A., Cravino, J. P., Santos, C. A., Cunha, A., Pinto, A., ... Branco, M. J. (2014). Constructing and using multimodal narratives to research in science education: Contributions based on practical classroom. *Research in Science Education, 44*(3), 415–438. doi:10.100711165-013-9381-y

Lopes, J. B., Silva, A. A., Cravino, J. P., Viegas, C., Cunha, A. E., Saraiva, E., ... Santos, C. A. (2012). Instrumentos de ajuda à mediação do professor para promover a aprendizagem dos alunos e o desenvolvimento profissional dos professores. *Revista Sensos, 2*, 125–171.

Lopes, J. B., Viegas, C., & Cravino, J. P. (2010). Improving the Learning of Physics and Development of Competences in Engineering Students. *International Journal of Engineering Education, 26*(3), 612–627.

Maheshwari, V. K. (2013). *Expository teaching – A direct instructional strategy*. Retrieved from http://www.vkmaheshwari.com/WP/?p=928

Osmond, C., & Barker, D. J. (2000). Fetal, infant, and childhood growth are predictors of coronary heart disease, diabetes, and hypertension in adult men and women. *Environmental Health Perspectives, 108*(Suppl), 545–553. PMID:10852853

Patterson, C., Stephens, M., Chiang, V., Price, A. M., Work, F., & Snelgrove-Clarke, E. (2017). The significance of personal learning environments (PLEs) in nursing education: Extending current conceptualizations. *Nurse Education Today, 48*, 99–105. doi:10.1016/j.nedt.2016.09.010 PMID:27744138

Pereira, F., Caetano, J., Moreira, J., & Ataíde, M. (2015). Práticas educativas em saúde na formação de acadêmicos de enfermagem. *Cogitare Enfermagem, 20*(2), 332–337. doi:10.5380/ce.v20i2.39767

Perrenoud, P. (2000). *Novas competências para ensinar*. Porto Alegre: Artmed.

Popil, I. (2011). Promotion of critical thinking by using case studies as teaching method. *Nurse Education Today, 31*(2), 204–207. doi:10.1016/j.nedt.2010.06.002 PMID:20655632

Prince, M. (2004). Does active learning work? A review of the research. *Journal of Engineering Education, 93*(3), 223–231. doi:10.1002/j.2168-9830.2004.tb00809.x

Rantavuori, J., Engeström, Y., & Lipponen, L. (2016). Learning actions, objects and types of interaction: A methodological analysis of expansive learning among pre-service teachers. *Frontline Learning Research in Learning Research, 4*(3), 1–27.

Redish, E. F. (2003). *Teaching Physics with the Physics Suite*. John Wiley &Sons, Inc.

Reiser, B. J. (2004). Scaffolding Complex Learning: The Mechanisms of Structuring and Problematizing Student Work. *Journal of the Learning Sciences, 13*(3), 273–304. doi:10.120715327809jls1303_2

Rua, M. (2011). *De aluno a enfermeiro. Desenvolvimento de competências em contexto de ensino clínico*. Lusociência.

Rua, M. (2012). A reflexão como estratégia de desenvolvimento de competências em contexto de ensino clinico. Vigo: Vicerreitoria de Alumnado Universidade de Vigo.

Rummel, N., Levin, J., & Woodward, M. (2003). Do pictorial mnemonic text-learning aids give students something worth writing about? *Journal of Educational Psychology, 95*(2), 327–334. doi:10.1037/0022-0663.95.2.327

Russell, A., Comello, R., & Lee Wright, D. (2007). Teaching strategies promoting active learning in healthcare education. *Journal of Education and Human Development, 1*(1), 1–8.

Sá-Chaves, I. (2002). *A construção de conhecimento pela análise reflexiva da praxis*. Lisboa: Edição Fundação CalousteGulbenkian, Fundação para a Ciência e Tecnologia.

Schon, D. (1983). *The reflective practitioner: How professionals think in action*. New York: Basic Books.

Sousa, L., Leal, A., & Sena, E. (2010). A importância da comunicação não-verbal do professor universitário no exercício de sua atividade profissional. *Revista CEFAC, 12*(5), 784–787. doi:10.1590/S1516-18462010005000088

Stanley, M. J. C., & Dougherty, J. P. (2010). Nursing education model. A paradigm shift in nursing education: A new model. *Nursing Education Perspectives, 31*(6), 378–380. Retrieved from http://10.0.4.19/1536-5026-31.6.378 PMID:21280445

Tavares, J. (2003). *Formação e inovação no ensino superior*. Porto: Porto Editora.

Thurlings, M., Vermeulen, M., Bastiaens, T., & Stijnen, S. (2013). Understanding feedback: A learning theory perspective. *Educational Research Review, 9*, 1–15. doi:10.1016/j.edurev.2012.11.004

Tomey, A. M. (2003). Learning with cases. *Journal of Continuing Education in Nursing, 34*(1), 34–38. Retrieved from http://www.ncbi.nlm.nih.gov/pubmed/12546132 PMID:12546132

Waterkemper, R., & Prado, M. (2011). Estratégias de ensino-aprendizagem em cursos de graduação em Enfermagem. *Avances En Enfermería, 39*(2), 234–246.

Yeigh, T. (2008). Quality teaching & professional learning: Uncritical reflections of a critical friend. *Australian Journal of Teacher Education, 33*(2), 1–15. doi:10.14221/ajte.2008v33n2.2

Zabalza, M. (2004). *Los diarios de classe. Un instrumento de investigación y desarrollo profissional*. Madrid: Nancea S. A. Ediciones.

Conclusion

This book, *Multimodal Narratives in Research and Teaching Practices*, focuses on Science and Technology Education, a relevant issue in the present society. It contributes to the field presenting an innovative way to capture the reality of what happens in the classroom and to make it public: the tool Multimodal Narratives (MN). This tool is explained from its origin, to its construction and illustration of how it can be used from several perspectives and purposes. The potential readers are primarily teachers, teacher trainers/students and researchers. This book shows how this tool can contribute to teachers' professional development and to educational research. It also approaches other contexts of usage (nursery and museums) as examples of other applications.

In brief, an MN is a descriptive and multimodal account that aggregates several types of data of professional practices in a real work context, which can be public and shareable. MN is characterized by their self-containment, reliability and representing public descriptions of professional practices in the real work context. MN tool is distinguished from others with similar purposes by: (a) having a verifiable chronological line, (b) being based on collected data of various types and the possibility to validate its accuracy, (c) incorporating descriptively the teachers' perspective, necessarily subjective, (d) being enriched by multimodal elements; (e) have a clear structure and similar characteristics; (f) present itself in a public register, easily consulted and open to different communities. These unique characteristics of MN allow them to be used by teachers, researchers and other professionals with different interests, visions and purposes taking advantage of the collection of validated MNs, in its present state, accessible on http://multimodal.narratives.utad.pt/ or even extend the collection.

All these characteristics are fully explained and described in the first chapter, which also shows the importance of the teaching practices become public and shareable and presents a tool –the MN– to do that. This chapter inscribes the work with MNs in a general trend of an Open Science perspective, gives the book a direction towards a strong coherence, and a fully assumed citizenship. In particular, explains in detail how MNs can be developed and validated and articulates the rationale of MN, its characteristics, the development process, its use in teacher training, professional development and research.

The book presents multifaceted contributions from several authors who over time have joined the team of researchers and teachers who developed the MN tool. Their seek to provide answers to questions that gradually emerged from the tool development process itself allowed to detail different views and lines of research and its acceptance in the professional and academic communities. Each perspective of the MN usage allowed the adjustment of this book for different target reading audiences closely aligned with each part of the book. By showing evidences of the MN's relevance and ways of usage through reports of cases, a useful mixture of theory and practice is made. In all chapters the respective authors had in mind the development of the teaching quality of Science & Technology and the students' learning quality.

The research developed and the collaborative work showed in this book involved teachers and researchers from Portugal, Brazil and Angola (with 11 Higher Education Institutions, 9 Basic and Secondary Schools directly involved). This allowed to verify that MNs are a rich and multifaceted tool that can:

1. Provide an integrative vision of a S&T teacher in action. In Section 2, chapters 2 to 6, showed how Teaching Practices in S&T can be seen through the MN, namely while exploiting specific issues of teacher mediation in different practices.
2. Be used by teachers in the context of their professional development or even in initial training. In Section 3, chapters 7 to 11, showed how the usage of MNs can aid improve the professional development of S&T teachers, articulating with research of teaching practices.
3. Be object of research. Since their foci are not in the "teaching subjects" but in the "teaching practices", MN from different contexts can be used to compare different practices and serve wider research purposes related to the improvement of S&T teaching or other professional practices. In Section 4, chapters 12 to 16, illustrates some of these possibilities.

In addition to the S&T teaching professionals and researchers' interest, NMs may also have a potential interest in professional development in other professional sectors that are based on interaction with people such as doctors, nurses, caregivers, etc. In Section 5, chapters 17 and 18 explored some possibilities in this direction, in particular, science communication in museums and nursing education.

The most values of this book are presenting, explaining, using and exploring the possibilities of use (researching, teaching and professional development) of a tool – MN – that may:

1. Enhance the possibilities of researching S&T teaching practices by other teams, using the currently MN collection available in a portal articulated with the book. It is possible to do large-scale research studies (using the MN diversity of the collection available) or in depth (using few MNs). This possibility is based on the structure and characteristics common to different NMs secured by a protocol for their elaboration. In addition, MN can be explored using different theoretical approaches and analysis techniques. Finally, the same MN can be used for different research purposes.
2. Expand the MN collection with other MNs from different professionals and contexts. Given their public character, MNs can be taken as a reference to improve the teaching practices of S&T. The MN reports and shows what happened, as it happened. There are unsuccessful situations and others with different degrees of success. The lessons to be taken from one or more MNs will be personal, perhaps even depending on the experience of each one. The same can no longer be said of research studies due to their aim of contributing to a better and systematic understanding of specific aspects of teacher mediation.
3. Disseminate the MNs' potential to larger audiences which can include not only academics, but also teachers and other professionals who daily contact with other people in their practice. MNs allow access to the point of view of the teacher. Since MNs reconcile objective data and subjective descriptions, and accepting them both, as fundamental, it is possible to focus on main points for professional practice or professional development. It is well known how important the implicit and subjective aspects are for the repertory of professional skills.

In light of the above, we are convinced that this book, while presenting the MNs and the different scenarios in which it can be used, can constitute the basis for many future and fruitful research lines as well as the empirical basis support for teachers in the development of their educational practices!

J. Bernardino Lopes
Universidade de Trás-os-Montes e Alto Douro, Portugal

Maria Clara Viegas
Politécnico do Porto, Portugal

José Alexandre Pinto
Politécnico do Porto, Portugal

Compilation of References

Adger, C. T., & Wright, L. J. (2015). Discourse in educational settings. In D. Tannen, H. Hamilton, & D. Schiffrin (Eds.), *The handbook of discourse analysis* (pp. 858–879). Chichester, UK: Wiley Blackwell.

Afonso, A. S., Afonso, S., & Rodrigues, F. (2017, August). *The role of language in building common ground in a science museum*. Poster presented at the 12th Conference of the European Science Research Association (ESERA), Dublin, Ireland.

Aho, E., Haverinen, H.-L., Juuso, H., Laukka, S. J., & Sutinen, A. (2010). Teachers' principles of decision-making and classroom management: A case study and a new observation method. *Procedia: Social and Behavioral Sciences, 9*, 395–402. doi:10.1016/j.sbspro.2010.12.171

Ainsworth, S. (2006). DeFT: A conceptual framework for considering learning with multiple representations. *Learning and Instruction, 16*(3), 183–198. doi:10.1016/j.learninstruc.2006.03.001

Alarcão, I., & Roldão, M. C. (2010). Supervisão: Um contexto de desenvolvimento profissional dos professores [Supervision: A context of professional development of teachers] (2nd ed.). Odivelas: Edições Pedago.

Alarcão, I., & Sá-Chaves, I. (1994). Supervisão de professores e desenvolvimento humano: Uma perspectiva ecológica [Teacher supervision and human development: an ecological perspective]. In J. Tavares (Ed.), Para intervir em educação: Contributos dos colóquios (pp. 201-232). Porto: Edições CIDInE.

Alarcão, I. (1996). Reflexão crítica sobre o pensamento de D. Schön e os programas de formação de professores [Critical reflection on D. Schön's thinking and teacher training programs]. In *Formação Reflexiva de Professores: Estratégias de Supervisão* [Reflective Teacher Training: Supervision Strategies] (pp. 9–40). Porto, Portugal: Porto Editora.

Alarcão, I. (1996). Reflexão crítica sobre o pensamento de D. Schön e os programas de formação de professores. In *Formação reflexiva de professores. Estratégias de supervisão*. Porto: Porto Editora.

Alarcão, I. (2001). Professor-investigador: Que sentido? Que função? [Teacher-researcher: What sense? What function?]. In B. P. Campos (Ed.), *Formação profissional de professores no ensino superior* [Professional training of teachers in higher education] (Vol. 1, pp. 21–31). Oporto, Portugal: Porto Editora.

Alibali, M. W., Bassok, M., Solomon, K. O., Syc, S. E., & Goldin-Meadow, S. (1999). Illuminating mental representations through speech and gesture. *Psychological Science, 10*(4), 327–333. doi:10.1111/1467-9280.00163

Allan, R. (2011). The role of water vapour in Earth's energy flows. *Surveys in Geophysics*, 1–8. doi:10.100710712-011-9157-8

Allen, L. B., & Crowley, K. (2014). Challenging beliefs, practices, and content: How museum educators change. *Science Education, 98*(1), 84–105. doi:10.1002ce.21093

Allen, S. (2002). Looking for learning in visitor talk: A methodological exploration. In G. Leinhardt, K. Crowley, & K. Knutson (Eds.), *Learning conversations in museums* (pp. 259–303). Mahwah, NJ: Lawrence Erlbaum Associates.

Alsop, S., Bencze, L., & Pedretti, E. (Eds.). (2005). *Analysing exemplary science teaching*. Maidenhead, UK: Open University Press.

Altunay, M., Avery, P., Blackburn, K., Bockelman, B., Ernst, M., Fraser, D., ... Livny, M. (2011). A science driven production cyberinfrastructure—The open science grid. *Journal of Grid Computing*, *9*(2), 201–218. doi:10.100710723-010-9176-6

Anderson, D., Cosson, A., & McIntoshosh, L. (2015). Foreword: Research informing the practice of museum educators: Diverse audiences, challenging topics, and reflective praxis. In D. Anderson, A. Cosson, & L. McIntoshosh (Eds.), *Research informing the practice of museum educators: Diverse audiences, challenging topics, and reflective praxis* (pp. vii–xii). Rotterdam: Sense Publishers.

Anema, M., & McCoy, J. (2010). *Competency-based nursing education*. New York: Springer Publishing Company, LLC.

Aranda, C., & Callejo, M. L. (2010). Construcción del concepto de dependencia lineal en un contexto de geometría dinámica: Un estudio de casos [Construction of the concept of linear dependence in a context of dynamic geometry: A case study]. *Relime*, *13*, 129–158.

Arcavi, A. (2003). The role of visual representations in the learning of mathematics. *Educational Studies in Mathematics*, *52*(3), 215–241. doi:10.1023/A:1024312321077

Arnold, I. (2016). Cheating online formative tests: Does it pay off? *The Internet and Higher Education*, *29*, 98–106. doi:10.1016/j.iheduc.2016.02.001

Artigue, M. (2001). What can we learn from educational research at the university level? In D. Holton (Ed.), *The teaching and learning of mathematics at university level: An ICMI study* (pp. 207–220). Dordrecht: Kluwer Academic.

Ash, D. B. (2007). Using video data to capture discontinuous science meaning making in non-school settings. In R. Goldman, R. Pea, B. Barron, & S. J. Derry (Eds.), *Video research in the learning sciences* (pp. 207–226). Mahwah, NJ: Erlbaum.

Ash, D. B., & Lombana, J. (2012). Methodologies for reflective practice and museum educator research. In D. Ash, J. Rahm, & L. M. Melber (Eds.), *Putting theory into practice: Tools for research in informal settings* (pp. 29–52). Rotterdam, The Netherlands: Sense Publishing. doi:10.1007/978-94-6091-964-0_4

Avalos, B. (2011). Teacher professional development in teaching and teacher education over ten years. *Teaching and Teacher Education*, *27*(1), 10–20. doi:10.1016/j.tate.2010.08.007

Badreddine, Z., & Buty, C. (2011). Discursive reconstruction of the scientific story in a teaching sequence. *International Journal of Science Education*, *33*(6), 773–795. doi:10.1080/09500693.2010.496475

Baird, D. (2004). *Thing knowledge: A philosophy of scientific instruments*. Berkeley, CA: University of California Press.

Bakkenes, I., Vermunt, J. D., & Wubbels, T. (2010). Teacher learning in the context of educational innovation: Learning activities and learning outcomes of experienced teachers. *Learning and Instruction*, *20*(6), 533–548. doi:10.1016/j.learninstruc.2009.09.001

Barbot, A. (2014). *Água, Energia, Sustentabilidade e Educação Sustentada* [Water, Energy, Sustainability and grouded Education] (Unpublished doctoral dissertation). University of Trás-os-Montes e Alto Douro, Vila Real, Portugal.

Barbot, A. (2017a). Integrating the transversal theme of water-energy in initial teacher training. In Teaching science: contributions of research for planning, practice and professional development (pp. 173-188). New York: Nova Science.

Barbot, A. (2017b). Problems and Questions: Elucidation and Relevance for Research and Teaching. In Teaching science: contributions of research for planning, practice and professional development (pp. 325-336). New York: Nova Science.

Barbot, A., Lopes, J. B., & Soares, A. (2010). Água, Energia, Sustentabilidade e Educação Sustentada [Water, Energy, Sustainability and grouded Education]. In Resumos do 20º encontro Ibérico para o ensino da Física 2010 [Iberian meeting for the teaching of Physics 2010] (pp. 311–312). Universidade de Trás-os-Montes e Alto Douro.

Barbot, A., Lopes, J. B., & Soares, A. A. (2011). Relating Water, Energy and Sustainability to Improve Teaching and Learning in Physical and Environmental Sciences in Teacher Education for Classes of 6 – 12 Age Levels. In *ICERI2011 Proceedings* (pp. 1254–1261). ICERI.

Bardin, L. (1997). *L'analyse de contenu* [Content analysis]. Paris: Presses Universitaires de France.

Bargagliotti, A. E., & Anderson, C. R. (2017). Using learning trajectories for teacher learning to structure professional development. *Mathematical Thinking and Learning, 19*(4), 237–259. doi:10.1080/10986065.2017.1365222

Barron, D. (2007). Video as a tool to advance understanding of learning and development in peer, family, and other informal learning contexts. In R. Goldman, R. Pea, B. Barron, & S. J. Derry (Eds.), *Video research in the learning sciences* (pp. 159–187). Mahwah, NJ: Erlbaum.

Bateman, J. A. (2016). From narrative to visual narrative to audiovisual narrative: The multimodal discourse theory connection (Invited Talk). *OASIcs-OpenAccess Series in Informatics, 53*(1), 1–11. doi:10.4230/OASIcs.CMN.2016.1

Bell, B., Lewenstein, B., Shouse, A., & Feder, M. (2009). *Learning science in informal environments: People, places, and pursuits*. Washington, DC: The National Academies Press.

Berger, J. L., & Karabenick, S. (2011). Motivation and students' use of learning strategies: Evidence of unidirectional effects in mathematics classrooms. *Learning and Instruction, 21*(3), 416–428. doi:10.1016/j.learninstruc.2010.06.002

Berland, L. K., Schwarz, C. V., Krist, C., Kenyon, L., Lo, A. S., & Reiser, B. J. (2016). Epistemologies in practice: Making scientific practices meaningful for students. *Journal of Research in Science Teaching, 53*(7), 1082–1112. doi:10.1002/tea.21257

Biggs, J. (1999). *Teaching for quality learning at university*. Buckingham, UK: Society for Research into Higher Education – O. U. Press.

Bird, F., & Yucel, R. (2015). Feedback codes and action plans: Building the capacity of first-year students to apply feedback to a scientific report. *Assessment & Evaluation in Higher Education, 40*(4), 508–527. doi:10.1080/02602938.2014.924476

Bishop, A. J. (1976). Decision-making, the intervening variable. *Educational Studies in Mathematics, 7*(1/2), 41–47. doi:10.1007/BF00144357

Black, P., & Wiliam, D. (2009). Developing the theory of formative assessment. *Educational Assessment, Evaluation and Accountability, 21*(1), 5–31. doi:10.100711092-008-9068-5

Black, P., & William, D. (1998). Assessment and classroom learning. *Assessment in Education: Principles, Policy & Practice, 5*(1), 7–64. doi:10.1080/0969595980050102

Black, P., & William, D. (2005). Lessons from around the world: How policies, politics and cultures constrain and afford assessment practices. *Curriculum Journal, 16*(2), 249–261. doi:10.1080/09585170500136218

Bogdan, R., & Biklen, S. (1994). *Investigação qualitativa em educação*. Porto, Portugal: Porto Editora.

Bogdan, R., & Biklen, S. (1994). *Investigação qualitativa em educação. Uma introdução à teoria e aos métodos* [Qualitative research in education. An introduction to theory and methods]. Porto: Porto Editora.

Bogdan, R., & Biklen, S. (1994). *Investigação Qualitativa em Educação: uma introdução à teoria e aos métodos* [Qualitative research in education: an introduction to the theory and methods]. Porto: Porto Editora.

Bogdan, R., & Biklen, S. K. (1997). *Qualitative research for education.* Boston, MA: Allyn & Bacon.

Böhm, A. (2004). Theoretical coding: Text analysis in grounded theory. In U. Flick, E. von Kardorff, & I. Steinke (Eds.), *A companion to qualitative research* (pp. 270–274). London: Sage.

Borko, H., Stecher, B., & Kuffner, K. (2007). *Using artifacts to characterize reform-oriented instruction: The scoop notebook and rating guide* (CSE Technical Report 707). Los Angeles, CA: The Regents of the University of California.

Borko, H., Roberts, S. A., & Shavelson, R. (2008). Teachers' decision making: From Alan J. Bishop to today. In P. Clarkson & N. Presmeg (Eds.), *Critical issues in mathematics education* (pp. 37–67). Boston, MA: Springer. doi:10.1007/978-0-387-09673-5_4

Bot, L., Gossiaux, P.-B., Rauch, C.-P., & Tabiou, S. (2005). "Learning by doing": A teaching method for active learning in scientific graduate education. *European Journal of Engineering Education, 30*(1), 105–119. doi:10.1080/03043790512331313868

Boud, D., Cohen, R., & Sampson, J. (1999). Peer learning and assessment. *Assessment & Evaluation in Higher Education, 24*(4), 413–426. doi:10.1080/0260293990240405

Boyd, R. (1979). Metaphor and theory change: What is 'metaphor' a metaphor for? In A. Ortony (Ed.), *Metaphor and thought* (pp. 356–408). Cambridge, UK: Cambridge University Press.

Bradforth, S. E., Miller, E. R., Dichtel, W. R., Leibovich, A. L., Martin, J. D., Bjorkman, K. S., ... Smith, L. T. (2015). Improve undergraduate science education: It is time to use evidence-based teaching practices at all levels by providing incentives and effective evaluations. *Nature, 523*(7560), 282–285. doi:10.1038/523282a PMID:26178951

Branco, M. J. (2018). *Desenvolvimento das práticas de mediação dos professores em contexto de ensino de Ciências Físicas - O papel de ferramentas de ajuda* [Development of the teachers' mediation practices in the context of Physics teaching: The role of tools for help] (Unpublished doctoral dissertation). Universidade de Trás-os-Montes e Alto Douro, Vila Real, Portugal.

Branco, M. J., Magalhães, R., Ferreira, A., Lopes, J. B., & Cravino, J. P. (2009). Narrações de aulas: Um estímulo ao desenvolvimento profissional? [Classroom narratives: A stimulus to professional development?]. In F. Paixão & F. R. Jorge (Eds.), Educação e Formação: Ciência, Cultura e Cidadania. Actas XIII Encontro Nacional de Educação em Ciências (pp. 740–749). Castelo Branco: Escola Superior de Educação, Instituto Politécnico de Castelo Branco.

Branco, M. J., Magalhães, R., Ferreira, A., Lopes, J., & Cravino, J. (2009). Narrações de aulas: um estímulo ao desenvolvimento profissional? [Narratives from lessons: A stimulus to professional development?]. In F. Paixão & F. Jorge (Eds.), *Proceedings XIII Encontro Nacional de Educação em Ciências* (pp. 740–749). Castelo Branco: Instituto Politécnico de Castelo Branco.

Bressem, J. (2012). *Repetitions in gesture: Structures, functions, and cognitive aspects* (Unpublished doctoral dissertation). European University Viadrina, Frankfurt, Germany.

Broadbent, J. (2017). Comparing online and blended learner's self-regulated learning strategies and academic performance. *The Internet and Higher Education, 33*, 24–32. doi:10.1016/j.iheduc.2017.01.004

Brownlee, J., Ferguson, L., & Ryan, M. (2017). Changing teachers' epistemic cognition: A new conceptual framework for epistemic reflexivity. *Educational Psychologist, 52*(4), 242–252. doi:10.1080/00461520.2017.1333430

Bruner, J. (1997). *Acts of meaning: for a cultural psychology*. Lisbon: 70 Editions.

Bruner, J. S. (1960). *The process of education*. Cambridge, MA: Harvard University Press.

Bryan, L. A. (2012). Research on science teacher beliefs. In B. J. Fraser, K. G. Tobin, & C. J. McRobbie (Eds.), *Second International Handbook of Science Education* (pp. 477–495). New York: Springer International Handbooks of Education. doi:10.1007/978-1-4020-9041-7_33

Bulman, C., & Schutz, S. (2013). Reflective practice in nursing. In C. Bulman & S. Schutz (Eds.), *Nurse education today* (5th ed.; Vol. 1). Oxford, UK: Wiley-Blackwell.

Buoyancy – project PhET. (n.d.). Retrieved from http://phet.colorado.edu/en/simulation/buoyancy

Burbank, M. D., & Kauchak, D. (2003). An alternative model for professional development: Investigations into effective collaboration. *Teaching and Teacher Education, 19*(5), 499–514. doi:10.1016/S0742-051X(03)00048-9

Buty, C., Badreddine, Z., & Régnier, J.-C. (2012). Didactique des sciences et interactions dans la classe: quelques lignes guidelines pour une analyze dynamique [Science teaching and interactions in the classroom: some guidelines for a dynamic analysis]. *Testing Research in Education in Science, 14*(1), 147-165.

Caldwell, L., & Grobbel, C. (2013). The importance of reflective practice in nursing. *International Journal of Caring Sciences, 6*(3), 319–326.

Cárcamo, A., Fortuny, J. M., & Gómez, J. (2017). Mathematical modelling and the learning trajectory: Tools to support the teaching of linear algebra. *International Journal of Mathematical Education in Science and Technology, 48*(3), 338–352. doi:10.1080/0020739X.2016.1241436

Carlson, D. (1993). Teaching linear algebra: Must the fog always roll in? *The College Mathematics Journal, 24*(1), 29–40. doi:10.1080/07468342.1993.11973503

Carlson, D., Johnson, C. R., Lay, D. C., & Porter, A. D. (1993). The Linear Algebra Curriculum Study Group recommendations for the first course in linear algebra. *The College Mathematics Journal, 24*(1), 41–46. doi:10.1080/07468342.1993.11973504

Carney, R. N., & Levin, J. R. (2002). Pictorial illustrations still improve students' learning from text. *Educational Psychology Review, 14*(1), 5–26. doi:10.1023/A:1013176309260

Carroll, J. (2015). *Tools for teaching in an educationally mobile world* (1st ed.). London: Routledge Edition.

Cazan, A. M. (2014, July). Self-regulated learning and academic achievement in the context of online learning environments. *The International Scientific Conference Elearning and Software for Education*, 90-95.

Center for Research on Learning and Teaching of the University of Michigan. (2016). *Introduction to active learning*. Retrieved from http://www.crlt.umich.edu/active_learning_introduction

Cerqueira, P., & Barbot, A. (2010). Ciclo Urbano da Água: Contributo para o seu tratamento no Ensino e na Educação [Urban Water Cycle: Contribution to its treatment in teaching and learning]. In Resumos do 20º encontro Ibérico para o ensino da Física 2010 [Iberian meeting for the teaching of Physics 2010] (pp. 339–340). Universidade de Trás-os-Montes e Alto Douro.

Christoforou, A., & Yigit, A. (2008). Improving teaching and learning in engineering education through a continous assessment process. *European Journal of Engineering Education, 33*(1), 105–116. doi:10.1080/03043790701746405

Cienki, A. (2012). Usage events of spoken language and the symbolic units we (may) abstract from them. In K. Kosecki & J. Badio (Eds.), *Cognitive processes in language* (pp. 149–158). Frankfurt am Main: Peter Lang.

Clark, C., & Peterson, P. (1986). Teachers' thought processes. In M. Wittrock (Ed.), *Handbook of research on teaching* (3rd ed.; pp. 255–296). New York: Macmillan.

Clark, H. H. (1996). *Using language*. Cambridge, UK: Cambridge University Press. doi:10.1017/CBO9780511620539

Clemente, M., & Ramírez, E. (2008). How teachers express their knowledge through narrative. *Teaching and Teacher Education, 24*(5), 1244–1258. doi:10.1016/j.tate.2007.10.002

Cochran-Smith, M., & Lytle, S. L. (1993). *Inside-outside: Teacher research and knowledge*. New York, NY: Teacher's College Press.

Cochran-Smith, M., & Lytle, S. L. (1993). *Inside-outside: Teacher researcher and knowledge*. New York, NY: Teacher's College Press.

Cohen, L., Manion, L., & Morrison, K. (2011). *Research methods in education* (7th ed.). London: Routledge.

Collins, A., Joseph, D., & Bielaczyc, K. (2004). Design research: Theoretical and methodological issues. *Journal of the Learning Sciences, 13*(1), 15–42. doi:10.120715327809jls1301_2

Cook, S., & Brown, J. (1999). Bridging epistemologies: The generative dance between organizational knowledge and organizational knowing. *Organization Science, 10*(4), 381–400. doi:10.1287/orsc.10.4.381

Cooper, J. (Ed.). (2014). Classroom teaching skills (10th ed.). Cengage Learning.

Corrigan, D., Dillon, J., & Gunstone, R. (Eds.). (2011). *The professional knowledge base of science teaching*. Dordrecht, The Netherlands: Springer. doi:10.1007/978-90-481-3927-9

Cortesão, L., & Stoer, S. R. (1997). Investigação-acção e a produção de conhecimento no âmbito de uma formação de professores para a educação inter/multicultural [Action research and the production of knowledge within a teacher training course for inter/multicultural education]. *Educação Sociedade & Culturas, 7*, 7–28.

Costa, M. (2011). *Mais saber, Melhor Enfermagem : A repercussão da formação na qualidade de cuidados*. Universidade Lusófona de Humanidades e Tecnologias.

Cotton, K. (2001). Classroom questioning. *School Improvement Research Series, 3*, 14. doi:10.1080/00221347308981289

Coutinho, C. P. S., Dias, A., Bessa, A., Ferreira, F., Vieira, M. J. R. C., & Regina, S. (2009). Investigação-acção: Metodologia preferencial nas práticas educativas [Action research: the preferred methodology within educational practice]. *Revista Psicologia. Education and Culture, 13*(2), 355–379.

Cox-Peterson, A., Marsh, D., Kiesel, J., & Melber, L. (2003). Investigation of guided school tours, student learning, and science reform recommendations at a museum of natural history. *Journal of Research in Science Teaching, 40*(2), 200–218. doi:10.1002/tea.10072

Craig, C. J. (2011). Narrative inquiry in teaching and teacher education. In J. Kitchen, D. C. Parker, & D. Pushor (Eds.), *Advances in research on teaching* (Vol. 13, pp. 19–42). Bingley, UK: Emerald Group Publishing.

Crisp, G. (2012). Integrative assessment: Reframing assessment practice for current and future learning. *Assessment & Evaluation in Higher Education, 37*(1), 33–43. doi:10.1080/02602938.2010.494234

Crouch, C., & Mazur, E. (2001). Peer instruction: Ten years of experience and result. *American Journal of Physics, 69*(9), 970–977. doi:10.1119/1.1374249

Cruz, E. (2012). *Da Avaliação do Impacte à Articulação da Investigação Práticas – O caso da Articulação na Formação Didáctica Pós-Graduada de Professores de Ciências e desafios futuros* [From impact evaluation to articulation of research practices – The case of articulation in post-graduate training of science teachers and future challenges] (Doctoral thesis). University of Aveiro, Portugal.

Cunha, A. E. (2015). *Construção de práticas de referência no Ensino da Física para o Ensino Secundário* [Construction of Physics Teaching practices of reference for Secondary Education] (Unpublished doctoral dissertation). Universidade de Trás-os-Montes e Alto Douro, Vila Real, Portugal.

Cunha, A. E. (2015). *Construção de práticas de referência no ensino da física para o ensino secundário.* Tese de doutoramento. Universidade de Trás-os-Montes e Alto Douro, Vila Real, Portugal.

Cunha, A. E., Branco, J., Lopes, J. B., Cravino, J. P., & Santos, C. (2009). Envolver os alunos de forma produtiva em aulas de ciências físicas: Evidências para uma proposta de uma ferramenta para ajudar os professores. In *Extra Nr. VIII Congresso Internacional sobre Investigación en Didáctica de las Ciencias* (pp. 1553–1558). Barcelona: Enseñanza de las Ciencias.

Cunha, A. E., Saraiva, E., Dinis, F., Anacleto, A., Pires, C., Agostinho, A., . . . Lopes, J. B. (2012). The influence of the teacher's research experience in his mediation of secondary school students learning using computer simulations. In R. Pintó, V. López, & C. Simarro (Eds.), Learning science in the society of computers (pp. 11–19). Barcelona, Spain: Centre for Research in Science and Mathematics Education (CRECIM).

Cunha, A. E. (2015). *Construção de práticas de referência no ensino da física para o ensino secundário. Tese de doutoramento.* Vila Real, Portugal: Universidade de Trás-os-Montes e Alto Douro.

Cunha, A. E. R. M. (2015). *Construção de práticas de referência no ensino da Física para o Ensino Secundário (Doutoramento).* Universidade de Trás-os-Montes e Alto Douro, Vila Real.

Cunha, A. E., Lopes, J. B., Cravino, J. P., & Santos, C. A. (2012). Envolver os alunos na realização de trabalho experimental de forma produtiva: O caso de um professor experiente em busca de boas práticas [Engage students in the performance of experimental work in a productive way: the case of an experienced teacher in search of good practices]. *Revista Electrónica de Enseñanza de las Ciencias, 11*(3), 635–659.

Cunha, A. E., Saraiva, E., Santos, C. A., Dinis, F., & Lopes, J. B. (2014). Teacher mediation actions and students' productive engagement during the use of computer simulations in physical science classrooms. *Procedia Technology, 13*, 76–85. doi:10.1016/j.protcy.2014.02.011

Dabbagh, N., & Kitsantas, A. (2012). Personal learning environments, social media, and self-regulated learning: A natural formula for connecting formal and informal learning. *The Internet and Higher Education, 15*(1), 3–8. doi:10.1016/j.iheduc.2011.06.002

Datnow, A., & Hubbard, L. (2016). Teacher capacity for and beliefs about data-driven decision making: A literature review of international research. *Journal of Educational Change, 17*(1), 7–28. doi:10.100710833-015-9264-2

Dennis, D. V., Gelfuso, A., & Sweeney, S. (2018). Trying and reflecting: Two novice teacher educators' first attempts at guiding pre-service teachers' reflection on literacy field experiences. *Education 3-13, 46*(4), 456-468.

Derry, S. J., Pea, R. D., Barron, B., Engle, R. A., Erickson, F., Goldman, R., ... Sherin, B. L. (2010). Conducting video research in the learning sciences: Guidance on selection, analysis, technology, and ethics. *Journal of the Learning Sciences, 19*(1), 3–53. doi:10.1080/10508400903452884

Desimone, L. M. (2009). Improving impact studies of teachers' professional development: Toward better conceptualizations and measures. *Educational Researcher, 38*(3), 181–199. doi:10.3102/0013189X08331140

Desimone, L. M., & Garet, M. S. (2015). Best practices in teachers' professional development in the United States. *Psychology. Social Education, 7*(3), 252–263.

Desimone, L., Garet, M., Birman, B., Porter, A., & Yoon, K. (2003). Improving teachers' in-service professional development in mathematics and science: The role of post-secondary institutions. *Educational Policy, 17*(5), 613–649. doi:10.1177/0895904803256791

Dias, V., Pitolli, A., Prudencio, C., & Oliveira, M. (2013). O Diário de Bordo como ferramenta de reflexão durante o Estágio Curricular Supervisionado do curso de Ciências Biológicas da Universidade Estadual de Santa Cruz – Bahia. S. Paulo: IX Encontro Nacional de Pesquisa em Educação em Ciências.

Dillon, J., DeWitt, J., Pegram, E., Irwin, B., Crowley, K., Haydon, R., . . . Xanthoudaki, M. (2016). A learning research agenda for natural history institutions. London: Natural History Museum.

diSessa, A. (2001). *Changing minds: Computers, learning, and literacy.* Cambridge, MA: MIT Press.

Ditcher, A. K. (2001). Effective teaching and learning in higher education, with particular reference to the undergraduate education of professional engineers. *International Journal of Engineering Education, 17*(1), 24–29.

Domínguez-García, S., García-Planas, M. I., & Taberna, J. (2016). Mathematical modelling in engineering: An alternative way to teach linear algebra. *International Journal of Mathematical Education in Science and Technology, 47*(7), 1076–1086. doi:10.1080/0020739X.2016.1153736

Dorier, J.-L. (1998). The role of formalism in the teaching of the theory of vector spaces. *Linear Algebra and Its Applications, 275/276*, 141–160. doi:10.1016/S0024-3795(97)10061-1

Dorier, J.-L. (2000). Epistemological analysis of the genesis of the theory of vector spaces. In J.-L. Dorier (Ed.), *On the teaching of linear algebra* (pp. 191–207). Dordrecht: Kluwer Academic.

Dorier, J.-L., Robert, A., Robinet, J., & Rogalski, M. (2000). The obstacle of formalism in linear algebra. In J.-L. Dorier (Ed.), *On the teaching of linear algebra* (pp. 85–124). Dordrecht: Kluwer Academic.

Dorier, J.-L., & Sierpinska, A. (2001). Research into the teaching and learning of linear algebra. In D. Holton (Ed.), *The teaching and learning of mathematics at university level: an ICMI study* (pp. 255–273). Dordrecht: Kluwer Academic.

Dorier, J.-L., & Sierpinska, A. (2001). Research into the teaching and learning of linear algebra. In D. Holton (Ed.), *The teaching and learning of mathematics at university level: An ICMI study* (pp. 255–273). Dordrecht: Kluwer Academic.

Draganoc, P., & Sanna, M. C. (2015). Avaliação da s competências dos professores de enfermagem para facilitar a aprendizagem de adultos. *Cogitare Enfermagem, 20*(3), 556–564. doi:10.5380/ce.v20i3.41104

Dreyfus, T. (1991). On the status of visual reasoning in mathematics and mathematics education. In F. Furinghetti (Ed.), *Proceedings of the 15th Conference of the International Group for the Psychology of Mathematics Education* (vol. 1, pp. 33–48). Assisi, Italy: Università di Genova.

Driver, R., Asoko, H., Leach, J., Scott, P., & Mortimer, E. (1994). Constructing Scientific Knowledge in the Classroom. *Educational Researcher, 23*(7), 5–12. doi:10.3102/0013189X023007005

Dubinsky, E. (1997). Some thoughts on a first course in linear algebra at the college level. In D. Carlson, C. R. Johnson, D. Lay, A. D. Porter, A. E. Watkins, & W. Watkins (Eds.), *Resources for teaching linear algebra* (Vol. 42, pp. 85–105). Washington, DC: Mathematical Association of America.

Duffy, A. (2007). A concept analysis of reflective practice: Determining its value to nurses. *British Journal of Nursing (Mark Allen Publishing)*, *16*(22), 1400–1407. doi:10.12968/bjon.2007.16.22.27771 PMID:18361389

Dufresne, R., & Gerace, W. (2004). Assessing-to-learn: Formative assessment in physics instruction. *The Physics Teacher*, *42*(7), 428–433. doi:10.1119/1.1804662

Dumas-Carré, A., & Goffard, M. (1997). *Rénover les activités de résolution de problèmes en physique. Concepts et démarches*. Paris: Armand Colin.

Dunn, K. E., Airola, D. T., Lo, W.-J., & Garrison, M. (2013). Becoming data driven: The influence of teachers' sense of efficacy on concerns related to data-driven decision making. *Journal of Experimental Education*, *81*(2), 222–241. do i:10.1080/00220973.2012.699899

Duschl, R. A., & Grandy, R. (2013). Two views about explicitly teaching nature of science. *Science & Education*, *22*(9), 2109–2139. doi:10.100711191-012-9539-4

Duschl, R. A., & Wright, E. (1989). A case study of high school teachers' decision making models for planning and teaching science. *Journal of Research in Science Teaching*, *26*(6), 467–501. doi:10.1002/tea.3660260602

Duval, R. (2006). Transformations de représentations sémiotiques et démarches de pensée en mathématiques [Transformations of semiotic representations and thought processes in mathematics]. In *Actes du XXXIIe Colloque COPIRELEM* (pp. 67–89). Strasbourg, France: ULP. IREM.

Duval, R. (1993). Registres de représentation sémiotique et fonctionnement cognitif de la pensée. *Annales de Didactiques et de Sciences Cognitives*, *5*, 37–65.

Duval, R. (2003). Registros de representação semiótica e funcionamento cognitivo da compreensão matemática [Registers of semiotic representation and cognitive functioning of mathematical understanding]. In *S. D. A. Machado (Ed.), Aprendizagem em matemática: registros de representação semiótica* [Learning in mathematics: Semiotic representation register] (pp. 11–33). Campinas, São Paulo: Papirus.

Duval, R. (2009). *Semiósis e pensamento humano: registros semióticos e aprendizagens intelectuais* (Vol. 1; L. F. Levy & M. R. A. da Silveira, Trans.). São Paulo: Livraria da Física.

Duval, R. (2017). Registers of semiotic representations and analysis of the cognitive functioning of mathematical thinking. In T. M. M. Campos (Ed.), *Understanding the mathematical way of thinking – The registers of semiotic representations* (pp. 45–71). Cham: Springer. doi:10.1007/978-3-319-56910-9_3

Earley, M. (2014). A synthesis of the literature on research methods education. *Teaching in Higher Education*, *19*(3), 242–253. doi:10.1080/13562517.2013.860105

Edwards, W., & Fasolo, B. (2001). Decision technology. *Annual Review of Psychology*, *52*(1), 581–606. doi:10.1146/annurev.psych.52.1.581 PMID:11148318

Eison, J., & Ph, D. (2010). *Using active learning instructional strategies to create excitement and enhance learning. East Fowler*. University of South Florida.

Engle, R. A., & Conant, F. R. (2002). Guiding principles for fostering productive disciplinary engagement: Explaining an emergent argument in a community of learners classroom. *Cognition and Instruction*, *20*(4), 399–483. doi:10.1207/S1532690XCI2004_1

Etkina, E. (2000). Weekly reports: A two way feedback tool. *Science Education*, *84*(5), 594–605. doi:10.1002/1098-237X(200009)84:5<594::AID-SCE3>3.0.CO;2-U

Evans, C. (2013). Making Sense of Assessment Feedback in Higher Education. *Review of Educational Research, 83*(1), 70–120. doi:10.3102/0034654312474350

Faria, Á., Faria, P., & Ramos, M. (2013). Interactive multimodal narrative as an approach to developing emergent literacy in early childhood education. *Sino-US English Teaching, 10*(4), 328–333.

Farrell, T., & Ives, J. (2015). Exploring teacher beliefs and classroom practices through reflective practice: A case study. *Language Teaching Research, 19*(5), 594–610. doi:10.1177/1362168814541722

Fecher, B., & Friesike, S. (2014). Open science: One term, five schools of thought. *Opening Science*, 17–47.

Felder, R. M., Woods, D. R., Stice, J. E., & Rugarcia, A. (2000). The Future of Engineering Education II. Teaching Methods That Work. *Chem. Engr. Education, 34*(1), 26–39.

Felder, R., Woods, D., Stice, J., & Rugarcia, A. (2000). The future of engineering education II. Teaching methods that work. *Chemical Engineering Education, 34*(1), 26–39.

Feucht, F. C., Brownlee, J. L., & Schraw, G. (2017). Moving beyond reflection: Reflexivity and epistemic cognition in teaching and teacher education. *Educational Psychologist, 52*(4), 234–241. doi:10.1080/00461520.2017.1350180

Fischer, C., Fishman, B., Dede, C., Eisenkraft, A., Frumin, K., Foster, B., ... McCoy, A. (2018). Investigating relationships between school context, teacher professional development, teaching practices, and student achievement in response to a nationwide science reform. *Teaching and Teacher Education, 72*, 107–121. doi:10.1016/j.tate.2018.02.011

Fischler, H. (2011). Didaktik: An appropriate framework for the professional work of science teachers? In D. Corrigan, J. Dillon, & R. Gunstone (Eds.), *The professional knowledge base of science teaching* (pp. 31–50). Dordrecht, The Netherlands: Springer. doi:10.1007/978-90-481-3927-9_3

Forman, E. A., Engle, R. A., Venturini, P., & Ford, M. J. (2014). Introduction to special issue: International examinations and extensions of the productive disciplinary engagement framework. *International Journal of Educational Research, 64*, 149–155. doi:10.1016/j.ijer.2013.07.007

Fosnot, C. (1996). *Construtivismo e Educação - Teoria, perspectivas e práticas. Maria João Batalha Reis (Trad.).* Lisboa: Instituto Piaget.

Freeman, S., Eddy, S. L., McDonough, M., Smith, M. K., Okoroafor, N., Jordt, H., & Wenderoth, M. P. (2014). Active learning increases student performance in science, engineering, and mathematics. *Proceedings of the National Academy of Sciences of the United States of America, 111*(23), 8410–8415. doi:10.1073/pnas.1319030111 PMID:24821756

G7. (2017). *Group of Seven Science Ministers Meeting | Declaration.* Retrieved from http://www.g7italy.it/sites/default/files/documents/G7%20Science%20Communiqu%C3%A9.pdf

Gardiner, W., & Lorch, J. (2015). From "outsider" to "bridge": The changing role of university supervision in an urban teacher residency program. *Action in Teacher Education, 37*(2), 172–189. doi:10.1080/01626620.2015.1004601

Georgakopoulou, A. (2006). Thinking big with small stories in narrative and identity analysis. *Narrative Inquiry, 16*(1), 122–130. doi:10.1075/ni.16.1.16geo

Gibbs, G. (2003). Improving student learning through assessment [Editorial]. *Journal of Geography in Higher Education, 27*(2), 123–132.

Gilbert, J. K. (2010). The role of visual representations in the learning and teaching of science: An introduction. *Asia-Pacific Forum on Science Learning and Teaching, 11*(1), 1–19.

Gilbert, J. K. (2005). Endpiece: Research and development on visualization in science education. In J. K. Gilbert (Ed.), *Visualization in science education* (pp. 333–335). Holland: Springer. doi:10.1007/1-4020-3613-2_16

Gilbert, J. K. (2013). Helping learning in science communication. In J. K. Gilbert & S. Stocklmayer (Eds.), *Communication and engagement with science and technology. Issues and dilemmas* (pp. 165–179). New York: Routledge.

Glickman, C. D., Gordon, S. P., & Ross-Gordon, J. M. (2001). *Supervision and instructional leadership: A developmental approach*. Boston: Pearson.

Goldin, G. A. (2008). Perspectives on representation in mathematical learning and problem solving. In L. English (Ed.), *Handbook of international research in mathematics education* (pp. 178–203). New York, NY: Routledge.

Goldin, G. A., & Kaput, J. J. (1996). A joint perspective on the idea of representation in learning and doing mathematics. In L. P. Steffe, P. Nesher, P. Cobb, G. A. Goldin, & B. Greer (Eds.), *Theories of mathematical learning* (pp. 397–430). Mahwah, NJ: Erlbaum.

Gonçalves, R. (2018). *A investigação em didática da álgebra linear: Uma perspetiva do seu contributo para o ensino por meio da reflexão sobre a prática* [Research on the teaching and learning of linear algebra: A perspective of its contribution to teaching through reflection on practice] (Unpublished doctoral dissertation). UTAD, Portugal.

Gonçalves, R. (2015). *Álgebra linear: Teoria e prática* [Linear algebra: Theory and practice]. Lisbon: Sílabo.

Gray, E. M., & Tall, D. O. (1994). Duality, ambiguity and flexibility: A proceptual view of simple arithmetic. *Journal for Research in Mathematics Education, 26*, 115–141.

Griffiths, G. (1985). Doubts, dilemmas and diary keeping: some reflections on teacher-based research. In R. G. Burgess (Ed.), *Issues in educational research – qualitative methods. London*: Falmer.

Guney, A., & Al, S. (2012). Effective learning environments in relation to different learning theories. *Procedia: Social and Behavioral Sciences, 46*, 2334–2338. doi:10.1016/j.sbspro.2012.05.480

Guskey, T. R. (2002). Professional development and teacher change. *Teachers and Teaching, 8*(3), 381–391. doi:10.1080/135406002100000512

Gutwill, J., & Allen, S. (2010). *Group inquiry at science museum exhibits: Getting visitors to ask juicy questions*. Walnut Creek: Left Coast Press.

Halliday, M. A. K., & Martin, J. R. (1993). *Writing science: Literacy and discursive power*. London: The Falmer Press.

Hamilton, L., Halverson, R., Jackson, S. S., Mandinach, E., Supovitz, J. A., Wayman, J. C., … Steele, J. L. (2009). *Using student achievement data to support instructional decision making*. United States Department of Education. Retrieved from http://repository.upenn.edu/gse_pubs/279

Handal, G., & Lauvås, P. (1987). *Promoting reflective teaching: Supervision in action*. Milton Keynes, UK: Open University Press.

Hannah, J., Stewart, S., & Thomas, M. (2013). Emphasizing language and visualization in teaching linear algebra. *International Journal of Mathematical Education in Science and Technology, 44*(4), 475–489. doi:10.1080/0020739X.2012.756545

Harel, G. (1997). The Linear Algebra Curriculum Study Group recommendations: moving beyond concept definition. In D. Carlson, C. R. Johnson, D. Lay, A. D. Porter, A. E. Watkins, & W. Watkins (Eds.), *Resources for teaching linear algebra* (Vol. 42, pp. 107–126). Washington, DC: Mathematical Association of America.

Harel, G. (1997). The Linear Algebra Curriculum Study Group recommendations: Moving beyond concept definition. In D. Carlson, C. R. Johnson, D. Lay, A. D. Porter, A. E. Watkins, & W. Watkins (Eds.), *Resources for teaching linear algebra* (Vol. 42, pp. 107–126). Washington, DC: Mathematical Association of America.

Hargreaves, A. (2007). Sustainable leadership and development in education: Creating the future, conserving the past. *European Journal of Education*, *42*(2), 223–233. doi:10.1111/j.1465-3435.2007.00294.x

Hattie, J. (2003). Teachers make a difference: What is the research evidence? In M. Meiers (Ed.), *Building teacher quality: What does the research tell us?* (pp. 1–17). Melbourne: Australian Council for Educational Research.

Hattie, J., & Timperley, H. (2007). The power of feedback. *Review of Educational Research*, *77*(1), 81–112. doi:10.3102/003465430298487

Heath, C., & vom Lehn, D. (2004). Configuring reception: Looking at exhibits in museums and galleries. *Theory, Culture & Society*, *21*(6), 43–65. doi:10.1177/0263276404047415

Heinze, A., Star, J. R., & Verschaffel, L. (2009). Flexible and adaptive use of strategies and representations in mathematics education. *ZDM*, *41*(5), 535–540. doi:10.100711858-009-0214-4

Hevner, A. (2007). A three-cycle view of design science research. *Scandinavian Journal of Information Systems*, *19*(2), 87–92.

Hickson, H. (2011). Critical reflection: Reflecting on learning to be reflective. *Reflective Practice: International and Multidisciplinary Perspectives*, *12*(6), 829–839. doi:10.1080/14623943.2011.616687

Hiebert, J., & Morris, A. K. (2012). Teaching, rather than teachers, as a path toward improving classroom instruction. *Journal of Teacher Education*, *63*(2), 92–102. doi:10.1177/0022487111428328

Hillel, J. (2000). Modes of description and the problem of representation in linear algebra. In J.-L. Dorier (Ed.), *On the teaching of linear algebra* (pp. 191–207). Dordrecht: Kluwer Academic.

Hillel, J. (2001). Computer Algebra Systems in the learning and teaching of linear algebra: Some examples. In D. Holton (Ed.), *The teaching and learning of mathematics at university level: an ICMI Study* (Vol. 7, pp. 371–380). Dordrecht: Kluwer Academic. doi:10.1007/0-306-47231-7_33

Hmelo-Silver, C. E. (2004). Problem-based learning: What and how do students learn? *Educational Psychology Review*, *16*(3), 235–266. doi:10.1023/B:EDPR.0000034022.16470.f3

Hooper-Greenhill, E. (2007). *Museums and education: Purpose, pedagogy, performance*. New York: Routledge. doi:10.4324/9780203937525

Hopkins, D. (2014). *A teacher's guide to classroom research*. Maidenhead, UK: McGraw-Hill Education.

Hundhausen, C. D. (2002). Integrating algorithm visualization technology into an undergraduate algorithms course: Ethnographic studies of a social constructivist approach. *Computers & Education*, *39*(3), 237–260. doi:10.1016/S0360-1315(02)00044-1

Ideaslab, A. N., & Paper, W. (2011). *Undersanding virtual pedagogies for contemporary teaching & learning*. Pascoe Vale Road: Richard Olsen & ideasLAB. Retrieved from http://www.ideaslab.edu.au/wp-content/uploads/2011/07/Understanding-Virtual-Pedagogies_CKC_ideasLAB.pdf

International Family Nursing Association (IFNA). (2015). *Position statement on generalist competencies for family nursing practice*. Mankato: IFNA.

International Family Nursing Association (IFNA). (2017). *IFNA Position statement on advanced practice competencies for family nursing*. Retrieved from https://internationalfamilynursing.org/2017/05/19/advanced-practice-competencies/

Jagger, S. L., & Yore, L. D. (2012). Mind the gap: Looking for evidence-based practice of science literacy for all in science teaching journals. *Journal of Science Teacher Education*, *23*(6), 559–577. doi:10.100710972-012-9271-6

Jamil, F. M., & Hamre, B. K. (2018). Teacher reflection in the context of an online professional development course: Applying principles of cognitive science to promote teacher learning. *Action in Teacher Education*, 1–17.

Järvelä, S., Kirschner, P., Panadero, E., Malmberg, J., Phielix, C., Jaspers, J., ... Järvenoja, H. (2015). Enhancing socially shared regulation in collaborative learning groups: Designing for CSCL regulation tools. *Educational Technology Research and Development*, *63*(1), 125–142. doi:10.100711423-014-9358-1

Jewitt, C. (2008). Multimodality and literacy in school classrooms. *Review of Research in Education*, *32*(1), 241–267. doi:10.3102/0091732X07310586

Jiménez-Aleixandre, M. P., & Bustamante, J. D. (1997, March). *Analysing classroom discourse: Practical work in the biology laboratory*. Paper presented at the American Educational Research Association (AERA) annual meeting, Chicago, IL.

Jiménez-Aleixandre, M. P., & Crujeiras, B. (2017). Epistemic practices and scientific practices in science education. In K. S. Taber & B. Akpan (Eds.), *Science education. New directions in mathematics and science education* (pp. 69–80). Rotterdam, The Netherlands: Sense Publishers. doi:10.1007/978-94-6300-749-8_5

Johannesburg Declaration on Sustainable Development. (2002, September). Retrieved from http://www.un.org/esa/sustdev/documents/WSSD_POI_PD/Spanish/WSSDsp_PD.htm

Johnson-Laird, P. N. (1983). *Mental models: Towards a cognitive science of language, inference, and consciousness*. Cambridge, MA: Harvard University Press.

Jonassen, D. H. (2012). Designing for decision making. *Educational Technology Research and Development*, *60*(2), 341–359. doi:10.100711423-011-9230-5

Jones, A., & Cowie, B. (2011). Moving beyond deconstruction and reconstruction: Teacher knowledge-as-action? In D. Corrigan, J. Dillon, & R. Gunstone (Eds.), *The professional knowledge base of science teaching* (pp. 51–63). Dordrecht, The Netherlands: Springer. doi:10.1007/978-90-481-3927-9_4

Jong, T., Lane, J., & Sharp, S. (2012). The efficacy of simulation as a pedagogy in facilitating pre-service teachers' learning about emotional self-regulation and its relevance to the teaching profession. *Australian Journal of Teacher Education*, *37*(3). doi:10.14221/ajte.2012v37n3.6

Kaufman, L., & Rousseeuw, P. J. (1990). *Finding groups in data: An introduction to cluster analysis*. Chichester, UK: John Wiley & Sons, Inc. doi:10.1002/9780470316801

Kelly, G. J., & Duschl, R. A. (2002, April). *Toward a research agenda for epistemological studies in science education*. Paper presented at the Annual meeting of the National Association for Research in Science Teaching, New Orleans, LA.

Kelly, G. J. (2008). Inquiry, activity, and epistemic practice. In R. Duschl & R. Grandy (Eds.), *Teaching Scientific Inquiry: Recommendations for Research and Implementation* (pp. 99–117). Rotterdam: Sense Publishers.

Kelly, G. J. (2014). Inquiry teaching and learning: Philosophical considerations. In M. R. Matthews (Ed.), *Handbook of Historical and Philosophical Studies in Science Education* (pp. 1363–1380). Springer.

Kelly, G. J., McDonald, S., & Wickman, P.-O. (2012). Science learning and epistemology. In B. Fraser & K. Tobin (Eds.), *Second international handbook of science education* (pp. 281–291). Dordrecht, The Netherlands: Springer. doi:10.1007/978-1-4020-9041-7_20

Kemmis, S. (1989). Research in action. In T. Husen & T. N. Postlehwaite (Eds.), *International Encyclopedia of Education* (Vol. 6, pp. 3330–3337). Barcelona: Vicens – Vives/MEC.

Kennedy, M. M. (1998). Education reform and subject matter knowledge. *Journal of Research in Science Teaching, 35*(3), 249–263. doi:10.1002/(SICI)1098-2736(199803)35:3<249::AID-TEA2>3.0.CO;2-R

Kennedy, M. M. (2006). Knowledge and vision in teaching. *Journal of Teacher Education, 57*(3), 205–211. doi:10.1177/0022487105285639

Kiely, R., Matt, D., & Wheeler, E. (2010). *Investigating Critical Learning Episodes: a practical guide for continuing professional development.* Academic Press.

King, H., & Tran, L. (2017). Facilitating deep conceptual learning: The role of reflection and learning communities. In P. G. Patrick (Ed.), *Preparing informal science educators: Lessons for science communication and education* (pp. 67–85). New York: Springer. doi:10.1007/978-3-319-50398-1_4

Kirschner, P., van Vilsteren, P., Hummel, H., & Wigman, M. (1997). The design of a study environment for acquiring academic and professional competence. *Studies in Higher Education, 22*(2), 151–171. doi:10.1080/03075079712331381014

Klasa, J. (2010). A few pedagogical designs in linear algebra with Cabri and Maple. *Linear Algebra and Its Applications, 432*(8), 2100–2111. doi:10.1016/j.laa.2009.08.039

Kleiner, I. (2007). *A history of abstract algebra.* Boston, MA: Birkhäuser. doi:10.1007/978-0-8176-4685-1

Klein, G. (2008). Naturalistic decision making. *Human Factors, 50*(3), 456–460. doi:10.1518/001872008X288385 PMID:18689053

Kok, K., Bergmann, K., Cienki, A., & Kopp, S. (2016). Mapping out the multifunctionality of speakers' gestures. *Gesture, 15*(1), 37–59. doi:10.1075/gest.15.1.02kok

Kok, K., & Cienki, A. (2016). Cognitive grammar and gesture: Points of convergence, advances and challenges. *Cognitive Linguistics, 27*(1), 67–100. doi:10.1515/cog-2015-0087

Korthagen, F. (2012). A prática, a teoria e a pessoa na formação de professores. *Educação, Sociedade & Culturas, 36,* 141-158.

Korthagen, F. (2010). How teacher education can make a difference. *Journal of Education for Teaching, 36*(4), 407–423. doi:10.1080/02607476.2010.513854

Korthagen, F., Loughran, J., & Russell, T. (2006). Developing fundamental principles for teacher education programs and practices. *Teaching and Teacher Education, 22*(8), 1020–1041. doi:10.1016/j.tate.2006.04.022

Kramp, M. K. (2004). Exploring life and experience through narrative inquiry. In M. Kathleen De & S. D. Lapan (Eds.), *Foundations for research: Methods of inquiry in education and the social sciences* (pp. 103–122). Mahwah, NJ: Erlbaum.

Kress, G., Jewitt, C., Ogborn, J., & Tsatsarelis, C. (2014). *Multimodal teaching and learning: The rhetorics of the science classroom.* London: Bloomsbury.

Krippendorff, K. (2004). *Content analysis: An introduction to its methodology.* Thousand Oaks, CA: SAGE Publications.

Krist, C. (2016). How a 6th grade classroom develops epistemologies for building scientific knowledge. In C.-K. Looi, J. Polman, U. Cress, & P. Reimann (Eds.), *Proceedings of the international conference of the learning sciences (ICLS 2016)* (Vol. 2, pp. 306–313). Singapore: International Society of the Learning Sciences.

Krist, C. (2016). How a 6th-grade classroom develops epistemologies for building scientific knowledge. In C.-K. Looi, J. Polman, U. Cress, & P. Reimann (Eds.), *Proceedings of the International Conference of the Learning Sciences (ICLS 2016)* (Vol. 2, pp. 306–313). Singapore: International Society of the Learning Sciences.

Kung, R. L., Kung, P., & Linder, C. (2005). Equipment issues regarding the collection of video data for research. Physical review special topics. *Physics Education Research, 1*(1), 010105-1/010105-9. doi:10.1103/PhysRevSTPER.1.010105

Labov, W. (2006). Narrative pre-construction. *Narrative Inquiry, 16*(1), 37–45. doi:10.1075/ni.16.1.07lab

Ladewig, S. (2011). Putting the cyclic gesture on a cognitive basis. *CogniTextes*, 6. Retrieved from http://journals.openedition.org/cognitextes/406

Lakoff, G., & Johnson, M. (1980). *Metaphors we live by*. Chicago: The University of Chicago Press.

Lakoff, G., & Nuñez, R. (2000). *Where mathematics comes from. How the embodied mind brings mathematics into being*. New York: Basic Book.

Lapp, D. A., Nyman, M. A., & Berry, J. S. (2010). Student connections of linear algebra concepts: An analysis of concept maps. *International Journal of Mathematical Education in Science and Technology, 41*(1), 1–18. doi:10.1080/00207390903236665

Larrivee, B. (2000). Transforming teaching practice: Becoming the critically reflective teacher. *Reflective Practice: International and Multidisciplinary Perspectives, 1*(3), 293–307. doi:10.1080/713693162

Laszlo, P. (2006). *Communicating science: A practical guide*. Berlin: Springer.

Latorre, A. (2003). *La investigación-acción* [Action research]. Barcelona: Graó.

Laurillard, D. (2002). *Rethinking university teaching: a conversational framework for the effective use of learning technologies* (2nd ed.). London: Routledge Falmer. doi:10.4324/9780203160329

Lay, D. C. (2012). *Linear algebra and its applications* (4th ed.). Boston, MA: Addison Wesley.

Lemke, J. L. (1998). Teaching all the languages of science: Words, symbols, images, and actions. In *Conference on Science Education in Barcelona*. Retrieved from http://www-personal.umich.edu/~jaylemke/papers/barcelon.htm

Little, J. W. (2007). Teachers' accounts of classroom experience as a resource for professional learning and instructional decision making. In P. Moss (Ed.), *Evidence and decision making* (pp. 217–240). Malden, MA: John Wiley & Sons. doi:10.1111/j.1744-7984.2007.00103.x

Lladó, A. (2014). Student perceptions of peer assessment: An interdisciplinary study. *Assessment & Evaluation in Higher Education, 39*(5), 592–610. doi:10.1080/02602938.2013.860077

Lloyd, C. A. (2017). Exploring the real-world decision-making of novice and experienced teachers. *Journal of Further and Higher Education*, 1–17. doi:10.1080/0309877X.2017.1357070

Loio, C., & Ançã, M. (2006). A dimensão pedagógica do gesto: Contributo para a formação de professores de Português. Aveiro: Universidade De Aveiro.

Lopes, B., & Cravino, J. (2017). *Práticas de Ensino e Tecnologia Acervo de Narrações Multimodais* (B. Lopes & J. Cravino, Eds.). VIla Real: Universidade de Trás-os-Montes e Alto Douro. Retrieved from http://multimodal.narratives.utad.pt

Lopes, J. B., Cravino, J. P., Branco, M. J., Saraiva, E., & Silva, A. (2008). Mediation of student learning: Dimensions and evidences in science teaching. *Problems of Education in the 21st Century, 9*, 42–52.

Lopes, J. B., Cravino, J. P., Branco, M., Saraiva, E., & Silva A. A. (2008). Mediation of student learning: Dimensions and evidences in science teaching. *Problems of Education in the 21st Century, 9*(9), 42–52.

Lopes, J. B., Cravino, J. P., Branco, M., Saraiva, E., & Silva, A. A. (2008). Mediation of student learning: Dimensions and evidences in science teaching. *PEC 2008: Problems of Education in the 21st Century, 9*(9), 42–52.

Lopes, J. B., Cravino, J. P., Branco, M., Saraiva, E., & Silva, A. A. (2008). Mediation of student learning: Dimensions and evidences in science teaching. *Problems of Education in the 21st Century, 9*(9), 42-52.

Lopes, J. B., Cravino, J. P., Branco, M., Saraiva, E., & Silva, A. A. (2008). Mediation of student learning: Dimensions and evidences in science teaching. *Problems of Education in the 21st Century, 9*, 42–52.

Lopes, J. B., Cravino, J. P., Silva, A. A., & Viegas, C. (2012, June). *The role of teacher mediation using computer simulations to improve students' epistemic competences in physical sciences: A theoretical framework.* Paper presented at the 10th International Conference on Computer Based Learning in Science (CBLIS), Barcelona, Spain.

Lopes, J. B., Silva A. A., Cravino, J. P., Viegas, C., Cunha, A. E., Saraiva, E., ... Santos, A. (2010). *Investigação sobre a Mediação de professores de Ciências Físicas em sala de aula* [Research on Teacher Mediation of Physical Science on classroom]. Vila Real: UTAD.

Lopes, J. B., Silva, A. A., Cravino, J. P., Viegas, C., Cunha, A. E., Saraiva, E., ... Santos, A. (2010). *Investigação sobre a mediação de professores de ciências físicas em sala de aula* [Research on physical sciences teacher mediation within classrooms]. Vila Real: UTAD.

Lopes, J. B., Silva, A. A., Cravino, J. P., Viegas, C., Cunha, A. E., Saraiva, E., ... Santos, C. A. (2010). Investigação sobre a Mediação de professores de Ciências Físicas em sala de aula [Research about teacher mediation of physical sciences in the classroom]. Vila Real, Portugal: Universidade de Trás-os-Montes e Alto Douro.

Lopes, J. B., Silva, A. A., Cravino, J. P., Viegas, C., Cunha, A. E., Saraiva, E., ... Santos, C. A. (2010). Investigação sobre a mediação de professores de ciências físicas em sala de aula [Research on physical sciences teachers' mediation in the classroom]. Vila Real, Portugal: Universidade de Trás-os-Montes e Alto Douro.

Lopes, J. B., Silva, A. A., Cravino, J. P., Viegas, C., Cunha, E., Saraiva, E., ... Santos, C. A. (2010). *Investigação sobre a mediação de professores de ciências físicas em sala de aula* [Research on physical science teachers' mediation in the classroom]. Vila Real: UTAD.

Lopes, J. B., Viegas, C., & Pinto, A. (2018). Melhorar Práticas de Ensino de Ciências e Tecnologia – Registar e Investigar com Narrações Multimodais [Improving Science and Technology Teaching Practices - Registering and Investigating Multimodal Narratives]. Lisboa: Edições Sílabo.

Lopes, J. B., Viegas, C., & Pinto, A. (Eds.). (2018). Melhorar práticas de ensino de ciências e tecnologia: Registar e investigar com narrações multimodais [Improving teaching practices in sciences and technology: Register and investigate with multimodal narratives]. Lisboa: Edições Sílabo.

Lopes, J. B. (2004). *Aprender e ensinar Física* [Learn and teach physics]. Lisbon: Fundação Calouste Gulbenkian.

Lopes, J. B. (2004). *Aprender e Ensinar Física.* Lisboa: Fundação Calouste Gulbenkian.

Lopes, J. B. (2017). Researching science teaching practices in the classroom over two decades: What matters? In J. B. Lopes, J. P. Cravino, E. S. Cruz, & A. Barbot (Eds.), *Teacher science: Contributions of research for planning, practice and professional development* (pp. 1–26). New York, NY: Nova Science Publishers.

Lopes, J. B. (2017). Researching Science Teaching Practices in the Classroom over Two Decades: What Matters? In J. B. Lopes, J. P. Cravino, E. S. Cruz, & A. Barbot (Eds.), *Teacher Science: Contributions of Research for Planning, Practice and Professional Development* (pp. 1–26). New York: Nova Science Publishers.

Lopes, J. B. (2017). Researching science teaching practices in the classroom over two decades: What matters? In J. B. Lopes, J. P. Cravino, E. Cruz, & A. Barbot (Eds.), *Teaching science: Contributions of research for planning, practice and professional development* (pp. 1–24). New York: Nova Science Publishers.

Lopes, J. B. (2017). Researching science teaching practices in the classroom over two decades: What matters? In J. B. Lopes, J. P. Cravino, E. Cruz, & A. Barbot (Eds.), *Teaching Science: Contributions of Research for Planning, Practice and Professional Development* (pp. 1–25). New York: Nova Science Publishers.

Lopes, J. B. (2017). Self-directed professional development to improve effective science teaching. In J. B. Lopes, J. P. Cravino, E. S. Cruz, & A. Barbot (Eds.), *Teaching science: Contributions of research for planning, practice and professional development* (pp. 247–264). New York: Nova Science Publishers.

Lopes, J. B., Branco, J., & Jiménez-Aleixandre, M. P. (2011). "Learning experience" provided by science teaching practice in a classroom and the development of students' competences. *Research in Science Education*, *41*(5), 787–809. doi:10.100711165-010-9190-5

Lopes, J. B., Cravino, J. P., Cruz, E., & Barbot, A. (2017). *Teaching science: Contributions of research for planning, practice and professional development.* New York: Nova Science Publishers.

Lopes, J. B., Cravino, J. P., & Silva, A. A. (2010). *A model for effective teaching in science and technology (Metilost).* New York: Nova Science Publishers, Inc.

Lopes, J. B., Cravino, J. P., & Silva, A. A. (2010). *Effective Teaching for Intended Learning Outcomes in Science and Technology (Metilost).* New York: Nova Science Publishers.

Lopes, J. B., & Cunha, A. E. (2017). Self-directed professional development to improve effective teaching: Keys points for a model. *Teaching and Teacher Education*, *68*, 262–274. doi:10.1016/j.tate.2017.09.009

Lopes, J. B., Cunha, A. E., Santos, C. A., Saraiva, E., Cravino, J. P., & Dinis, F. (2012). Envolver produtivamente os alunos em aulas de Física e Química durante o uso de simulações computacionais: Dois professores com mediações distintas e uso distinto [Productively engaging students in physics and chemistry lessons during the use of computer simulations: Two teachers with different mediations and different use]. *SENSOS*, *2*(2), 121–137.

Lopes, J. B., Cunha, A. E., Santos, C. A., Saraiva, E., Cravino, J. P., & Dinis, F. (2013). Envolver os alunos produtivamente em aulas de física e química durante o uso de simulações computacionais: Dois professores com mediações distintas e uso distinto das simulações. *Sensos*, *4*(2), 121–137.

Lopes, J. B., Silva, A. A., Cravino, J. P., Viegas, C., Cunha, A. E., Saraiva, E., & Santos, C. A. (2010). *Investigação sobre a mediação de professores de Cências Físicas em sala de aula.* Vila Real, Portugal: Universidade de Trás-os-Montes e Alto Douro.

Lopes, J. B., Silva, A. A., Cravino, J. P., Viegas, C., Cunha, A. E., Saraiva, E., ... Santos, C. A. (2012). Instrumentos de ajuda à mediação do professor para promover a aprendizagem de alunos e o desenvolvimento profissional dos professores [Tools to help teacher mediation to promote student learning and teacher professional development]. *Sensos*, *2*(1), 125–171.

Lopes, J. B., Silva, A. A., Cravino, J. P., Viegas, C., Cunha, A. E., Saraiva, E., ... Santos, C. A. (2012). Instrumentos de ajuda à mediação do professor para promover a aprendizagem dos alunos e o desenvolvimento profissional dos professores. *Revista Sensos*, *2*, 125–171.

Lopes, J. B., Silva, A. A., Cravino, J. P., Viegas, C., Cunha, A. E., Saraiva, E., ... Santos, C. A. (2012). Instrumentos de ajuda à mediação do professor para promover a aprendizagem dos alunos e o desenvolvimento profissional dos professores. *Sensos, 2*(1), 125–171.

Lopes, J. B., Silva, A., Cravino, J., Viegas, C., Cunha, A., Saraiva, E., & Santos, C. (2010). *Investigação sobre a Mediação de professores de ciências físicas em sala de aula* [Research on science teachers' mediation in the classroom]. Vila Real, Portugal: Universidade de Trás-os-Montes e Alto Douro.

Lopes, J. B., Viegas, C., & Cravino, J. P. (2010). Improving the learning of physics and development of competences in engineering students. *International Journal of Engineering Education, 26*(3), 612–627.

Lopes, J. B., Viegas, C., & Cravino, J. P. (2010). Improving the Learning of Physics and Development of Competences in Engineering Students. *International Journal of Engineering Education, 26*(3), 612–627.

Lopes, J., Silva, A., Cravino, J., Santos, C., Cunha, A., Pinto, A., ... Branco, M. (2014). Constructing and using Multimodal Narratives to Research in Science Education: Contributions Based on Practical Classroom. *Research in Science Education, 44*(3), 415–438. doi:10.100711165-013-9381-y

Loughran, J. (2012). *What expert teachers do: Enhancing professional knowledge for classroom practice*. New York: Routledge. doi:10.4324/9780203851470

Love, B., Hodge, A., Grandgenett, N., & Swift, A. W. (2014). Student learning and perceptions in a flipped linear algebra classroom. *International Journal of Mathematical Education in Science and Technology, 45*(3), 317–324. doi:10.1080/0020739X.2013.822582

Luebke, J. F., & Matiasek, J. (2013). An exploratory study of zoo visitors' exhibit experiences and reactions. *Zoo Biology, 32*(4), 407–416. doi:10.1002/zoo.21071 PMID:23740472

Luft, J. A., & Hewson, P. W. (2014). Research on teacher professional development programs in science. Handbook of Research on Science Education, 2, 889-909.

Luke, N., & Rogers, C. (2015). Responding to uncertainty: Teacher educator professional development through coteaching and collaborative reflection. *Learning Landscapes, 8*, 245–259.

Luke, N., & Rogers, C. (2015). Responding to uncertainty: Teacher educator professional development through coteaching and collaborative reflection. *Teacher Research: Integrating Action, Observation, and Reflection, 8*(2), 245–260.

Lumpe, A., Vaughn, A., Henrikson, R., & Bishop, D. (2014). Teacher professional development and self-efficacy beliefs. In R. Evans, J. Luft, C. Czerniak, & C. Pea (Eds.), *The role of science teachers' beliefs in international classrooms* (pp. 49–63). Rotterdam, The Netherlands: SensePublishers.

Lustick, D. (2009). The Failure of Inquiry: Preparing Science Teachers with an Authentic Investigation. *Journal of Science Teacher Education, 20*(6), 583–604. doi:10.100710972-009-9149-4

Maheshwari, V. K. (2013). *Expository teaching – A direct instructional strategy*. Retrieved from http://www.vkmaheshwari.com/WP/?p=928

Martínez, J. F., Borko, H., & Stecher, B. M. (2012). Measuring instructional practice in science using classroom artifacts: Lessons learned from two validation studies. *Journal of Research in Science Teaching, 49*(1), 38–67. doi:10.1002/tea.20447

Mason, J. (2002). *Researching your own practice: Discipline of noticing*. New York: Routledge.

Mason, J. (2002). *Researching your own practice: The discipline of noticing*. New York: Routledge.

Masuda, A. M., Ebersole, M. M., & Barrett, D. (2013). A Qualitative Inquiry: Teachers' attitudes and willingness to engage in professional development experiences at different career stages. *International Journal of Progressive Education, 79*(2), 6–14.

Maturano, C., Aguilar, S., & Nuñez, G. (2009). Conversion de imagenes al linguaje escrito. Un desafio para el estudiante de ciencias naturales. *Revista Eureka sobre Enseñanza y Divulgación de las Ciencias, 6*(1), 63–78.

Mazur, E. (1997). *Peer instruction: a user's manual.* Prentice Hall.

McDonald, B., & Boud, D. (2003). The impact of self-assessment on achievement: The effects of self-assessment training on performance in external examinations. *Assessment in Education: Principles, Policy & Practice, 10*(2), 209–220. doi:10.1080/0969594032000121289

McGee, D. L., & Martinez-Planell, R. (2014). A study of semiotic registers in the development of the definite integral of functions of two and three variables. *International Journal of Science and Mathematics Education, 12*(4), 883–916. doi:10.100710763-013-9437-5

McIntyre, D. (2005). Bridging the gap between research and practice. *Cambridge Journal of Education, 35*(3), 357–382. doi:10.1080/03057640500319065

McNeill, D. (1992). *Hand and mind: What gestures reveal about thought.* Chicago: University of Chicago Press.

McNeill, D. (2015). *Why we gesture: The surprising role of hand movements in communication.* Cambridge, UK: Cambridge University Press.

McNeill, K. L., & Krajcik, J. (2009). Synergy between teacher practices and curricular scaffolds to support students in using domain-specific and domain-general knowledge in writing arguments to explain phenomena. *Journal of the Learning Sciences, 18*(3), 416–460. doi:10.1080/10508400903013488

McNeill, K. L., Lizotte, D. J., Krajcik, J., & Marx, R. W. (2006). Supporting students' construction of scientific explanations by fading scaffolds in instructional materials. *Journal of the Learning Sciences, 15*(2), 153–191. doi:10.120715327809jls1502_1

McNiff, J., & Whitehead, J. (2006). *All you need to know about action research.* London: Sage Publications.

Meirink, J., Meijer, P., Verloop, N., & Bergen, T. C. M. (2009). Understanding teacher learning in secondary education: The reactions of teacher activities to changed beliefs about teaching and learning. *Teaching and Teacher Education, 25*(1), 89–100. doi:10.1016/j.tate.2008.07.003

Melville, W. (2008). Mandated curriculum change and a science department: A superficial language convergence? *Teaching and Teacher Education, 24*(5), 1185–1199. doi:10.1016/j.tate.2007.03.004

Mestre, J. (2001). Implications of research on learning for the education of prospective science and physics teachers. *Physics Education, 36*(1), 44–51. doi:10.1088/0031-9120/36/1/308

Millar, R., & Osborne, J. (2009). Research and practice: A complex relationship? In M. C. Shelley, L. D. Lore, & B. Hand (Eds.), *Quality research in literacy and science education: International perspectives and gold standards* (pp. 41–61). Dordrecht, The Netherlands: Springer. doi:10.1007/978-1-4020-8427-0_3

Montenegro, P., Campos, H., & Aires, A. P. (2017). Tasks involving perimeter and area of plane figures: Analyses of a mathematical textbook. In L. Gómez Chova, A. López Martínez, & I. Candel Torres (Eds), *EDULEARN2017: Proceedings 9th International Conference on Education and New Learning Technologies* (pp. 2469–2475). Barcelona, Spain: IATED Academy. 10.21125/edulearn.2017.1518

Montenegro, P., Costa, C., & Lopes, B. (2017a). Transformações de representações visuais na exploração de múltiplos e divisores de um número [Transformations of visual representations in the exploration of multiples and divisors of a number]. *Comunicações Piracicaba, 24*(1), 55–68.

Montenegro, P., Costa, C., & Lopes, B. (2017b). The use of visual representations in mathematics classes: A comparative case study. In L. Gómez Chova, A. López Martínez, & I. Candel Torres (Eds.), *INTED2017: Proceedings of 11th International Technology, Education and Development Conference* (pp. 6845–6852). Valencia, Spain: IATED Academy. 10.21125/inted.2017.1590

Montenegro, P., Costa, C., & Lopes, B. (2017c). Multiple representations in problem solving with divisors and multiples. In L. Gómez Chova, A. López Martínez, & I. Candel Torres (Eds.), *INTED2017: Proceedings of 11th International Technology, Education and Development Conference* (pp. 6884–6890). Valencia, Spain: IATED Academy. 10.21125/inted.2017.1599

Montenegro, P., Costa, C., & Lopes, B. (2017d). Visual representation (table) to recreate a routine mathematical exercise. In L. Gómez Chova, A. López Martínez, & I. Candel Torres (Eds.), *EDULEARN2017: Proceedings 9th International Conference on Education and New Learning Technologies* (pp. 2514–2520). Barcelona, Spain: IATED Academy. 10.21125/edulearn.2017.1525

Montenegro, P., Costa, C., & Lopes, J. B. (2018). Transformations in the visual representation of a figural pattern. *Mathematical Thinking and Learning, 20*(2), 91–107. doi:10.1080/10986065.2018.1441599

Montero, L. (2001). *A construção do conhecimento profissional docente. Armando Pereira da Silva (Trad.).* Lisboa: Instituto Piaget.

Morgan, C., Craig, T., Schuette, M., & Wagner, D. (2014). Language and communication in mathematics education: An overview of research in the field. *ZDM Mathematics Education, 46*(6), 843–853. doi:10.100711858-014-0624-9

Morris, A. K., & Hiebert, J. (2011). Creating shared instructional products: An alternative approach to improving teaching. *Educational Researcher, 40*(1), 5–14. doi:10.3102/0013189X10393501

Morris, C., & Chikwa, G. (2016). Audio versus written feedback: Exploring learners' preference and the impact of feedback format on students' academic performance. *Active Learning in Higher Education, 17*(2), 125–137. doi:10.1177/1469787416637482

Mortensen, M. F. (2011). Analysis of the educational potential of a science museum Learning environment: Visitors' experience with and understanding of an immersion exhibit. *International Journal of Science Education, 33*(4), 517–545. doi:10.1080/09500691003754589

Mortimer, E. F., & Araújo, A. O. (2014). Using productive disciplinary engagement and epistemic practices to evaluate a traditional Brazilian high school chemistry classroom. *International Journal of Educational Research, 64*, 156–169. doi:10.1016/j.ijer.2013.07.004

Mortimer, E. F., Mol, G., & Duarte, L. P. (1994). Regra do octeto e teoria da ligação química no ensino médio: Dogma ou ciência? [Octet rule and chemical bond theory in high school: Dogma or science?]. *Quimica Nova, 17*(2), 243–252.

Mugur-Schächter, M. (2006). *Sur le tissage des connaissances* [On the weaving of knowledge]. Paris: Hermès Science Publishing – Lavoisier.

Mulkay, M. (1972). *The social process of innovation.* London: Macmillan. doi:10.1007/978-1-349-01450-7

Mushyikwa, E., & Lubben, F. (2009). Self-directed professional development – Hope for teachers working in deprived environments. *Teaching and Teacher Education, 25*(3), 375–432. doi:10.1016/j.tate.2008.12.003

National Council of Teachers of Mathematics (NCTM). (2008). *Princípios e normas para a Matemática escolar.* Lisbon, Portugal: APM.

Nicol, D., & Macfarlane-Dick, D. (2004). *Rethinking formative assessment in HE: a theoretical model and seven principles of good feedback practice.* York, UK: HEA. Retrieved from http://www.enhancementthemes.ac.uk/pages/docdetail/docs/workshop/rethinking-formative-assessment-a-theoretical-model-and-seven-principles-of-good-feedback-practice-paper

Nicol, D. (2008). *Transforming assessment and feedback: enhancing integration and empowerment in the first year.* Quality Assurance Agency.

Nicol, D., & Macfarlane-Dick, D. (2006). Formative assessment and self-regulated learning: A model and seven principles of good feedback practice. *Studies in Higher Education, 31*(2), 199–218. doi:10.1080/03075070600572090

Nonaka, I., & Takeuchi, H. (1995). *The knowledge-creating company: How Japanese companies create the dynamics of innovation.* New York: Oxford University Press.

Norton, A. H., & McCloskey, A. (2008). Teaching experiments and professional development. *Journal of Mathematics Teacher Education, 11*(4), 285–305. doi:10.100710857-008-9076-x

Novak, J. D., & Gowin, D. B. (1984). *Learning how to learn.* Cambridge University Press. doi:10.1017/CBO9781139173469

Nóvoa, A. (2009). *Para uma formação de professores construída dentro da profissão* [For a teacher training built within the profession]. Retrieved from http://www.revistaeducacion.educacion.es/re350/re350_09por.pdf

Nzau, D. K. (2010). *Das concepções dos slunos sobre força ao desenvolvimento de estratégias de ensino fundamentadas num modelo didáctico construtivista.* Tese de Doutoramento. Vila Real: Universidade de Trás-Os-Montes e Alto Douro.

Nzau, D. K., & Costa, N. (1992). As concepções alternativas dos alunos sobre força como medida de interacção dos corpos. In 8ª Conferência Nacional de Física e 2º Encontro Ibérico para o Ensino de Física (pp. 570–571). Vila Real: Universidade de Trás-os-Montes e Alto Douro.

Nzau, D. K. (2013). *Formação de professores de física na perspectiva construtivista. Proposta didáctica para a introdução do conceito da força.* Saarbrücken: Novas Edições Acadêmicas.

Nzau, D. K., Lopes, J. B., & Costa, N. (2012). Formação Continuada de Teacheres de Física, em Angola, com base num modelo didáctico para o campo conceptual de força. *Revista Brasileira de Ensino de Física, 34*(3), 3402. doi:10.1590/S1806-11172012000300014

Oliveira, D. K. B., Mendonça, P. C. C., & Justi, R. (2012). Argumentar para aprender ligações químicas: contribuições de uma atividade investigativa [Argumentation to learn chemical bonds: Contributions of a research activity]. XVI Encontro Nacional de Ensino de Química (XVI ENEQ), Salvador – Bahia.

Oliveira, I., & Serrazina, L. (2002). A reflexão e o professor como investigador [Reflection and teacher as a researcher]. In APM-GTI (Ed.), Reflectir e investigar sobre a prática profissional (pp. 29-42). Lisbon: APM.

Osborne, J. (2014). Teaching scientific practices: Meeting the challenge of change. *Journal of Science Teacher Education, 25*(2), 177–196. doi:10.100710972-014-9384-1

Osmond, C., & Barker, D. J. (2000). Fetal, infant, and childhood growth are predictors of coronary heart disease, diabetes, and hypertension in adult men and women. *Environmental Health Perspectives, 108*(Suppl), 545–553. PMID:10852853

Ozkan, B. C. (2004). Using NVivo to analyze qualitative classroom data on constructivist learning environments. *Qualitative Report, 9*(4), 589–603.

Panadero, E., Jonsson, A., & Strijbos, J. W. (2016). Scaffolding self-regulated learning through self-assessment and peer assessment: Guidelines for classroom implementation. In *Assessment for learning: Meeting the challenge of implementation* (pp. 311–326). Cham, Switzerland: Springer International Publishing.

Patel, V. L., Kaufman, D. R., & Arocha, J. F. (2002). Emerging paradigms of cognition in medical decision-making. *Journal of Biomedical Informatics*, *35*(1), 52–75. doi:10.1016/S1532-0464(02)00009-6 PMID:12415726

Patrick, P. G. (2017). Introduction. In P. G. Patrick (Ed.), *Preparing informal science educators: Lessons for science communication and education* (pp. 1–9). New York: Springer. doi:10.1007/978-3-319-50398-1_1

Patterson, C., Stephens, M., Chiang, V., Price, A. M., Work, F., & Snelgrove-Clarke, E. (2017). The significance of personal learning environments (PLEs) in nursing education: Extending current conceptualizations. *Nurse Education Today*, *48*, 99–105. doi:10.1016/j.nedt.2016.09.010 PMID:27744138

Pattison, S. A., & Dierking, L. D. (2012). Exploring staff facilitation that supports family learning. *Journal of Museum Education*, *37*(3), 69–80. doi:10.1080/10598650.2012.11510743

Pattison, S. A., & Dierking, L. D. (2013). Staff-mediated learning in museums: A social interaction perspective. *Visitor Studies*, *16*(2), 117–143. doi:10.1080/10645578.2013.767731

Pazinato, M. S. (2016). Ligações químicas: investigação da construção do conhecimento no ensino médio [Chemical bonds: Investigation of the construction of knowledge in high school] (Tese de Doutorado). Universidade Federal de Santa Maria, Santa Maria.

Pedretti, E. (2012). The medium is the message: Unraveling visitors' experiences of Body Worlds and the Story of the Heart. In E. Davidsson & A. Davidsson (Eds.), *Understanding interactions at science centers and museums – A socio-cultural perspective* (pp. 45–62). Sense Publishers. doi:10.1007/978-94-6091-725-7_4

Pedrosa-de-Jesus, H., & Watts, M. (2014). Managing affect in learners' questions in undergraduate Science. *Studies in Higher Education*, *39*(1), 102–116. doi:10.1080/03075079.2011.646983

Peirce, C. S. (2005). *Semiótica* [Semiotics] (3rd ed.). São Paulo: Perspectiva.

Pereira, A., Mendes, A. Q., Morgado, L., Amante, L., & Bidarra, J. (2008). *Universidade Aberta's pedagogical model for distance education: A university for the future*. Lisbon: Universidade Aberta.

Pereira, F., Caetano, J., Moreira, J., & Ataíde, M. (2015). Práticas educativas em saúde na formação de acadêmicos de enfermagem. *Cogitare Enfermagem*, *20*(2), 332–337. doi:10.5380/ce.v20i2.39767

Perrenoud, P. (1998). La transposition didactique à partir de pratiques: Des savoirs aux compétences [Didactic transposition from practices: From knowledge to skills]. *Revue des Sciences de l'Education*, *24*(3), 487–514. doi:10.7202/031969ar

Perrenoud, P. (2000). *Novas competências para ensinar*. Porto Alegre: Artmed.

Pinto, A., Barbot, A., Viegas, C., Silva, A. A., Santos, C. A., & Lopes, J. B. (2012). Teacher education using computer simulations pre and in-service primary school teacher training to teach science. In R. Pintó, V. López, & C. Simarro (Eds.), Learning science in thes of computers (pp. 28–36). Barcelona, Spain: Centre for Research in Science and Mathematics Education (CRECIM).

Pinto, A., Barbot, A., Viegas, C., Silva, A. A., Santos, C. A., & Lopes, J. B. (2012). Teacher education using computer simulations – pre and in-service primary school teacher training to teach science. *10th International Conference on Computer Based Learning in Science (CBLIS)*.

Pinto, A., Barbot, A., Viegas, C., Silva, A. A., Santos, C. A., & Lopes, J. B. (2014). Teaching science with experimental work and computer simulations in a primary teacher education course: What challenges to promote epistemic practices? *Procedia Technology, 13*, 86–96. doi:10.1016/j.protcy.2014.02.012

Pinto, A., & Nzau, D. K. (2017). Training teachers how to plan for teaching science and technology. In J. B. Lopes, J. P. Cravino, E. Cruz, & A. Barbot (Eds.), *Teaching science: Contributions of research for planning, practice and professional development* (pp. 61–77). New York: Nova Science Publishers.

Ponte, J. P. (2002). Investigar a nossa própria prática [Investigate our own practice]. In Grupo de Trabalho sobre Investigação (Ed.), Reflectir e investigar sobre a prática profissional [Reflect and investigate on professional practice] (pp. 5-28). Lisbon: Associação de Professores de Matemática.

Ponte, J. P. (1994). O desenvolvimento profissional do professor de Matemática [The professional development of the Mathematics teacher]. *Educación Matemática, 31*, 9–12, 20.

Ponte, J. P. (2004). Investigar a nossa prática: Uma estratégia de formação e de construção do conhecimento profissional [Investigating our practice: A strategy of training and building professional knowledge]. In *Actas do VIII Simpósio do SEIEM*. Coruña: Universidad de la Coruña.

Popil, I. (2011). Promotion of critical thinking by using case studies as teaching method. *Nurse Education Today, 31*(2), 204–207. doi:10.1016/j.nedt.2010.06.002 PMID:20655632

Prachagool, V., Nuangchalerm, P., Subramaniam, G., & Dostál, J. (2016). Pedagogical decision making through the lens of teacher preparation program. *Journal for the Education of Gifted Young Scientists, 4*(2), 41–52.

Prince, M. (2004). Does active learning work? A review of the research. *Journal of Engineering Education, 93*(3), 223–231. doi:10.1002/j.2168-9830.2004.tb00809.x

Qualitative Research. (2016). Editorial. *Qualitative Research, 16*(1), 3–8.

Radford, L., Edwards, L., & Arzarello, F. (2009). Introduction. *International Journal of Mathematics Education, 70*, 91–95.

Rantavuori, J., Engeström, Y., & Lipponen, L. (2016). Learning actions, objects and types of interaction: A methodological analysis of expansive learning among pre-service teachers. *Frontline Learning Research in Learning Research, 4*(3), 1–27.

Redish, E. (2003). *Teaching physics with the physics suite*. New York: John Wiley & Sons.

Redish, E. F. (2003). *Teaching Physics with the Physics Suite*. John Wiley &Sons, Inc.

Reinholz, D. (2016). The assessment cycle: A model for learning through peer assessment. *Assessment & Evaluation in Higher Education, 41*(2), 301–315. doi:10.1080/02602938.2015.1008982

Reiser, B. J. (2004). Scaffolding complex learning: The mechanisms of structuring and problematizing student work. *Journal of the Learning Sciences, 13*(3), 273–304. doi:10.120715327809jls1303_2

Rennie, L. J. (2014). Learning science outside of school. In N. G. Lederman, & S. K. Abell (Eds.), Handbook of research on science education (Vol. 2, pp. 120–144). Routledge.

Rennie, L. J. (2013). The practice of science and technology communication in science museums. In J. K. Gilbert & S. Stocklmayer (Eds.), *Communication and engagement with science and technology. Issues and dilemmas* (pp. 197–211). New York: Routledge.

Rettie, R., Robinson, H., Radke, A., & Ye, X. (2008). CAQDAS: A supplementary tool for qualitative market research. *Qualitative Market Research, 11*(1), 76–88. doi:10.1108/13522750810845568

Rich, P. J., & Hannafin, M. J. (2008). Decisions and reasons: Examining preservice teacher decision-making through video self-analysis. *Journal of Computing in Higher Education*, *20*(1), 62–94. doi:10.1007/BF03033432

Richter, D., Kunter, M., Klusmann, U., Lüdtke, O., & Baumert, J. (2014). Professional development across the teaching career. In S. Krolak-Schwerdt, S. Glock, & M. Böhmer (Eds.), *Teachers' Professional Development. The Future of Education Research* (pp. 97–121). Rotterdam: Sense Publishers. doi:10.1007/978-94-6209-536-6_7

Rodrigues, F., & Oliveira, P. (2014). A system for formative assessment and monitoring students' progress. *Computers & Education*, *76*, 30–41. doi:10.1016/j.compedu.2014.03.001

Rogalski, M. (2000). The teaching experiment in Lille. In J.-L. Dorier (Ed.), *On the teaching of linear algebra* (pp. 133–149). Dordrecht: Kluwer Academic.

Roque, N. F., & Silva, J. L. P. B. (2008). A linguagem química e o ensino da química orgânica [The chemical language and the teaching of organic chemistry]. *Quimica Nova*, *31*(4), 921–923. doi:10.1590/S0100-40422008000400034

Roth, W.-M., Masciotra, D., & Boyd, N. (1999). Becoming-in-the-classroom: A case study of teacher development through coteaching. *Teaching and Teacher Education*, *15*(7), 771–784. doi:10.1016/S0742-051X(99)00027-X

Rowan, B., Camburn, E., & Correnti, R. (2004). Using teacher logs to measure the enacted curriculum: A study of literacy teaching in third-grade classrooms. *The Elementary School Journal*, *105*(1), 75–101. doi:10.1086/428803

Rua, M. (2011). *De aluno a enfermeiro. Desenvolvimento de competências em contexto de ensino clínico*. Lusociência.

Rua, M. (2012). A reflexão como estratégia de desenvolvimento de competências em contexto de ensino clinico. Vigo: Vicerreitoria de Alumnado Universidade de Vigo.

Rummel, N., Levin, J., & Woodward, M. (2003). Do pictorial mnemonic text-learning aids give students something worth writing about? *Journal of Educational Psychology*, *95*(2), 327–334. doi:10.1037/0022-0663.95.2.327

Russell, A., Comello, R., & Lee Wright, D. (2007). Teaching strategies promoting active learning in healthcare education. *Journal of Education and Human Development*, *1*(1), 1–8.

Sabella, M., & Redish, E. (2007). Knowledge organization and activation in physics problem solving. *American Journal of Physics*, *75*(11), 1017–1029. doi:10.1119/1.2746359

Sá-Chaves, I. (2002). *A construção de conhecimento pela análise reflexiva da praxis*. Lisboa: Edição Fundação CalousteGulbenkian, Fundação para a Ciência e Tecnologia.

Sá-Chaves, I. (2000). *Formação, conhecimento e supervisão: Contributos nas áreas de formação de professores e de outros profissionais* [Training, knowledge and supervision: Contributions in the areas of teacher training and other professionals]. Aveiro: Universidade de Aveiro.

Sadler, R. (1989). Formative assessment and the design of instructional systems. *Instructional Science*, *18*(2), 119–144. doi:10.1007/BF00117714

Sambell, K. (2016). Assessment and feedback in higher education: Considerable room for improvement? *Student Engagement in Higher Education Journal*, *1*(1), 2–14.

Sandoval, W. A. (2005). Understanding students' practical epistemologies and their influence on learning through inquiry. *Science Education*, *89*(4), 634–656. doi:10.1002ce.20065

Sandoval, W. A., & Millwood, K. A. (2005). The quality of students' use of evidence in written scientific explanations. *Cognition and Instruction*, *23*(1), 23–55. doi:10.12071532690xci2301_2

Sandoval, W. A., & Reiser, B. J. (2004). Explanation-driven inquiry: Integrating conceptual and epistemic scaffolds for scientific inquiry. *Science Education, 88*(3), 345–372. doi:10.1002ce.10130

Santagata, R., & Yeh, C. (2016). The role of perception, interpretation, and decision making in the development of beginning teachers' competence. *ZDM Mathematics Education, 48*(1-2), 153–165. doi:10.100711858-015-0737-9

Santos, C. A., Lopes J. B., & Cravino, J. P. (2011). Patterns of didactic decisions made by teachers in physical sciences classrooms. *Problems of Education in the 21st Century, 36*, 76–89.

Santos, C. A., Lopes, J. B., & Cravino, J. P. (2011). Patterns of didactic decisions made by teachers in physical sciences classroom. *Problems of Education in the 21st Century, 36*, 67–89.

Santos, C. A., Lopes, J. B., & Cravino, J. P. (2011). Patterns of didactic decisions made by teachers in physical sciences classrooms. *Problems of Education in the 21st Century, 36*, 76–89.

Santos, C. A., Lopes, J. B., & Cravino, J. P. (2011). Patterns of didactic decisions made by teachers in physical sciences classrooms. *Problems of Education in the 21th Century, 36*, 76–89.

Santos, C. A., Lopes, J. B., Saraiva, E., & Cunha, A. E. (2015). Teacher decisions and didactical intentionality using computer simulations in physical science classrooms. In L. G. Chova, A. L. Martínez, & I. C. Torres (Eds.), *9th International Technology, Education and Development Conference* (pp. 5347–5356). Madrid: IATED Academy.

Santos, G. L. S. (2014). Os registros de representação semiótica mobilizados por acadêmicos de um curso de ciências contábeis em resolução de problemas [Semiotic representation registers mobilized by academics from a course in accounting sciences in problem solving] (Tese de Doutorado). Universidade Estadual de Londrina, Londrina.

Santos, C. A. (2017). The role of teacher decisions in the classroom to develop students' epistemic activity. In J. B. Lopes, J. P. Cravino, E. S. Cruz, & A. Barbot (Eds.), *Teaching science: Contributions of research for planning, practice and professional development* (pp. 119–134). Nova Science Publishers.

Santos, L. (2016). A articulação entre a avaliação sumativa e a formativa, na prática pedagógica: Uma impossibilidade ou um desafio? [Combining summative and formative assessment in pedagogical practice. An impossibility or a challenge?]. *Avaliação e Políticas Públicas em Educação, 24*(92), 637–669.

Saraiva, E. (2017). *Estudo da Linguagem gráfica no contexto da mediação da aprendizagem em Ciências Físicas* [Study of graphical language in the context of learning mediation in physical sciences] (Unpublished doctoral dissertation). Universidade de Trás-os-Montes e Alto Douro, Vila Real, Portugal.

Saraiva, E. (2017). *Estudo do papel da representação visual no contexto da mediação dos professores de ciências físicas* [Study of the role of visual representation in the context of the mediation of teachers of physical sciences] (Unpublished doctoral dissertation). Universidade de Trás-os-Montes e Alto Douro, Vila Real, Portugal.

Saraiva, E. (2017a). *Estudo do papel da representação visual no contexto da mediação dos professores de ciências físicas* [Study of the role played by visual representation in the context of teacher mediation of physical sciences] (Doctoral thesis). Universidade de Trás-os-Montes e Alto Douro, Vila Real, Portugal.

Saraiva, E., Lopes, J. B., Cravino, J. P., & Santos, C. A. (2012). How do teachers of physical sciences with different professional experience use visual representations with epistemic functions in the classroom? *PEC 2012: Problems of Education in the 21st Century, 42*(42), 97–114.

Saraiva, E., Lopes, J. B., Cravino, J. P., & Santos, C. A. (2012). How do teachers of physical sciences with different professional experiences use visual representations with epistemic functions in the classroom? *Problems of Education in the 21st Century, 42*, 97–114.

Saraiva, E., Lopes, J. B., Cravino, J. P., & Santos, C.A. (2012). How do teachers of physical sciences with different professional experiences use visual representations with epistemic functions in the classroom. *Problems of Education in the 21st Century, 42*, 97–114.

Saraiva, E., Lopes, J. B., Cravino, J. P., Santos, C., & Cunha, A. E. (2015). Teacher mediation of students learning using visual representations displayed by computer simulations. In *INTED2015 Proceedings* (pp. 4996–5005). Madrid, Spain: IATED.

Saraiva, E. (2017b). Visual representations in the classroom when teaching physical sciences. In J. B. Lopes, J. P. Cravino, E. S. Cruz, & A. Barbot (Eds.), *Teaching science: Contributions of research for planning, practice and professional development* (pp. 227–246). New York, NY: Nova Publishers.

Saraiva, E., Cunha, A. E., Santos, C. A., Lopes, J. B., & Cravino, J. P. (2012). Papel da mediação do professor na promoção de trabalho epistémico dos alunos durante o uso de simulações computacionais. In C. Portela & T. Peña (Eds.), *Física 2012: Atas do XXII Encontro Ibérico para o Ensino da Física* (pp. 51–58). Lisboa, Portugal: Sociedade Portuguesa de Física.

Schoenfeld, A. H. (2015). How we think: A theory of human decision-making, with a focus on teaching. In S. J. Cho (Ed.), *The Proceedings of the 12th International Congress on Mathematical Education* (pp. 229–243). Seoul, South Korea: Springer International Publishing.

Schoenfeld, A. H. (2015). How we think: A theory of human decision-making, with a focus on teaching. In S. J. Cho (Ed.), *The Proceedings of the International Congress on Mathematical Education* (pp. 229–243). Seoul, South Korea: Springer International Publishing.

Schoenfeld, A. H., Thomas, M., & Barton, B. (2016). On understanding and improving the teaching of university mathematics. *International Journal of STEM Education, 3*(4), 1–17.

Schon, D. (1983). *The reflective practitioner: How professionals think in action*. New York: Basic Books.

Schön, D. (1987). *Educating the reflective practitioner*. San Francisco: Jossey-Bass Publishers.

Schön, D. (2000). *Educando o profissional reflexivo: Um novo design para o ensino e a aprendizagem* [Educating the reflective professional: a new design for teaching and learning]. Porto Alegre, Brazil: ARTMED.

Schwarz, G. (2001). Using teacher narrative research in teacher development. *Teacher Educator, 37*(1), 37–48. doi:10.1080/08878730109555279

Sedig, K., & Sumner, M. (2006). Characterising interaction with visual mathematical representations. *International Journal of Computers for Mathematical Learning, 11*(1), 1–55. doi:10.100710758-006-0001-z

Seidel, T., Stürmer, K., Blomberg, G., Kobarg, M., & Schwindt, K. (2011). Teacher learning from analysis of videotaped classroom situations: Does it make a difference whether teachers observe their own teaching or that of others? *Teaching and Teacher Education, 27*(2), 259–267. doi:10.1016/j.tate.2010.08.009

SEPUP. (n.d.a). *Electrolysis Simulation*. Retrieved from http://www.sepuplhs.org/high/hydrogen/electrolysis_sim.html

SEPUP. (n.d.b). *Fuel Cell Animation*. Retrieved from http://www.sepuplhs.org/high/hydrogen/fuelcell_sim.html

Shagrir, L. (2010). Professional development of novice teacher educators: Professional self, interpersonal relations and teaching skills. *Professional Development in Education, 36*(1–2), 45–60. doi:10.1080/19415250903454809

Shavelson, R. J. (1973). *The basic teaching skill: Decision making*. Stanford, CA: Stanford University School of Education, Center for Research and Development in Teaching.

Shavelson, R. J., & Stern, P. (1981). Research on teachers' pedagogical thoughts, judgments, decisions, and behavior. *Review of Educational Research, 51*(4), 455–498. Retrieved from http://rer.aera.net. doi:10.3102/00346543051004455

Sherr, R. (2008). Gesture analysis for physics education researchers. *Physics Education Research, 4*(1), 1–9.

Shulman, L. (1986). Those who understand: Knowledge growth in teaching. *Educational Researcher, 15*(2), 4–14. doi:10.3102/0013189X015002004

Shute, V. J., & Kim, Y. J. (2014). Formative and stealth assessment. In *Handbook of research on educational communications and technology* (pp. 311–321). New York: Springer. doi:10.1007/978-1-4614-3185-5_25

Sierpinska, A. (2000). On some aspects of students' thinking in linear algebra. In J.-L. Dorier (Ed.), *On the teaching of linear algebra* (pp. 209–246). Dordrecht: Kluwer Academic.

Sierpinska, A., Trgalová, J., Hillel, J., & Dreyfus, T. (1999). Teaching and learning linear algebra with Cabri. In O. Zaslavsky (Ed.), *Proceedings of the 23rd Conference of the International Group for the Psychology of Mathematics Education* (Vol. 1, pp. 119–134). Haifa, Israel: Program Committee.

Silva, A. A. (1998). Archimedes' law and potential energy: Modelling and simulation with a spreadsheet. *Physics Education, 33*(2), 87–92. doi:10.1088/0031-9120/33/2/010

Sindelar, N. (2015). *Assessment powered teaching*. New York: Skyhorse Publishing.

Smyth, J. (1989). Developing and sustaining critical reflection in teacher education. *Journal of Teacher Education, 40*(2), 2–9. doi:10.1177/002248718904000202

Soebari, T. S., & Aldridge, J. M. (2015). Using student perceptions of the learning environment to evaluate the effectiveness of a teacher professional development programme. *Learning Environments Research, 18*(2), 163–178. doi:10.100710984-015-9175-4

Sondergaard, H., & Mulder, R. (2012). Collaborative learning through formative peer review: Pedagogy, programs and potential. *Computer Science Education, 22*(4), 343–367. doi:10.1080/08993408.2012.728041

Sousa, L., Leal, A., & Sena, E. (2010). A importância da comunicação não-verbal do professor universitário no exercício de sua atividade profissional. *Revista CEFAC, 12*(5), 784–787. doi:10.1590/S1516-18462010005000088

Sparks, D., & Loucks-Horsley, S. (1989). Five models of staff development. *Journal of Staff Development, 10*(4), 40–57.

Speer, N., Smith, J. III, & Horvath, A. (2010). Collegiate mathematics teaching: An unexamined practice. *The Journal of Mathematical Behavior, 29*(2), 99–114. doi:10.1016/j.jmathb.2010.02.001

Stake, R. E. (2013). *Multiple case study analysis*. New York, NY: Guilford Press.

Stanley, M. J. C., & Dougherty, J. P. (2010). Nursing education model. A paradigm shift in nursing education: A new model. *Nursing Education Perspectives, 31*(6), 378–380. Retrieved from http://10.0.4.19/1536-5026-31.6.378 PMID:21280445

Starkey, L., Yates, A., Meyer, L. H., Hall, C., Taylor, M., Stevens, S., & Toia, R. (2009). Professional development design: Embedding educational reform in New Zealand. *Teaching and Teacher Education, 25*(1), 181–189. doi:10.1016/j.tate.2008.08.007

Stein, S. J., Isaacs, G., & Andrews, T. (2004). Incorporating authentic learning experiences within a university course. *Studies in Higher Education, 29*(2), 239–258. doi:10.1080/0307507042000190813

Stenhouse, L. (1988). Case study methods. In J. P. Keeves (Ed.), *Educational research, methodology and measurement: an international handbook* (pp. 49–53). Oxford, UK: Pergamon.

Stevens, R., & Hall, R. (1998). Disciplined perception: Learning to see in technoscience. In M. Lampert & M. L. Blunk (Eds.), *Talking mathematics in school: Studies of teaching and learning* (pp. 107–150). New York: Cambridge University Press. doi:10.1017/CBO9780511571251.007

Stewart, S., & Thomas, M. (2009). A framework for mathematical thinking: The case of linear algebra. *International Journal of Mathematical Education in Science and Technology*, *40*(7), 951–961. doi:10.1080/00207390903200984

Stigler, J. W., & Hiebert, J. (2009). *The teaching gap: Best ideas from the world's teachers for improving education in the classroom*. New York: Simon and Schuster.

Stocklmayer, S. (2013). Engagement with science: Models of science communication. In J. K. Gilbert & S. Stocklmayer (Eds.), *Communication and engagement with science and technology. Issues and dilemmas* (pp. 19–38). New York: Routledge.

Stocklmayer, S., & Rennie, L. J. (2017). The attributes of informal science education: A science communication perspective. In P. G. Patrick (Ed.), *Preparing informal science educators: Lessons for science communication and education* (pp. 527–544). New York: Springer. doi:10.1007/978-3-319-50398-1_26

Talbert, R. (2014). Inverting the linear algebra classroom. *PRIMUS Problems, Resources, and Issues in Mathematics Undergraduate Studies*, *24*, 361–374.

Tal, T., & Morag, O. (2007). School visits to natural history museums: Teaching or enriching? *Journal of Research in Science Teaching*, *44*(5), 747–769. doi:10.1002/tea.20184

Tan, A. L., & Towndrow, P. A. (2009). Catalyzing student-teacher interactions and teacher learning in science practical formative assessment using digital video technology. *Teaching and Teacher Education*, *25*(1), 61–67. doi:10.1016/j.tate.2008.07.007

Tardif, M. (2000). Os saberes profissionais dos professores e conhecimentos universitários [The professional knowledge of teachers and university knowledge]. *Revista Brasileira de Educação*, *13*, 5–24.

Tavares, J. (2003). *Formação e inovação no ensino superior*. Porto: Porto Editora.

Thompson, I. (Ed.). (2015). *Designing tasks in secondary education: Enhancing subject understanding and student engagement*. New York: Routledge.

Thurlings, M., Vermeulen, M., Bastiaens, T., & Stijnen, S. (2013). Understanding feedback: A learning theory perspective. *Educational Research Review*, *9*, 1–15. doi:10.1016/j.edurev.2012.11.004

Tiberghien, A. (1996). Construction of prototypical situations in teaching the concept of energy. In G. Welford, J. Osborne, & P. Scott (Eds.), *Research in science education in Europe* (pp. 100–114). London: The Falmer Press.

Tiberghien, A., & Buty, C. (2007). Studying science teaching practices in relation to learning: Time scales of teaching phenomena. In R. Pintó & D. Couso (Eds.), *Contributions from science education research* (Vol. 2, pp. 59–75). Dordrecht, The Netherlands: Springer. doi:10.1007/978-1-4020-5032-9_5

Tlili, A., Cribb, A., & Gewirtz, S. (2006). What becomes of science in a science centre? Reconfiguring science for public consumption. *Review of Education, Pedagogy & Cultural Studies*, *28*(2), 203–228. doi:10.1080/10714410600739921

Tomey, A. M. (2003). Learning with cases. *Journal of Continuing Education in Nursing*, *34*(1), 34–38. Retrieved from http://www.ncbi.nlm.nih.gov/pubmed/12546132 PMID:12546132

Toulmin, S. (1977). *Human understanding: the collective use and evolution of concepts*. Princeton, NJ: University Press.

Tran, L. U. (2007). Teaching science in museums: The pedagogy and goals of museum educators. *Science Education, 91*(2), 278–297. doi:10.1002ce.20193

Tran, L. U., & King, H. (2007). The professionalization of museum educators: The case in science museums. *Museum Management and Curatorship, 22*(2), 129–147.

Tran, L. U., Werner-Avidon, M., & Newton, L. R. (2013). What is successful professional learning for informal educators and how do we get there? *Journal of Museum Education, 38*(3), 333–348. doi:10.1080/10598650.2013.11510785

Trigueros, M., & Possani, E. (2013). Using an economics model for teaching linear algebra. *Linear Algebra and Its Applications, 438*(4), 1779–1792. doi:10.1016/j.laa.2011.04.009

Tucker, A. (1993). The growing importance of linear algebra in undergraduate mathematics. *The College Mathematics Journal, 24*(1), 3–9. doi:10.1080/07468342.1993.11973500

Tytler, R. (2003). A window for a purpose: Developing a framework for describing effective science teaching and learning. *Research in Science Education, 33*(3), 273–298. doi:10.1023/A:1025423704068

Uhlig, F. (2003). A new unified, balanced and conceptual approach to teaching linear algebra. *Linear Algebra and Its Applications, 361*, 147–159. doi:10.1016/S0024-3795(02)00319-1

Valente de Oliveira, L. (Ed.). (2018). O papel dos museus [The role of museums]. Porto: Amigos do Museu Nacional Soares dos Reis.

Van Eekelen, I. M., Vermunt, J. D., & Boshuizen, H. P. A. (2006). Exploring teachers' will to learn. *Teaching and Teacher Education, 22*(4), 408–423. doi:10.1016/j.tate.2005.12.001

Vanassche, E., & Kelchtermans, G. (2016). Facilitating self-study of teacher education practices: Toward a pedagogy of teacher educator professional development. *Professional Development in Education, 42*(1), 100–122. doi:10.1080/19415257.2014.986813

Vergnaud, D. (1987). In J. Piaget, P. Mounoud, & P. J. Brankart (Eds.), *Encyclopédie de la Pléiade Psychologie* (pp. 821–844). Paris: Gallimard.

Vescio, V., Ross, D., & Adams, A. (2008). A review of research on the impact of professional learning communities on teaching practice and student learning. *Teaching and Teacher Education, 24*(1), 80–91. doi:10.1016/j.tate.2007.01.004

Viegas, C., Lopes, J. B., & Cravino, J. P. (2009). Incremental innovations in a physics curriculum for engineering undergraduates. In W. Aung, K.-S. Kim, J. Mecsi, J. Moscinski, & I. Rouse (Eds.), Innovations 2009: World innovations in engineering education and research (pp. 175–186). Arlington, VA: iNEER.

Viegas, C., Lopes, J. B., & Cravino, J. P. (2010). Learning through Assessment and Feedback In and Out of Class. In Innovations 2010 – World Innovations in Engineering Education and Research (pp. 33-46). Potomac, MD: iNEER.

Viegas, C., Lopes, J. B., & Cravino, J. P. (2010b). Learning through assessment and feedback in and out of class. In W. Aung, K. Kim, J. Mecsi, J. Moscinski, & I. Rouse (Eds.), Innovations 2010: World innovations in engineering education and research (pp. 33–46). Arlington, VA: iNEER.

Viegas, C., Lopes, J. B., & Cravino, J. P. (2011). Reflection on teachers' practice to improve teaching and learning of introductory physics in engineering education. In W. Aung, V. Ilic, J. Moscinski, & J. Uhomoibhi (Eds.), Innovations 2011: World innovations in engineering education and research (pp. 211–225). Potomac, MD: iNEER.

Viegas, C., Lopes, J., & Cravino, J. (2010a). Introductory physics teaching in engineering education: How can the reflection on teaching practice be used to improve teaching and learning? In *ICEE 2010 Proceedings: International conference on engineering education*. Gliwice, Poland: iNEER. Retrieved from http://www.ineer.org/Events/ICEE2010/papers/R9A/Paper_961_1175.pdf

Viegas, C., Alves, G., & Lima, N. (2015). Formative Assessment Diversity to Foster Students' Engagement. In *Proceedings of ICL2015 – International Conference on Interactive Collaborative Learning* (pp. 929-935). Florence, Italy: Academic Press. 10.1109/ICL.2015.7318152

Vieira, F. (2011). Entre o privado e o público: O diário como lugar de transformação. In M.A. Moreira (Eds.), Narrativas dialogadas na investigação, formação e supervisão de professores (pp. 115−136). Mangualde: Pedágio.

Vieira, F., & Moreira, M. A. (2011). *Supervisão e avaliação do desempenho docente: Para uma abordagem de orientação transformadora* [Supervision and evaluation of teacher performance: Towards a transformative orientation approach.]. Lisbon: Ministério da Educação. Retrieved from http://www.ccap.min-edu.pt/docs/Caderno_CCAP_1-Supervisao.pdf

Vieira, F. (1993). Observação e supervisão de professores [Observation and supervision of teachers]. In F. Sequeira (Ed.), *Dimensões da Educação em Língua Estrangeira* [Dimensions of Foreign Language Education] (pp. 69–90). Braga: Universidade do Minho.

Viennot, L. (2002). *Enseigner la physique*. Bruxelles: Éditions De Boeck Université.

Vinner, S. (1997). Scenes from linear algebra classes. In D. Carlson, C. R. Johnson, D. Lay, A. D. Porter, A. E. Watkins, & W. Watkins (Eds.), *Resources for teaching linear algebra* (Vol. 42, pp. 155–171). Washington, DC: Mathematical Association of America.

vom Lehn, D. (2006). Embodying experience: A video-based examination of visitors' conduct and interaction in museums. *European Journal of Marketing*, *40*(11/12), 1340–1359. doi:10.1108/03090560610702849

vom Lehn, D., Heath, C., & Hindmarsh, J. (2001). Exhibiting interaction: Conduct and collaboration in museums and galleries. *Symbolic Interaction*, *24*(2), 189–216. doi:10.1525i.2001.24.2.189

Waight, N., & Abd-El-Khalick, F. (2007). The impact of technology on the enactment of "inquiry" in a technology enthusiast's sixth grade science classroom. *Journal of Research in Science Teaching*, *44*(1), 154–182. doi:10.1002/tea.20158

Wallace, M. J. (1991). *Training foreign language teachers*. Cambridge, UK: CUP.

Waterkemper, R., & Prado, M. (2011). Estratégias de ensino-aprendizagem em cursos de graduação em Enfermagem. *Avances En Enfermería*, *39*(2), 234–246.

Weber, R. P. (1990). *Basic content analysis* (2nd ed.). Newbury Park, CA: Sage Publications. doi:10.4135/9781412983488

Weil-Barais, A. (1995). La formation des connaissances en sciences expérimentales. In D. Gaonach & C. Golder (Eds.), *Manuel de Psychologie pour l'Enseignement* (pp. 410–437). Paris: Hachette Éducation.

Wei, R. C., Darling-Hammond, L., Andree, A., Richardson, N., & Orphanos, S. (2009). *Professional learning in the learning profession: A status report on teacher development in the United States and abroad*. Dallas, TX: National Staff Development Council.

Williams, G. (2001). *Linear algebra with applications* (4th ed.). Sudbury, MA: Jones and Bartlett.

Winch, C., Oancea, A., & Orchard, J. (2015). The contribution of educational research to teachers' professional learning: Philosophical understanding. *Oxford Review of Education*, *41*(2), 202–216. doi:10.1080/03054985.2015.1017406

Woolhouse, C. (2017). Multimodal life history narrative. *Narrative Inquiry*, *27*(1), 109–131. doi:10.1075/ni.27.1.06woo

Wright, S. P., Horn, S. P., & Sanders, W. L. (1997). Teacher and classroom context effects on student's achievement: Implications for teacher evaluation. *Journal of Personnel Evaluation in Education*, *11*(1), 57–67. doi:10.1023/A:1007999204543

Yeigh, T. (2008). Quality teaching & professional learning: Uncritical reflections of a critical friend. *Australian Journal of Teacher Education*, *33*(2), 1–15. doi:10.14221/ajte.2008v33n2.2

Yeo, J., & Gilbert, J. K. (2014). Constructing a scientific explanation – A narrative account. *International Journal of Science Education*, *36*(11), 1902–1935. doi:10.1080/09500693.2014.880527

Yin, R. K. (2010). *Estudo de caso—Planejamento e métodos* [Case study—Planning and methods] (4th ed.). Porto, Portugal: Porto Editora.

Zabalza, M. (2004). *Los diarios de classe. Un instrumento de investigación y desarrollo profissional*. Madrid: Nancea S. A. Ediciones.

Zacharia, Z. T., & De Jong, T. (2014). The effects on students' conceptual understanding of electric circuits of introducing virtual manipulatives within a physical manipulatives-oriented curriculum. *Cognition and Instruction*, *32*(2), 101–158. doi:10.1080/07370008.2014.887083

Zangori, L., Forbes, C. T., & Schwarz, C. V. (2015). Exploring the effect of embedded scaffolding within curriculum tasks on third-grade students' model-based explanations about hydrologic cycling. *Science & Education*, *7*(24), 957–981. doi:10.100711191-015-9771-9

Zeichener, K. M. (1993). *Reflective teacher training: ideas and practices*. Lisbon: Educa.

Zeichner, K., & Liston, D. (2014). *Reflective teaching: An introduction* (2nd ed.). New York: Routledge.

Zeichner, K., Payne, K., & Brayko, K. (2015). Democratizing teacher education. *Journal of Teacher Education*, *66*(2), 122–135. doi:10.1177/0022487114560908

Zhai, J., & Dillon, J. (2014). Communicating science to students: Investigating professional botanic garden educators' talk during guided school visits. *Journal of Research in Science Teaching*, *51*(4), 407–429. doi:10.1002/tea.21143

Zimmerman, B. J. (2013). From cognitive modeling to self-regulation: A social cognitive career path. *Educational Psychologist*, *48*(3), 135–147. doi:10.1080/00461520.2013.794676

Zimmerman, B. J., & Moylan, A. R. (2009). Self-regulation: Where metacognition and motivation intersect. In D. J. Hacker, J. Dunlosky, & A. C. Graesser (Eds.), *Handbook of metacognition in education* (pp. 299–315). New York: Routledge.

Zodik, I., & Zaslavsky, O. (2008). Characteristics of teachers' choice of examples in and for the mathematics classroom. *Educational Studies in Mathematics*, *69*(2), 165–182. doi:10.100710649-008-9140-6

Zompero, A. F., & Laburú, C. E. (2010). As relações entre aprendizagem significativa e representações multimodais [The relationships between meaningful learning and multimodal representations]. *Revista Ensaio*, *12*(3), 31–40.

About the Contributors

J. Bernardino Lopes holds a PhD and Aggregation in Physics Education. Associate Professor of UTAD (Universidade de Trás-os-Montes e Alto Douro). Researcher of Research Centre CIDTFF - Research Centre "Didactics and Technology in Education of Trainers and coordinator of a research lab of this centre. Former President of Scientific Council, Pedagogical Council and Head of Department at UTAD. Currently is Director of doctoral program: "Science and Technology Education Research" at UTAD. Invited Editor of scientific journals and referee in JCR journals. Consultant in a professional development center. Author/editor of books (international edition) and papers in international scientific journals with peer review. Present research interests include Sciences and Technological Education Research, in particular: i) Curriculum design and evaluation of teaching; ii) Teacher mediation of student learning; iii) Teaching practices (questions, tasks, inquiry, epistemic practices, decisions, engagement, assessment, CTS contexts, interactions, practical work); iv) Conceptual learning; v) Tools for data collection in classroom; vi) computer simulation; vii) tools (beneficial and epistemic) to enhance C&T learning viii) Teacher education and Teacher professional development; ix) Philosophy of Science.

Maria Clara Viegas holds a PhD in Science and Technology and a MSc in Mechanical Engineering. Currently she is an Adjunct Professor in Polytechnic of Porto – School of Engineering (ISEP), Porto, Portugal. Researcher in CIETI Research Centre - Centre for Innovation in Engineering and Industrial Technology. Former member of the Pedagogical Council and Vice-Head of Physics Department at ISEP. Participant in several national and international research projects. Invited chair in International Conferences and referee in JCR journals. Author of several papers in international scientific journals, book chapters and international conferences with peer review. Her present research interests are: Engineering Education; Physics Didactics; Science and Technological Education Research, in particular Teacher mediation of student learning and Teaching practices Teacher professional development; Remote Labs Learning Outcomes.

José Alexandre Pinto holds a PhD in Didactics of Science and technology and a MSc in Science Teaching Supervision. Currently he is an Adjunct Professor in Polytechnic of Porto – School of Education, Porto, Portugal. He is currently the Vice-President of the School of Education. Previously, he was director of the department of Mathematics, Science and Technology (between 2016-2018). He is also a researcher at the Research Centre "Didactics and Technology in Education of Trainers" (CIDTFF) at the University of Aveiro. His research interests are in the areas of Teacher Education, Science Teaching Supervision and STEM Education (Science, Technology, Engineering and Mathematics Education). He has participated in several funded research projects and is author or coauthor of books, book chapters, peer-reviewed papers and conference communications.

***

Ana Sofia Afonso is Assistant Professor in Science Education at the University of Minho, Portugal, and a researcher at the Research Centre on Education, Institute of Education, University of Minho. She holds a first degree in Physics (University of Porto) and a Master degree in Physics Education (University of Minho). She completed her PhD in 2005 at the University of Reading, UK. Her dissertation focused on learning in science museums. Since then, her research has explored learning in other informal environments, particularly with popular science books. She is a member of the editorial board of "Revista Portuguesa de Educação".

Susana Afonso is Senior Lecturer in Portuguese at the University of Exeter. She has completed her PhD in 2008 at the University of Manchester. Her dissertation is an empirical study of Impersonal constructions in European Portuguese from a Cognitive Linguistics perspective. Since then, her research interests widened to encompass multilingualism and language contact from a Cognitive Sociolinguistics perspective, particularly with reference to non-standard varieties of Portuguese around the world. More recently, Susana is involved in collaborative work on communication of Science in informal contexts, researching the discourse of museum guides and its effectiveness in communicating scientific phenomena.

Carla Aguiar dos Santos is a student of the PhD course in Physical Sciences - didactics, at the Universidade de Trás-os-Montes e Alto Douro. Carla has a degree in "Teaching of Physics and Chemistry" from the Universidade de Trás-os-Montes e Alto Douro. Research interests are in the area of science education, in particular in the study of the pedagogical decisions that teachers adopt in the classroom, using multimodal narratives as a research tool.

António Barbot was born on November 4, 1979 in Porto, Portugal. He is graduated in Biological Engineering and has a postgraduate degree in Environmental Technologies from University of Minho, Braga, Portugal. He holds a PhD in Physical Sciences Education from University of Trás-os-Montes and Alto Douro, Vila Real, Portugal, where he dealt with topics related to the relationships between Water and Energy in Education. He is currently an Assistant Teacher at the School of Education of the Polytechnic Institute of Porto in the department of Mathematics, Science and Technology. He is a member of the Research Centre "Didactics and Technology in Education of Trainers" (CIDTFF), Aveiro, Portugal. His main research interests are in the areas of Physical Science Education in Initial Teachers Training courses, and in the use of Information and Communication Technologies (ICT) in the classroom.

Maria Cecília Rosas Pereira Peixoto da Costa was born in Porto, Portugal, in 1966. She holds a BSc in Mathematics by University of Porto, a MSc in Educational Sciences by University of Coimbra, a PhD in Mathematics and Agregation in Science and Technology Education Research, both by the University of Trás-os-Montes and Alto Douro (UTAD). She teaches in the Department of Mathematics of UTAD since 1994, mainly linear algebra and didactics of mathematics. She is member of CIDTFF - Research Center "Didactics and Technology in Education of Trainers". Her interests as a researcher are Mathematics Education, Didactics of Linear Algebra and Geometry, Ethnomathematics, History of Mathematics and its Teaching, and Teaching practices. She has several national and international publications.

Nilza Costa is a full professor in Education in the Department of Education and Psychology (DEP), University of Aveiro (UA)/Portugal, being responsible in DEP for the area of Didactics and Educational Technology. She is a founding member of the Research Centre "Didactics and Technology in Teacher Education" (CIDTFF), allocated in DEP/UA, and external financed by the Portuguese Science and Technology Foundation (FCT), since its creation in 1994. She was the coordinator of CIDTFF, between 2012-2016, leading 80 researchers from UA and other Higher Education Institutions. She was, during five years, the Coordinator of the Integrated Centre of Techer Education in UA. Her main research interests are: teacher education and staff development from secondary to higher education; science/physics education; and evaluation and assessment in education. In the last thirty years she has been responsible for several research projects (at a national and international level), an author and/or co-author of several publications and a supervisor of more than 20 PhD students (mainly from Portugal, but also from Brazil and Africa). She has been invited to give key lectures in many Universities in Europe, Africa and Brazil.

José Paulo Cravino (http://orcid.org/0000-0002-5376-6128) has an academic background in Physics and holds a PhD in Physics/Physics Education Research. He is Assistant Professor of Physics at the Physics Department of the School of Science and Technology of the Universidade de Trás-os-Montes e Alto Douro. He is currently the director of the department (2017-2020). Previously, he was Vice-President of the School of Science and Technology (between 2009 and 2013). He is also a researcher at the Research Centre "Didactics and Technology in Education of Trainers" (CIDTFF) at the University of Aveiro. His research interests are in the areas of Physics Education, STEM Education (Science, Technology, Engineering and Mathematics Education), Educational Technology and Teacher Education. He has participated in several competitively funded research projects and is author or coauthor of several books, book chapters, peer-reviewed papers and conference communications.

Eliane de Souza Cruz graduated with a degree in Physics Education from the University of São Paulo in São Paulo, Brazil. She has a postgraduate degree in Science Education from the University of Aveiro in Aveiro, Portugal. She holds a PhD in Education, Didactics and Teacher Training Research. She is currently an Adjunt Professor at the Federal University of São Paulo and a researcher at CEFE (Centre of Educators Training of Nonhigher Education) in Diadema, Brazil. Her main research interest is "Articulation between Research and Practices in Nonhigher and Higher Education". Member of the Education Regional Forum of the ABC that discusses Brazil's educational policies. She is coordinator of the Program ARTICUL@ÇÕES that includes 4 projects (articul@ events, articul@ training, articul@ schools and articul@ popular courses).

Gonçalo Cruz is a PhD student in Didactics of Science and Technology and a research fellow at the University of Trás-os-Montes and Alto Douro (UTAD), in Portugal. In 2011, Gonçalo achieved his Master degree in Educational Sciences at the University of Coimbra. With several publications in journals, conferences and book chapters, his main research interests are Distance Learning, Social Learning, Technology Enhanced Learning, Cooperative Learning, Critical Thinking, and Teacher Training.

Ana Edite Cunha is a professor of Physics and Chemistry at the Secondary School. She has a master degree in Physics of University of Minho in 2002 and a PhD degree in Sciences and Technology Didactics at the University of Trás-os-Montes and Alto Douro in 2015. Her research interests are professional development, teacher mediation and students engagement in science education. She has several publications of papers in national and international journals and in books. In her professional life besides teaching subjects of physics and chemistry, she holds several teaching positions as Class Director, Cycle Coordinator and member of the School's Pedagogical Council.

Ricardo Jorge Castro Gonçalves was born in Fafe, Portugal, in 1975. He holds a BSc in Mathematics by University of Aveiro, a MSc in Mathematics Education by Faculty of Science of University of Porto, and a PhD in Science and Technology Education Research by University of Trás-os-Montes and Alto Douro. He teaches in Polytechnic Institute of Cávado and Ave, mainly linear algebra, since 2009. His research interests include mathematics education, in particular on teaching and learning of linear algebra, having some publications in this area. He is also author of a linear algebra textbook. When he is not teaching, Ricardo Gonçalves enjoys hiking and travel.

Rita Maria Ferreira Leal has a Master's Degree in Family Health Nursing, a Master's Degree in Business Administration and Accounting, is Coordinator, Registered Nurse-Midwife and Family Nurse Practitioner at a Primary Health Care Unit in Aveiro, Portugal (Unidade de Cuidados na Comunidade Aveiro) and has been a Part-time Visiting Professor at the School of Health, University of Aveiro, Portugal (since 2016). Areas of scientific interest: Family health, Midwifery, Nursing and Education.

Carolina José Maria has a Bachelor's degree in Sciences - Qualification in Chemistry. She is a Master in Education for Science and has a PhD in Education. She began her teaching career in Basic Education as a teacher of Chemistry and Sciences. She is currently a teacher and coordinator of the Chemistry-Licentiate Course at the Methodist University of Piracicaba. She also works as an education researcher, working especially with semiotic registers in Science/Chemistry lessons, besides guiding Scientific Initiation projects in this line. She has experience in the area of Education in Chemistry Teaching, mainly working in the subjects of teacher's mediation in the teaching of Chemistry, didactic resources for the teaching of Chemistry, semiotic registers and teacher's training.

Cristina Marques has a degree in Chemistry from the University of Coimbra and she has a PhD in Didactics of Chemistry from the University of Trás-os-Montes e Alto Douro since 2011. She is an assistant professor in the Department of Chemistry of this university since 1993. Her research interests are related with the area of physical sciences and with curricular development, problem solving, practical work, work by project, design of educational tools, teacher mediation and teacher training.

Maria Paula Montenegro Vieira Cardoso was born in Espinho, Portugal, in 1965. She holds a BSc in Mathematics and Science by Instituto Piaget, a MSc in Mathematics and Science by University of Trás-os-Montes and Alto Douro (UTAD) and she is a PhD student in Science and Technology Education Research in University of Trás-os-Montes and Alto Douro (UTAD). She is a science and mathematics teacher in the 2nd cycle of basic education. Her research interest lies on the specifics of each one of verbal, symbolic and visual representation systems, used together, providing experiences with multiples representations. She considers that systems of visual representation can have a significant impact in teaching and learning of Mathematics, in middle school, reason why she developed investigation work in this area, based on her teaching experience and together with other investigators. She has several national and international publications on this topic.

Leonel Morgado is Assistant Professor with Habilitation, at Universidade Aberta (Portuguese Open University), a public university focusing on e-learning, where he lectures on programming and the use of virtual worlds, and develops his research as a senior researcher at INESC TEC, a research laboratory rated as Excellent by the national research assessment authorities. His main research interest is the use and development of virtual worlds and immersive technologies as tools for learning and business, which he pursues since 2000, focusing on multi-user platforms since 2006. He authored over 100 papers, in journals, conferences, and as book chapters, most related to this field of interest. Leonel is a board member of the Immersive Learning Research Network. He is also the President of the Board of the Portuguese Association of Videogame Sciences.

Domingos Kimpolo Nzau is an Angolan and career professor. He graduated in Physics from the Higher Institute of Educational Sciences (ISCED) in Lubango - Angola in 1985 and a PhD in Physics Didactics from the University of Trás-os-Montes e Alto Douro-Portugal in 2010. He was a Teacher of Primary School from 1970 to 1977 in Cabinda - Angola, teacher of Mathematics and Natural Sciences in Secondary Schools in Cabinda and Lubango from 1979 to 1985; a university professor from 1985 until the present time. Actually, he is Associate Professor and General Director of the Higher Institute of Educational Sciences (ISCED) of Uíge - Angola as well as a researcher of the Department of Exact Sciences of the same Institution. His research concern is focused on students' learning in exact, natural, and technological sciences, notably in physics teaching practices, teachers' training, and students' learning assessment.

Daniela Pedrosa is a Post-Doc Researcher at the University of Trás-os-Montes and Alto Douro (UTAD), and Research Centre on Didactics and Technology in the Education of Trainers (CIDTFF), in Portugal. She holds a PhD in Science and Technology Didactics at UTAD, specializing in Didactics of Informatics. Her doctoral research project funded by a PhD grant awarded by the Foundation for Science and Technology (FCT), whose theme was "Self-Regulation and Co-Regulation Learning in Higher Education: Strategies Adopted by Computer Programming Students." With several publications in journals, conferences and book chapters, her scientific interest are Self-regulation and Co-regulation of Learning, Formative Assessment, Distance Learning, Technology Enhanced Learning, Teaching Training, and research and intervention methodologies. She collaborated in the organization of events (conferences, seminars, workshops). She organizes a special track in international conference.

Marília dos Santos Rua has a PhD in Health Sciences, a Master's Degree in Nursing Sciences, and is currently Professor at the School of Health of the University of Aveiro (EESUA). Areas of scientific interest: Family health, Education, Supervision.

Elisa Saraiva was born in Oporto – Portugal in 1974. She studied Chemisty in University of Aveiro where sucessefully completed a degree in Chemestry. After she continued her studies in University of Trás-os-Montes e Alto Douro where she completed a master degree in teaching (Physics and Chemistry) and a Ph.D. in Physical Sciences. She has worked as physics and chemistry teacher at secondary school in Portugal for 20 years. She also has experience as trainer in teacher training programs (initial and continuous).

Index

F

H

I

J

L

M

N

O

P

Q

R

S

T

U

V

W

Premier Reference Source
Emerging GIS Applications for Emergency and Disaster Management

Premier Reference Source
Managerial Strategies and Green Solutions for Project Sustainability

Premier Reference Source
Comparative Approaches to Using R and Python for Statistical Data Analysis

Premier Reference Source
Solutions for High-Touch Communications in a High-Tech World

www.ingramcontent.com/pod-product-compliance
Lightning Source LLC
LaVergne TN
LVHW062111300926
849161LV00010B/2432